T0192883

Basics of Precision Engineering

Basics of Precision Engineering

Edited by
Richard Leach
Stuart T. Smith

CRC Press
Taylor & Francis Group
Boca Raton London New York

CRC Press is an imprint of the
Taylor & Francis Group, an **informa** business
A CHAPMAN & HALL BOOK

CRC Press
Taylor & Francis Group
6000 Broken Sound Parkway NW, Suite 300
Boca Raton, FL 33487-2742

First issued in paperback 2020

© 2018 by Taylor & Francis Group, LLC
CRC Press is an imprint of Taylor & Francis Group, an Informa business

No claim to original U.S. Government works

ISBN-13: 978-1-4987-6085-0 (hbk)
ISBN-13: 978-0-367-78139-2 (pbk)

Library of Congress Cataloging-in-Publication Data
Names: Leach, R. K., editor.	Smith, S. T. (Stuart T.), 1961- editor.
Title: Basics of precision engineering / [edited by] Richard Leach & Stuart T. Smith.
Description: Boca Raton : Taylor & Francis, a CRC title, part of the Taylor & Francis imprint, a member of the Taylor & Francis Group, the academic division of T&F Informa, plc, [2018]	Includes bibliographical references.
Identifiers: LCCN 2017041321	ISBN 9781498760850 (hardback : acid-free paper)	ISBN 9781498760874 (ebook)
Subjects: LCSH: Machine design.	Tolerance (Engineering)	Machine-tools.	Metrology.	Engineering instruments.
Classification: LCC TJ233 .B36 2018	DDC 621.8/15–dc23
LC record available at https://lccn.loc.gov/2017041321

Visit the Taylor & Francis Web site at
http://www.taylorandfrancis.com

and the CRC Press Web site at
http://www.crcpress.com

Contents

Preface

Advances in engineering precision have tracked with technological progress for hundreds of years. Over the last few decades, precision engineering has been the specific focus of research on an international scale. The outcome of this effort has been the establishment of a broad range of engineering principles and techniques that form the foundation of precision design.

Today's precision manufacturing machines and measuring instruments represent highly specialised processes that combine deterministic engineering with metrology. Spanning a broad range of technology applications, this involves scientific disciplines such as mechanics, materials, optics, electronics, control, thermomechanics, dynamics and software engineering.

This book provides a collection of these ideas in a single source. Each topic is presented at a level suitable for both undergraduate students and precision engineers in the field. Also included is a wealth of references and example problems to consolidate ideas and help guide the interested reader to more advanced literature on specific implementations.

Acknowledgements

The editors thank Professor Pat McKeown for providing useful material for Chapter 1. Thanks also to Patrick Bointon, Danny Sims-Waterhouse and Luke Todhunter (University of Nottingham) for running through some of the exercises and offering suggestions for improvements. Shah Karim thanks Teguh Santosa (University of Nottingham) for help with figures. Harish Cherukuri thanks Dr Mohammed Hassan (University of North Carolina at Charlotte) for help with drawing the figures in Table 3.1 and Daniel Skoog (MaplePrimes.com) for providing help with generating normal distribution tables. Richard Leach, Shah Karim and Waiel Elmadih thank the UK Engineering and Physical Sciences Research Council (grant EP/M008983/1) for funding.

The editors also thank the National Physical Laboratory for the cover figure.

About the Editors

Richard Leach is a professor in Metrology at the University of Nottingham and heads the Manufacturing Metrology Team. Prior to this position, he was at the National Physical Laboratory from 1990 to 2014. His primary love is instrument building, from concept to final installation, and his current interests are the dimensional measurement of precision and additive manufactured structures. His research themes include the measurement of surface topography, development of methods for measuring 3D structures, development of methods for controlling large surfaces to high resolution in industrial applications and x-ray computed tomography. He is a leader of several professional societies and a visiting professor at Loughborough University and the Harbin Institute of Technology.

Stuart T. Smith has been working in engineering for four decades starting in 1977 with a factory maintenance apprenticeship with Miles Redfern Limited. He is now a Professor of Mechanical Engineering and leads the Instrument Development Group at the University of North Carolina at Charlotte. Throughout the years, his main focus has been the development of instrumentation and sensor technologies primarily aimed towards the challenges of atomic scale discrimination, manipulation and manufacture with applications in the fields of optical, biological and mechanical processes.

About the Contributors

Patrick Baird is an R&D scientist at Gnosys in the Surrey Research Park, where he works primarily in spectroscopy and software development for multivariate analysis. He received his PhD degree from Brunel University in 1996 for work in CMM probe metrology at the National Physical Laboratory. For some years, he carried out postdoctoral research in scanning probe microscopy at the University of Surrey and Imperial College.

Niels Bosmans is the founder of Innovate Precision, a start-up specialised in innovations of precision motion systems and metrology. Previously, he was the R&D manager at Wielandts UPMT, which develops ultra-precision machining technologies aiming at drastically increasing the precision of mass-produced polymer lenses. He obtained his PhD at KU Leuven in 2016 while developing mechatronic positioning and metrology systems for ultra-precision machine tools. He has a passion for design and development of mechatronic systems and has a strong background in the combination of precision engineering principles with advanced control.

Eric S. Buice is an engineer at the Lawrence Berkeley National Laboratory (LBNL) with 15 years of experience in the field of precision engineering. Prior to joining LBNL, he had appointments at Carl Zeiss Semiconductor Manufacturing Technology, Lawrence Livermore National Laboratory, Technical University of Delft and the University of North Carolina at Charlotte. His primary interests are in the design and development of precision instruments. Eric is an active member of the American Society for Precision Engineering (ASPE) and was elected as the president of the society for the 2018 term.

Harish P. Cherukuri is a professor in the Department of Mechanical Engineering and Engineering Science at the University of North Carolina (UNC) at Charlotte. He obtained his PhD degree in 1994 from the Department of Theoretical and Applied Mechanics at the University of Illinois at Urbana-Champaign, and has been with the UNC Charlotte since 1995. His research expertise is in solid mechanics, wave propagation in elastic solids, modelling of manufacturing and metal-forming processes, computational mechanics and particle-based methods. His current research interests include modelling of shot peening processes and material removal processes using particle-based methods, and irradiation-induced creep and swelling phenomena in metals.

Derek G. Chetwynd has almost 50 years' experience in high-precision engineering, instrument systems design and metrology, first at Rank Taylor Hobson (as it then was called) and in academia. Later, he led both the Precision Engineering Group and Mechanical Engineering at the School of Engineering, University of Warwick, where he is now a Professor Emeritus. Consistently promoting a cross-disciplinary approach to instrument design, he has contributed to topics as varied as roughness and roundness metrology, small-scale tribology, mechanisms and x-ray interferometry. In the last decade, he has concentrated mostly on developing instruments and methods for characterising and better understanding mechanical behaviour at or close to solid surfaces, and on kinematics. He holds visiting professor posts at the Harbin Institute of Technology and at Tianjin University.

Waiel Elmadih is a manufacturing metrology researcher at the University of Nottingham. He received a BSc degree in mechanical engineering at the University of Khartoum and an MSc degree in manufacturing engineering and management from the University of Nottingham in 2016. He worked on lean manufacturing and continuous improvement projects during his time as a manufacturing engineer in Rolls-Royce. His current research includes the investigation of additively manufactured lattice structures for vibration and thermal isolation of precision engineering instruments.

Massimiliano Ferrucci has been working in the field of dimensional metrology for the past 8 years. From 2008 to 2013, he worked as a physicist at the National Institute of Standards and Technology. He then worked as a research scientist at the National Physical Laboratory from 2013 to 2016. Currently, Massimiliano is a researcher in Engineering Technology at KU Leuven. Massimiliano's research has been largely focused on development of geometrical calibration procedures for modern coordinate measuring systems, more specifically large-volume laser scanners and x-ray computed tomography.

Han Haitjema obtained his MSc degree in physics at Utrecht University in 1985 and his PhD degree at Delft University of Technology in 1989. Then he joined the NMi Van Swinden Laboratory, the National Metrology Institute of the Netherlands. In 1997, he moved to the Eindhoven University of Technology as an assistant professor in the Precision Engineering Group of Professor Schellekens. In 2004, he was appointed as the director of Mitutoyo Research Center Europe. During almost three decades of research work, he has developed calibration and instrument concepts in the field of dimensional metrology that are in daily use.

Shah Karim obtained an MSc degree in manufacturing systems engineering and worked as an engineer and consultant at Bristol-based manufacturing companies to carry out 'continuous improvement' projects. He received an MPhil degree in manufacturing from the University of the West of England, Bristol. In 2015, he joined the Manufacturing Metrology Team of the University of Nottingham as a researcher in the field of ultra-precision engineering and is passionate about the kinematics of machines.

Stephen Ludwick directs the mechatronic research group at Aerotech Inc., where he has been employed since 1999. This group specialises in the development of modelling, identification and control techniques for precision automation systems, and has ample opportunity to test new ideas on a wide range of motion systems in order to identify those with the broadest appeal. He is also an adjunct associate professor at the University of Pittsburgh, a leader in the American Society for Precision Engineering, and an editor-in-chief for *Precision Engineering*, and he enjoys building robots with middle-school students.

Jimmie Miller has worked at the Center for Precision Metrology (CPM) at the University of North Carolina at Charlotte for more than 25 years. He holds degrees in engineering (PhD, MS), mathematics (BS), physics (BS) and electronics (AAS). He teaches a graduate course in machine tool metrology as well as supports the research and development activities of the CPM. He serves (2016–2018) on the board of the American Society for Precision Engineering (ASPE) as the director-at-large and chairs of the metrology systems committee.

Marwène Nefzi received a diploma and PhD degree in mechanical engineering from RWTH Aachen University, and a diploma in engineering management from the University of Hagen. He worked for more than 5 years in the Department of Mechanism Theory and Dynamics of Machines of RWTH Aachen University. He is now a systems engineer at Carl Zeiss. His main focus is in the development of high-precision optomechanical systems for photolithographic machines needed to enable the manufacturing of microchips.

Dominiek Reynaerts received the MSc and PhD degrees in mechanical engineering from KU Leuven. In 1986, he joined the Department of Mechanical Engineering, KU Leuven, as a research assistant and became an assistant professor in 1997. He is currently a full professor of mechanical engineering at KU Leuven, where he chaired the Department of Mechanical Engineering from 2008 to 2017. He is performing research and teaching activities in micro- and precision engineering, focusing on advanced machine tool components and medical robotics. He is a member of the euspen, IEEE and ASME. He is also a member of Flanders Make, the Flemish strategic research centre for the manufacturing industry.

Richard M. Seugling has been working in precision engineering for over twelve years at Lawrence Livermore National Laboratory (LLNL) supporting the design and manufacture of experiments at the National Ignition Facility. He has a passion for the discipline of precision engineering and its application to manufacturing in support of large-scale science experiments. Richard is currently the deputy program manager at LLNL, and his research interests include meso-scale metrology, advanced manufacture of precision structures, x-ray metrology, precision forming and diamond machining.

Ulrich Weber has been working for over 18 years at Carl Zeiss SMT GmbH in the field of mechanical design for semiconductor lithography lenses and wafer inspection microscopes. Besides all mechanical aspects in the development of mountings and manipulators for optics with nanometre accuracy in small series production, his main focus is directed to reliable designs often driven by kinematics during assembly and operation. He has a passion for compliant mechanisms, especially the development of parametric simulation models allowing rapid optimisation and tolerance analysis.

1

Introduction to Precision

Richard Leach and Stuart T. Smith

CONTENTS

ABSTRACT Ultimately, the goal of precision engineering design is to create a process for which the outcomes are deterministic and controllable over a range of operation, with unpredictable deviations from a desired result being as small as is physically and economically possible. This book outlines concepts that might be considered good practice in precision engineering, concentrating on the basic principles and how to use them as part of the design, development and characterisation of the precision process in question.

Many conceptual tools are discussed throughout the book and have been collected in this introductory chapter. Because these ideas are only briefly explained here, it is recommended that this be reviewed both before and after reading the rest of this book. To introduce this topic, this chapter discusses some general ideas of what constitutes precision engineering as a field of study and concludes with an outline of fundamental limits to precision.

1.1 Introduction

Precision engineering has been, and continues to be, one of the disciplines needed to enable future technological progress. Being always at the edge of technological capability and pushing towards the limits imposed by physical laws, the drive for increased precision is, and always will be, an intellectually demanding pursuit, and brings with it the benefit of being pivotally involved in some of the most exciting of human endeavours.

It is clear that technology is changing the world in many ways, but both its impact and progress is difficult to summarise into a single equation, chart or graph. However, it is readily apparent that the advent of the transistor has played a large part in recent technological advances. Gordon Moore (1965) ably illustrated progress in this field by plotting the number of transistors on a chip as a function of time, showing that this number was doubling every two or so years, a relation now called Moore's law. Over time, the quoted number of components on a chip has changed, but the overall trend has stayed relatively consistent for more than six decades, although it appears to have slowed a little over the last decade or so to a doubling every two and a half years. However, Moore's plot does not contain the effect of increasing clock speeds, and newer roadmaps of this technology now incorporate this to reflect the rate at which information can be transferred and processed, typically plotting a measure of the number of calculations per second over time (see also Chapter 2, Section 2.2). Notwithstanding these details, a simplified and modified Moore's law plot is shown in Figure 1.1. As usual, time is shown as the horizontal axis, while there are two lines showing the number of components on a chip as the primary vertical axis and

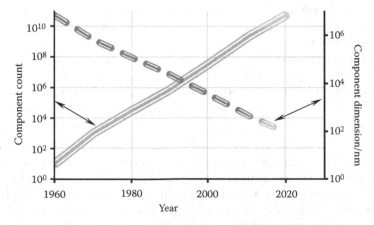

FIGURE 1.1
Simplified Moore's law plot with estimates of component count on a chip and size of these components assuming a square chip of dimension 20 mm and components forming a square array in this area.

component size (assuming that the chip is square and 20 mm on a side). Typically, for a chip consisting of an array of transistors, the features on this component will be around one quarter of the size. A glance at this plot shows the number of components increasing exponentially, with the size of components correspondingly reducing. For example, with a component size of 100 nm, the individual feature dimensions will be of the order 25 nm. A reduction in component size to 10 nm in the mid-2020s indicates features to have dimensions of the order of only a few nanometres. Interestingly, when the dimensions of electrical circuits approach atomic scales, quantum effects will significantly influence the nature of conduction and place constraints on the motion of electrons. Fundamentally, a wire must comprise a conductor surrounded by an insulating barrier. However, on a quantum level, the barrier only affects the probability of the electron being located in the region of the conductor and insulator, with an exponentially decaying probability that the electron will be found at a specific distance into the insulator. However, if the insulator is sufficiently thin and there is another conductor adjacent, the probability that electrons from one conductor will 'appear' in the adjacent one, and vice versa, becomes significant and there will be a tunnelling of currents between conduction paths. Additionally, the wave functions of the conducting electrons will overlap, resulting in interference effects that might be exploited in the form of quantum-based computing devices. There are clearly challenges and opportunities for the future as technology pushes against the basic laws of physics.

Moore's law is an excellent map to illustrate technological change: its motivation (the information age) and limits to progress (manufacturing precision). Another map developed by Norio Taniguchi (1974) provides insight into the response of manufacturing industries to address the limitations imposed by the need for increased precision to facilitate technological progress. Figure 1.2 plots the achievable accuracy of manufacturing machines and processes

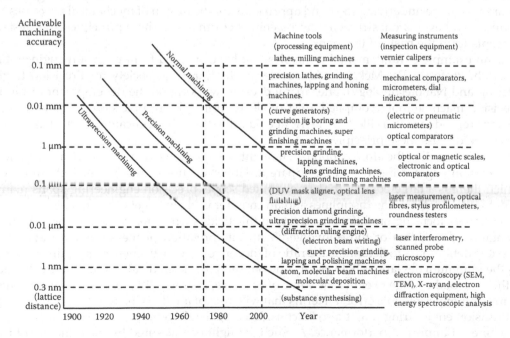

FIGURE 1.2
Taniguchi plot indicating the evolution of precision manufacturing machines and instruments as a function of time. (Adapted from Taniguchi N, 1974, On the basic concept of nanotechnology, *Proceedings of the International Conference on Production Engineering Tokyo* (Tokyo ASPE), Part 2:18–23.)

as a function of time, starting at the end of the nineteenth century. Within this graph, three lines indicate representative values for the achievable accuracy of manufacturing machines. To the right of the plot are two columns indicating the types of machines, and measuring instruments that are capable of the indicated performance of the state-of-the-art in terms of precision at that time. A major and profound conclusion from Figure 1.2 is that the trend showed that, with the current rate of progress, the resolution of state-of-the-art precision manufacturing equipment in the 21st century will provide the capability to deterministically process materials at atomic levels. This new capability he termed 'nanotechnology' (Taniguchi 1974), although it is not entirely clear that he was the first to use this term. Most of the processes evaluated in Taniguchi's study involve manufacture by either adding or removing materials to pattern relatively large areas, often termed a top-down approach. Nature, on the other hand, constructs life by building structures at a molecular level, a so-called bottom-up approach. A merging of these two approaches clearly creates exciting opportunities for future technologies.

Much of the preceding discussion has placed an emphasis on the (very important) semiconductor industry and its impact on technology. However, similar demands for increasing precision have also evolved in almost all other industrial sectors, ranging from planes, trains and automobiles, to printers (printing electronics and optical surfaces), scientific and analytical instruments (microscopes and telescopes to particle accelerators), medical and surgical tools, and traditional and renewable power generation. At the root of all of these technologies are increasingly advanced machines and control systems. This ubiquitous need for increased precision has resulted in the growth of an international community with precision engineering societies in Asia, Europe and America gathering to exchange information and ideas focused on conceptual and technological solutions for the creation of increasingly precise processes. An outcome of these societal activities has been a growing awareness of the underlying tools and approaches for the design of mechanical processes to optimise and evaluate instrument and machine performance. These principles and design concepts form the basis of this book.

As an example of fundamental principles that have evolved from precision engineering research, Professor Pat Mckeown (a founder of the European Society for Precision Engineering and Nanotechnology), after many decades working on the development of ultra-precision machine tools, compiled a list of eleven principles of precision machine design that are reproduced in Table 1.1. The right column of this table indicates which chapters of this book discuss these principles.

Considering all of the aforementioned, it might be supposed that a simple, clear definition of precision engineering would be apparent. However, unlike many fields of study for which there is a targeted and, therefore, limited scope, precision engineering spans many scientific and engineering disciplines and an enquirer is likely to get different answers from practitioners focusing on different aspects in this field, such as machine designers, instrumentation developers or those involved with interpretation of measurement data. One possible definition is to determine the ratio of the range of an instrument or machine divided by the minimum deviations between its outcome and the desired value. Quantitatively, this is the ratio of the range to either the value of accuracy, repeatability, resolution, uncertainty or maximum permissible error (all terms that will be explained in Chapter 2 and Chapter 5).

Precision engineering might also be considered as the drive to provide instruments and machines of improved performance. As such, it might be measured by the difficulty of the

TABLE 1.1

Eleven Principles of Precision Machine Design Enumerated by Pat Mckeown

Principle	Considerations	Chapter
1. Structure	Symmetry, dynamic stiffness, high damping, stability, thermal stability, independent foundation, seismic isolation	1, 7, 12, 13
2. Kinematic/semi-kinematic design	Rigid body kinematics, three-point support	6
3. Abbe principle	Alignment	10
4. Direct displacement transducers	Scale or laser interferometer(s)	5
5. Metrology frames	Isolate measuring system from force paths and machine distortion	11
6. Bearings	High accuracy, high averaging/low rumble, low thermal effects, low limiting friction	7
7. Drives/carriages	Through axes of reaction, non-influencing couplings and clamps	7
8. Thermal effects	Eliminate/minimise thermal inputs and drift, stabilisation, compensation	2, 12, 13
9. Servo drives and control (CNC)	High stiffness, response, bandwidth, zero following errors, dynamic position loop, synchronisation	7, 14
10. Error budgeting	Geometrical-angular, straightness and orthogonal error motion, thermal-loop expansions, deformations	8, 9
11. Error compensation	Linear, planar, volumetric, quasi-static and dynamic	2, 8

endeavour or availability of existing technologies and solutions to satisfy specific needs. One method of assessing difficulty is to map the performance of existing instruments or machines (that, in a competitive economic environment, will already be optimised for performance using all available technologies) and judge the technological difficulty by looking at the region on this map to be occupied by the proposed new design. One particularly useful mapping is to plot the limits of performance of machines in terms of the height of features that can be realised (or measured) to the length or area over which they can be evaluated. For example, this might be related to the shapes and sizes of precision optics manufacture (Franks et al. 1988), or, more commonly, the performance of surface measuring instruments, as presented by Stedman (1987). Such maps are now referred to as 'Stedman diagrams' (see Chapter 5, Figure 5.30).

Precision engineering has been promoted as the pursuit of determinism in manufacturing processes (Bryan 1993). In this view, the lack of precision that often comes from a lack of repeatability is considered to be a lack of attention to causal effects within the process. Consequently, precision engineering might be viewed as an endeavour to identify and quantify the causes leading to non-repeatability and utilise the principles of precision engineering to reduce, and preferably eliminate, the effect. For example, with machine tools, common sources of non-repeatability include thermal effects, loose joints, vibration, limited measurement and actuator resolution, contamination, friction, ground loops in electrical circuits, rolling elements and variations in electricity, air and vacuum. Again approaches for addressing these limitations to precision form the basis for many of the topics discussed in this book. However, it is realised that, at a fundamental level, there will always be noise, with the sources and estimates of their magnitude being briefly discussed in Section 1.5.

1.2 Foundational Concepts for Precision Process Design and Evaluation

The following sections form a list of concepts or activities that should be an intrinsic part of the design process, particularly for the development of precision instruments and machines. Not all of these are design concepts, for example, 'performance measures' are typically used to determine the quantitative limitations of any particular design, but nevertheless are intrinsic to design and process development. On a first pass, many of these concepts may appear to be abstractions from considerations otherwise accepted as 'good practice'. If, after having studied the rest of this book, the relevance and importance of these ideas take on more significance upon revisiting this section, then one major aim of this book has been achieved.

These concepts represent a guide for the designer or design teams at the early creative stages when trying to conceive of possible solutions to engineering challenges where precision performance is a necessary requirement.

Most important, it is necessary to try to avoid subjective bias when judging designs. It is often difficult to throw away an idea after giving it much thought and easy to reject those ideas that appear to lack feasibility at first glance. Often, after a number of ideas have been considered, a new solution will show obvious benefits over all others considered. When this happens, there is often a 'sense' that 'the' solution has been created, at which point, further creative thinking can come to a stop (in fact, there is only ever 'a' solution). Even 'final' solutions need to be reviewed critically and studied to recognise how the thinking about the problem has been changed by 'a' new solution. With this continual process of reviewing, further new, and even better, solutions often start to emerge and the design team will evolve how to think about the nature of the problems and acceptability of solutions for consideration.

After the design concept has been decided upon, it becomes necessary to plan the stages of process development and identify critical performance requirements. Two useful concepts for subsequent process evaluation and measurement, namely Ishikawa diagrams and 'concept of operations' planning, are discussed in Section 1.4.

1.2.1 Analysis Is Not Design Synthesis

No amount of analysis will change a bad design. The outcome of an analysis of a bad design will be an optimised bad design.

1.2.2 Design Specifications and Other Requirements

Requirements should always be stated, where possible, in terms of quantitative measures that are independent of the design of the subsequent instrument/machine. For example, specifying that a load measuring cantilever beam must be capable of withstanding a force of 100 N while not deflecting greater than 1 mm is a design-dependent specification. Requiring that loads covering a range of 100 N with deformations as a result of this load being less than 1 mm is design independent.

1.2.3 Symmetry

Symmetry within mechanisms and supporting structures is easily spotted and results in numerous benefits, mainly those of improving performance and simplifying analysis in terms of stresses and dynamic and thermal response (see Chapter 7, Section 7.1.2). For example, it is not necessary to use any analysis to deduce that forces generated by a machine can be exactly balanced in a given direction by attaching it to a second identical machine to create a plane of symmetry.

1.2.4 Identify and Eliminate, Where Possible, Bending Moments

Bending moments can be considered as force multipliers resulting in correspondingly large stresses and strains. Deformation is the integral of strain and typically results in rotations about the axis of bending, with displacements due to this deformation being increased by the length of the lever arm that is responsible for the bending moment. In many instances, failure of a structural element will occur where bending moments are the greatest. Stresses in these regions are often exacerbated by sharp changes in geometry (stress concentrations).

When there are bending moments, major distortion errors are likely to occur. Where these cannot be avoided, the designer should try to compensate or null the effect. For a common lathe slideway, such as that shown in Chapter 7, Figure 7.16, when a force is applied using a feedscrew drive, it is difficult to eliminate bending moments due to the frictional forces caused by the sliding surfaces. The moments created by these forces can result in small pitch and yaw error motions of the moving carriage. However, by joining two slideways together (see Figure 7.19) and driving the moving carriage through a line of symmetry, moments will be substantially reduced.

1.2.5 Loops

Newton's third law states that for every action there is an equal and opposite reaction. Hence, for any forces applied to a mechanism that is in equilibrium, there will be a loop of structural elements around which this force is transmitted, called the force loop. Forces produce distortion of the structural elements around this loop. When measuring relative displacements or locations of points in a structure, there will also be a loop of structural elements connecting the measurement scale to the point of interest, called the metrology loop. To eliminate measurement errors, the metrology loop should ideally remain rigid. Hence, because force loops distort and metrology loops should be rigid, the structural elements of these two loops should be separated as far as is possible. These and other force loop issues are discussed in detail in Chapter 11.

Other rules for loops include: keep elements stiff; keep forces small and moments smaller; maintain stable temperatures and avoid temperature gradients, especially in the direction for which the structure has no symmetry; and ensure resonant frequencies are high and keep loops small (for high frequency response, small bending moments, low self-weight distortion and fast thermal response).

1.2.6 Stiffness

Most machines, whether for measurement or processing, are required to deliver precisely controlled work and power. Stiffness is a direct measure of the ability of a machine to maintain precision while doing mechanical work and accelerating (and decelerating) structural elements, and is of particular importance when the components of metrology and force loops coincide.

For many precision applications, the requirement for stiff structures results in small strains within the force loop components so that strength need not be considered. However, the large structures required for high stiffness tend to have high mass and reduce the dynamic response, which can be a significant problem.

It is difficult to predict, ahead of time, the nature of the forces that might be produced in complex mechanisms, so that it is often necessary to consider the effect on a structure of forces and moments in many coordinate directions. For many mechanisms, it is necessary to consider the three-dimensional stiffness of a design subject to all possible forces and moments.

In some cases, low stiffness is a desired goal. One example might be a force controlled machining process such as polishing. For such a process, the goal is to apply a constant (and often, uniform) force between the specimen and polishing lap. Gravity is often utilised and provides an effective zero-stiffness force vector.

1.2.7 Compensation

When calibrating instruments or machines, it is often found that errors will repeat. Errors are typically constant or can be predicted if they follow a systematic or causal model. Consequently, repeatable errors can be minimised; at least to within any random or unpredictable deviations from calibration. Known errors are often compensated directly within computers almost routinely with modern machine tools and measuring instruments. Causal errors, such as thermal expansion effects with temperature changes, predictable wear rates or variation of optical wavelength with temperature, pressure and humidity, can be compensated using predictive models, examples of which are given throughout the book.

1.2.8 Null Control

Null control can be used to remove undesirable errors typically, in the case of measurement, by restoring an instrument to a null condition and using the restoring force or displacement to deduce the measured value. A simple and familiar null method is used in the design of the assay balance or common mass balance. Assay balances comprise two balance pans, where a mass to be measured is balanced by calibrated masses. In the balanced state, the mechanism is restored to its configuration before the masses were applied, whereupon any errors due to tilting or bending of the scale arm will compensate about the plane of symmetry of the mechanism.

Sometimes active control can be used with feedback. For example, the proposed new definition of the kilogram balances the gravitational force on a mass with electromagnetic forces in an instrument called the Watt or Kibble balance (Haddad et al. 2016). Another null-based design is that of scanned probe microscopes that measure surface topography using a sharp tipped surface sensor: the 'probe' (see Chapter 5, Section 5.7.4, and Chapter 4). This probe determines the location of a small point on a specimen surface by monitoring the interaction of the tip as it is moved into close proximity with the surface. Today's scanning probes either measure the force of interaction or, for the highest resolutions, the tunnelling of electrons between conducting tips and specimens at tip-to-specimen surface gaps of less than 1 nm. The tip is servo-controlled to keep this interaction constant as it is traversed over the surface (see Chapter 4). Subsequently, the motion of the tip is used as a measure of the surface topography and, in the case of tunnelling probes, is capable of resolving the presence of electron orbitals in individual atoms, a feat that earned research scientist Gerd Binnig and co-workers the 1986 Nobel Prize in Physics. In many such implementations, the null is used as a means of transferring the localised measurement into the mechanism where high precision measurements can be readily achieved.

Zerodur is a two-phase ceramic with proportions of these phases chosen so that the positive expansion of a glass phase is compensated by a negative expansion of a crystalline phase for variations around room temperature, and might be considered a null or compensation approach to the problem of thermal instabilities (see Chapter 12).

1.2.9 Error Separation

Error separation might be implemented in any repeatable machine or process in which the errors can be considered to be the direct sum of two or more effects. If one or more of the effects can be reversed, it is, in principle, possible to separate them to determine their influence or to compensate the effects independently. One example is the measurement of cylinders (or spheres) using a spindle and a probe that measures the radial variations of the cylinder surface as the spindle rotates. This measurement will include both the radial error motion of the spindle and the radial deviations from roundness of the cylinder. If the probe and cylinder are rotated 180° and the measurement is repeated, the errors of the cylinder will be the same, whereas spindle error motions that result in a peak during the first measurement (i.e. the spindle moving towards the probe) will produce a valley when measured from the other side. By adding these two measurements together, the spindle errors will cancel leaving only the out-of-roundness of the cylinder; hence these two errors can be separated. Specific examples of error separation and self-calibration are discussed in Section 5.8.

1.2.10 Self-Correction/Calibration

As with error separation, self-correction uses the ability to utilise the configuration of a process to correct for errors. One application is the lapping of flat surface plates; at least for smaller plates, larger, metre-sized plates are not produced this way. In this process, three plates can be successively lapped against each other, so that flatness errors from one flat will be selectively removed by the form of another, until all three converge to a flat surface (see Chapter 7, Section 7.2.1.2; Evans et al. 1996).

1.2.11 Kinematic Design

Kinematic design represents a methodology aimed towards the design of theoretically ideal and deterministic mechanisms by considering the effect of connecting rigid bodies using joints that are designed to constrain specific degree of freedom of motion between them (see Chapter 6). In particular, it can be readily proved that a point contact between two bodies will constrain a single degree of freedom and the total number of contacts, suitably arranged, will be equal to the total number of constraints.

Exact constraint design considers complete mechanisms for which, if there is to be total determinism, all bodies will be constrained in all degrees of freedom at all times. For example, the moving carriage of a linear motion stage will be free to move along its axis until an actuator is added to constrain it to a specific location in all degrees of freedom.

1.2.12 Psuedo-Kinematic Design

Psuedo-kinematic design is often necessary when load-bearing capability is required. At higher loads, the stresses of 'point' contacts become impractical and these contacts are replaced by conformal surfaces in the form of bearings, hinges, mechanical joints and couplings. Larger area contacts provide higher load capacity but deviate from pure kinematic principles. However, the ideas from kinematic design are not abandoned altogether and joints are still classified, and assemblies evaluated, in terms of the degrees of freedom that they impart or constrain. Kinematic and pseudo-kinematic design, as well as the subsequent 'mobility' of mechanism assemblies, are presented in Chapter 6.

1.2.13 Elastic Design and Elastic Averaging

All manufactured components necessarily deviate from their ideal geometry, often arbitrarily, to within the specified manufacturing tolerances. Correctly toleranced parts can be fabricated into assemblies and will function to specified performance requirements. Depending on the constraints imposed on the assembly (minimised by applying exact constraint and pseudo-kinematic design principles), for components to fit together, some limited deformation is required. This might be considered the exploitation of elastic deformation to provide the functional accommodation of manufacturing tolerances in a mechanism. For example, the assembly of rolling element bearings relies upon elastic deformation to accommodate the small deviations of the rolling elements. If many elements are used in the bearing, the effects of these geometric deviations can average out to provide deviations of the inner and outer race of the bearing that are smaller than that of the individual rolling elements. Elastic averaging is often considered a form of elastic design.

1.2.14 Plastic Design

Permanent deformation can be exploited to provide conformal contacts or component alignment. 'Running-in' of sliding surfaces is a combination of plastic design and elastic averaging. Other processes that result in precision forming or conformal surfaces include lapping and honing, the use of 'white metals' (chosen for embeddability, conformability and tribological compatibility), compliant nuts, slideway scraping and sphere manufacture using tumbling.

Dial gauges are used for many precision measurements and typically comprise a spherical contact on a sliding or rotating shaft. In some applications, such as a gauge block comparator (see Chapter 5), repeated contact with flat and polished surfaces will preferentially wear micrometre-scale asperities until the surface of the sphere in the contact region is itself flattened to a large radius and 'burnished', so that plastic stresses are reduced to elastic contact conditions, resulting in more repeatable measurements.

Most fasteners rely on the plastic deformation of surface texture asperities to develop a sufficient contact area. Whether the surface contact conditions will be elastic or plastic is outlined in Chapter 7, Section 7.2.1. Other examples of elastic and plastic design are presented throughout Chapter 7.

1.2.15 Reduction

It is surprising how often, with a little design ingenuity, many sub-assemblies within a mechanism can be replaced with a single component and sometimes these can be eliminated altogether. Replacement of complex assemblies with fewer components will always simplify analysis. Many assemblies, especially those requiring many fasteners or requiring specific tolerances to be functional, are resistant to precise and/or deterministic theoretical models.

Another issue with complex assemblies is that of tolerance stack-up. As the number of components in an assembly increases, the deviation of the assembled mechanism from the desired geometry will increase, often to the detriment of its overall functionality. The role of reduction as a simplifying principle when considering assembly alignment is discussed, with a detailed case study, in Chapter 10, Section 10.1.5.

Reduction can often be used when looking at the evolution of engineering solutions as technologies progress. For example, carburettors have given way to fuel injectors and there is now a slow evolution towards electric motors as a replacement for the internal combustion engine.

Related to reduction is the principle of, whenever possible, using direct rather than derived measurements. For example, the pressure of a known gas can be obtained from measurements of volume and temperature. Although both ideally provide the same result, the direct pressure measurement will half the number of measurements, thereby reducing the number of influence factors in an uncertainty determination.

1.2.16 Cosine and Abbe Errors

Monitoring of linear or rotational motions within instruments and machines often uses scales in the form of a bar or disk having regularly patterned lines that can be sensed either electronically or optically. With the scale fixed to one part of a machine, and the sensing head attached to a moving element (a carriage say), the reading is obviously the motion of the sensor head relative to the scale. However, the design intent is to measure the motion at one point on the machine relative to another, for example, the motion of a cutting tool relative to the workpiece that is being machined. For practical reasons, the sensor and scale are often at different locations on the machine than the line of interest. This difference in location between the line (or axis) of interest and the measurement results in alignment (cosine) and offset (Abbe) errors that can be significant sources of measurement or position uncertainty, even for small motion errors of the machine (see Chapter 10, Section 10.1).

1.2.17 Design Inversion

Sometimes the mechanism of a system can be constructed in an inverted configuration. If this is so, consideration of the differences between the designs often illuminates the relative merits and limitations of different approaches to solving a problem and, sometimes, the nature of the problem.

Examples of inversion include:

- Heat pumps or refrigerators
- Power cables (mount high and move vehicles under, or mount below ground and drive over)
- Suspension bridges (wires strong in tension)
- Arch bridges (concrete strong in compression)
- Turning tools (reverse spindle and mount the tool upside down so that the chips fall away from the cutting zone)

- One large motor driving many small machines or a single small motor driving each machine
- Brake cables on bicycles routed outside or inside the tube (one easier, the other more streamlined and does not snag)
- Drum brake pads push away from one another but are symmetric about the axis, and disc brake pads push towards one another but apply moments that produce a torque on the supporting axle

In most machine tools (milling, grinding), the tool goes in the spindle and the part moves. In a lathe, the part goes in the spindle and the tool moves.

1.2.18 Energy Dissipation

Systems without damping will oscillate endlessly when subject to disturbing forces. Adding damping often reduces the amplitude of vibrations and makes control easier, but also generates heat and slows things down. Consequently, getting damping right always represents a compromise between speed and precision.

1.2.19 Test and Verify

Most designs work on paper and look good in a solid model. Models, however, do not reveal all aspects of the functioning device. Many large corporations have suffered losses, both financially and in terms of reputation, by over-reliance on models. For a design to be successful, a lot of interacting influences will come to play, including social, physical and economic. Hence, prototypes help to address all issues of fabricating the process and getting a product to market. Rapidly producing and testing prototypes helps to identify major issues at the early stages of development.

1.2.20 Occam's Razor

Briefly stated: 'simpler is better'. When there are competing solutions to a problem, each of which satisfies the requirements at comparable cost and performance, it is best to choose the simplest. Essentially, Occam's razor is used to make a decision between competing designs that appear to solve the same problem with equal fidelity.

1.3 Performance Measures

When a process is developed, it is essential that the performance is evaluated and specified with terminology that engineers can understand, to determine whether it will provide a solution to a specific need. Consequently, the customer is interested in all relevant parameters, such as range of operation, speed (or bandwidth) and repeatability. For these terms to be unambiguously understood, they must be specifically defined. For the purpose of defining terms used in metrology, there is a guide called the *International Vocabulary of Metrology* (VIM), which is presented in Chapter 2.

Other terms of particular relevance, and also detailed in Chapter 2, are traceability of measurements to international or national standards as a means of maintaining accuracy for interchangeability, specification standards that provide specific guidelines for the experimental measurement of performance parameters, as well as algorithms and protocols for the reporting of results. By defining how performance is measured, manufacturers in competing fields are able to provide performance measures that can be directly compared. When producing products that may be used in safety critical applications or are capable of environmental damage, adherence to specification standards is important to ensure safe practice and protect from legal liability. Finally, at the heart of precision engineering is calibration, a process that can be considered as the transfer of physical standards to the measurements within a processes, to within the limits of the calibration processes (see Chapter 2). Adherence to all of the principles for determining performance parameters is essential for the realisation of deterministic processes.

When reporting the performance measures as well as any other measurements, it is necessary to evaluate the limits to which the values are considered to correspond to the actual quantity to be measured, called the measurand in the VIM. For a direct measured value (such as the length of a rod), it is required that an estimate of the uncertainty of the value is always determined and recorded. Again, there are developed procedures for determination of uncertainty, which are described in the *Guide to the Expression of Uncertainty in Measurement* (GUM). Adherence to the GUM is of fundamental importance for all scientific measurements and is discussed in detail in Chapter 9.

Evaluating machine performance is complicated by the fact that different customers will use the machine for processing of components having almost arbitrary composition and geometry. Because it is often not possible to know what is being processed, it is not possible to specify performance in terms of an uncertainty (in fact uncertainty should only be used when reporting the outcome of a measurement). Consequently, performance measures are usually specified in terms of measured values during performance testing when measuring a standard part (typical for measuring machines) or, for machine tools, when the cutter is not removing material ('cutting air'). Hence, the performance measures (with their associated uncertainty) in this case are termed the maximum permissible errors (MPEs) of the machine (see Chapter 5).

1.4 Development of Precision Processes

Ultimately, the precision engineering principles outlined throughout this book will be incorporated in the creation of new technologies and processes. Such processes start as problems or ideas, the solutions of which are envisaged to provide a worthy benefit. Having decided upon a solution path, a process team is assembled with individual members addressing specific components of the solution. This team effort must then be planned and managed. To manage a process or machine development, it is desirable to map out both the major components of the development and schedule the required operations for its full implementation. Two methods that are frequently used for these purposes are briefly outlined in the following two sections. The first of these, Ishikawa diagrams, represents a method for visualising all elements of process development in a single diagram. The second, concepts of operations, is a detailed guide for process or operations planning.

1.4.1 Ishikawa Diagrams for Precision

Originally called a cause and effect diagram, now named after its creator, an Ishikawa diagram contains a central horizontal arrow pointing to a target goal at the right-hand side, as shown in Figure 1.3 (Ishikawa 1990). For a precision engineering process, the major outcomes might be the creation of a machine having an MPE, and measurement with known uncertainty or a machine tool that satisfies desired performance specifications. Above and below this central arrow are primary arrows representing major categories that must be addressed and understood to achieve the stated goal. The Ishikawa diagram shown in Figure 1.3 indicates the major categories for a generic process development project. Other arrows connecting to these primary arrows indicate specific factors that will influence the outcome of the process. For the generic example of Figure 1.3, the major categories for a mechanical process are the materials being processed, the processing operations, measurements, machines and team activities. In practice, while these categories might be different, it is important that all factors that can influence the process outcome are included in the diagram. Once all of the factors have been identified (including sub-factors shown as arrows above and below the arrows of the factors), it is then possible to evaluate the tasks that will need to be addressed to create a deterministic process of the required precision. These tasks can then be incorporated into the project management and systems engineering plans outlined in the following section.

An example Ishikawa diagram for a micro-droplet deposition instrument is shown in Figure 1.4. The goal of this instrument is to deposit droplets onto a specimen surface using an automatic syringe, with positioning of the syringe relative to the specimen achieved using three orthogonal linear translation stages. Once the droplets have been deposited, they are imaged from the side using a camera and the contact angle between the spherical droplet (ignoring gravitational distortions) and the flat surface is determined using image processing techniques. To market this instrument as a metrology tool, it is necessary to provide customer information on the appropriate protocol for evaluating uncertainty of subsequent measurements. Being a small team, development of the roles and structuring of team members is more readily resolved at the project planning stage. From Figure 1.4 it can be seen that the process has been split into four categories. These categories are (1) surface and fluid controls that will depend on the physical interactions and interaction dynamics between the droplet and surface for specific syringe needles; (2) measurement of the contact angle from the camera image based on its orientation relative to the droplet and surface; (3) parameters of the droplet deposition procedure determined by the speed of the fluid coming out of the syringe and the approach and retraction of the needle from the surface; and, finally, (4) the performance of the positioning system.

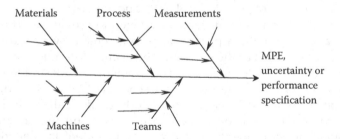

FIGURE 1.3
Ishikawa diagram for process development.

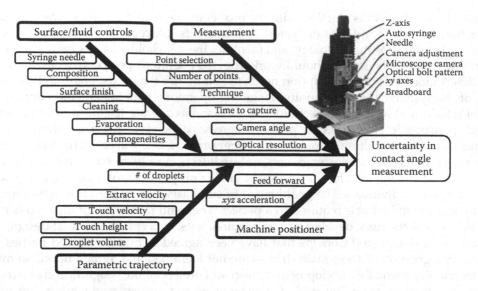

FIGURE 1.4
Ishikawa diagram for a syringe-based micro-droplet deposition process.

1.4.2 Introduction to Project Management and System Engineering

A project in its simplest form will have a project sponsor (customer), project manager and a technical team. The line of communication that is required for a project to run effectively along with the ownership between the project sponsor, project manager and technical team is shown in Figure 1.5. The project sponsor defines the scope of the project in terms of outcome/goals and provides financial support. The project manager interfaces with the project sponsor and technical team to define the required process steps (required documentation, milestones, project reviews etc.), schedule and budget. The technical team is typically led by a lead systems engineer (also referred to as the lead engineer at some

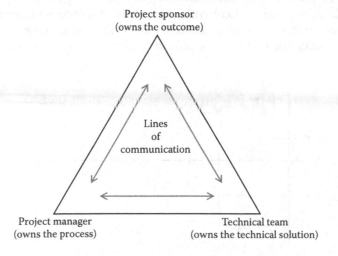

FIGURE 1.5
A triangular depiction of the communication lines that is required between the project sponsor, project manager and technical team.

organisations) who works together with the project manager and project sponsor to define the technical solution in terms of system requirements, concept of operations (ConOps), design and verification methodology, and identifies the stakeholders. ConOps is a document describing the way the system should work from the operator's perspective and includes a user description. This user description needs to summarise the needs, goals and characteristics of the system's user community, including operators, maintainers and support personnel (Haskins et al. 2007). Stakeholders may also incorporate technical teams from the project sponsor side. How in detail a project progresses depends on the size of the project and individual corporate culture. Typically, for larger projects, this will entail having a product and project manager, systems engineers and architects, and team members with specific technical expertise required for the project. In smaller projects, the roles may be combined to a single or few individuals. The goal here is to emphasise the importance as an engineer to fully understand what is required in a project. The main takeaway for an engineer is to identify the stakeholders, develop system requirements, conceptual and final design, verification methodology and ConOps that have been agreed to by all involved parties, and preferably signed off. If these main drivers are not identified and agreed upon, no matter how technically sound the development is, the risk of project failure is significantly increased. For further reading regarding project management and system engineering, the reader is referred to Gruebl and Welch (2013), Haskins (2006), Montesanti (2014) and PMI (2013).

1.5 Limits of Precision

Considering precision engineering to be the pursuit of determinism, it is informative to determine the fundamental limits of a measurement process that, in turn, provides a measure of the degree to which a process can be controlled (see Chapter 14 for more information on control systems).

Consider a displacement sensor (such as a capacitance gauge, Chapter 5) that outputs a voltage that can be plotted as displacement x against time. Figure 1.6 shows such a measurement signal, where the vertical axis represents displacement and the horizontal time axis intercepts at the actual displacement x_t, and the signal noise is assumed to vary about this value. Analogue voltages are almost invariably converted into binary number

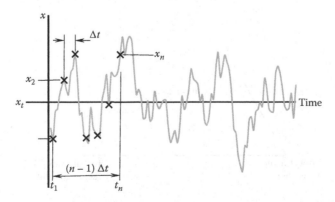

FIGURE 1.6
Displacement signal from an electrical sensor as a function of time. Markers indicate samples of the signal.

representations, using analogue-to-digital converters (ADCs). Consequently, ADCs are used to collect samples of the signal where these samples are represented by the markers in Figure 1.6. To improve an estimate of the displacement, an obvious approach would be to collect a number of samples and then average the readings. If the original signal has a standard deviation σ_{x_i}, then taking a total of n samples and averaging these, then a measure of the displacement x_m and its standard deviation can be computed from

$$x_m = \frac{1}{n} \sum_{i=1}^{n} x_i$$

$$\sigma_{x_m}^2 = \frac{\sigma_{x_i}^2}{n}. \tag{1.1}$$

Based on Equation 1.1, it would seem reasonable to record a very large number of samples to reduce the deviation of the measured values. However, if all of the samples are collected in a short time interval, then, in the limit of zero time interval, they will all be the same value and are said to be correlated. Consequently, for the successive sample to be uncorrelated, it is necessary to use a finite sampling interval, Δt in Figure 1.6. To determine this interval, it is possible to compute the autocorrelation function of the signal by simply multiplying the signal by a copy of itself that is shifted a finite distance τ, integrating this product and averaging over the total time. Mathematically, the autocorrelation function can be computed from

$$R_{xx}(\tau) = \frac{1}{T - \tau} \int_{0}^{T-\tau} x(t)x(t - \tau)dt, \tag{1.2}$$

where T is the total time of the signal. Certain features of this function can be determined without resort to computation. When the time shift is zero, Equation 1.2 is the variance of the signal. Similarly, if the time shift is large, then, for these random signals, the products of the two signals will also be random, so that the average integral will tend to zero as can be seen by the autocorrelation function of a random signal in Figure 1.7. It is also shown in

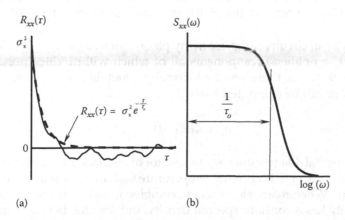

(a) (b)

FIGURE 1.7
Correlation and frequency content of a random signal, (a) autocorrelation function, (b) power spectral density.

Figure 1.7 that the autocorrelation function for this signal can be modelled as an exponential decay of the form

$$R_{xx}(\tau) = \sigma_x^2 e^{-\frac{\tau}{\tau_0}}, \tau > 0$$
$$= \sigma_x^2 e^{\frac{\tau}{\tau_0}}, \tau < 0,$$

(1.3)

where τ_0 is the correlation length and might be considered a minimum time for sampling from the signal if successive measurements are to be considered uncorrelated. The Fourier transform (the equivalent of the Fourier series for signals containing a continuous range of frequencies and therefore not being periodic, see Chapter 3) of the autocorrelation function produces the power spectral density $S_{xx}(\omega)$ in the form of the noise power per unit frequency in the signal. Because the autocorrelation function is even, the power spectral density is given by

$$S_{xx}(\omega) = \frac{\sigma_x^2}{\pi} \int_0^\infty e^{-\frac{\tau}{\tau_0}} \cos(\omega\tau) d\tau = \frac{\sigma_x^2}{\pi\tau_0} \frac{1}{1+(\omega\tau_0)^2} = \frac{S_o}{1+(\omega\tau_0)^2}.$$

(1.4)

Equation 1.4 is plotted in Figure 1.7b and shows that the signal has a constant frequency spectrum with a bandwidth equal to the reciprocal of the correlation length. Returning to the measurement, it is possible to now write the number of samples in terms of the sampling time

$$\sigma_{x_m} = \frac{\sigma_{x_t}\sqrt{\tau_0}}{\sqrt{T}} = \frac{\sigma_{x_t}}{\sqrt{(BW)_s}}\sqrt{(BW)_m} = P_o\sqrt{(BW)_m},$$

(1.5)

where P_o [m $\cdot$ Hz$^{-1/2}$] is the noise measured in metres per square root of bandwidth and BW is the bandwidth, with the subscripts s and m indicating the original signal and measurement, respectively. Equation 1.5 indicates the fundamental relationship between speed and precision. Clearly, to increase measurement precision given the available signal, the inherent measurement deviations can only be reduced by correspondingly reducing the measurement bandwidth. The square root term indicates a diminishing return, for example, reducing the bandwidth by a factor of 100 will only produce a tenfold improvement in precision.

In practice, electrical circuits comprise many passive (resistors, inductors, capacitors) and active (transistors, amplifiers) components all of which will produce noise. For even the simplest passive elements, there will be thermally generated noise, called Johnson noise, which for a resistor can be computed from

$$\sigma_V^2 = 4RkT(BW)_{HZ},$$

(1.6)

where V is the potential difference across the resistor of resistance R, k is Boltzmann's constant (equal to 1.38×10^{-3} J$\cdot$K^{-1}), T is absolute temperature and, in this equation, the bandwidth is measured in hertz. When many elements are combined, each will contribute to a complete noise that typically has a constant spectral density, but also tends to increase as frequency reduces. Such low frequency noise is modelled as being inversely proportional to frequency and is called $1/f$, pink or flicker noise. Hence, a typical noise spectral density will be of the form

$$G(f) = \frac{C}{f} + G_0 \quad [\text{V}^2\,\text{Hz}^{-1}], \tag{1.7}$$

where $G(f)$ is the one-sided spectral density, f is frequency in hertz, C is the flicker noise constant and G_0 is the white noise spectral density. Integration of the one-sided spectral density gives the mean square noise

$$\sigma_V^2 = \int_{f_1}^{f_2} G(f)df = C\ln\frac{f_2}{f_1} + G_0(f_2 - f_1) \quad [\text{V}^2]. \tag{1.8}$$

Equation 1.8 indicates the temporal nature of precision, in particular, the difficulty of maintaining precision over long timescales. Electrical circuits also carry currents via transmission of electrons that also contain temperature fluctuations, often modelled as a Poisson statistic (see Chapter 3). Noise currents are assumed to be a white noise spectrum with current deviations σ_i predicted from

$$\sigma_i = \sqrt{2qI(BW)}, \tag{1.9}$$

where q is the charge on an electron (1.602×10^{-19} C) and I is the dc value of the current. Current noise will show up as a voltage, as the current is transmitted through the impedance of circuitry, such as a resistor. Current noise played through a loud speaker sounds a little like a metal shot falling on a drum and is often referred to as 'shot noise'. Other sources of electronic or sensor noise can be attributed to specific physical effects, such as burst noise (also called 'popcorn') typically attributed to defects in oxide layers in transistors.

Today's clocks can measure time with unprecedented precision. Atomic clock and, publically available, global positioning system (GPS, not to be confused with geometrical product specification described in Chapter 2) signals have stability and accuracy measured as parts in 10^{11} or more, at frequencies ranging from 10 MHz to 1.57542 GHz. The precision of these clocks can be exploited by converting sensor voltages to signals that encode these voltages in a time signature. This has the twofold advantages that, once converted, the signal can be read with the precision of the clock and also that, being based on time alone, changes in amplitude of the signal at the receiver have no effect on the value. There are a number of ways that voltages can be converted to a time-based signal. Possibly, the most common method is to use a voltage controlled oscillator (VCO) and then measure the frequency of the transmitted signal. This is commonly used in radio transmission, with high frequency transmission being wirelessly transmitted. Typically, at the receiving end, the frequency is followed using a phase locked loop (PLL), in which a similar VCO is used to follow the frequency variations by keeping the phase between the received signal and an internal VCO constant (Best 2007). Because phase is the integral of frequency, this is equivalent to an integral-based control (the 'loop' in the initialism PLL refers to the closed 'loop' control, see Chapter 14).

Another method, called pulse width modulation (PWM), converts a voltage into a periodic on–off cycle of constant period. The value of the voltage is proportional to the ratio of the time spent in the on state to the time in the off state.

For measuring phase shifts in laser interferometers (and radio communications), one common method is to create two laser beams having a constant frequency difference (called a heterodyne method), typically around 3 MHz to 80 MHz. These two beams are configured in a Michelson interferometer with each frequency going around separate paths before they are recombined (see Chapter 5 for the basics of interferometry). When the two beams combine, they produce a beat in intensity equal to the frequency difference, with the phase of these beats being the same as the phase shift produced by a single frequency (or homodyne) interferometer. Converting the periodicity of the beat of the interferometer output into a square wave (or on–off cycle), this can then be compared with the original beat frequency of the two beams before entering into the interferometer optics (see Chapter 5, Section 5.4.5). Hence, the phase shift of the interferometer can be determined by timing the 'on' transition between the reference and output signals.

Using these time-based methods to measure sensor outputs has the potential to exploit the precision of clocks. However, the conversion from analogue to digital signals invariably involves using circuits to compare voltage transitions and these circuits will be subject to the fundamental noise limitations mentioned earlier. Typically, this leads to a noise in the transition timing (called 'jitter') that, in turn, shows up as noise in the measured value.

For mechanical mechanisms of reasonable size, there does not appear to be any significant limits on position resolution. However, as the scale of mechanism or mechanism stiffness reduces, thermal noise is detectable and represents a fundamental limit to position resolution. In general for any state of a system, there will be a noise energy of magnitude one half Boltzmann's constant times absolute temperature. As a simple example, a point particle of mass m and stiffness s (s being chosen to avoid confusion with Boltzmann's constant) constrained to move only in a single linear degree of freedom x will, over time, exhibit average displacements $\bar{x}$ and velocities $\dot{\bar{x}}$ given by

$$m\dot{\bar{x}}^2 = kT, \tag{1.10}$$

$$s\bar{x}^2 = kT. \tag{1.11}$$

This rather surprising result incorporates no knowledge of how this mechanism is implemented and is an example of Einstein's principle of equipartition (Bohm 1951) that can be deduced for any mechanism. As an example, consider the equation for the free displacement of a string given in Chapter 3, Section 3.7.11,

$$y(x,t) = \sum_{n=1}^{\infty} \left[\sin\left(\frac{n\pi x}{l}\right) (a_n \cos(\omega_n t) + b_n \sin(\omega_n t)) \right] = \sum_{n=1}^{\infty} \phi_n(x) q_n(t), \tag{1.12}$$

where the constants a_n and b_n depend on initial displacement and velocity of the string and

$$\omega_n = \frac{n\pi}{l} c. \tag{1.13}$$

Within this summation, the term on the left is the dimensionless mode shape φ and represents half of an odd Fourier series representation of period $2l$ that can be used to model an arbitrary deformation shape of the string between the bridge supports at each end over which the string is stretched. The term on the right-hand side has units of length and can be used to represent the infinite number of 'normal' coordinates q of this system. Derivatives of Equation 1.12 with respect to x and t are

$$\frac{dy}{dx} = \sum_{s=1}^{\infty} \left(\frac{n\pi}{l}\right) \cos\left(\frac{n\pi x}{l}\right) q_n$$

$$\frac{dy}{dt} = \sum_{n=1}^{\infty} \phi_n \dot{q}_n. \tag{1.14}$$

In this case, the potential energy V and kinetic energy T_{KE} are given by

$$V = \frac{F_o}{2} \int_0^l \left(\frac{dy}{dx}\right)^2 dx = \frac{F_0}{2} \int_0^l \left(\sum_{n=1}^{\infty} \left(\frac{n\pi}{l}\right) \cos\left(\frac{n\pi x}{l}\right) \right)^2 dx$$

$$= \frac{F_o}{4} \sum_{n=1}^{\infty} \left(\frac{n\pi}{l}\right)^2 q_n^2, \tag{1.15}$$

$$T_{KE} = \frac{\rho l}{4} \sum_{n=1}^{\infty} \dot{q}_n^2,$$

where F_o is the tension in the string and ρ is its mass per unit length.

From Equation 1.15, it is apparent that the potential and kinetic energy terms are in simple quadratic forms representing independent states of the string. Because of this, the coordinates of this system can be used with Lagrange's equation to derive equations of motion in terms of an infinite sum of single degree of freedom systems (a technique called modal analysis, not covered in this book). For each individual normalised coordinate, the mean square motion due to thermal excitation will be

$$\frac{F_o l}{4} \left(\frac{n\pi}{l}\right)^2 \overline{q_n^2} = \frac{kT}{2}$$

$$\frac{\rho l}{4} \overline{\dot{q}_n^2} = \frac{kT}{2}. \tag{1.16}$$

Because of the orthogonality of the mode shape for this, and any other, vibrating systems, the mean square of the displacement, given by the upper line of Equation 1.16, is

$$\overline{y(x,t)^2} = \left\langle \left(\sum_{n=1}^{\infty} \phi_n(x) q_n(t) \right)^2 \right\rangle = \sum_{n=1}^{\infty} \phi_n^2(x) \overline{q_n^2(t)} = \sum_{s=1}^{\infty} \sin^2\left(\frac{n\pi x}{l}\right) \overline{q_n^2(t)}$$

$$= \left(\frac{2kTl}{F_o \pi^2}\right) \sum_{n=1}^{\infty} \frac{1}{n^2} \sin^2\left(\frac{n\pi x}{l}\right). \tag{1.17}$$

Observing the motion of the midpoint of the string, the sine term disappears for even values of s and is equal to unity for odd values. Hence, Equation 1.17 becomes

$$\overline{y(x,t)^2} = \left(\frac{2kTl}{F_o\pi^2}\right) \sum_{n=1,3,5}^{\infty} \frac{1}{n^2} = \left(\frac{2kTl}{F_o\pi^2}\right) \frac{\pi^2}{8} = \frac{l}{F_o 4} kT. \tag{1.18}$$

Consider now a string subject to a deflection δ at its midspan. This will give rise to a triangular shape with an angle of $2\delta/l$ that must be balanced by a force of magnitude

$$F = 2F_o \sin\left(\frac{2\delta}{l}\right) \approx \frac{4F_o\delta}{l}. \tag{1.19}$$

The stiffness s of the string at the midpoint is given by

$$s = \frac{dF}{d\delta} = \frac{4F_o}{l}. \tag{1.20}$$

Substituting Equation 1.18 into Equation 1.20 yields the Brownian motion of the midpoint of the string

$$\overline{y(x,t)^2} = \frac{kT}{s}. \tag{1.21}$$

Equation 1.21 is identical with Equation 1.11 and demonstrates Einstein's principle of equipartition for the case of a string. However, this principle applies to all conservative mechanisms and provides a means to avoid lengthy derivations based on states of a system. Hence, for a more analytically involved case of thermal motion of a cantilever beam in the lateral direction at its free end, it is possible to again use Equation 1.21, only, in this case, the stiffness in this direction is given in Equation 3.108 (see Chapter 3).

For small particles, such as pollen dust on a water surface, this thermally induced motion can be, and was, observed under a microscope by Scottish scientist Robert Brown in 1827 and is now called Brownian motion. For large-sized objects, such as those typically encountered in engineering processes, this ever-present Brownian motion is too small to be detected. For most engineering processes, the major source of mechanical disturbances are due to vibration. Vibrations come from a variety of sources; these mainly being ground-borne, acoustic and self-generated by other moving components in the machine, and methods of isolating mechanisms from such disturbances are discussed in Chapter 13.

References

Best R E, 2007. *Phase-locked loops*. McGraw-Hill.

Bohm D, 1951. *Quantum theory*. Dover Publications.

Bryan J B, 1993. The deterministic approach in metrology and manufacturing. The *ASME* 1993 International Forum on Dimensional Tolerancing and Metrology, June 17–19, Dearborn, Michigan.

Evans C J, Hocken R J, Estler T W, 1996. Self-calibration: Reversal, redundancy, error separation and 'absolute testing'. *Ann. CIRP* **45**:617–635.

Franks A, Gale B, Stedman M, 1988. Grazing incidence optics: Mapping as a unified approach to specification, theory, and metrology. *Applied Optics* **27**(8):1508–1517.

Gruebl T, Welch J, 2013. *Bare knuckled project management: How to succeed at every project.* Gameplan Press.

Haddad D, Seifert F, Chao L S, Newell D B, Pratt J R, Williams C, Schlamminger S, 2016. Invited paper: A precise instrument to determine the Planck constant and future kilogram. *Review of Scientific Instruments* **87**:061301.

Haskins C, 2006. *Systems engineering handbook: A guide for system life cycle processes and activities.* International Council on Systems Engineering (INCOSE).

Ishikawa K, 1990. *Introduction to quality control.* 3A Corporation.

Montesanti R C, 2014. Process for developing stakeholder-driven requirements and concept of operations. *Proceedings of ASPE, Boston, November 9–14,* 329–334.

Moore G E, 1965. Cramming more components onto integrated circuits. *Electronics* **38**(8):114–117.

PMI, 2013. *A guide to the project management body of knowledge: PMBOK® guide.* Project Management Institute (PMI).

Stedman M, 1987. Basis for comparing the performance of surface-measuring machines. *Precision Engineering* **9**:149–152.

Taniguchi N, 1974. On the basic concept of nanotechnology. *Proceedings of the International Conference on Production Engineering Tokyo* (Tokyo ASPE), Part 2:18–23.

2

Metrology

Jimmie Miller

CONTENTS

ABSTRACT Measurements enable engineers and scientists to quantify, model, specify, control and verify component and systemic parameters, both statically and dynamically. In this chapter, concepts of measurement accuracy, trueness, precision, calibration, traceability and uncertainty are examined. The foundation and methods of metrology are discussed from accepted definitions of quantities and units with their physical realisations to the international bodies that provide consensus on related matters.

2.1 Introduction

Metrology, 'the science of measurement' (JCGM 200:2012), comprises all theoretical and practical aspects of processes to obtain numerical values for parametric properties of systems and components. This includes the general principles of metrology, physical standards and documentary standards, along with measurement devices and systems, and their calibration, traceability and related uncertainties. A practical knowledge of metrology is essential because 'progress in science is founded upon the results of measurements, whether carried out primarily for the increase of natural knowledge, or for the improvement of technical processes and apparatus used in all branches of applied science' (Rayner 1923). Although a primary function of measurement is to establish that what has been made is what was intended to be made (within tolerances) and will function as intended, the role of measurement from design through production is to make certain that at each stage, processes are performed in such a way that the specified end result is ensured.

2.2 Metrology and Precision Engineering

Measurement is interwoven throughout all stages of the engineering life cycle. Before the design stage, applied metrology has provided the quantitative foundation for describing the modelled and parameterised capabilities of various engineered materials. The functional engineering specifications require well-defined measureable parameters. These definitions and the methods for measuring them are commonly the subjects of documentary (written) metrology standards. During design, metrology-based simulations are used to predict system performance to evaluate the efficacy of the proposed solution. Culminating the design, metrology information provides the basis for communicating the physical design intent through geometric dimensioning and tolerancing along with other materials specifications. After design, metrology provides the means for construction (accurate manufacturing production, including assembly) via the production system's internal supervision and quality control, as well as process capability studies. Process capability evaluation can be used to provide information to modify overall system parameters to optimise performance and verify that the system is performing to the required specifications. After any necessary refinements (repairs and design and assembly modifications), original operational specifications are confirmed through measurement, with some systems requiring documentation that complies with specification standards. Metrology is essential to the total engineering process and is the foundation for many of the discussions in this book.

The precision engineer takes the measured and modelled values for each of the component parameters along with their associated possible variations (uncertainty) and determines their effect on the ability of the design to meet the specifications within stated tolerances. Furthermore, where the tolerances are not met, precision engineers determine the most influential parametric variations and adjust the materials and/or processes to decrease their uncertainties, so that the specification can be achieved.

The precision engineer, through measurement and modelling, rationally chooses and controls the deterministic properties of the components and their system assembly arrangement to purposefully, predictably, repeatably, accurately and optimally operate according to functionally related, numerically toleranced specifications. A secondary challenge for the precision engineer is to enhance manufacturing capabilities in all areas. An example of this, the increasing demand placed upon precision engineers by the manufacture of integrated circuits (ICs), is found in an observation first shown over fifty years ago (Moore 1965) that the number of components in a microprocessor integrated circuit (IC) approximately doubles every year. This was revised in 1975 to be a doubling every two years (Moore 1975). This exponential growth has come to be known as Moore's law. Moore's law reveals a continual tightening of size specifications on integrated circuits to the point where the interconnecting traces and allowable layer overlay requirements are below 10 nm and approaching 1 nm for some circuits. These feature requirements are less than one part in a million of the IC's chip size dimension. The transistor count increase is made possible by a concurrent increase in the areal density of transistors on each chip. The density increase also positively contributes to the computation

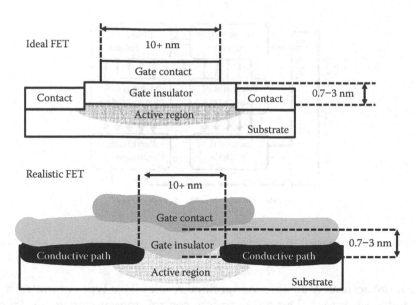

FIGURE 2.1
Drawing of a field-effect-transistor (FET), both theoretical design and a more realistic sketch resembling an as-made FET.

performance doubling every eighteen months, because a smaller transistor area has smaller electrical capacitance, allowing a quicker response. Common transistor gate thicknesses, illustrated in Figure 2.1, are currently below the 10 nm mark, which is approximately 20 times the interatomic spacing of atoms in a silicon crystal. It is at these dimensions that atomic defects can crucially alter the performance of the circuit. Metrology plays a critical role in reaching these manufacturing dimensions with minimal defects. In-process metrology provides the feedback for controlling each process step, whereas post-process metrology ascertains the variations from specification of the final product. The material deposition processes that build the numerous stacked integrated circuit layers require in-process metrology systems which supervise the thickness deposition with atomic layer resolution. A futuristic look at the metrology needs and other technical roadblocks of the semiconductor industry has been provided for nearly two decades by the combined international Semiconductor Industries Associations' International Technology Roadmap for Semiconductors (ITRS), ITRS 2.0 (2015). Recently, the ITRS has entered into cooperation with the Institute of Electrical and Electronics Engineers Rebooting Computing initiative (IEEE-RC) to 'create a new roadmap to successfully restart computer performance scaling' (Conte and Gargini 2015). This roadmap is called the International Roadmap for Devices and Systems (IRDS).

2.3 Metrology and Control

To optimise the operation of a designed system through control of its parameters, accurate metrology having both trueness and precision is a requirement. Manufacturing process control, as illustrated in Figure 2.2 is the means, both analytical and physical, used to analyse the quantitative values obtained through measurement and alter the operation of

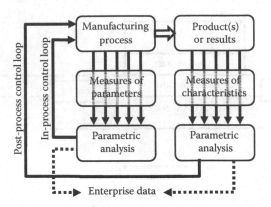

FIGURE 2.2
Alternate process control techniques.

a system to more accurately achieve the specified results. Two types of control are shown: in-process and post-process. During in-process control, measurements are taken which, with analysis and feedback, concurrently influence the manufacturing process. Post-process control uses statistical measurements of the final product characteristics, including both manufacturing specifications and performance, analyses them and adjusts the system to influence the product manufactured at a later time. Both in-process and post-process measurements can be recorded and used for comparative state analysis at any level of the entire manufacturing enterprise.

In-process control is further illustrated for a single parameter control loop in Figure 2.3. The controller of the manufacturing process determines the nominal reference value (set-point) for parameter A of the process. Measurements are continually being taken within the process and compared, usually by calculating the difference from the set-point. This difference, called the error, is then used to make a change in the system actuation affecting parameter A to attempt to maintain A at its desired valued. For example, the controller may be attempting to keep the temperature of an integrated-circuit gate-oxide growing furnace at 1200 °C (the set-point). By way of a pyrometer, thermocouple or other sensor, the temperature is determined to be low by 10 °C. The analysis components change the power to the heating coils by an amount that is related to the error. Chapter 14 gives more

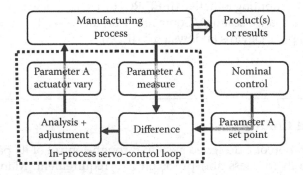

FIGURE 2.3
Single parameter process control loop.

background to the design and use of control approaches in precision engineering that focuses more on machine and instrument dynamics.

2.4 Accuracy, Precision and Trueness

The value of the measurements to benefit controlled improvement is largely dependent on their accuracy (Rayner 1923). Accuracy is, qualitatively, the ability to achieve with precision the true measurement quantity of a product characteristic or related process parameter. Precision is the ability to obtain the same measured quantity value over multiple attempts, whereas trueness is the closeness of the average of repeated measurements to the measurand value that is consistent with its definition. Thus improvement or progressive societal benefit is dependent on control and control requires measurement, and the value of those measurements depends on their accuracy including both trueness and precision (ISO 5725-1:1994).

The concepts of accuracy, trueness and precision are illustrated in Figure 2.4 as attempts by an archer to hit the bull's eye (true value) of a target from a specified distance. If the attempts are well grouped (i.e. repeatable), the archer has demonstrated 'precision'. The average position over a sufficient number of tries compared with the true value determines the 'trueness' of the attempts. The accuracy of any attempt is dependent on both the trueness and the precision. The left target shows a grouping of attempts whose precision is much better than their trueness. The target on the right shows attempts which on average are near to the true value but individually are dominated by the inconsistency of the archer to hit the same place. There are other aspects which may also play a role when considering the variations of the archer's attempts. If the archer was standing 10 m away from the left target, but 30 m away from the right target, the angular precision (skill) in combination with the distance of the archer changes the position precision of the grouping. This illustrates the range-dependence of precision. Perhaps, the bow used for the attempts on the right target may have better aligned sights than the bow used for the attempts on

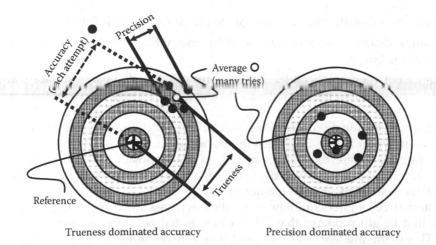

FIGURE 2.4
Classic target illustration of accuracy, trueness and precision.

the left target, or perhaps the bows have differing full-draw stored energies; this is equipment variation. Perhaps the attempts at the right target were influenced by a windy day; this is environmentally induced variation. If there were two different archers, then one may have greater skill than the other; this illustrates operator variation. These factors are possible contributing influences when trying to assign causes for the variability of inaccurate processes. Detailed process specifications along with the measurements of related parameters can be used to determine the sources of process variation. Note that the 'true value' does not exist in reality and can only be approximated. There will always be some form of variation of the true value (error) and ultimately the laws of physics dictate a fundamental limit (i.e. Heisenberg's uncertainty principle and Brownian motion at finite temperatures).

In the case of machines and instruments, precision is also inherently related to the quality of their construction and the conditions of their use. Calibration, discussed later, can be used to make a precision process more accurate but it has no influence on precision.

2.5 Measurement Basics

The *International Vocabulary of Metrology (VIM)* (JCGM 200:2012) serves to harmonise terminology used in the metrology and related disciplines. Some VIM terms are listed in Table 2.1.

To illustrate the vocabulary, consider a cylindrical bar. Some of the property quantities, with associated arbitrary example quantity values that can be attributed to the bar, are diameter (25.4 mm), length (200.0 mm), mass (3.3 kg), density (8.05 $g \cdot cm^{-3}$), coefficient of thermal expansion (17.3×10^{-6} K^{-1}), linear elastic modulus (198 GPa), shear modulus (77.2 GPa), Poisson's ratio (0.30) and fracture toughness (50 $MPa \cdot m^{1/2}$). These are just some of the mechanical properties; the bar also has chemical, electrical and optical properties.

Suppose the quantity value is to be determined for the length property of the bar. The length of the bar is then considered the measurand. Due to the fact that the end surfaces of the bar are not perfectly flat or parallel (see Figure 2.5), different quantity values may be determined depending on how the measurements are performed. For example, the length could be defined alternatively as the

1. distance between the circular faces of the bar at the bar axis intersection points;
2. minimum distance between two parallel planes which totally enclose the circular faces of the bar; or
3. distance between points on the least-squares calculated average surface planes of each circular face of the bar determined along the centre-line axis of the bar.

Other length determinations are also possible. The effective difference in plane descriptions is also shown in Figure 2.5. When the planes are not parallel, there is a line at some extended distance where they intersect and, in the opposite direction, they ultimately become infinitely separated. A well-defined measurand is, therefore, necessary for a proper unambiguous measurement of its quantity value.

In the alternative length definitions given, there is an assumed criterion that the length of the bar is that length property that it has when its temperature property quantity value is 20 °C. This is the internationally defined standard reference temperature for the specification of geometrical and dimensional properties (ISO 1:2016) (see Chapter 5).

TABLE 2.1

Definitions of Metrology-Related Terms

Metrology Term	Definition
Quantity	Property of a phenomenon, body, or substance, where the property has a magnitude that can be expressed as a number and a reference, for example height may be expressed as 3.01 metres.
Quantity value	Number and reference representing a given property, for example with 3 μm, 3 is the number and μm the reference.
Measurand	Quantity intended to be measured, for example mass, length, voltage.
True quantity value	Quantity value consistent with the definition of a quantity, generally unknowable but estimable within a stated uncertainty (clarified elsewhere). For example the length of a reference bar may be 10.001 mm according to its definition but the estimated value may be 10.002 mm with a stated uncertainty of 0.005 mm.
Measurement	Process of experimentally obtaining one or more quantity values that can reasonably be attributed to a quantity.
Indication	Quantity value provided by a measuring instrument or a measuring system.
Measurement precision	Closeness of agreement between indications or measured quantity values.
Certified reference material	Reference material, accompanied by documentation issued by an authoritative body and providing one or more specified property values with associated uncertainties and traceabilities, using valid procedures.
Reference quantity value	Quantity value used as a basis for comparison with values of quantities of the same kind.
Measurement trueness	Closeness of agreement between the average of an infinite number of replicate measured quantity values and a reference quantity value.
Measurement accuracy	Closeness of agreement between a measured quantity value and a true quantity value of a measurand.
Measurement repeatability	Measurement precision under a set of repeatability conditions of measurement (same conditions might include procedure, operators, measuring system, operating conditions, location).
Measurement reproducibility	Measurement precision under reproducibility conditions of measurement (different conditions include locations, operators, measuring systems, and replicate measurements on the same or similar objects).
Systematic error	Component of measurement error that in replicate measurements remains constant or varies in a predictable manner.
Instrumental bias	Average of replicate indications minus a reference quantity value.
Measurement uncertainty	Non-negative parameter characterizing the dispersion of the quantity values being attributed to a measurand, based on the information used.
Metrological traceability	Property of a measurement result whereby the result can be related to a reference through a documented unbroken chain of calibrations, each contributing to the measurement uncertainty of comparisons.
Traceability chain	Sequence of measurement standards and calibrations that is used to relate a measurement result to a reference.

Source: JCGM 200:2012, International vocabulary of metrology – Basics and general concepts and associated terms (VIM), Bureau International des Poids et Mesures.

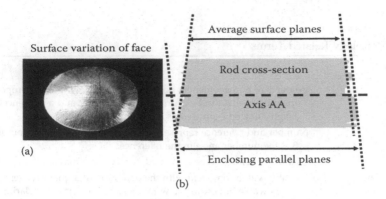

FIGURE 2.5
(a) Photograph of the end of steel bar revealing process-induced variations in roughness and form. (b) Illustration of cross-section showing variation in end surfaces and planes for possible measurand quantification choices.

Based on the quantity description, there is a true quantity value for the length which is never exactly determinable but is estimated using a measuring instrument. A single reported value by the measuring instrument is called an indication. An estimation of the true quantity value for the length can be made using an instrument such as a coordinate measuring system (CMS), as shown in Figure 2.6 and discussed in detail in Chapters 5 and 11. A CMS moves a probe, such as a stylus ball attached to a displacement sensor, through a 3D work zone to contact the surfaces of an object to determine the object's spatial

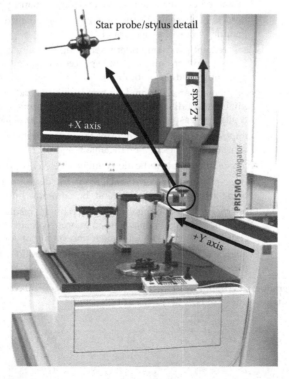

FIGURE 2.6
Coordinate measurement system.

geometry and calculate deviations from the nominal values. By contacting points on both ends near the axis centre-line, the CMS can determine the length of the bar that is consistent with its quantity definition. Arbitrary quantity value data for the length showing the deviation from the nominal for ten indications are plotted in Figure 2.7. For a finite number n of indications V_i, measurement precision p can be estimated by a one statistical deviation through the variance $p^2 = \frac{1}{n-1} \sum_{i=1}^{n} (V_i - V_{avg})^2$ from the average indication value $V_{avg} = \frac{1}{n} \sum_{i=1}^{n} V_i$. The measurement precision of the example centreline point-to-point mea-surement data is 3.3 μm. Since the procedure or measuring system did not change for the indications, this is also an estimate of measurement repeatability using the CMS. This measurement precision/repeatability is a statistically bounded, but unpredictably varying, contributor to the accuracy of a point-to-point length indication from the CMS. The average deviation from nominal of the length indications is 3.47 μm. If the specified tolerance for the length was 5 μm, then the bar would appear to be within tolerance. However, another contributor to the accuracy of measurement is the systematic measurement error which includes the instrument bias. This bias is estimated during calibration. By measuring an appropriate reference artefact that has a known reference quantity value of low uncertainty, the CMS's bias can be determined and accounted for in the measurement. Suppose the reference artefact was a gauge block having a 200.000 0 mm reference quantity value for its length (see Chapter 5 for information about gauge blocks). After repeated measurements, the calculated average measurement value is 199.993 9 mm. By subtracting the reference value from the average, the instrument bias is found to be –0.006 1 mm. So 6.1 μm must be added to measured lengths to account for the bias. Correcting the bar length average for this bias yields a corrected average length deviation from specification of 9.57 μm, which is beyond the allowable 5 μm tolerance value.

Calibration for any instrument is much more complicated than the simple illustration given. For example, the instrument bias can be a function of the magnitude of the measured value or of the environmental temperature, or other factors. If the present bias is not determined, it must be included as a component of uncertainty contributing to a measured quantity. Geometrically, for the CMS, each motion axis of the machine contributes six

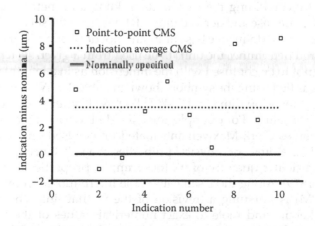

FIGURE 2.7
Example data from point-to-point length measurement along the axis.

positionally dependent degrees of error motion to the model. For the simplest three axis Cartesian CMS, this corresponds to eighteen positionally dependent parameters (Hocken and Periera 2012) plus three alignment errors (see Chapter 5). After measurement and correction by an error model, CMS's are verified for accuracy performance by measuring step gauges with multiple reference surfaces and reference ball-bars placed in multiple positions and orientations (see Chapters 5 and 11). The uncertainty of the reference standard(s) used contributes to uncertainty in instrument calibration, which, in turn, must be combined with the uncertainty estimates in measurement indications. Chapter 9 gives detailed information on methods and procedures for determining measurement uncertainty.

2.6 Measurement Units and Reference Standards

Civilized society has from antiquity demanded just and equitable trade, and established it upon reference quantities called weights and measures. As length, measured in cubits, had the forearm for a reference, each weight and measure needs some comparison measurement unit, 'a real scalar quantity, defined and adopted by convention, with which any other quantity of the same kind can be compared to express the ratio of the two quantities as a number' (JCGM 200:2012). This number, in conjunction with the measurement unit, is the quantity value previously discussed. Historically, there have been many units and systems used to provide comparison. These reference units of measurement allow the determination of a specific quantity as a number with respect to the reference unit as the measured quantity value. In this case the base quantity is called length.

A collection of units of measurement is called a system. The most common systems of units used today are the imperial system and the Système International d'Unites or simply the SI system (SI 2017). In this book, the SI system is used unless otherwise stated.

Within a system, a base quantity having a base unit is a quantity in a conventionally chosen subset of a given system of quantities, where no subset quantity can be expressed in terms of the others (JCGM 200:2012). The set of base quantities in the SI with their units, symbols and definitions are given in Table 2.2. Derived units are products of powers of base units. They include no numerical factor other than unity. For example, the newton, a unit for the quantity of force, is derived by combining the base units of kilogram, metre and second into the derived unit $kg \cdot m \cdot s^{-2}$. The base and derived units of the SI form the set of coherent SI units (SI 2017). It is only necessary to define the base units and not derived units. Proper definition of what is to be measured determines the units to be used from a given system. The dimension (dim) of a quantity (not to be confused with the dimension as in a length) is independent of the system and is identified using the symbols shown in Table 2.2. With a specific measurand quantity (Q) having dimension $\dim Q = L^{\alpha} M^{\beta} T^{\gamma} I^{\delta} \Theta^{\varepsilon} N^{\zeta} J^{\eta}$ with $\alpha, \beta, \gamma, \delta, \varepsilon, \zeta$ and η being generally small integer values. For example, force would have a dimension $L^{1} M^{1} T^{-2}$.

Lord Kelvin and James Clerk Maxwell anticipated units based on 'natural standards of pieces of matter such as atoms', or 'physical properties of a substance without the necessity of specifying any particular quantity of it', for example, properties of light in a vacuum (Thomson and Tait 1879). Along these same lines, the International Committee for Weights and Measures (CIPM) is pursuing a 'revision of the SI that links the definitions of the kilogram, ampere, kelvin, and mole to exact numerical values of the Planck constant h, elementary charge e, Boltzmann constant k, and Avogadro constant N_A, respectively'. The CIPM also plans revising 'the way the SI is defined including the wording of the definitions

TABLE 2.2

SI Set of Base Quantities and Units of the SI

Dimension:: Quantity:: SI Unit:: Symbol:: Definition
L:: length:: metre or metre:: m:: The length of the path travelled by light in vacuum during a time interval of 1/299 792 458 of a second.
M:: mass:: kilogram:: kg:: The mass of the international prototype of the kilogram. Imminent redefinition of this will be in terms of setting an exact value for the Plank constant $h = 6.6260 \ldots \times 10^{-34}$ J·s (J·s = s^{-1}·m^2·kg, so that kg = h·m^{-2}·s with metres and seconds being explicitly defined).
T:: time:: second:: s:: The duration of 9 192 631 770 periods of the radiation corresponding to the transition between the two hyperfine levels of the ground state of the caesium 133 atom.
I:: electric current:: ampere:: A:: That constant current which, if maintained in two straight parallel conductors of infinite length, of negligible circular cross-section, and placed 1 metre apart in vacuum, would produce between these conductors a force equal to 2×10^{-7} newton per metre of length.
Θ:: thermodynamic temperature:: kelvin:: K:: The fraction 1/273.16 of the thermodynamic temperature of the triple point of water.
N:: amount of substance:: mole:: mol:: The amount of substance of a system which contains as many elementary entities as there are atoms in 0.012 kilogram of carbon 12.
J:: luminous intensity:: candela:: cd:: The luminous intensity, in a given direction, of a source that emits monochromatic radiation of frequency 540×101^2 hertz and that has a radiant intensity in that direction of 1/683 watt per steradian.

Source: SI, 2017, The International System of Units (SI), Bureau International des Poids et Mesures, Organisation Intergouvernementale de la Convention du Mètre.

of the SI units for time, length, mass, electric current, thermodynamic temperature, amount of substance, and luminous intensity so that the reference constants on which the SI is based are clearly apparent' (CGPM 2014).

2.6.1 Unit of Length

Although competing with the length of the pendulum on a clock having a 2 s period, the metre was first defined in 1791 by the French Academy of Sciences as being one ten-millionth of the length of the quadrant (pole-equator) of the Earth through Paris. Although actually short of its definition, the metre was later physically embodied in platinum-iridium reference bars in 1874 and 1889. In 1927, the premier prototype reference bar became the definition, albeit at the melting point of ice. Based on interferometric techniques using single frequency light sources, there was also a supplementary relation at that time linking the prototype metre to 1 533 164 13 waves of the red spectroscopic line of cadmium at 15 °C. This was an intermediary step which eventually led to the redefinition of the metre so that it was based on a natural standard rather than a man-made one. In 1960, the metre was redefined to be 1 650 763.73 wavelengths of the orange-red spectroscopic line of krypton 86 under specified conditions (NBS 1960). This was an improvement over earlier definitions, but still required establishing the proper temperature, humidity and pressure. At the 17th General Conference on Weights and Measures, the metre received its current definition in 1983 as 'the length of the path travelled by light in vacuum during a time interval of 1/299 792 458 of a second' (CGPM 1984). Thus, length is currently defined by relating a natural physical constant, i.e. the speed of light in a vacuum, and time. Since the definition relates to a vacuum, the embodiment does not depend on temperature as material embodiments do. By this definition, the speed of light is also fixed as 299 792 458 m·s^{-1} and a summary of these definitions is provided in Table 2.3.

TABLE 2.3

Summary of Milestones in the Definition of the Metre

Year	Definition
1791	Ten-millionth of the Earth's quadrant, reference bars in 1799, 1874, 1889
1927	BIPM platinum-iridium reference bar length
1960	1 650 763.73 orange-red spectral wavelengths of krypton 86
1983	1/1299 792 458 of 1 s travel distance of light in vacuum

2.7 Physical Reference Standards

Measurement standards are the 'realization of the definition of a given quantity, with stated quantity value and associated measurement uncertainty, used as a reference' (JCGM 200:2012). Manufacturing production often depends upon hard gauges, a type of measurement standard, by which a comparison between or a 'fit' check is made to a feature to determine whether it is in tolerance. 'If a gauge is going to check one-thousandth—which is not extreme accuracy—then that gauge itself must be accurate to at least one ten-thousandth of an inch and it must be kept accurate' (Ford et al. 1931). This is the traditional gauge-maker's guideline of a 10:1 ratio between the specification being checked and the accuracy of the verifying reference gauge. Due to metrology capability limitations for high tolerance work, this guideline has been relaxed to 4:1 (ASME B89.7.3.1 2001), but can result in the necessity of parts being manufactured to statistical tolerances a little tighter than their required specifications to pass inspection. Currently, generalised standards provide flexibility in setting formal rules for acceptance and rejection of parts, for example ISO 14253-1 (2013) and JCGM 106 (2012).

Introduced at the turn of the 19th and 20th centuries (Dixie 1907), the gauge block set is a reference embodiment that has enabled growth in precision production. A gauge block set (see Chapter 5, Figure 5.3) originally consisted of 102 rectangular steel blocks that were hardened to reduce wear under continuous use. These blocks can be 'wrung' together, as shown in Figure 2.8, to achieve over one hundred thousand different lengths (Bryan 1993) and close to any desired calibration specification up to that limited by the set (see Chapter 5 for a more detailed description of gauge blocks and their use, including wringing). See also ASME B89.1.2 (2002).

Rulers, metre sticks, tape measures, gauge blocks, plug gauges, ball gauges, go/no-go gauges, step gauges and ball bars are also embodiments of measurement standards, having various accuracies, for length comparison or measurement system calibration. Rulers and tape measures are measurement standards that provide a sub-millimetre level uncertainty from a visual direct comparison, while the uncertainty of a CMS can be on the order of 1 μm and even less for high-end systems. CMSs use internal reference scales to provide displacement resolution and realise the length metric. Mills and lathes used in precision manufacturing also have similar scales that are used to control position of the workpiece. Diamond turning machines are available with linear encoder scale resolutions of 1 nm and even smaller if laser interferometric measurement is used.

Measurement standards are also commonly discriminated by their accuracy and for what purpose they are used, as identified in Table 2.4.

The working standard either calibrates an instrument for measurement or may be used to physically verify according to a decision rule (see ISO/TR 14253-6:2012) that the geometric

FIGURE 2.8
'Wrung' gauge blocks in vertical suspension.

dimension of a workpiece is within tolerance by a geometric fitting or other comparative process. Figure 2.9 shows comparative go–no go gauges used for comparative tolerance checking. The working standard is calibrated by a more accurate local (building or shop) reference standard which is in turn calibrated to a standard having lower uncertainty. There is a traceable chain of verification from working standard to reference standard to secondary standard to the primary standard that ultimately links a measurement to its unit definition.

2.8 Traceability, Verification and Uncertainty

In manufacturing, hand-held calipers and micrometres (gauges), as well as more accurate automated measuring systems, capable of diverse form and surface measurement, are

TABLE 2.4

Categories of Reference Standards

Measurement Standard Class	Realisation or Purpose
Primary (lowest uncertainty)	'Established using a primary reference measurement procedure, or created as an artefact, chosen by convention'. These are typically developed and maintained in government-funded standards laboratories.
Secondary (lower uncertainty)	'Established through calibration with respect to a primary measurement standard for a quantity of the same kind'.
Reference (low uncertainty)	'Designated for the calibration of other measurement standards for quantities of a given kind in a given organization or at a given location'.
Working (decision rule uncertainty)	'Used routinely to calibrate or verify measuring instruments or measuring systems'.
Travelling	'Sometimes of special construction, intended for transport between different locations'.

Source: JCGM 200:2012, International vocabulary of metrology – Basics and general concepts and associated terms (VIM), Bureau International des Poids et Mesures.

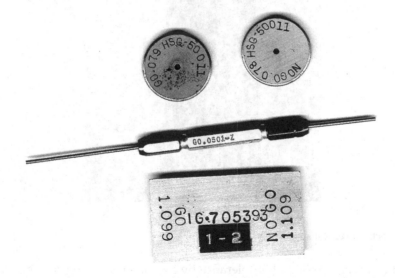

FIGURE 2.9
Various go–no-go gauges used for tolerance verification.

commonly used for verifying dimensional tolerances of components. The quantity values produced by these gauges and systems are verified using measurement standards as previously discussed. As a practical example, Figure 2.10 shows a micrometre being verified for a 0.9 in measurement using a gauge block as a working reference standard. Every properly performed measurement has metrological traceability through a traceability chain (defined in Table 2.1 and see Chapter 5). The traceability chain for a micrometre using gauge blocks as references is illustrated in Figure 2.11. M1 represents a primary measurement standard (system) that directly embodies the definition of length. For a gauge block this may be a gauge block interferometre (Doiron and Beers 1995). Gauge block B1 is placed in this system and the distance L1 between the faces is determined and reported with

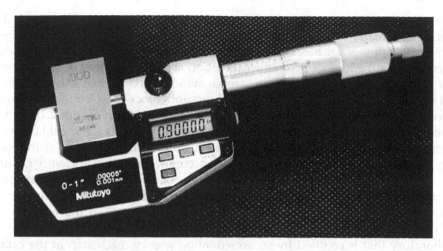

FIGURE 2.10
Micrometre measuring a gauge block for measurement process verification.

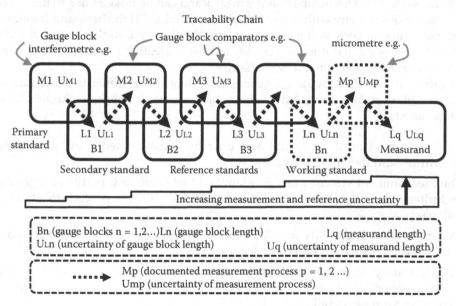

FIGURE 2.11
Links in the traceability chain for micrometre calibration and measurement.

an uncertainty U_{L1} that is due to the inability of the system to ascertain the true value. Thus, B1 has been calibrated and is considered a secondary measurement standard. In the same building with B1 is another gauge B2 that will be used at another location for verifying other gauge blocks. B1 is compared to B2 in measurement system M2, likely a gauge block comparator built for that purpose and may in some respects resemble the gauge block interferometre M1. M2 directly measures the difference $\Delta L2$ between the lengths of B1 and B2 with an uncertainty of U_{M2}. The estimated quantity value for the length of L2 is L1 + $\Delta L2$. This calculated value can be considered a part of the measurement process. If the measurement is a comparative measurement, the uncertainty U_{L2} can be calculated by $(U_{L2})^2 = (U_{L1})^2 + (U_{M2})^2$.

If the measurement is not a direct comparison but B1 was used to calibrate M2, then the uncertainty has already been included in U_{M1}, so that $U_{M1} = U_{L2}$. Since B2 is going to be used to calibrate other gauge blocks, it is considered a reference measurement standard. B2 is returned to its home location from the calibration laboratory where it then becomes the reference for calibrating block B3 as a reference standard. B3 is then used as a reference standard to calibrate another block and so forth up until block Bn. Gauge block Bn is used as the working reference to calibrate the micrometre as shown in Figure 2.10. This series of referenced measurements is known as a calibration hierarchy. The calibration of the micrometre verifies its operation if the result of the measurement made of the length of the gauge block is within the micrometre's stated accuracy, usually sufficiently greater than that of the reference block, so that the 10:1 gauge maker's guideline is satisfied. The micrometre (Mp in Figure 2.11) is now ready to verify the width of components near 0.9 inches and whose specified tolerance is ten times greater than the uncertainty of a measurement taken using the micrometre system.

In a laboratory that is certified by an accreditation agency, each step in the calibration chain is verified so that the calibration certificate for the measurand is linked to the micrometre, the micrometre to gauge Bn, Bn to B3 via M4, B3 to B2 via M3, and so on, until Bn can be traced to M1. This means that the measurand can be linked back to the realisation of the definition of the metre with a determinable uncertainty. If the base unit is defined as a physical embodiment, such as the kilogram, then the final link in the traceability chain is that embodiment. Note that it is not possible to have traceability without uncertainty; the two concepts are intimately linked (see Chapter 9).

Specification standards provide guidelines for establishing sufficient corporate metrological traceability, such as those listed next, which include those from the B89.7.5 length traceability standard:

1. The measurand must be clearly and unambiguously stated. This may include algorithms and data filters to be used.

2. The measurement system must be identified including any pertinent algorithms and filters which determine the measurand and/or written procedural standards followed.

3. Documentation traceability for that reference standard(s) used for comparison or calibration of the measurement system must exist.

4. An uncertainty statement of the measurement result in harmony with the *Guide to the Expression of Uncertainty in Measurement* (JCGM 100:2008) must be provided. See Chapter 9 for more detail.

5. An uncertainty budget describing and quantifying any significant contributors and their effect on uncertainty must be available.

6. A measurement system program that verifies the conditions of the measurement are valid to ensure the stated uncertainty must be in operation.

A calibration certificate for a 0.9 inch gauge block is shown in Figure 2.12. Note that, along with the serial number of the gauge block being calibrated, there is the serial number given for the master set by which calibration comparison was made. This master set would also have a calibration certificate as well as the set that was used to calibrate that master set. In this manner, the chain is maintained through the realisation of the definition of the metre. Notice also that the tests were carried out in accordance with the ASME B89.1.9 standard (ASME 2002), which in this case is considered a part of the total measurement system.

FIGURE 2.12
Example gauge block calibration certificate.

2.9 Metrology Institutes, Standards Bodies and Performance Test Codes

Because of the importance of trade, organised nations have established governmental entities for the supervision and administration of reference standards for metrology. Preeminent metrology entities are national metrology institutes (NMIs) and organisations called standards bodies for developing and distributing specification standards. NMIs conduct research aimed at the establishment, verification, maintenance and distribution of physical standards. Internationally, the impartial 'intergovernmental organization through which Member States act together on matters related to measurement science and measurement standards' is the Bureau International des Poids et Mesures (BIPM). In addition to assisting member states with scientific activities, the BIPM's goal in regard to primary standards is to 'coordinate the realization and improvement of the world-wide measurement system to ensure it delivers accurate and comparable measurement results'.

Some major NMI locations are identified in Figure 2.13. The contact information for these and other NMIs are available through the BIPM website (www.bipm.org).

Standards bodies are entities that oversee the development and dissemination of specification 'standards' also known as codes. Examples of these standards bodies are the International Organization for Standardization (ISO), the British Standards Institution (BSI), the American National Standards Institute (ANSI) and the International Electrotechnical Commission (IEC).

Specification standards can be described as a set of technical definitions and guidelines, or how-to instructions for designers, manufacturers and users. They provide straightforward product comparison, consumer protection, and corporate liability protection and also promote safety, reliability, productivity and efficiency in almost every industry that relies on engineered components or process equipment (ASME 2016).

Specification standards should be repeatable, enforceable, definite, realistic, authoritative, complete, clear and consistent but have a specific limited scope ASME (2016). There are even specification standards for producing standards, for example ISO/IEC (2016) and ASME PTC-01 (2015).

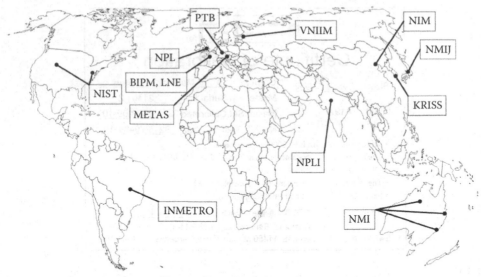

BIPM – International Bureau of Weights and Measures
NIST – United States – National Institute of Standards and Technology
NPL – United Kingdom – National Physical Laboratory
LNE – France – Laboratoire National de Métrologie et D'essais
PTB – Germany – Physikalisch-Technische Bundesanstalt
METAS – Switzerland – Federal Institute of Metrology
NMIJ – Japan – National Metrology Institute of Japan
NIM – P.R. China – National Metrology Institute
VNIIM – Russian Federation – D.I. Mendeleev All-Russian Institute for Metrology
KRISS – Republic of Korea – Korea Research Institute of Standards and Science
NPLI – India – National Physical Laboratory of India
NMI – Australia – National Measurement Institute
INMETRO – Brazil – Instituto Nacional de Metrologia, Qualidade e Tecnologia

FIGURE 2.13
Selected National Metrology Institute (NMI) locations. (Map is a derivative of map retrieved from d-maps.com
/carte.php?num_car=13181&lang=en under permissions allowed per d-maps.com/conditions.php?lang=en.)

Specification standards for procedural instruction are necessary because poor procedures increase reproducibility variations in the data or as the Portas principle states 'random results are the consequence of random procedures' (Bryan 1984). Ambiguity can also provide 'loop holes' for presenting misleading data. For example, if the number of equally spaced data points is not specified to determine the roundness of a cylinder, then using only three will allow the reporting of the roundness as 'perfect' because three points are the minimally sufficient number to define a perfectly round circle. Specification standards can run from a few paragraphs to hundreds of pages, and are written by committees of professionals with knowledge and expertise spanning a particular field. The committees are made up of individuals from industry, government and academia, who have a vested interest in the technical scope of the standards. The typical sequence stages for creation of an ISO speci-fication standard is the preliminary stage, the proposal stage, the preparatory stage, the committee stage, the enquiry stage, the approval stage and the publication stage.

In the preliminary stage the topic is developed and approved for a work item. In the proposal stage an outline or a first working draft is developed for discussion and a project leader is nominated. In the preparatory stage a working group prepares a fully developed working draft suitable for circulation to the members of the technical committee or subcommittee as a first committee draft (CD). The key stage in which comments from national bodies are considered, with the goal of a reaching a technical content consensus, is the committee stage. During the enquiry stage, national standards bodies are allowed to comment for a set period from eight to twelve weeks. These comments are compiled and given consideration by the committee. If consensus is not reached, then a revised committee draft is prepared for distribution and comment. If consensus is reached, then the draft is registered for the enquiry stage in which the national bodies are given twelve weeks to submit a vote. Negative votes are not counted unless they have been submitted with a justification. The committee is required to respond to the technical reasons for the negative votes. After distribution to the national bodies, including the committee responses, another vote is taken during the approval stage, at which time the standard can be adopted with the reasons for the technical no-votes to be discussed at the next meeting of the committee.

Consensus is the general agreement, characterised by the absence of sustained opposition to substantial issues, by any important part of the concerned interests and by a process that involves seeking to take into account the views of all parties concerned and to reconcile any conflicting arguments. Consensus need not imply unanimity (ISO/IEC 2016).

Specification standards are considered voluntary because they serve as guidelines, but do not of themselves normally have the force of law. However, standards do become legally binding when they have been incorporated into a business contract or incorporated into governmental regulations. Examples are building codes for safety and accessibility adopted by municipalities.

For both manufacturers and users of fabrication and measurement equipment to have a level playing field, performance test standards known as codes are developed. These can be referenced in purchase specifications as well as used to periodically assess the operation of that equipment. These standards represent the state-of-the art knowledge in obtaining a consistent accurate assessment while taking into account the cost and informational value of testing.

An example of performance test standards useful to precision engineers is the ISO 230 'Test code for machine tools' series. This series is useful for assessing the ability of a machine to position a tool with respect to a part. The various components of the standard are shown in Table 2.5.

TABLE 2.5

Components of ISO 230 Standard for Machine Tool Performance

ISO 230-1:2012	Part 1: Geometric accuracy of machines operating under no-load or quasi-static conditions
ISO 230-2:2014	Part 2: Determination of accuracy and repeatability of positioning of numerically controlled axes
ISO 230-3:2007	Part 3: Determination of thermal effects
ISO 230-4:2005	Part 4: Circular tests for numerically controlled machine tools
ISO 230-5:2000	Part 5: Determination of the noise emission
ISO 230-6:2002	Part 6: Determination of positioning accuracy on body and face diagonals (diagonal displacement tests)
ISO 230-7:2015	Part 7: Geometric accuracy of axes of rotation
ISO/TR 230-8:2010	Part 8: Vibrations
ISO/TR 230-9:2005	Part 9: Estimation of measurement uncertainty for machine tool tests according to series ISO 230, basic equations
ISO 230-10:2016	Part 10: Determination of the measuring performance of probing systems of numerically controlled machine tools
ISO/DTR 230-11	Part 11: Measuring instruments suitable for machine tool geometry tests

TABLE 2.6

ISO 15530 Standards for Geometrical Product Specifications (GPS) – Coordinate Measuring
Machines (CMM): Technique for Determining the Uncertainty of Measurement

ISO/TS 15530-1:2013	Part 1: Overview and metrological characteristics
ISO 15530-3:2011	Part 3: Use of calibrated workpieces or measurement standards
ISO/TS 15530-4:2008	Part 4: Evaluating task-specific measurement uncertainty using simulation

Another useful set of standards as applied to a metrology system is the ISO 10360
'Acceptance and reverification tests for coordinate measuring machines (CMM)', which are
individually listed in Chapter 5.

There are standards such as the ISO 15530, identified in Table 2.6, which are helpful in
determining measurement uncertainties.

To communicate the geometric dimensioning and tolerancing requirements, the ISO 129-
1:2004(en) and the ASME Y14.5 standards offer basic guidelines, whereas there are others
for diverse but specific applications such as for optical systems (ISO 10110) and ship-
building (ISO 129-4). There are also standards for calculating specific parameters, such as
those related to surface texture in ISO 25178-2 and ASME B46.1 (see Chapter 5).

2.10 Measurement Realisation

The physical realisation for an estimation of the quantity value follows a predetermined
path as shown in Figure 2.14. The first step to take is to unambiguously define the
measurand. This may involve specifying the exact steps taken and method of analysis used
to obtain the value. Second, a sensor must be used that has a parameter S_p which is sensitive

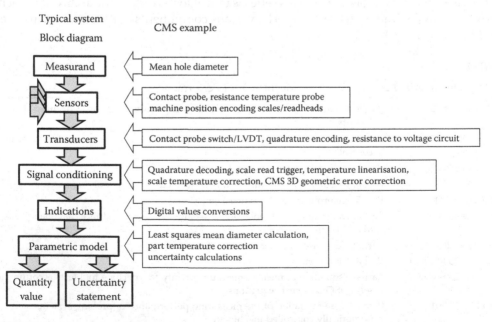

FIGURE 2.14
Typical measurement system block diagram.

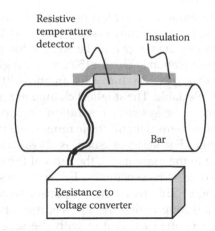

FIGURE 2.15
Surface temperature sensor attached to a cylindrical rod.

to variations in the measurand. Third, there must be an optical, mechanical or electrical transducer which provides an analogue or digital value corresponding to S_p that can be used to obtain an indication. A mechanical transducer example is the dial indicator on a pair of callipers. An electronic circuit may also operate with the sensor to provide a voltage or current which changes with the measurand. Fourth, the output of the transducer may be further altered by electronic or digital circuits with filters to reduce the noise, amplify the signal or linearise the system response. In many instrumentation texts this is referred to as the signal conditioning component of the circuit. Fifth, the indication is read visually or digitised using an electronic analogue-to-digital converter. As a sixth step, it may be necessary to combine indications according to a parametric model via calculations provided by computer programming to produce the reported quantity value. The completing seventh step is a measurement uncertainty evaluation that includes all of the known factors that may have influenced the measurement (see Chapter 9).

Suppose it is desired to measure the temperature of the aforementioned steel rod. The measurand could be defined as the surface temperature of the steel bar measured at a position halfway between the end faces. A resistance temperature detector (RTD) could be chosen as the sensor because it has a resistance which linearly varies with temperature. By attaching the RTD on the surface with an appropriate adhesive, and covering other faces of the RTD with thermal insulation (making sure that the effective heat flow change is minimised), the temperature of the bar and RTD become linked. The wire leads of the RTD can be connected to an electronic bridge circuit which has an output voltage which is proportional to the change in the resistance of the RTD. By recording the voltage and using a linear model whose slope $(K \cdot V^{-1})$ and offset have been determined by calibration, an indication of the temperature of the bar can be obtained (see Figure 2.15).

2.11 Measurement System Parameters

A measurement system is also describable using associated system parameters. Consider a simple displacement gauge, which may be based on any electrical, optical or mechanical principle. There is a limit to the amount of displacement that it can detect. This total possible change in displacement from the minimum possible reading to the maximum possible reading is the range. For example, the range may be 300 µm for a capacitance gauge. Often it is expressed as a bipolar specification such as ±150 µm. A single probe system may have varying range selections, for example ±200 µm, ±40 µm and ±10 µm. These ranges may be software or electronic switch selectable. The smallest change in a measured parameter that a measuring system can detect is the system's resolution (not to be confused with optical resolution; see Chapter 3). For systems with multiple ranges, the resolution usually depends on the chosen operating range. For systems, such as those with voltage outputs corresponding linearly to a change in the measurand, the ratio of the change in indication to the change in the quantity measured is its sensitivity. For some systems, the sensitivity may change as the quantity changes. This nonlinear response, which increases the system's uncertainty, is often corrected through calibration procedures and system modelling using calibration curves. Sensitivity is often associated with the sensor element. For example,

TABLE 2.7

Metrology System Specification Parameters

System Parameter	Description
Range(s)	Absolute value of the difference between the extreme quantity values available for any control setting of an instrument
Resolution(s)	Smallest change in a quantity being measured that causes (or would cause) a perceptible change in the corresponding indication
Sensitivity	Quotient of the change in an indication of a measuring system and the corresponding change in a value of a quantity being measured
Selectivity	Property of a measuring system such that the values of each measurand are independent of other possible measurands
Stability	Property of a measuring instrument, whereby its metrological properties remain constant in time (regardless of allowable variations in temperature humidity or air pressure and so son and thus provide indications having a favourable repeatability)
Instrument drift	Continuous or incremental change over time in indication, due to changes in metrological properties of a measuring instrument
Instrumental measurement uncertainty	Component of measurement uncertainty arising from a measuring instrument or measuring system in use
Maximum permissible measurement error	Extreme value of measurement error, with respect to a known reference quantity value, permitted by specifications or regulations for a given measurement, measuring instrument, or measuring system
Calibration diagram	Graphical expression of the relation between indication and corresponding measurement result
Calibration curve	Expression of the relation between indication and corresponding measured quantity value
Influence quantity variation	Difference in indication for a given measured quantity value when an influence quantity assumes successively two different quantity values (for example when the air pressure changes while measuring length)
Reference operating condition	Operating condition prescribed for evaluating the performance of a measuring instrument or measuring system or for comparison of measurement results
Rated operating condition	Operating condition that must be fulfilled during measurement in order that a measuring system performs as specified

Source: After JCGM 200:2012, International vocabulary of metrology – Basics and general concepts and associated terms (VIM), Bureau International des Poids et Mesures.

a linear variable differential transformer (LVDT; see Chapter 5) requires an applied alternating current input which is multiplied by its gain. The gain is a function of the relative displacement of the core inductor with respect to a null position. The change in gain per core displacement is its sensitivity so that, for example, the sensitivity may be 0.01 per micrometre. The consistency in the sensitivity and null position over time is considered stability. System stability will influence the mean time between calibrations. It is the precision engineer's task in designing the system to select the applied input to match the chosen LVDT which produces an output that can be evaluated with a resolution defined by the system specification. VIM definitions for these and other parameters are given in Table 2.7. Many other system specific parameters are also defined and included in related specification standards.

2.12 Summary

Measurements and the related systems and instruments that provide them along with their specification standards are critical in maintaining global economic commerce, societal infrastructure, and enabling engineering and manufacturing processes. The precision engineer must be concerned with an understanding of how well measured values reflect the reality of the parameters that they represent. The precision engineer, through measurement and modelling, rationally chooses and controls deterministic properties of components to purposefully, predictably, accurately and optimally build and operate systems according to functionally related, numerically toleranced specifications. The precision engineer also enhances state-of-the-art manufacturing capabilities through improving metrology-enabled control within all areas of engineering and related manufacturing.

Exercises

1. Choose a specific engineering material (perhaps an alloy), then list six material properties and their values with references for those values. For two of the properties, list a reported variant value for one of the properties and describe why you think there is discrepancy. What effect will the variations in values have on an engineered design?
2. Which kind of measurement standard would be used while performing duties on a shop floor to calibrate hand tools? Research the classifications for the quality grades for gauge blocks and their related tolerances. Find three manufacturers of gauge block sets. What materials are commercially available for gauge blocks? Why would differing materials be necessary?
3. Find three coordinate measuring system (CMS) manufacturers and compare the range, resolution and/or accuracy (or uncertainty) specifications given for their most accurate machines. Note that CMS is a relatively new term adopted by ISO; you may want to use coordinate measuring machine (CMM) in your search.
4. Name four kinds of sensors that are available on CMSs for geometric measurement of components? Briefly describe the operational principles for two of them.
5. A reference ring gauge (10.003 01 mm reference diameter having very low uncertainty) is used to estimate the repeatability and bias of a CMS's measurement. Calculate the

statistical standard deviation representing the measurement precision if separate indications of quantity values for the diameter taken in millimetres are 9.9931, 10.017, 9.9953, 9.9875, 10.003, 9.9853, 9.9853, 9.9918, 10.012, 10.001, 9.9921, 9.9935, 9.9998, 10.005, 9.9978, 10.010, 9.9908, 9.9981, 9.9985, 9.9995. What is the estimated instrument bias determined by the ring gauge measurement above?

6. Identify a set of specification standards that applies to an industry of your choice. What are seven of the parametric terms and their definitions associated with that standard? (Hint: ISO.org has partial preview documents available that include the terms and definitions.)

7. Suppose you wanted to make a measurement of the coefficient of thermal expansion of a particular material. What are the indications that would need to be measured? What parameter would need to be controlled to make low uncertainty measurements? Search relevant journals or patents and describe briefly two methods and/or instruments used for that purpose. One term used to identify these instruments is dilatometer.

8. Use other resources to find and describe the operation principle and mathematical modelling for three of the following: RTDs, thermistors, thermocouples, temperature integrated circuits, pyrometers and liquid thermometers.

9. Measure the diameter of a stock (as purchased) cylindrical bar with a set of calipers. First take ten measurements at the same location then repeat at several different locations. Determine the diameter and the precision (statistical variation) of the diametrical measurement of the cylindrical bar. Compare these results with those of your classmates or others who are willing to repeat the experiment. What do the comparisons suggest about the reproducibility of your measurement results?

10. What are three reference points from the ITS-90 (1990) would be best for calibrating thermistors to measure from below the freezing point of water to above the boiling point of water?

References

ASME 2016. Standards and Certification FAQs. Retrieved June 3, 2016, from www.asme.org,www.asme.org/about-asme/who-we-are/standards/about-codes-standards.

ASME B89.1.2 – 2002. *Gage blocks*. American Society of Mechanical Engineers.

ASME B89.7.3.1 – 2001 (R2011). Guidelines for Decision Rules: Considering Measurement Uncertainty in Determining Conformance to Specifications. American Society for Mechanical Engineers.

ASME B89.7.5 – 2006. Metrological traceability of dimensional measurements to the SI unit of length. American Society of Mechanical Engineers.

ASME PTC-01-2015. General instructions, performance test codes American Society of Mechanical Engineers.

Bryan, F. R. 1993. *Henry's Lieutenants*. Wayne State University Press, Detroit, MI.

Bryan, J. B. 1984. The power of deterministic thinking in machine tool accuracy. UCRL 91531 Lawrence Livermore National Lab. e-reports-ext.llnl.gov/pdf/197002.pdf.

CGPM. 1984. Comptes Rendus de la 17th Conférence Générale des Poids et Mesures (CGPM). www.bipm.org/utils/common/pdf/CGPM/CGPM17.pdf.

CGPM. 2014. Resolution 1 of the 25th CGPM. www.bipm.org/en/CGPM/db/25/1/.

Conte, T. M., and Gargini, P. A. 2015. On the foundation of the new computing industry beyond 2020. Preliminary IEEE RC-ITRS report, IEEE September 2015.

Dixie, E. A. 1907. A new Swedish combination gaging system. *American Machinist*, September 19, 393–396.

Doiron, T., and Beers, J. 1995. The gauge block handbook. National Institute of Standards and Technology. www.nist.gov/calibrations/upload/mono180.pdf.

Ford, H., Crowther, S., and Johansson, C. E. 1931. Millionth of an inch. In: *Moving Forward*. Doubleday, Duran and Co.

Hocken, R. J., and Periera, P. H. 2011. *Coordinate Measuring Machines and Systems*. CRC Press.

ISO 1:2016. Geometrical product specifications (GPS) – Standard reference temperature for the specification of geometrical and dimensional properties. International Organisation for Standardization.

ISO 5725-1:1994. Accuracy (trueness and precision) of measurement methods and results – Part 1: General principles and definitions. International Organization for Standardization.

ISO 10360-1:2000. Geometrical Product Specifications (GPS) – Acceptance and reverification tests for coordinate measuring machines (CMM). International Organization for Standardization.

ISO 14253-1:2013. Geometrical product specifications (GPS) – Inspection by measurement of workpieces and measuring equipment – Part 1: Decision rules for proving conformity or nonconformity with specifications. International Organization for Standardization.

ISO/IEC. 2016. Directives, Part 1 Consolidated ISO Supplement – Procedures specific to ISO and Part 2 Principles and rules for the structure and drafting of ISO and IEC documents.

ISO/TR 14253-6:2012. Geometrical product specifications (GPS) – Inspection by measurement of workpieces and measuring equipment – Part 6: Generalized decision rules for the acceptance and rejection of instruments and workpieces.

ITRS 2.0 2015. International Technology Roadmap for Semiconductors 2.0 Executive Summary. URL: www.itrs2.net.

ITS-90 1990. The International Temperature Scale of 1990. The International Committee for Weights and Measures. (Available through www.BIPM.org.)

JCGM 100:2008. Evaluation of measurement data – Guide to the expression of uncertainty in measurement. Bureau International des Poids et Mesures (BIPM).

JCGM 106:2012. Evaluation of measurement data – The role of measurement uncertainty in conformity assessment. Bureau International des Poids et Mesures (BIPM).

JCGM 200:2012. International vocabulary of metrology – Basics and general concepts and associated terms (VIM). Bureau International des Poids et Mesures (BIPM).

Moore, G. E. 1965. Cramming more components onto integrated circuits. *Electronics* **38**(8), 114ff. Reprint available from IEEE Solid-State Circuits Society Newsletter, doi: 10.1109/N-SSC.2006 .4785860.

Moore, G. E. 1975. Progress in digital integrated electronics. Proceedings of IEDM Technical Digest, pp. 11–13.

National Bureau of Standards (NBS). 1960. Units of weights and measures. National Bureau of Standards Misc Pub 233. December 20. United States Department of Commerce.

Rayner, E. H. 1923. The scheme for a journal of scientific instruments. *Journal of Scientific Instruments* 1, 2–4. iopscience.iop.org/article/10.1088/0950-7671/1/0/301.

SI. 2017. *The International System of Units (SI)*. Bureau International des Poids et Mesures, Organisation Intergouvernementale de la Convention du Mètre.

Thomson, W. and Tait, P. 1879. *Treatise on Natural Philosophy*, Vol. **1**, Part 1. Cambridge Press.

3

Background Principles

Harish P. Cherukuri

CONTENTS

ABSTRACT   In this chapter, some mathematical preliminaries and a review of solid mechanics and optics are presented. The mathematics sections cover the topics of elementary trigonometry, linear algebra, vector algebra, Taylor series, Fourier series and statistics. The solid mechanics section introduces the concepts of stress, strain, linear elastic constitutive relationships, elementary beam theory and Mohr's circle. In addition, the Hertz contact theory is also briefly reviewed. In the optics section, the wave representation of light is presented along with the laws of reflection, refraction, the thin-lens formula and diffraction. While the chapter is self-contained and meant to serve as a quick reference for many of the mathematical and physical concepts used in the rest of the book, the reader interested in learning these topics in a more thorough fashion is referred to the references provided at the end of the chapter.

3.1 Some Mathematical Preliminaries

In this section, a few concepts and basic identities from trigonometry, linear algebra and vector algebra are summarised. The important topics of Taylor series, Fourier series and statistics are covered separately in Sections 3.2 to 3.4. The coverage in this chapter is necessarily brief. The readers will find references Kreyszig (2007) and Riley et al. (2006) useful for more in-depth understanding of the topics discussed and as more comprehensive references to the mathematical techniques used throughout this book.

3.1.1 Review of Trigonometric Functions and Identities

Consider a circle of radius r centred at the origin in a two-dimensional space as shown in Figure 3.1. The equation of the circle is $x^2 + y^2 = r^2$. The point P lies on the circle and has the coordinates (x, y). Using the coordinates of this point, the six trigonometric functions can be defined as follows:

$$\sin \theta = \frac{y}{r}, \cos \theta = \frac{x}{r}, \tan \theta = \frac{y}{x}, \csc \theta = \frac{r}{y}, \sec \theta = \frac{r}{x}, \text{ and } \cot \theta = \frac{x}{y}. \tag{3.1}$$

From $r^2 = x^2 + y^2$, the following Pythagorean identities can be derived:

$$\sin^2 \theta + \cos^2 \theta = 1, 1 + \tan^2 \theta = \sec^2 \theta, \text{ and } 1 + \cot^2 \theta = \csc^2 \theta. \tag{3.2}$$

Many other identities are satisfied by trigonometric functions. Some of the most useful identities are

$$\sin(\theta_1 + \theta_2) = \sin \theta_1 \cos \theta_2 \pm \sin \theta_2 \cos \theta_1, \tag{3.3}$$

$$\cos(\theta_1 \pm \theta_2) = \cos \theta_1 \cos \theta_2 \mp \sin \theta_1 \sin \theta_2, \tag{3.4}$$

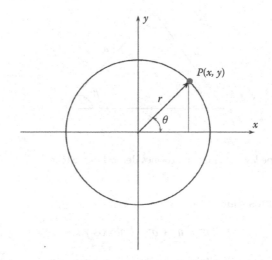

FIGURE 3.1
A circle of radius with centre at the origin. The six trigonometric functions can be defined using the points on the circle.

and

$$\tan(\theta_1 \pm \theta_2) = \frac{\tan \theta_1 \pm \tan \theta_2}{1 \mp \tan \theta_1 \tan \theta_2}. \qquad (3.5)$$

From these equations, the following identities involving double angles can be derived:

$$\sin 2\theta = 2 \sin \theta \cos \theta,\ \cos 2\theta = \cos^2 \theta - \sin^2 \theta \text{ and } \tan 2\theta = \frac{2 \tan \theta}{1 - \tan^2 \theta}. \qquad (3.6)$$

Another set of very useful identities involves sums and differences of sines and cosines:

$$\sin \theta_1 \pm \sin \theta_2 = 2 \sin \frac{\theta_1 \pm \theta_2}{2} \cos \frac{\theta_1 \mp \theta_2}{2}, \qquad (3.7)$$

$$\cos \theta_1 + \cos \theta_2 = 2 \cos \frac{\theta_1 + \theta_2}{2} \cos \frac{\theta_1 - \theta_2}{2} \qquad (3.8)$$

and

$$\cos \theta_1 - \cos \theta_2 = -2 \sin \frac{\theta_1 + \theta_2}{2} \sin \frac{\theta_1 - \theta_2}{2}. \qquad (3.9)$$

In addition to the preceding, trigonometric functions relate the sides and angles of a triangle through the laws of sines and cosines. Consider a triangle with sides a, b and c as shown in Figure 3.2. Let the angles opposite these sides be α, β, and γ respectively.

The law of sines states that

$$\frac{\sin \alpha}{a} = \frac{\sin \beta}{b} = \frac{\sin \gamma}{c} = \frac{abc}{2A/(abc)}, \qquad (3.10)$$

where A is the area of the triangle.

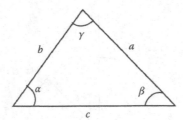

FIGURE 3.2
The nomenclature used for the laws of sines and cosines defined in the text.

The law of cosines states that

$$c^2 = a^2 + b^2 - 2ab \cos \gamma .$$ (3.11)

Often, approximations to the trigonometric functions are used when the angles (in radians) are small compared to unity. The three most used approximations are

$$\sin \theta \approx \theta, \cos \theta \approx 1 \text{ and } \tan \theta \approx \theta$$ (3.12)

when $\theta \ll 1$.

Finally, Euler's formula relating the sine and cosine functions to the exponential function is an important tool for studying wave phenomena. It is given by

$$e^{i\theta} = \cos \theta + i \sin \theta,$$ (3.13)

where $i = \sqrt{-1}$. The very useful de Moivre's formula follows from Equation 3.13:

$$(\cos \theta + i \sin \theta)^n = \cos n\theta + i \sin n\theta .$$ (3.14)

3.1.2 Linear Algebra

A common occurrence in engineering problems is the solution to a system of linear algebraic equations in several independent variables. The determination of voltages and currents in an electrical network analysis involves the solution to a system of linear equations. The nodal displacements in the finite element solution to structural problems involves the solution of a large number of linear equations. The mode shapes of a structure are often calculated using the solution to a system of linear equations.

A general system of n equations in n unknowns, $x_1, x_2, ..., x_n$, can be written as

$$A_{11}x_1 + A_{12}x_2 + A_{13}x_3 + \cdots + A_{1n}x_n = b_1,$$

$$A_{21}x_1 + A_{22}x_2 + A_{23}x_3 + \cdots + A_{2n}x_n = b_2,$$

$$A_{31}x_1 + A_{32}x_2 + A_{33}x_3 + \cdots + A_{3n}x_n = b_3,$$ (3.15)

$$\vdots \quad + \quad \vdots \quad + \quad \vdots \quad + \cdots + \quad \vdots \quad = \vdots,$$

$$A_{n1}x_1 + A_{n2}x_2 + A_{n3}x_3 + \cdots + A_{nn}x_n = b_n .$$

Here, the coefficients A_{ij} are either real or complex numbers. For the purposes of this chapter, it is assumed that these coefficients are real. A concise way of representing these equations is through the concepts of matrices and vectors in linear algebra.

A matrix is a collection of numbers (real or complex) arranged in a rectangular array consisting of a fixed set of rows and columns. Thus, if there are m rows and n columns in a matrix $\mathbf{A}$, the matrix is said to be of order $m \times n$. When $m = n$, the matrix is said to be a square matrix of size n. A matrix with one column and m rows is called a column vector of order m. A matrix with one row and n columns is called a row vector of order n.

For example, the matrices $\mathbf{A}$, $\mathbf{x}$ and $\mathbf{y}$ given by

$$\mathbf{A} = \begin{bmatrix} 5 & 7 & 8 \\ 6 & 8 & 4 \\ 3 & 2 & 1 \end{bmatrix}, \mathbf{x} = \begin{bmatrix} 7 \\ 4 \\ 9 \end{bmatrix} \text{ and } \mathbf{y} = [5 \ 4 \ 2 \ 12 \ 10] \tag{3.16}$$

are respectively a 3×3 matrix, a three-element column vector and a five-element row vector.

The element in the ith row and jth column of a matrix $\mathbf{A}$ is denoted by A_{ij}. Thus, the indices i and j take on values $1, 2, 3, \ldots, m$ and $1, 2, 3, \ldots, n$ respectively.

The transpose of a matrix $\mathbf{A}$ is often denoted by $\mathbf{A}^T$ and obtained by setting $A_{ij}^T = A_{ji}$. Thus, the transpose of an $m \times n$ matrix is of order $n \times m$. The transpose of a row vector is a column vector, and vice versa.

A square matrix $\mathbf{A}$ is said to be symmetric when $\mathbf{A}^T = \mathbf{A}$, that is $A_{ij}^T = A_{ji} = A_{ij}$. For square matrices (not necessarily symmetric), the elements with $i = j$ are called the diagonal elements.

A diagonal matrix $\mathbf{D}$ is a square matrix with $D_{ij} = 0$ for $i \neq j$. An identity matrix $\mathbf{I}$ is a diagonal matrix with $I_{ij} = 1$ for $i = j$. Finally, a zero matrix is a matrix with all zero components.

Two matrices $\mathbf{A}$ and $\mathbf{B}$ are said to be equal when both have the same order and $A_{ij} = B_{ij}$ for all i and j. Addition (subtraction) of matrices $\mathbf{A}$ and $\mathbf{B}$ of the same order is accomplished by adding (subtracting) corresponding elements of the two matrices. That is the matrix $\mathbf{C}$ in $\mathbf{C} = \mathbf{A} \pm \mathbf{B}$ is obtained from $C_{ij} = A_{ij} \pm B_{ij}$.

Suppose that a matrix $\mathbf{A}$ is of the order $m \times p$ and a second matrix $\mathbf{B}$ of order $q \times n$. Then, the product $\mathbf{C} = \mathbf{AB}$ is possible when $\mathbf{A}$ and $\mathbf{B}$ are compatible or conformable, that is when $p = q$. In other words, for matrix multiplication to be carried out, the number of columns in $\mathbf{A}$ must be equal to the number of rows in $\mathbf{B}$. The resulting product $\mathbf{C}$ will be of the order $m \times n$. The component C_{ij} is given by:

$$C_{ij} = \sum_{k=1}^{p} A_{ik} B_{kj}. \tag{3.17}$$

When $\mathbf{x}$ is a column vector of size n and $\mathbf{A}$ is a square matrix of size n, then the product $\mathbf{Ax}$ is a column vector of size n. Note that if $\mathbf{y}$ is a row vector of size n, then, $\mathbf{Ay}$ is not defined, since $\mathbf{A}$ and $\mathbf{y}$ are not compatible for matrix multiplication. However, $\mathbf{Ay}^T$ and $\mathbf{yA}$ are defined. In the former case, a column vector of size n is obtained and in the latter case, a row vector of size n is obtained. Another important result is that the product of a row vector of size n with a column vector of size n is a scalar.

By the definition of matrix multiplication, it should be obvious that for two square matrices of the same size n, $\mathbf{AB} \neq \mathbf{BA}$, in general. Furthermore, $\mathbf{A(B + C)} = \mathbf{AB} + \mathbf{AC}$ and $\mathbf{(A + B)C} = \mathbf{AC} + \mathbf{BC}$.

Upon combining the definitions of the transpose of a matrix and matrix multiplication, it can also be shown that for conformable matrices $\mathbf{A}$ and $\mathbf{B}$, $(\mathbf{AB})^T = \mathbf{B}^T\mathbf{A}^T$.

With the help of the foregoing discussion on matrices, the system of equations given in Equation 3.15 can be succinctly represented as

$$\mathbf{Ax} = \mathbf{b}. \tag{3.18}$$

with $\mathbf{A}$ being an $n \times n$ matrix with components A_{ij}, and $\mathbf{x}$ and $\mathbf{b}$ being column vectors of size n with components x_i and b_i respectively. The ith equation in the system of equations can be written as

$$\sum_{j=1}^{n} A_{ij}x_j = b_i, \text{ with } i = 1, 2, 3, \ldots, n. \tag{3.19}$$

The solution $\mathbf{x}$ to the system of equations (3.18) requires the concept of the inverse of a matrix. The inverse of a square matrix $\mathbf{A}$ of size n is (when it exists) also a square matrix of size n and denoted by $\mathbf{A}^{-1}$. It satisfies the properties

$$\mathbf{AA}^{-1} = \mathbf{A}^{-1}\mathbf{A} = \mathbf{I}. \tag{3.20}$$

When $\mathbf{A}^{-1}$ exists, the matrix $\mathbf{A}$ is said to be invertible or non-singular. If the inverse does not exist, the matrix $\mathbf{A}$ is said to be singular.

To obtain the solution $\mathbf{x}$ to the system of Equation 3.18, both of sides of Equation 3.18 are pre-multiplied with $\mathbf{A}^{-1}$. Then,

$$\mathbf{A}^{-1}\mathbf{Ax} = \mathbf{A}^{-1}\mathbf{b} \text{ or } \mathbf{x} = \mathbf{A}^{-1}\mathbf{b} \tag{3.21}$$

since $\mathbf{A}^{-1}\mathbf{A} = \mathbf{I}$ and $\mathbf{Ix} = \mathbf{x}$. Thus, the solution $\mathbf{x}$ is simply the product of $\mathbf{A}^{-1}$ with $\mathbf{b}$. However, this solution makes sense only as long as $\mathbf{A}^{-1}$ exists. It can be shown that the matrix $\mathbf{A}$ is invertible if its columns are linearly independent or equivalently the determinant of the matrix $\mathbf{A}$, denoted by det $\mathbf{A}$, is non-zero. For a square matrix $\mathbf{A}$ of order 2 with components A_{ij} with $i,j = 1,2$, the determinant is simply $A_{11}A_{22} - A_{12}A_{21}$. For a square matrix of order 3, the determinant is given by

$$\det \mathbf{A} = A_{11}(A_{22}A_{33} - A_{32}A_{23}) - A_{12}(A_{21}A_{33} - A_{31}A_{23}) + A_{13}(A_{21}A_{32} - A_{31}A_{22}). \tag{3.22}$$

For higher order matrices, the following equation can be used to calculate the determinant:

$$\det \mathbf{A} = \sum_{i=1}^{n} A_{i,j}\mathbf{C}_{ij} \text{ or } \sum_{j=1}^{n} A_{i,j}\mathbf{C}_{ij}. \tag{3.23}$$

Here, $\mathbf{C}_{ij}$ is the co-factor of $A_{i,j}$ given by $\mathbf{C}_{ij} = (-1)^{i+j}M_{i,j}$. $M_{i,j}$ is the minor of $A_{i,j}$ and equals the determinant of the matrix obtained by omitting the ith row and jth column of $\mathbf{A}$. The matrix $\mathbf{C}$ with components $\mathbf{C}_{ij}$ is called the co-factor matrix of $\mathbf{A}$. Then, it can be shown that the inverse $\mathbf{A}^{-1}$ of $\mathbf{A}$ is given by

$$A^{-1} = \frac{1}{\det A} C^{T}.$$
(3.24)

Clearly, A is non-singular when $\det A$ is non-zero and consequently, the system of Equation 3.18 will have a unique solution x. In practice, the inverse of a matrix is easy to calculate only when the order of the matrix is small. For matrices of large sizes, the solution x is rarely calculated by first finding the inverse of A, since the computational cost of inverse calculations tends to be high. Instead, a numerical approach is often taken to find an approximate solution to the system of equations. The most commonly used methods are based on the Gaussian elimination technique. A discussion of these methods is beyond the scope of this chapter. The interested reader is referred to Ascher and Greif (2011) and Kreyszig (2007).

An important application of linear algebra is in the determination of natural frequencies and mode shapes of structures. The governing equations for determining the mode shapes reduce to finding x such that

$$(K - \omega^2 M)x = 0.$$
(3.25)

The matrices K and M are known as the stiffness and mass matrices. They are square matrices of the same order (say, n). The vector x is the mode shape corresponding to the natural frequency ω. In linear algebra, the problem of determining x is known as the generalised eigen value problem. An obvious solution is the trivial solution $x = 0$. However, a non-trivial solution exists if and only if the determinant of $K - \omega^2 M$ is zero, that is

$$\det(K - \omega^2 M) = 0.$$
(3.26)

This equation leads to an nth-order polynomial in ω^2 that can be used to solve for the natural frequencies ω. Once the natural frequencies are determined, Equation 3.25 is solved for each of the natural frequencies to determine the corresponding mode shape. When $M = I$, the generalized eigenvalue problem reduces to the standard eigenvalue problem and x and $\lambda = \omega^2$ are called the eigenvector and the corresponding eigenvalue respectively of K.

3.1.3 Vector Algebra

Many physical quantities, such as the mass of an object or the length of a bar, can be represented by a single number and, therefore, are called scalar quantities. On the other hand, there are many other physical quantities which can only be described by both magnitude and direction. Examples of such quantities are the force acting on a body, the displacement of a particle and the velocity of a projectile; vectors are used to describe these quantities.

A vector can be thought of as a quantity with a direction and a magnitude. Suppose that a is a vector. The magnitude of a is typically represented by $\| a \|$. Vector $-a$ has the same magnitude as a but points in the opposite direction to that of a.

Two vectors a and b are said to be equal if they have the same direction and same magnitude, that is $\| a \| = \| b \|$. The addition of two vectors a and b is performed by placing the tail of b at the head of a. The direction of the resulting vector is from the tail of a to the head of b.

An alternative and equivalent approach to adding vectors is by making use of the component representation of vectors. Consider a three-dimensional space with a rectangular Cartesian coordinate system. Let x_1, x_2 and x_3 be the coordinate axes and let e_1, $i = 1, 2, 3$ be the unit vectors pointing along the positive coordinate directions. A unit vector has a length of one. Then, any vector a can be written as

$$\mathbf{a} = a_1\mathbf{e}_1 + a_2\mathbf{e}_2 + a_3\mathbf{e}_3 \ . \tag{3.27}$$

The quantities a_1, a_2 and a_3 are scalars and called the components of $\mathbf{a}$ along each of the coordinate directions. Often, the vector $\mathbf{a}$ is represented by the column vector consisting of (a_1, a_2, a_3) as its components:

$$\mathbf{a} = \begin{bmatrix} a_1 & a_2 & a_3 \end{bmatrix}^T \tag{3.28}$$

or simply a_i, with the implicit understanding that i takes on values from 1 to n, where n is the dimension of the space. The addition or subtraction of two vectors $\mathbf{a}$ and $\mathbf{b}$ is performed by adding or subtracting the corresponding components:

$$\mathbf{a} = \begin{bmatrix} a_1 & a_2 & a_3 \end{bmatrix}^T \ . \tag{3.29}$$

In terms of the components, the length $\| \mathbf{a} \|$ of the vector $\mathbf{a}$ is $\sqrt{a_1^2 + a_2^2 + a_3^2}$. Thus, if $\mathbf{a}$ is divided by its length, the resulting vector will have a unit length and the vector $\mathbf{a}$ is said to be normalised.

Two vectors $\mathbf{a}$ and $\mathbf{b}$ are equal if and only if the corresponding components are equal, that is $a_1 = b_1$, $a_2 = b_2$ and $a_3 = b_3$.

Multiplication of a vector by a scalar scales the magnitude of the vector by the scalar. If the scalar is positive, the direction remains the same, whereas if the scalar is negative, the direction is reversed. Furthermore, each of the components is also scaled by the same factor.

The dot product of a vector $\mathbf{a}$ with another vector $\mathbf{b}$, denoted by $\mathbf{a} \cdot \mathbf{b}$ is a scalar defined by

$$\mathbf{a} \cdot \mathbf{b} = a_1 b_1 + a_2 b_2 + a_3 b_3 = \| \mathbf{a} \| \ \| \mathbf{b} \| \cos \theta, \tag{3.30}$$

where θ is the angle between the two vectors. Thus, if two vectors are parallel to each other, $\theta = 0$ and the dot product equals the product of the lengths of the two vectors. On the other hand, if the two vectors are perpendicular to each other, the dot product equals zero since $\theta = \dfrac{\pi}{2}$.

From the definition of the dot product of two vectors, it is clear that $\mathbf{a} \cdot \mathbf{b} = \mathbf{b} \cdot \mathbf{a}$. Furthermore, $\mathbf{a} \cdot (\alpha \mathbf{b} + \beta \mathbf{c}) = \alpha \mathbf{a} \cdot \mathbf{b} + \beta \mathbf{a} \cdot \mathbf{c}$ for two scalars α and β.

The cross-product of two vectors $\mathbf{a}$ and $\mathbf{b}$, denoted by $\mathbf{a} \times \mathbf{b}$ is also a vector defined by

$$\mathbf{a} \times \mathbf{b} = \| \mathbf{a} \| \ \| \mathbf{b} \| \sin \theta \mathbf{n} = \det \begin{bmatrix} \mathbf{e}_1 & \mathbf{e}_2 & \mathbf{e}_3 \\ a_1 & a_2 & a_3 \\ b_1 & b_2 & b_3 \end{bmatrix} \ . \tag{3.31}$$

Here, $\mathbf{n}$ is the unit vector normal to the plane of $\mathbf{a}$ and $\mathbf{b}$ and determined by using the right-hand rule. According to the right-hand rule, if the index finger of the right hand is pointed towards $\mathbf{a}$ and the middle finger (of the right hand) is pointed towards $\mathbf{b}$, then the thumb (of the right hand) points in the direction of the vector $\mathbf{n}$. As with the dot product, θ is the angle between the two vectors $\mathbf{a}$ and $\mathbf{b}$. Clearly, when two vectors are parallel to each other, the cross-product of the two vectors is the zero vector. Furthermore, the magnitude of the cross-product of the two vectors represents the area of the parallelogram formed by the two vectors $\mathbf{a}$ and $\mathbf{b}$.

The components of a vector depend on the underlying coordinate frame. If the coordinate frame is rotated to a new frame with rectangular Cartesian coordinate axes denoted by

x'_1, x'_2, x'_3, the components a'_i of a vector **a** with respect to the new coordinate frame can be shown to be given by

$$a'_i = \sum_{j=1}^{3} Q_{ij} a_j, \quad i = 1, 2, 3. \tag{3.32}$$

The quantities $Q_{ij} = \cos(x'_i, x_j), i, j = 1, 2, 3$ are the components of a 3×3 direction cosine (or rotation) matrix denoted by **Q**. The notation $\cos(x'_i, x_j)$ represents the cosine of the angle between x'_i-axis and x_j-axis. The matrix **Q** has the interesting property that $\mathbf{Q}\mathbf{Q}^T = \mathbf{Q}^T\mathbf{Q} = \mathbf{I}$, that is the inverse of **Q** is its transpose. Such matrices are called orthogonal matrices in linear algebra. In matrix form, the aforementioned transformation rule can be expressed as

$$\begin{Bmatrix} a'_1 \\ a'_2 \\ a'_3 \end{Bmatrix} = \begin{bmatrix} Q_{11} & Q_{12} & Q_{13} \\ Q_{21} & Q_{22} & Q_{23} \\ Q_{31} & Q_{32} & Q_{33} \end{bmatrix} \begin{Bmatrix} a_1 \\ a_2 \\ a_3 \end{Bmatrix}. \tag{3.33}$$

In some applications, the coordinate frame is fixed while the vector is rotated. Suppose that a vector **a** is rotated through an angle θ to obtain a new vector **b**. Then, the rotated vector is related to the initial vector through

$$b_i = \sum_{j=1}^{3} Q_{ji} a_j, \quad i = 1, 2, 3. \tag{3.34}$$

The matrix **Q** in Equation 3.34 is the same as the matrix defined for the case of rotation of coordinate frames. In direct notation, Equation 3.34 can be written as $\mathbf{b} = \mathbf{Q}^T\mathbf{a}$.

3.1.4 Cylindrical and Spherical Coordinate Systems

When there is a cylindrical or spherical symmetry present in geometry, it is often more convenient to work in cylindrical or spherical coordinate systems. Suppose that a point P in three-dimensional space is identified by the rectangular coordinates (x_1, x_2, x_3). In a cylindrical coordinate system, the same point is identified by the coordinates (r, θ, z) (Figure 3.3a). The radial coordinate $r(\geq 0)$ is the distance (from the origin O) of the projection of the point P onto the $x_1 - x_2$ coordinate. The coordinate z is the same as x_3 and the azimuthal angle θ varies over a range of 2π, typically between 0 and 2π or from $-\pi$ to π.

The unit basis vectors in the three coordinate directions (r, θ, z) are denoted by $\mathbf{e}_r, \mathbf{e}_\theta$ and $\mathbf{e}_z$ respectively. Unlike in the rectangular coordinate system, the unit vectors $\mathbf{e}_r$ and $\mathbf{e}_\theta$ are not constant vectors. Instead, they change with the azimuthal coordinate θ.

In a spherical coordinate system, the point P is identified by the coordinates (r, θ, ϕ) (Figure 3.3b). The azimuthal angle θ again varies over a range of 2π and the polar angle ϕ varies over a range of π. It is common to assume $0 \leq \theta \leq 2\pi$ or $-\pi \leq \theta \leq \pi$ and $0 \leq \phi \leq \pi$. The unit basis vectors in the three coordinate directions (r, θ, ϕ) are denoted by $\mathbf{e}_r, \mathbf{e}_\theta$ and $\mathbf{e}_\phi$ respectively. Just as in the cylindrical coordinate system, these three vectors are functions of the spherical coordinates θ and ϕ.

It may be recalled that a position vector **v** has the representation $v_1\mathbf{e}_1 + v_2\mathbf{e}_2 + v_3\mathbf{e}_3$ in a rectangular coordinate systems. In a cylindrical coordinate system, the same vector has the representation $v_r\mathbf{e}_r + v_z\mathbf{e}_z$. Here, $\mathbf{e}_r$ is a unit vector in the direction of the radial distance of

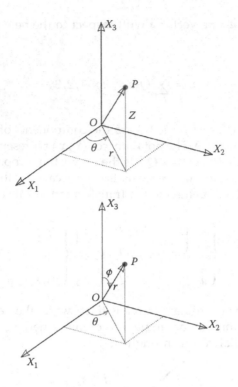

FIGURE 3.3
Cylindrical and spherical coordinates and the coordinate nomenclatures used to locate a point P in these systems.

the projection of P onto the x_1–x_2 plane and $\mathbf{e}_z = \mathbf{e}_3$ is the unit vector in the axial direction. v_r and v_z are the components of the vectors in the radial and axial directions respectively.

In a spherical coordinate system, the position vector of P has the representation $v_r\mathbf{e}_r$ with $\mathbf{e}_r$ being the unit vector in the radial direction. As before, v_r is the component of the vector in the radial direction or simply the length of the vector.

3.2 Taylor Series

Taylor series approximation of a function in the neighbourhood of $x = a$ is one of the most commonly used mathematical tools in engineering. The Taylor series of a function $f(x)$ about $x = a$ is given by

$$f(x) = f(a) + f'(a)(x - a) + f''(a)\frac{(x - a)^2}{2!} + \cdots + f^{(n)}(a)\frac{(x - a)^n}{n!} + \cdots . \qquad (3.35)$$

In Equation 3.35, $f^{(n)}(a)$ is the nth derivative of $f(x)$ at $x = a$ and $n!$ is the factorial of n, i.e. $n! = 1 \times 2 \times 3 \times \ldots n$ with the understanding that $0! = 1$. Then, the Taylor series representation of $f(x)$ near a can be written as

$$f(x) = \sum_{n=0}^{\infty} f^{(n)}(a) \frac{(x-a)^n}{n!}. \tag{3.36}$$

For this to make sense, the derivatives of all orders of the function $f(x)$ must exist. The special case corresponding to $a = 0$ goes by the name McLaurin series.

More generally, Taylor's theorem states that if the first $n + 1$ derivatives of $f(x)$ exist at $x = a$, then the function $f(x)$ for some x in the open interval containing x can be expressed as

$$f(x) = f(a) + f'(a)(x-a) + f''(a)\frac{(x-a)^2}{2!} + \cdots + f^{(n)}(a)\frac{(x-a)^n}{n!} + f^{(n+1)}(\xi_n)\frac{(x-a)^{n+1}}{(n+1)!}, \tag{3.37}$$

where ξ_n is between a and x. The Taylor polynomial is the nth-order polynomial on the right-hand side of Equation 3.37. It approximates the function $f(x)$ in a neighbourhood of a using an nth order polynomial in x. The term involving ξ_n on the right-hand side is called the remainder and is the error of approximation of $f(x)$ using the Taylor polynomial.

For example, when $n = 1$, the Taylor polynomial leads to a linear approximation of $f(x)$ and when $n = 2$, $f(x)$ is approximated using a quadratic polynomial.

As an example the Taylor series of the exponential function $e^{\alpha x}$ about $x = 0$ is

$$f(x) = 1 + \alpha x + \frac{\alpha^2 x^2}{2!} + \frac{\alpha^3 x^3}{3!} + \cdots = \sum_{n=0}^{\infty} \frac{\alpha^n x^n}{n!}. \tag{3.38}$$

Similarly, the Taylor series for $\sin x$ and $\cos x$ are given by

$$\sin x = x - \frac{x^3}{3!} + \frac{x^5}{5!} - \cdots = \sum_{n=0}^{\infty} \frac{(-1)^n x^{2n+1}}{(2n+1)!} \tag{3.39}$$

and

$$\cos x = 1 - \frac{x^2}{2!} + \frac{x^4}{4!} - \cdots = \sum_{n=0}^{\infty} \frac{(-1)^n x^{2n}}{2n!} \tag{3.40}$$

respectively.

3.3 Fourier Series

The Fourier series method is a widely used technique in engineering with applications in obtaining solutions to partial differential equations governing various physical problems, such as heat conduction and propagation, signal processing, image processing, optics and approximation of functions. The method is used to express a periodic function in terms of an infinite series consisting of sine and cosine functions.

A function $f(x)$ is said to be periodic with a period $T(>0)$ if $f(x+T) = f(x)$ for all real x. The least possible value of T for which this relation holds true is called the fundamental period (or simply, the period). For example, the function $\sin x$ has a fundamental period of 2π. Suppose that a function $f(x)$ has a period of 2π. Then, it can be shown that $f(x)$ has the representation

$$f(x) = \frac{a_0}{2} + \sum_{n=1}^{\infty}(a_n \cos nx + b_n \sin nx), \qquad (3.41)$$

where the coefficients $a_n, n = 0,1,2\ldots$ and $b_n, n = 1,2,3\ldots$ are called the Fourier coefficients. The Fourier coefficients are obtained from the fact that $\sin nx$ and $\cos nx$ are orthogonal over the interval $-\pi \leq x \leq \pi$, that is

$$\int_{-\pi}^{\pi} \sin mx \sin nx \, dx = 0, \quad m \neq n,$$

$$\int_{-\pi}^{\pi} \cos mx \cos nx \, dx = 0, \quad m \neq n, \qquad (3.42)$$

$$\int_{-\pi}^{\pi} \sin mx \cos nx \, dx = 0, \quad all \, m, n$$

and

$$\int_{-\pi}^{\pi} \sin mx \sin mx \quad dx = \pi \text{ and } \int_{-\pi}^{\pi} \cos mx \cos nx \quad dx = \pi, \quad m = 1,2,3,\cdots. \qquad (3.43)$$

With the help of these properties, it can be shown that the Fourier coefficients are given by

$$a_n = \frac{1}{\pi}\int_{-\pi}^{\pi} f(x)\cos nx dx, \quad n = 0,1,2,\cdots$$

$$b_n = \frac{1}{\pi}\int_{-\pi}^{\pi} f(x)\sin nx dx, \quad n = 1,2,\cdots. \qquad (3.44)$$

When $f(x)$ is periodic over an interval $2L$ instead of 2π, then the corresponding Fourier series expansion of $f(x)$ is given by

$$f(x) = \frac{a_0}{2} + \sum_{n=1}^{\infty}\left(a_n \cos\frac{n\pi x}{L} + b_n \sin\frac{n\pi x}{L}\right) \qquad (3.45)$$

with

$$a_n = \frac{1}{L}\int_{-L}^{L} f(x)\cos\frac{n\pi x}{L} dx, \quad n = 0,1,2,\cdots,$$

$$b_n = \frac{1}{L}\int_{-L}^{L} f(x)\sin\frac{n\pi x}{L} dx, \quad n = 1,2,3,\cdots. \qquad (3.46)$$

The conditions under which the Fourier series of $f(x)$ converges to $f(x)$ can be found in Tolstov (2012). It is worth pointing out though that at a point of discontinuity in the interval $[-L \leq x \leq L]$, the Fourier series of $f(x)$ converges to the average value of $f(x)$ obtained from the left and right limits of $f(x)$ at this point. Often, the Fourier series converges fast enough that only the first few terms of the infinite series are used as a good approximation to $f(x)$.

The concepts of even and odd functions play a key role in the application of the Fourier series to the solution of partial differential equations. A function is said to be an even function if $f(-x) = f(x)$. The cosine function and x^{2n} for $n = 0,1,2,\ldots$ are examples of even functions. A function is said to be an odd function if $f(-x) = -f(x)$. The sine function and x^{2n+1} for $n = 0,1,2,\ldots$ are examples of odd functions. The Fourier series of an odd function will consist only of sine terms (i.e. $a_n = 0$ for all $n = 0,1,2,\ldots$), whereas the Fourier series of an even function will consist only of the cosine terms (i.e. $b_n = 0$ for all $n = 1,2,3,\ldots$). Suppose that a function $f(x)$ is given on the interval $0 \leq x \leq L$. Then, a convenient way to obtain a Fourier series is to extend the function $f(x)$ as an even function or an odd function to the interval $[-L,0]$ so that the extended function has a period of $2L$. The reader is referred to Kreyszig (2007) for further information on this approach. It is also worth pointing out that a given function $f(x)$ can always be represented by the sum of an even function defined by $(f(x)+f(-x))/2$ and an odd function defined by $(f(x)-f(-x))/2$.

Two examples are considered to illustrate the representation of a function by a Fourier series. The functions considered are the square and triangle functions shown in Figure 3.4. The square wave function has the Fourier series

$$f(x) = \frac{4}{\pi} \sum_{n=0}^{\infty} \frac{1}{2n+1} \sin 2(2n+1)\pi x. \tag{3.47}$$

The function is periodic with a period of one and is an odd function. Therefore, the Fourier expansion consists of only the sine functions. In Figure 3.5, the square wave function along with the Fourier series with only the first two, five and ten terms considered are shown. The oscillations seen in the Fourier series approximations are called the Gibbs phenomenon and typically occur near the points of discontinuity of $f(x)$.

The triangle wave equation has the Fourier series representation given by

$$f(x) = \sum_{n=0}^{\infty} \frac{8(-1)^n}{\pi^2(2n+1)^2} \sin(2n+1)\pi x. \tag{3.48}$$

Again, since $f(x)$ is an odd function (with a period of 2), the Fourier representation consists only of sine terms. The approximations obtained by truncating the Fourier series after the first, second and third terms are shown in Figure 3.6 along with the exact function $f(x)$. Clearly, the more terms that are used in the Fourier series, the better the approximation of $f(x)$.

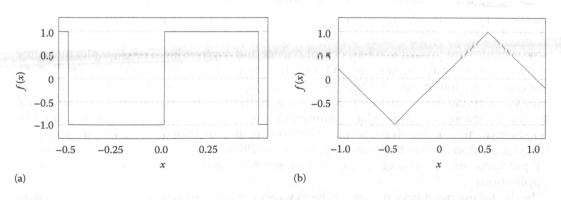

(a) (b)

FIGURE 3.4
Two functions considered to illustrate Fourier series representation of periodic functions: (a) a square wave function and (b) a triangular wave function.

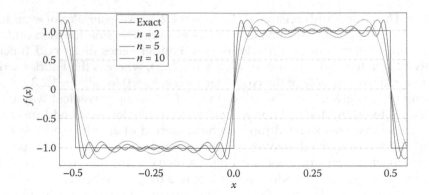

FIGURE 3.5
Fourier series representation of the square wave function of Figure 3.4 for various values of the maximum number of terms retained in the series. The oscillations near the points of discontinuity are called the Gibbs phenomenon.

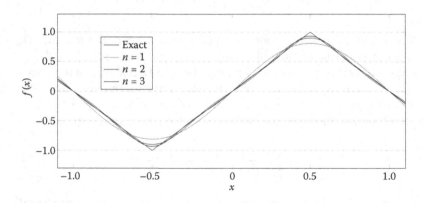

FIGURE 3.6
Fourier series representation of the triangular wave function of Figure 3.4 for various values of the maximum number of terms retained in the series.

3.4 Statistics

Uncertainties are present in all aspects of science and engineering. For example, measurements of temperatures, lengths and material properties involve uncertainties due to instrument limitations (see Chapter 9 for an in depth discussion of measurement uncertainty). Uncertainties in measurements are also introduced by the conditions under which the measurements are made. Manufacturing processes can introduce uncertainties into part geometries and material properties. Statistics is the mathematical tool employed to understand and analyse these uncertainties. In addition, statistics is used in quality control of products and to predict service life of machine components amongst many other applications.

In the following, a brief review of the key concepts of statistics is presented. The references Kreyszig (2007), Riley et al. (2006) and Holický (2013) are suggested for a more comprehensive understanding of the concepts presented here.

3.4.1 Populations and Samples

In statistical analysis, a population is the set of all possible values for a quantity. The population can be finite or infinite depending on the number of elements in it. Since it is often impossible to analyse the data based on the population, the characteristics of the population are often inferred using a smaller set of data sampled from the population. This set is called a sample. When all the elements of the population have the same chance of being sampled, the samples taken from such a population are called random samples. Random samples must be of sufficient size to capture the characteristics of the population. Too small a size can lead to misleading conclusions and too large a size can be unwieldy.

As an example, in the experiments carried out to measure the diameter of a component, the population is the interval $[L, U]$ for some known values L and U and a sample is the result of a finite number of measurements.

3.4.2 Discrete and Continuous Random Variables

Random variables are variables whose values are based on the outcomes of random events (Holický 2013). When a random variable assumes a discrete set of distinct values, then the variable is said to be a discrete random variable. A continuous random variable is a random variable that assumes values in a specified interval. The diameter of a bar or the tensile strength of a material is an example of a continuous random variable. The number of tails obtained in 100 throws of a coin and the number of defective parts in a given production cycle are examples of discrete random variables.

If X is a discrete random variable that attains the values $x_i, i = 1, 2, \ldots, n$ and p_i denotes the probability $P(x_i)$ of x_i, then a probability density function (often abbreviated as PDF) $f(x)$ can be defined as follows:

$$f(x) = P(X = x) = \begin{cases} p_i & \text{if } x = x_i, \\ 0 & \text{otherwise}. \end{cases} \tag{3.49}$$

Note that $\sum_{i=1}^{n} p_i = \sum_{i=1}^{n} f(x_i) = 1$. A cumulative distribution function (often abbreviated as CDF), or simply a distribution function $F(x)$ of the discrete random variable X, is defined as the probability that $X \leq x$, so that

$$F(x) = P(X \leq x) = \sum_{x_i \leq x} f(x_i). \tag{3.50}$$

From the preceding, it is clear that the probability of X in an interval $a < x \leq b$ is given by

$$P(a < X \leq b) = F(b) - F(a). \tag{3.51}$$

For a continuous random variable X, the probability density function $f(x)$ is defined so that $f(x)dx$ gives the probability of X in the interval $(x, x + dx]$, that is

$$P(x < X \leq x + dx) = f(x)dx. \tag{3.52}$$

Clearly, $f(x) \geq 0$. The corresponding cumulative distribution function $F(x)$ is defined as

$$F(x) = \int_{-\infty}^{x} f(u)du. \tag{3.53}$$

Note that $F(\infty) = 1$ and $f(x) = F'(x)$. The probability of $a < X \le b$ is given by

$$P(a < X \le b) = F(b) - F(a) = \int_a^b f(u)du . \tag{3.54}$$

3.4.3 Parameters of Random Variables

The probability density function $f(x)$ and the cumulative distribution function $F(x)$ are often characterised by moments and central moments of the distribution. The kth moment of a distribution is given by

$$\mu_k = \sum_{i=1}^n x_i^k f(x_i) \tag{3.55}$$

for a discrete random variable and

$$\mu_k = \int_{-\infty}^\infty x^k f(x)dx \tag{3.56}$$

for a continuous random variable. The special case $k = 1$ defines the mean of the distribution, that is

$$\mu = \mu_1 = \begin{cases} \sum_{i=1}^n x_i f(x_i), & \text{for discrete cases,} \\ \int_{-\infty}^\infty x f(x)dx, & \text{for continuous cases.} \end{cases} \tag{3.57}$$

The central moment of order k is given by

$$\lambda_k = \sum_{i=1}^n (x_i - \mu)^k f(x_i) \tag{3.58}$$

for a discrete random variable and

$$\lambda_k = \int_{-\infty}^\infty (x - \mu)^k f(x)dx \tag{3.59}$$

for a continuous random variable. The special case $k = 2$ defines the variance, σ^2, of the distribution, that is

$$\sigma^2 = \lambda_1 = \begin{cases} \sum_{i=1}^n (x_i - \mu)^2 f(x_i), & \text{for discrete cases,} \\ \int_{-\infty}^\infty (x - \mu)^2 f(x)dx, & \text{for continuous cases.} \end{cases} \tag{3.60}$$

Furthermore, it is worth noting that the central moment corresponding to $k = 1$ is zero. The quantity σ itself is called the standard deviation of the distribution.

The central moments of order 3 and 4 normalized by σ^3 and σ^4 are called the skewness and kurtosis respectively. Skewness characterises the symmetry of the distribution about the mean, and kurtosis defines the shape of the peak of the distribution as compared to a normal distribution. Note that when the distribution is symmetric about the mean μ, skewness is zero and for normal distributions, kurtosis is 3.

It is common to express the distribution functions in terms of a standardised random variable Z denoted by

$$Z = \frac{X - \mu}{\sigma}. \tag{3.61}$$

The variable Z has the special property that it has a mean of zero and a variance of unity. Distribution functions are often tabulated in terms of Z. If a value z of Z is known, the corresponding value x of X can be found from

$$x = \mu + \sigma z. \tag{3.62}$$

3.4.4 Distributions

For brevity sake, only one distribution for each type of random variable (discrete and continuous) is discussed. Suppose that an event occurs with a probability p. Then, $q = 1 - p$ denotes the probability that the event does not occur. If n independent trials are carried out, the random variable X of interest is the number of times the event occurs. Therefore, X can assume the discrete values $0, 1, 2, \ldots, n$. Then, the distribution function $f(x)$ for X, called the binomial or Bernoulli distribution, can be shown to be

$$f(x) = \binom{n}{x} p^x q^{n-x} = \frac{n!}{x!\,(n-x)!} p^x q^{n-x}, \qquad x = 0, 1, 2, \ldots, n. \tag{3.63}$$

For the Bernoulli distribution, by using the definitions of the moments and central moments, it can be shown (Riley et al. 2006) that

$$\mu = np \text{ and } \sigma = npq. \tag{3.64}$$

Furthermore, the normalised skewness and excess kurtosis can also be shown to be $\dfrac{q - p}{\sqrt{npq}}$ and $\dfrac{1 - 6pq}{npq}$ respectively. Here, excess kurtosis is defined as kurtosis $- 3$.

The most important distribution function for continuous random variables is the normal or Gaussian distribution that is defined by two parameters: the population mean μ and the standard deviation σ. The probability density function corresponding to the normal distribution is defined by

$$f(x) = \frac{1}{\sigma\sqrt{2\pi}} \exp\left[-\frac{1}{2}\left(\frac{x - \mu}{\sigma}\right)^2 \right] \tag{3.65}$$

with $\sigma > 0$. Clearly, the probability density function is symmetric about $x = \mu$ and, therefore, the skewness value is zero. In addition, the excess kurtosis value can also be shown to be zero for normal distributions. The corresponding cumulative distribution function $F(x)$ is given by

$$F(x) = \frac{1}{\sigma\sqrt{2\pi}} \int_{-\infty}^{x} \exp\left[-\frac{1}{2}\left(\frac{u-\mu}{\sigma}\right)^2\right] du. \tag{3.66}$$

In terms of the standardised normal variable z, the probability density function and the cumulative distribution function have the forms

$$f_n(z) = \frac{1}{\sqrt{2\pi}} e^{-\frac{z^2}{2}} \text{ and } F_n(z) = \frac{1}{\sqrt{2\pi}} \int_{-\infty}^{z} e^{-\frac{v^2}{2}} dv. \tag{3.67}$$

The subscript n is used to denote that these are defined in terms of z. The two functions are shown in Figure 3.7. Typically, the function values $F_n(z)$ are found from tables or numerically. If the corresponding $F(x)$ is desired, the relation

$$F(x) = F_n\left(\frac{x-\mu}{\sigma}\right) \tag{3.68}$$

is to be used. From this, it follows that the probability of a random variable X, assuming a value x in the interval $(a,b]$, is given by

$$P(a < X \le b) = F(b) - F(a) = F_n\left(\frac{b-\mu}{\sigma}\right) - F_n\left(\frac{a-\mu}{\sigma}\right). \tag{3.69}$$

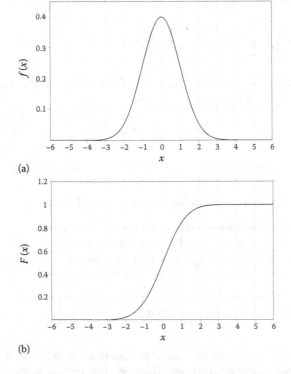

(a)

(b)

FIGURE 3.7
(a) Probability density function of the standardized normal distribution and (b) the corresponding cumulative distribution function.

3.4.5 Sampling Distributions

The distributions presented in Section 3.2.4 are for a population. However, typically, the parameters of a population (mean, variance, etc.) are unknown and are estimated from a sample of size n. For this reason, an understanding of the distributions of the sample statistics (such as the sample mean and variance) is needed. In this section, distributions for a random sample of size n are discussed. The two main sample statistics of interest are the sample mean $\bar{x}$ defined as

$$\bar{x} = \frac{1}{n}(x_1 + x_2 + \dots + x_n) \tag{3.70}$$

and the variance of the sample denoted by s calculated from

$$s = \frac{1}{n-1} \sum_{i=1}^{n} (x_i - \bar{x})^2 . \tag{3.71}$$

It should be noted that $\bar{x}$ is in general different from the population μ. However, the mean of the sampling distribution of the sample means can be shown to be equal to μ. Furthermore, for a large enough sample size n, the standard deviation of the sample mean (also called the standard error) and denoted by $\sigma_{\bar{x}}$ is equal to $\frac{\sigma}{\sqrt{n}}$, where σ is the population variance. When σ is known, it can be shown that the standardised normal variable Z, defined by

$$Z = \frac{\bar{x} - \mu}{\sigma_{\bar{x}}} = \frac{\bar{x} - \mu}{\dfrac{\sigma}{\sqrt{n}}}, \tag{3.72}$$

has a normal distribution.

3.4.6 Student's *t*-Distribution

The Student's *t*-distribution is used to estimate confidence intervals (discussed in Section 3.4.9) for the population mean μ, when the standard deviation of the population is unknown, the population distribution is normal or approximately normal and the sample size is reasonably large ($n \geq 30$). To estimate the confidence interval for μ, a new variable t (called the *t*-statistic) is defined as

$$t = \frac{\bar{x} - \mu}{\dfrac{s}{\sqrt{n}}} . \tag{3.73}$$

This variable can be shown to have Student's *t*-distribution function with $v = n - 1$ degrees of freedom given by (Riley et al. 2006)

$$f(t) = \frac{1}{\sqrt{v\pi}} \frac{\Gamma\left(\dfrac{v+1}{2}\right)}{\Gamma\left(\dfrac{v}{2}\right)} \left(1 + \frac{t^2}{v}\right)^{-(v+1)/2} . \tag{3.74}$$

The *t*-distribution function is symmetric, has a mean of zero and a variance equal to $\dfrac{v}{v-2}$ with $v \geq 2$. The skewness parameter is also zero, whereas the excess kurtosis parameter equals

$\dfrac{6}{v-4}$ for $v > 4$. Clearly, as v approaches infinity, the variance approaches unity. Furthermore, as $n \to \infty$, this distribution approaches the standardised normal distribution.

The corresponding distribution or the cumulative distribution function $F(t)$ is given by

$$F(t) = \frac{1}{\sqrt{v\pi}} \frac{\Gamma\left(\dfrac{v+1}{2}\right)}{\Gamma\left(\dfrac{v}{2}\right)} \int_{-\infty}^{t} \left(1 + \frac{u^2}{v}\right)^{-(v+1)/2} du . \tag{3.75}$$

The probability density function and the (cumulative) density function for the t-distribution are shown in Figure 3.8 for three different values of v.

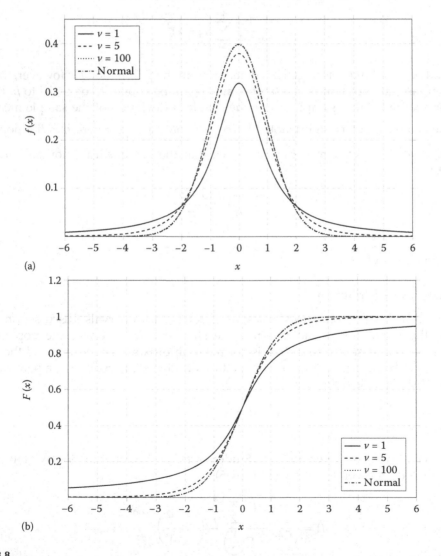

FIGURE 3.8
(a) Probability density function of t-distribution for three different degrees of freedom and (b) the corresponding cumulative distribution function. Note that the density function of the standardized normal distribution (also shown) is on top of the $v = 100$ curve.

3.4.7 χ^2-Distribution

The χ^2-distribution is used when confidence intervals for the population variance are desired.

The probability density function for a χ^2-distribution with $v = n - 1$ degrees of freedom is given by (Riley et al. 2006)

$$f(y) = \begin{cases} 0, & y < 0, \\ A_v y^{\frac{v}{2}-1} e^{-\frac{y}{2}}, & y \geq 0 \text{ with } A_v = \dfrac{1}{\Gamma\left(\dfrac{v}{2}\right) 2^{\frac{v}{2}}}. \end{cases} \tag{3.76}$$

Here, $\Gamma(\frac{v}{2})$ is the Gamma function (Abramowitz and Stegun 1972). The corresponding cumulative distribution function is

$$F(y) = \begin{cases} 0, & y < 0, \\ \int_0^y A_v u^{\frac{v}{2}-1} e^{-\frac{u}{2}} du, & y \geq 0. \end{cases} \tag{3.77}$$

Clearly, the χ^2-distribution is a one-parameter family of distributions obtained by varying v, the number of degrees of freedom. Furthermore, the distribution can be shown to have the statistical parameters (Weisstein 2017, Holický 2013)

$$\mu = v, \ \sigma = \sqrt{2v}, \ \alpha = 2\sqrt{\frac{2}{v}} \text{ and } \varepsilon = \frac{12}{v}. \tag{3.78}$$

Here, α and ε are the skewness and excess Kurtosis respectively.

The probability density function is not symmetric. However, it approaches a normal distribution as $v \to \infty$. The probability density function and the (cumulative) density function for the χ^2-distribution are shown in Figure 3.9 for four different values of v.

3.4.8 F-Distribution

The F-distribution is another important distribution in statistics. Suppose that Y_1 and Y_2 are two independent random variables, with χ^2-distributions drawn from two separate normally distributed populations. Let n and m be the corresponding degrees of freedom. Then, the F-statistic is defined as

$$X = \frac{Y_1/n}{Y_2/m}. \tag{3.79}$$

Alternatively, the F-statistic can also be defined as

$$X = \frac{s_1^2/\sigma_1^2}{s_2^2/\sigma_2^2}. \tag{3.80}$$

Here, s_i is the standard deviation of the sample i and σ_i is the standard deviation of the population i. Then, it can be shown that the probability density function for the distribution of F-statistics is given by (Weisstein 2017)

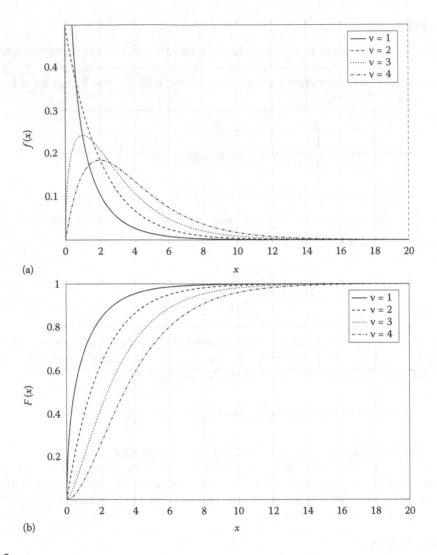

FIGURE 3.9
(a) Probability density function of χ^2-distribution for three different degrees of freedom and (b) the corresponding cumulative distribution function.

$$f_{n,m}(x) = \frac{\Gamma\left(\dfrac{n+m}{2}\right) m^{(m/2)}}{\Gamma\left(\dfrac{n}{2}\right)\Gamma\left(\dfrac{m}{2}\right)} \frac{n^{n/2}}{m} \frac{x^{n/2(n+m)=2}}{\left(1+\dfrac{nx}{m}\right)^{(n/2+m/2)}}, x \geq 0, \qquad (3.81)$$

and the corresponding cumulative distributive function is given by

$$F_{n,m}(x) = \int_0^x f_{n,m}(u)du . \qquad (3.82)$$

The probability density function and the (cumulative) density function for the F-distribution are shown in Figure 3.10 for four different (n,m) values.

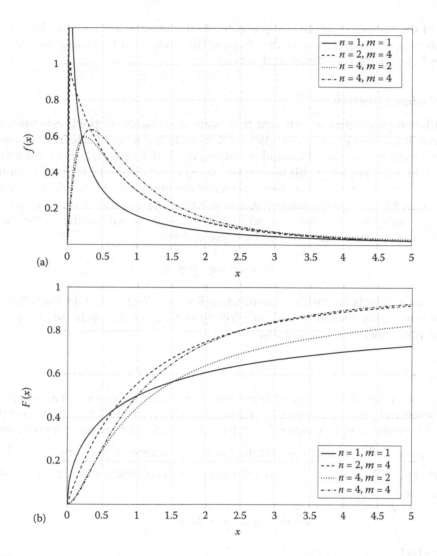

FIGURE 3.10

(a) Probability density function of F-distribution for three different degrees of freedom and (b) the corresponding cumulative distribution function.

The F-distribution is ≥ 0, a two-parameter function that depends on n and m and is positively skewed. Furthermore, it is important to note that $f_{n,m}(x) \neq f_{m,n}(x)$. The mean and variance for the F-distribution are given by (Weisstein 2017, Holický 2013)

$$\mu = \frac{m}{m-2}, m > 2 \text{ and } \sigma2 = \frac{2m^2(m+n-2)}{n(m-2)^2(m-4)}, m > 4. \tag{3.83}$$

Note that the mean does not depend on n. The reader is referred to Holický (2013) for the expressions for skewness and kurtosis. An additional property worthy of note is that, if X has an F-distribution given by $F_{n,m}$, then $1/X$ has a distribution given by $F_{m,n}$.

Tables of the various distributions presented in this chapter are listed in Tables 3.A.1 to 3.A.6 in the Appendix at the end of the chapter. The reader will find these useful in solving the example problems considered in this book.

3.4.9 Confidence Intervals

The population parameters determined from sample statistics are point estimates of these parameters. It is more useful in practice to determine the level of uncertainty associated with these parameters; for additional examples applied to estimation of uncertainty in measurement, see Chapter 9. This is achieved using confidence intervals for a parameter q of interest (for example, the mean and variance of the population). The confidence interval for q is specified by three parameters γ, q_1 and q_2 such that $q_1 \le q \le q_2$ with a high probability of γ. γ is typically chosen to be 90%, 95% or 99% and is known as the confidence level. Alternatively, the confidence level can be expressed as

$$P(q_1 \le q \le q_2) = \gamma. \tag{3.84}$$

In terms of a cumulative distributive function $F(x)$, $\gamma = F(q_2) - F(q_1)$. In the following, the two most commonly used tests for establishing confidence intervals for the population mean and population variance are discussed.

3.4.10 *t*-Test

As mentioned earlier, the Student's *t*-test is used when a confidence interval on μ is desired, σ is unknown and $\gamma = 1 - \alpha$ is given. First, the sample mean $\bar{x}$ and the sample variance s are calculated from the sample of size n. From the *t*-distribution tables, the value of t for which $F(t) = 1 - \dfrac{\alpha}{2}$ is found from the tables for the number of degrees of freedom $v = n - 1$. Let this be denoted by $t_{\frac{\alpha}{2}}$. Then, following the definition of *t*-statistic, the confidence interval for μ is determined as

$$\bar{x} - \delta \le \mu \le \bar{x} + \delta \text{ with } \delta = \frac{t_{\frac{\alpha}{2}} s}{\sqrt{n}}. \tag{3.85}$$

3.4.11 χ^2-Test

The χ^2-test is used when a confidence interval on the variance is desired. First, a random variable χ^2,

$$\chi^2 = \frac{(n-1)s^2}{\sigma^2}. \tag{3.86}$$

is defined. This variable has a χ^2-distribution with $v = n - 1$ degrees of freedom. Suppose again that $\gamma = 1 - \alpha$ is given. Then, from the χ^2-distribution tables (i.e. the tables for the cumulative distribution function $F(y)$ for χ^2-distribution), the values of $y = \chi^2$, for which $F(y) = 1 - \dfrac{\alpha}{2}$ and $F(y) = \dfrac{\alpha}{2}$ are determined. Let these be denoted by $\chi^2_{\frac{\alpha}{2}}$ and $\chi^2_{1-\frac{\alpha}{2}}$ respectively. From the earlier definition of χ^2, the confidence interval for the variance is easily established as

$$\frac{(n-1)s^2}{\chi^2_{\frac{\alpha}{2}}} \le \sigma^2 \le \frac{(n-1)s^2}{\chi^2_{1-\frac{\alpha}{2}}}. \tag{3.87}$$

3.4.12 *F*-Test

The *F*-test is used when a comparison of the variances σ_1 and σ_2 of two normally distributed populations is needed. Recall that the *F*-statistic is defined by

$$X = \frac{Y_1/n}{Y_2/m} \text{ or } X = \frac{s_1^2/\sigma_1^2}{s_2^2/\sigma_2^2}. \tag{3.88}$$

In general, the parameters with the subscript 1 correspond to the sample with a higher variance. Suppose that the confidence interval $\gamma = 1 - \alpha$ is given. From the *F*-distribution tables, the values of X corresponding to $F_{n,m} = 1 - \frac{\alpha}{2}$ and $F_{n,m} = 1 - \frac{\alpha}{2}$ are determined. Let these be denoted by $F_{n,m,1-\frac{\alpha}{2}}$ and $F_{n,m,\frac{\alpha}{2}}$ respectively. The confidence interval for the ratio $\frac{\sigma_1^2}{\sigma_2^2}$ is then established as

$$\frac{s_1^2/s_2^2}{F_{n,m,\frac{\alpha}{2}}} \le \frac{\sigma_1^2}{\sigma_2^2} \le \frac{s_1^2/s_2^2}{F_{n,m,1-\frac{\alpha}{2}}}. \tag{3.89}$$

The *F*-values are obtained from published tables or using statistical software. *F*-tables provide the *x* values such that $P(F \le x) = 1 - \alpha$. Typically, α tends to be small with commonly used numbers being about 0.01 or 0.025 or 0.05. Thus, in Equation 3.89, when $F_{n,m,1-\frac{\alpha}{2}}$ is needed, the important relationship

$$F_{n,m,1-\frac{\alpha}{2}} = \frac{1}{F_{m,n,\frac{\alpha}{2}}} \tag{3.90}$$

is used.

3.5 Some Concepts from Solid Mechanics

In the following, a brief review of the mechanics of solids is presented. The readers interested in learning the mechanics of solids in detail will find references Crandall et al. (1978) and Slaughter (2002) useful.

When a body is subjected to external loads, internal forces develop in the body and the body deforms. The deformation is characterised by the concept of strain, whereas the internal force distribution is quantified by the concept of stress. Stress and strain are related through constitutive relations that account for material behaviour due to the external loads. In this chapter, the material behaviour is assumed to be purely elastic and isotropic. Isotropy implies that the material response is the same in all directions.

3.5.1 Strain

Consider a three-dimensional body as shown in Figure 3.11. The current configuration of the body (after the application of external loads) is also shown. Suppose that any material point on the body in the initial configuration is denoted by its position with respect to a rectangular coordinate frame. Thus, a material point P_0 is identified by its coordinates (x_1, x_2, x_3). The position of the point P in the current configuration is denoted by P. Then, (u_1, u_2, u_3) denote the components of the displacement vector from P_0 to P.

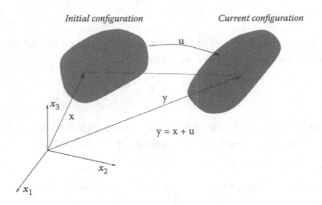

FIGURE 3.11
Motion of a body: a point at **x** moves to a new location **y** given by **u** = **y** − **x**.

The deformation of a body due to external loads is characterised by the strain tensor, which in component form is given by

$$\varepsilon_{ij} = \frac{1}{2}\left(\frac{\partial u_i}{\partial x_j} + \frac{\partial u_j}{\partial x_i}\right) \text{ with } i,j = 1,2,3 . \qquad (3.91)$$

Note that $\varepsilon_{ij} = \varepsilon_{ji}$ and, therefore, only six of the nine components are needed. The components $\varepsilon_{11}, \varepsilon_{22}$, and ε_{33} are called the normal strains and represent changes in lengths per unit length in each of the three coordinate directions. The components ε_{ij} with $i \neq j$ are called the shear strains and represent changes in angles (in radians) between two initially perpendicular lines aligned with the i and j axes.

Suppose that the coordinate frame is rotated to obtain a new coordinate frame with the axes (x_1', x_2', x_3'). The displacement components with respect to the rotated coordinate frame, denoted by u_i', are given by

$$u_i' = \sum_{j=1}^{3} Q_{ij} u_j , \qquad (3.92)$$

where Q_{ij} is the $(i,j)^{th}$ component of the 3×3 direction cosine matrix, that is Q_{ij} denotes the cosine of the angle between the x_i' and x_j axes.

Similarly, if ε_{ij}' denotes the components of the strain tensor in the rotated coordinate frame, then the transformation rule for these components is given by

$$[\varepsilon_{ij}'] = \mathbf{Q}[\varepsilon_{ij}]\mathbf{Q}^T . \qquad (3.93)$$

Here, **Q** is the 3×3 direction cosine matrix and the notation ε_{ij} represents a 3×3 matrix of ε_{ij}.

The coordinate frame in which the off-diagonal terms are zero is called the principal frame and each of the coordinate axis directions of this particular frame is called a principal direction. The diagonal terms are the normal strains and are also called the principal strains.

3.5.2 Stress

In addition to deformation, external forces also induce internal forces in the body. The distribution of these internal forces at any cut is characterised by the stress vector. The stress vector denoted by **s** is the force per unit area at any point in the body (see Figure 3.12). Then, the total force vector acting at a cut in the body is given by

$$F = \int_A s \, dA, \tag{3.94}$$

where A is the area of the cut. The component of the stress vector normal to the cut surface is called the normal stress and the component tangential to the surface is the tangential stress.

Suppose that three mutually perpendicular planes that are parallel to the coordinate planes pass through a point of interest (see Figure 3.13). The stress vector on the cut surface with a unit outward normal **n** at the point of interest is related to the stress vectors on these three mutually perpendicular planes through the relation (known as the Cauchy stress theorem)

$$\mathbf{s} = \begin{Bmatrix} s_1 \\ s_2 \\ s_3 \end{Bmatrix} = \begin{bmatrix} S_{11} & S_{12} & S_{13} \\ S_{21} & S_{22} & S_{23} \\ S_{31} & S_{32} & S_{33} \end{bmatrix} \begin{Bmatrix} n_1 \\ n_2 \\ n_3 \end{Bmatrix}. \tag{3.95}$$

The square matrix is the stress tensor in component form. The column i in the square matrix represents the components of the stress vector on the plane the is perpendicular to the ith coordinate axis. The diagonal component S_{ii} represents the normal stress in the ith direction. The off-diagonal terms $S_{ij}, i \neq j$ represents the shear stress acting in the jth direction on a plane that is perpendicular to the ith direction. Furthermore, the principle of balance of angular momentum leads to the symmetry of the stress tensor, that is $S_{ij} = S_{ji}$.

If the coordinate frame is rotated to another coordinate frame with the coordinate axes (x_1', x_2', x_3'), then the components S_{ij}' of the stress tensor with respect to this rotated coordinate system are given by the equation

$$[S_{ij}'] = \mathbf{Q}[S_{ij}]\mathbf{Q}^T. \tag{3.96}$$

As for the case of strain, the coordinate frame in which the shear stresses are zero (the off-diagonal components of the stress tensor are zero) is called the principal frame for stress. Thus, in the principal frame for stress, the stress tensor components are only the normal

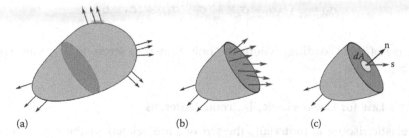

(a)   (b)   (c)

FIGURE 3.12
Stress vector is a measure of the internal force distribution in a body. (a) A three-dimensional body subjected to external loads. (b) A schematic of the internal force distribution at a cut in the body. (c) The stress vector at a point on the surface resulting from the cut. **n** is the unit outward normal to the surface.

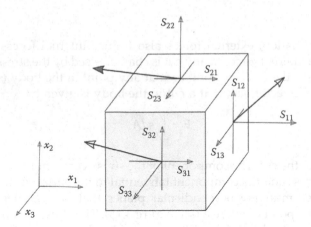

FIGURE 3.13
Components of the stress vectors acting on three planes perpendicular to the coordinate axes.

stresses. These are known as the principal stresses and the coordinate axis directions of the principal frame are called the principal directions. The principal stresses point along these principal directions. The principal stresses are commonly denoted by $\sigma_1, \sigma_2, \sigma_3$ with the algebraically largest and smallest values being σ_1 and σ_3 respectively.

Often in applications, the maximum shear stress and the von Mises stress are of interest. The maximum shear stress at a point is given by

$$\tau_{max} = \frac{|\sigma_1 - \sigma_3|}{2}, \tag{3.97}$$

which acts on the planes at an angle of 45° to the directions of the maximum and minimum principal stresses.

The von Mises stress σ_M is given by

$$\sigma_M = \frac{1}{\sqrt{2}} \sqrt{(S_{11} - S_{22})^2 + (S_{22} - S_{33})^2 + (S_{33} - S_{11})^2 + 6\left(S_{12}^2 + S_{23}^2 + S_{31}^2\right)}. \tag{3.98}$$

In terms of the principal stresses, Equation 3.98 simplifies to

$$\sigma_M = \frac{1}{\sqrt{2}} \sqrt{(\sigma_1 - \sigma_2)^2 + (\sigma_2 - \sigma_3)^2 + (\sigma_3 - \sigma_1)^2}. \tag{3.99}$$

For the case of axial loading, where the only non-zero stress is S_{11}, von Mises stress reduces to S_{11}.

3.5.3 Hooke's Law for Linear Elastic, Isotropic Materials

For linear elastic, isotropic materials, the stresses are related to the strains through the generalised Hooke's law:

$$\varepsilon_{ij} = \frac{1}{E}\left[(1 + v)S_{ij} - vS_{kk}\delta_{ij}\right] \text{ with } S_{kk} = S_{11} + S_{22} + S_{33} \text{ and } i, j = 1, 2, 3. \tag{3.100}$$

The quantities v and E are material properties and known as Poisson's ratio and Young's modulus respectively. Furthermore, δ_{ij} is equal to one when $i = j$ and zero when $i \neq j$. Poisson's ratio has values between -1 and 0.5. Most metallic materials have a value of around 0.3 for Poisson's ratio. Young's modulus has the same units as stress (pascal). It is approximately equal to 200 GPa for steels.

The inverse relationship relating stresses to strains is given by

$$S_{ij} = \frac{E}{(1 - 2v)(1 + v)} \left[(1 - 2v)\varepsilon_{ij} + ve\delta_{ij} \right],\qquad(3.101)$$

where $e = \varepsilon_{11} + \varepsilon_{22} + \varepsilon_{33}$ represents the change in volume per unit volume and is known as the dilatation or volumetric strain.

Two commonly used simplifications of the general three-dimensional elasticity are the plane stress and plane strain cases. Both of these are applicable to prismatic cylindrical bodies subjected to lateral loads/constraints that are independent of the axial coordinate x_3.

For plane stress problems (valid for prismatic disks where the lateral dimensions are much larger than the height of the disk), S_{13}, S_{23} and S_{33} are taken to be zero. The generalized Hooke's law implies that the non-zero strains are $\varepsilon_{11}, \varepsilon_{12}, \varepsilon_{22}$ and ε_{33}.

For plane strain problems (valid for prismatic long cylinders), $u_3 = 0$ and all the stresses and strains are independent of x_3. In this case, the non-zero strains are $\varepsilon_{11}, \varepsilon_{12}$ and ε_{22}. The corresponding non-zero stresses are the in-plane stresses S_{11}, S_{12}, S_{12} and the out-of-plane stress S_{33}.

The obvious advantage of the aforementioned two simplifications is that the general three-dimensional elasticity problem reduces to a planar problem involving only the cross-section of the prismatic cylinder. Many solution methodologies (including the very powerful complex analysis methods) exist for solving two-dimensional problems.

A further simplification is the beam theory applicable to long, slender bars of arbitrary cross-sections carrying transverse loads. This theory is summarised in the following for the special case when the cross-section is symmetric about a vertical axis, the material is homogeneous and the transverse load is symmetrically applied with respect to the surface bisecting the beam along the axial direction.

3.5.4 Beam Theory

A structural member is called a beam if the cross-sectional dimensions are significantly smaller than the length of the member and the loading is transverse. One such example of a beam is the cantilever beam as shown in Figure 3.14.

In the following, it is assumed that a given beam is subjected to a distributed load $w(x)$ (force per unit length) along the length of the beam. In addition, the beam may be subjected to concentrated transverse forces as well as concentrated moments. Suppose that the longitudinal direction of the beam is denoted by x.

The applied loads are resisted by the internal forces. At each cross-section of the beam, the net effect of these internal forces is a shear force V and a bending moment M both of which can vary along the length of the beam depending on the nature of the applied loads. The actual force distribution at each cross-section is quantified by the normal stress S_{11} and the shear stress S_{12}. These two stresses are more commonly denoted by σ_x and τ_{xy} respectively with y being the direction parallel to the direction of the transverse load. The two stresses are calculated from the equations

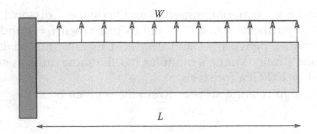

FIGURE 3.14
A schematic of a cantilever beam subjected to a distributed load.

$$\sigma_x = -\frac{My}{I_z} \text{ and } \tau_{xy} = \frac{VQ}{I_z b}. \tag{3.102}$$

In Equation 3.102, y is the y-coordinate of any point on a cross-section, with the origin being the centroid of the cross-section, b is the width of the cross-section at y, I_z is the second moment of the cross-sectional area about the z-axis (neutral axis) and Q is the first moment of the entire cross-sectional area either above or below the point of interest. For a positive bending moment, the normal stress is compressive for $y > 0$ and tensile for $y < 0$. The shear stress points in the direction of the shear force at the cross-section of interest.

For a rectangular cross-section of width w and height h, the quantities Q and I_z are given by

$$Q = \frac{b}{2}\left(\frac{h^2}{4} - y^2\right) \text{ and } I_z = \frac{bh^3}{12} \tag{3.103}$$

and the maximum normal and shear stresses are given by

$$\sigma_x^{max} = \frac{6M}{bh^2} \text{ and } \tau_{xy}^{max} = \frac{4V}{3bh} \tag{3.104}$$

with the understanding that M and V are the maximum possible values in the beam.

For a more general cross-section with a vertical axis of symmetry, the maximum normal stresses (in tension and compression) can be obtained from the relations

$$\sigma_x^{max/min} = -\frac{Mc_1}{I_z} \text{ and } \sigma_x^{min/max} = -\frac{Mc_2}{I_z}. \tag{3.105}$$

Here, c_1 and c_2 are the extremal values of y on the cross-section from the z-axis. The actual signs of the stresses depend on the sign of the bending moment.

The deflection $v(x)$ of the beam due to applied loads is calculated from the second-order differential equation

$$EI_z \frac{d^2v}{dx^2} = M(x). \tag{3.106}$$

Alternatively, the following fourth-order differential equation can be used:

$$EI_z \frac{d^4v}{dx^4} = w(x). \tag{3.107}$$

Either of these equations, along with appropriate boundary conditions, can be used to find $v(x)$. For example, for a cantilever beam with the fixed end at $x = 0$, the deflection and slope $(v'(x))$ are zero at $x = 0$.

In practice, the maximum deflection and stiffness of the beam are often of interest. For a cantilever beam of length L subjected to a point load P at the free-end, the maximum deflection (often denoted by δ) can be shown to be equal to

$$\delta = |v|_{max} = \frac{PL^3}{3EI_z}. \tag{3.108}$$

Then, the stiffness, defined as the ratio of the applied load and the maximum deflection, equals $\frac{3EI_z}{L^3}$.

3.5.5 Mohr's Circle

For two-dimensional problems, Mohr's circle offers a convenient method for calculating the stresses with respect to a rotated coordinate system. Suppose that the stress-state at a point A is given by the normal stresses S_{11} and S_{22} and the shear stress S_{12} with respect to $x_1 - x_2$ axes. Let $x_1' - x_2'$ denote the primed axes obtained by rotating the unprimed axes through an angle θ. The stresses at A with respect to these rotated coordinate axes are denoted by S_{11}', S_{22}' and S_{12}'.

A convenient way to visualise the stress-state at A is by imagining an infinitesimal square surrounding A with sides parallel to the coordinate axes. The stresses with respect to $x_1 - x_2$ axes are shown on the left element in Figure 3.15 and the stresses with respect to $x_1' - x_2'$ axes are shown on the right element in Figure 3.15. The two representations are equivalent to each other since one can be obtained from the other through the transformation 3.96.

Next, consider a two-dimensional stress plane defined by the coordinates (σ, τ), where σ represents the normal stress and τ, the shear stress. The stresses on each of the planes M_1, M_2, N_1 and N_2 are represented by a point on this stress plane. For convenience, the plane labels are also used to identify the points on the stress plane. The locus of N_1 on the stress plane as θ changes is a circle called Mohr's circle (see Figure 3.16). The centre C of the circle is at $(\sigma_{ave}, 0)$ with $\sigma_{ave} = \frac{S_{11} + S_{22}}{2}$. Since M_1 and M_2 correspond to $\theta = 0$ and $\frac{\pi}{2}$ respectively, the stresses on each of these planes are represented by points M_1 and M_2 on the circle. Then, the distances CM_1 and CM_2 are equal to R, the radius of the circle. Thus, to draw the Mohr's circle for the stress-state at A, the centre of the circle and one of the points M_1 with coordinates (S_{11}, S_{12}) or M_2 with coordinates S_{22}, S_{12} are located. The radius of the circle given by CM_1 or CM_2 is then calculated and the circle is drawn. In locating points M_1 or M_2 on the stress plane, it is assumed that shear stress is positive if it produces counterclockwise rotation of an element and negative otherwise. Furthermore, tensile stresses are taken to be positive with compressive stresses as negative.

The stresses on plane N_1 are obtained by starting at M_1 on Mohr's circle and moving through 2θ in the same direction as plane N_1 is from plane M_1. The coordinates of N_1 on the stress plane, calculated using trigonometry, are the stresses acting on plane N_1. The stresses on the plane N_2 can be calculated from the fact that N_2 is diametrically opposite to N_1 on Mohr's circle.

The two extreme points P_1 and P_2 on the σ-axis represent the principal stresses σ_1 and σ_2 respectively. The shear stress at these points is clearly zero. The maximum shear stress is equal to the radius of the circle and denoted by points S_1 and S_2 on the circle. The normal

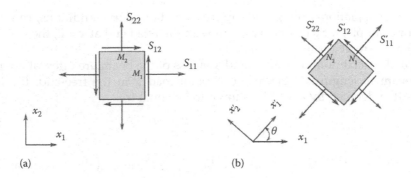

FIGURE 3.15
Representation of stress-state at a point: (a) with respect to the unprimed coordinate axes and (b) with respect to primed coordinate axes.

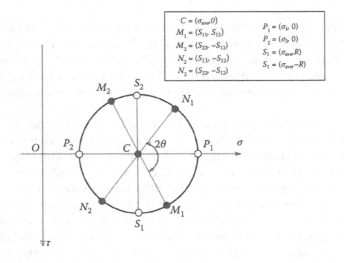

FIGURE 3.16
Mohr's circle for plane stress problems.

stress on the maximum shear planes is equal to σ_{ave}. Clearly, the maximum shear stress planes make an angle $\pm\dfrac{\pi}{4}$ to the principal planes.

3.5.6 Hertz Contact Theory

An important consideration in the design of precision instruments and machines is the nature of the interaction during contact between various components during operation. When two components come into contact with each other under the action of an external force, the resulting (contact) stresses in the contact area can be significant. The stiffness of the overall system may also be affected by the contact. Consequently, the stresses and deformations due to contact must be accounted for during the design process. Heinrich Hertz in 1881 was the first to provide a solution to the contact between two elastic curved bodies. By making certain assumptions, including frictionless contact, that stresses are localised near the contact area and that the contact surfaces are described by quadratic surfaces, Hertz was able to obtain the pressure distribution in the contact area. Hertz's solution to the contact problem is still the basis on which many components are designed.

In the following, a brief review of the solutions to the contact problems involving spherical and cylindrical bodies is provided. For more details on the derivation of the provided solutions, the interested reader is referred to Johnson (1987).

3.5.6.1 Contact between Two Spheres

Suppose that a sphere of radius R_1 made of material 1 is in contact with another sphere of radius R_2 made of material 2. The contact between the two spheres is due to the action of an external force F_n as shown in Table 3.1. Clearly, the contact area in this case will be circular. The radius a of this area, the contact pressure $p(r)$ and the relative displacement between the two centres of the spheres δ are given by

$$p(r) = p_0 \left(1 - \frac{r^2}{a^2}\right)^{\frac{1}{2}} \text{ with } p_0 = \left(\frac{6F_n \bar{E}^2}{\pi^3 R^2}\right)^{\frac{1}{3}} \tag{3.109}$$

and

$$a = \left(\frac{3F_n R}{4\bar{E}}\right)^{\frac{1}{3}} \text{ and } \delta = \frac{a^2}{R} = \left(\frac{9F_n^2}{16\bar{E}^2 R}\right)^{\frac{1}{3}} \tag{3.110}$$

respectively. Here, the quantities $\bar{E}$ and R are defined by

$$\frac{1}{\bar{E}} = \frac{1 - v_1^2}{E_1} + \frac{1 - v_2^2}{E_2} \text{ and } \frac{1}{R} = \frac{1}{R_1} + \frac{1}{R_2}. \tag{3.111}$$

The subscripts 1 and 2 on Young's modulus E and Poisson's ratio v correspond to materials 1 and 2 respectively. The quantities $\bar{E}$ and R are known as the contact modulus and reduced radius of curvature respectively. The quantity p_0 is the maximum contact pressure and occurs at $r = 0$.

A quantity of practical importance is the contact stiffness K_n, which can be defined as $\frac{dF_n}{d\delta}$. From Equation 3.110, the expression for K_n is found as

$$K_n = \frac{dF_n}{d\delta} = 2\bar{E}\sqrt{\delta R}. \tag{3.112}$$

Two important results can be derived from the expressions for contact radius and δ. When one of the spheres (say 2) has an infinite radius (i.e. a flat surface), $R_2 = \infty$. The second case corresponds to when sphere 1 contacts sphere 2 on the inner surface. In this case, R_2 is taken to be negative.

The expressions for the stresses are complex and the complete fields can be found in the book by Sackfield et al. (2013). On the axial line $r = 0$ these expressions have much simpler form. The non-zero components are the radial, hoop and axial stresses with radial stress being equal to hoop stress:

$$S_{rr} = S_{\theta\theta} = -p_0(1 + v)\left[1 - \frac{z}{a}\arctan\frac{a}{z}\right] + \frac{a^2 p_0}{2(z^2 + a^2)}, \tag{3.113}$$

$$S_{zz} = -\frac{p_0 a^2}{z^2 + a^2}.$$

TABLE 3.1

Contact Pressure and Area for Hertz-Type Contact between Various Geometries of Practical Interest

Case	Contact Pressure	Contact Width/Radius
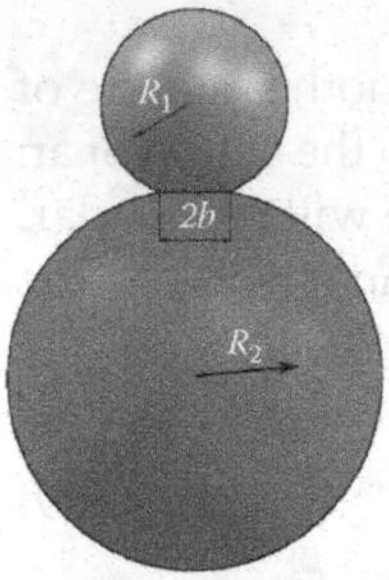 Sphere on sphere	$p(r) = p_0\left(1 - \dfrac{r^2}{a^2}\right)^{\frac{1}{2}}$ with $\dfrac{1}{R} =$ $p_0 = \left(\dfrac{6F\bar{E}^2}{\pi^3 R^2}\right)^{\frac{1}{3}}$ $\dfrac{1}{R_1} + \dfrac{1}{R_2}$	$a = \left(\dfrac{3FR}{4\bar{E}}\right)^{\frac{1}{3}}$
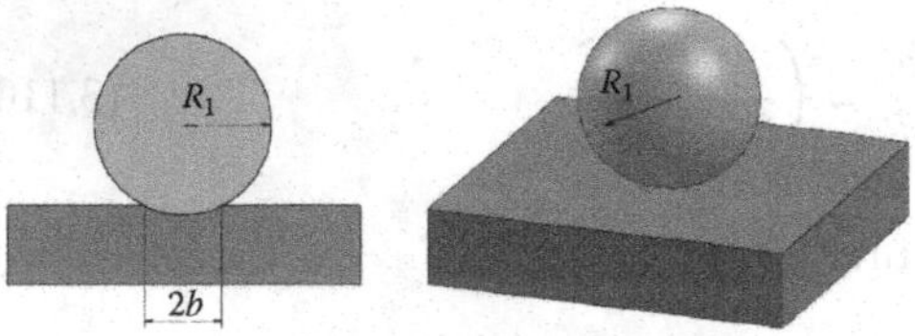 Sphere on flat surface	Same as case 1 with $R_2 \to \infty$	Same as case 1 with $R_2 \to \infty$
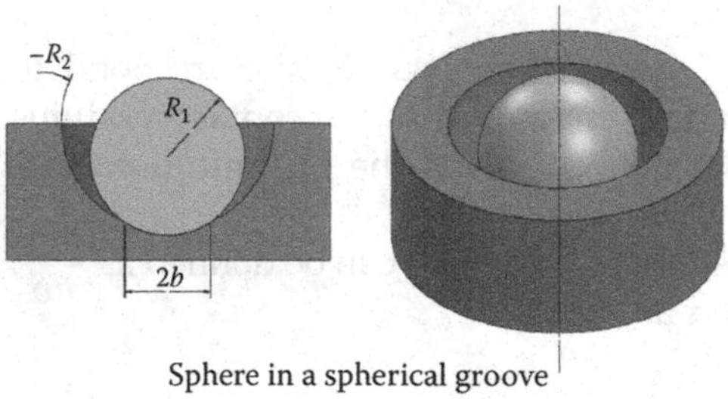 Sphere in a spherical groove	Same as case 1 with R_2 negative	Same as case 1 with R_2 negative

(Continued)

TABLE 3.1 (CONTINUED)

Contact Pressure and Area for Hertz-Type Contact between Various Geometries of Practical Interest

Case	Contact Pressure	Contact Width/Radius
Cylinder on cylinder with parallel axes	$p(x) = p_0 \left(1 - \dfrac{r^2}{b^2}\right)^{\frac{1}{2}}, p_0 = \dfrac{2F}{\pi b L}$	$b = \left(\dfrac{4FR}{\pi \bar{E} L}\right)^{\frac{1}{2}}$
Cylinder on a flat surface	Same as case 4 with $R_2 \to \infty$	Same as case 4 with $R_2 \to \infty$
Cylinder in a cylindrical groove	Same as case 4 with R_2 negative	Same as case 4 with R_2 negative

The maximum shear stress along the axial line equals

$$\tau_{max} = \frac{1}{2}|S_{zz} - S_{rr}| = \frac{p_0}{2}\left|\frac{3}{2}\frac{a^2}{z^2 + a^2} - (1 + \nu)\left(1 - \frac{z}{a}\arctan\left(\frac{a}{z}\right)\right)\right|. \qquad (3.114)$$

From this, it can be easily deduced that the maximum shear stress occurs in the subsurface and not at the contact interface. The locations of the maximum shear stress and its magnitude along the axial line for various values of the Poisson's ratio are listed in Table 3.2.

3.5.6.2 Contact between Two Parallel Cylinders

Next, consider two circular cylinders of radii R_1 and R_2 brought into contact with each other with their axes parallel by a force F_n (see Table 3.1). Let $2b$ and L be the width and length respectively of the resulting contact area. Then, the half-width b and the contact pressure are given by

$$b = \left(\frac{4F_n R}{\pi \bar{E} L}\right)^{\frac{1}{2}} \text{ and } p(x) = p_0\left(1 - \frac{r^2}{b^2}\right)^{\frac{1}{2}}. \qquad (3.115)$$

The maximum contact pressure p_0 and the mean contact pressure p_m are

$$p_0 = \frac{2F_n}{\pi b L} = \left(\frac{F_n \bar{E}}{\pi R L}\right)^{\frac{1}{2}} \text{ and } p_m = \frac{F_n}{2bL} = \frac{\pi p_0}{4}. \qquad (3.116)$$

The corresponding decrease in the distance between the axes of the two cylinders is (Puttock and Thwaite 1969)

$$\delta = \frac{F_n}{\pi L \bar{E}}\left[1 + \ln\frac{\pi L^3 \bar{E}}{PR}\right]. \qquad (3.117)$$

As with the spherical case, the case of the contact between a cylinder in an outer cylindrical cavity can be obtained by attaching a negative sign to the radius of curvature of the cavity.

It should be noted that some researchers use the following expression for δ (Hale 1999):

$$\delta = \delta_1 + \delta_2 \qquad (3.118)$$

TABLE 3.2

Maximum Shear Stress Locations for Spherical Bodies in Hertzian Contact

ν	$\dfrac{z}{a}$	$\tau_{\{max\}}/p_0$
0.0	0.38	0.385
0.1	0.41	0.359
0.2	0.45	0.333
0.3	0.48	0.310
0.32	0.49	0.305
0.4	0.51	0.288
0.5	0.55	0.266

with

$$\delta_1 = \frac{F_n}{\pi L}\left(\frac{1-v_1^2}{E_1}\right)\left[2\ln\frac{4R_1}{b}-1\right] \text{ and } \delta_2 = \frac{P_n}{\pi L}\left(\frac{1-v_2^2}{E_2}\right)\left[2\ln\frac{4R_2}{b}-1\right]. \tag{3.119}$$

This expression appears to be based on the results presented in Johnson (1987) for the problem of compression of a cylinder by two surfaces located along diametrically opposite generators of the cylinder.

The results presented in this section are summarised in Table 3.1 for the reader's convenience.

3.5.7 Tangential Loading of Two Spheres in Contact

The Hertz contact problem is concerned with the contact between two bodies subjected to normal (compressive) loads. The response of contact bodies due to loads applied tangential to the contact area is of equal importance. It is widely regarded that the pioneering work on this problem was due to Cattaneo (1938) and Mindlin (1949). Mindlin and Deresiewicz (1953) further extended this work to include various loading types. For a concise presentation of their results, the reader is referred to Vu-Quoc et al. (2001) and for a more detailed presentation, to Johnson (1987). In the following, the case of two spheres of the same radius and same material is considered. For the expressions for the relative tangential displacement of the spheres and the tangential contact stiffness when the spheres are made of dissimilar materials, the reader is referred to Johnson (1987).

Consider two spherical bodies that are pressed together with a normal force F_n. Suppose that F_n is held constant and a tangential force F_t is now applied starting from zero and increasing monotonically. To avoid the possibility of infinite shear stresses in the contact area, the contact area is assumed to consist of an outer annulus of slip region ($c \leq r \leq a$) and an inner circular area of stick region ($0 \leq r \leq c$). In the slip region, the shear stress equals $\mu_f p$ where μ_f is the friction coefficient. In the stick region, the shear stress is less than $\mu_f p$ and there is no relative motion between the surfaces in this region. As the tangential load F_t increases, the stick region decreases and eventually becomes zero when F_t reaches the critical value $F_t = \mu_f F_n$. At this point, the entire contact surface is the slip region and sliding motion takes place between the two spheres. The stick and slip zones along with the corresponding shear stress distribution in the contact area are shown in Figure 3.17.

The expression for the location of the stick–slip interface was derived by Cattaneo (1938) and Mindlin (1949) as

$$c = a\left(1 - \frac{F_t}{\mu_f F_n}\right)^{\frac{1}{3}}. \tag{3.120}$$

The corresponding shear stress in the contact region has the expressions

$$\tau = \begin{cases} \dfrac{3\mu_f F_n}{2\pi a^2}\left[1-\dfrac{r^2}{a^2}\right]^{\frac{1}{2}}, & c \leq r \leq a, \\[4mm] \dfrac{3\mu_f F_n}{2\pi a^2}\left[\left(1-\dfrac{r^2}{a^2}\right)^{\frac{1}{2}} - \dfrac{c}{a}\left(1-\dfrac{r^2}{c^2}\right)^{\frac{1}{2}}\right], & 0 \leq r \leq c. \end{cases} \tag{3.121}$$

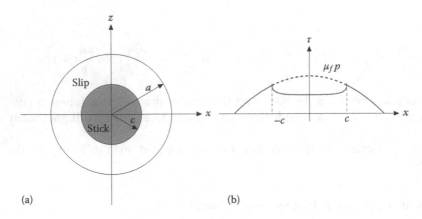

FIGURE 3.17
Tangential loading of two spheres in contact. (a) Slip and stick regions in the contact area. (b) Tangential stress distribution in the contact area.

In addition, Mindlin (1949) derived the relative displacement δ_t of the spheres (assuming they are made of the same material and are of the same radii) relative to the stick region in the contact area as

$$\delta_t = \frac{3(2-v)\mu_f F_n}{16\mu a}\left[1 - \left(1 - \frac{F_t}{\mu_f F_n}\right)^{\frac{2}{3}}\right]. \tag{3.122}$$

The load–displacement curve for this loading phase follows the path AB shown in Figure 3.18. From these expressions, the tangential stiffness K_t can be found as

$$K_t = \frac{dF_t}{d\delta_t} = \frac{8\mu a}{2-v}\left(1 - \frac{F_t}{\mu_f F_n}\right)^{\frac{1}{3}}. \tag{3.123}$$

It is important to note that this expression is valid for the case of increasing F_t while F_n is held constant.

If the tangential force is increased to a value $F_t^* < \mu_f F_n$ and then decreased, Mindlin and Deresiewicz (1953) proposed that a region of counterslip (slip in the direction opposite to the slip due to increasing F_t) must exist to prevent τ from going to infinity. If $b(c \le b \le a)$ is the inner radius of this counterslip region, the shear stress in the contact region during unloading of F_t can be shown to be

$$\tau = \begin{cases} -\dfrac{3\mu_f F_n}{2\pi a^2}\left[1 - \dfrac{r^2}{a^2}\right]^{\frac{1}{2}}, & b \le r \le a, \\[4mm] -\dfrac{3\mu_f F_n}{2\pi a^2}\left[\left(1 - \dfrac{r^2}{a^2}\right)^{\frac{1}{2}} - \dfrac{2b}{a}\left(1 - \dfrac{r^2}{b^2}\right)^{\frac{1}{2}}\right], & c \le r \le b, \\[4mm] -\dfrac{3\mu_f F_n}{2\pi a^2}\left[\left(1 - \dfrac{r^2}{a^2}\right)^{\frac{1}{2}} - \dfrac{2b}{a}\left(1 - \dfrac{r^2}{b^2}\right)^{\frac{1}{2}} + \dfrac{c}{a}\left(1 - \dfrac{r^2}{c^2}\right)^{\frac{1}{2}}\right], & 0 \le r \le c. \end{cases} \tag{3.124}$$

The corresponding relative tangential displacement δ_t is given by

$$\delta_t = \frac{3(2-v)\mu_f F_n}{16\mu a} \left[\left(1 - \frac{F_t^* - F_t}{2\mu_f F_n}\right)^{\frac{2}{3}} - \left(1 - \frac{F_t}{\mu_f F_n}\right)^{\frac{2}{3}} - 1 \right]. \tag{3.125}$$

The quantity b (termed as the depth of penetration of the counterslip by Mindlin and Deresiewicz, 1959) is calculated from the equation

$$b = a \left(1 - \frac{F_t^* - F_t}{2\mu_f F_n} \right)^{\frac{1}{3}}. \tag{3.126}$$

It is clear from this expression that as long as $F_t > -F_t^*$, $c \le b \le a$. Furthermore, when $F = -F_t^*$, $b = c$. During this unloading stage, the load–displacement curve follows path BD in Figure 3.18. If F_t is further decreased beyond this value, the curve follows the dashed line in the third quadrant of Figure 3.18. On the other hand, if F_t is increased from $-F_t^*$ to F_t^*, the load–displacement curve will follow path DEB as shown in Figure 3.18. Thus, the full load cycle leads to a hysteresis loop. The area enclosed by the loop $ABCDEB$ is the energy lost due to friction (see Mindlin and Deresiewicz, 1953, for an expression for the energy loss). The stiffness during the unloading phase can be readily shown to be

$$K_t = \frac{dF_t}{d\delta_t} = \frac{8\mu a}{2-v} \left(1 - \frac{F_t^* - F_t}{2\mu_f F_n} \right)^{\frac{1}{3}}. \tag{3.127}$$

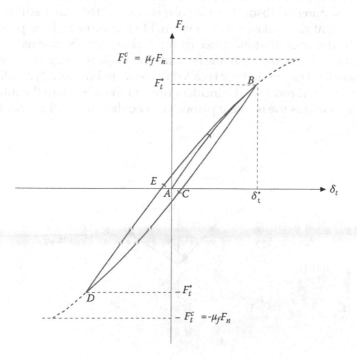

FIGURE 3.18
Tangential loading of two spheres in contact. The tangential force versus tangential relative displacement of the two spheres exhibits hysteresis. When the tangential load is increased from zero (in the presence of a constant normal compressive force) to a value less than the critical value for full sliding and then decreased to zero, the curve shows a residual relative displacement AC.

Another interesting observation is that when $F_t = 0$ during the unloading phase, there is a residual tangential relative displacement (AC in Figure 3.18).

The results presented in this section are for the case when the normal load is held constant and the tangential load is varied. For other types of loads, such as when both F_n and F_t are varying, the work by Mindlin and Deresiewicz (1953) and Johnson (1987) are good references.

3.6 Optics

Optical methods, such as laser interferometry, are commonly used for the accurate measurement of flatness, length, pressure and velocity (see examples in Chapter 5). Some of the fundamental concepts of optics needed to understand these methods are summarised next. A more thorough presentation of these concepts can be found in Hecht (2002).

Geometrical optics is that branch of optics where the path of light is treated as rectilinear and the wave nature is ignored. Physical or wave optics is the field where light is treated as a wave. The former approach leads to the laws of reflection and refraction, whereas the latter leads to the explanation of the phenomena of diffraction and interference that are important in measurement science.

Consider a ray of light traveling through medium 1 (see Figure 3.19). When this ray encounters the interface between medium 1 and medium 2, a part of it is reflected back into medium 1 and (assuming no absorption) the rest is transmitted into medium 2. Let θ_i be the angle that the incident ray makes with the normal to the surface at the point of incidence. Similarly, let θ_r be the angle that the reflected ray makes with the normal and θ_t, the angle that the transmitted wave makes with the normal. The angle θ_t is in general different from θ_i and for this reason, the transmitted ray is called the refracted ray and θ_t is called the angle of refraction. The plane that contains the incident ray and the normal to the interface between the two media is known as the incident plane. It can be shown that the reflected ray and the

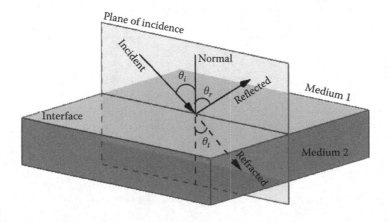

FIGURE 3.19
The laws of light reflection and refraction.

refracted ray also lie in the same plane. Furthermore, the angles θ_i, θ_r and θ_t obey the laws of reflection and refraction.

3.6.1 Law of Reflection

The angle of reflection is equal to the angle of incidence, that is $\theta_i = \theta_r$. For a smooth surface, the reflected rays will concentrate in one direction, called the specular direction. On the other hand, for a rough surface, the reflected rays propagate in many directions leading to what is called a diffuse reflection.

3.6.2 Law of Refraction (Snell's Law)

Let n_1 be the refractive index of medium 1 and n_2, the corresponding refractive index of medium 2. Then, the angle of incidence and the angle of refraction are related through Snell's law, which states that

$$n_1 \sin \theta_i = n_2 \sin \theta_t. \tag{3.128}$$

The refractive index of a medium is dimensionless number that describes how light is propagated through the medium, and is always greater than unity since $n = \dfrac{c}{v}$, where c is the speed of light in vacuum and v is the speed of light in the medium. Furthermore, v, the speed of light in the medium, is also given by the relation $v = f\lambda$, where f is the frequency and λ is the wavelength (this is known as the dispersion relation). The frequency does not change as light passes from one medium to another. However, the wavelength λ changes. Therefore, $f = \dfrac{v_1}{\lambda_1} = \dfrac{v_2}{\lambda_2}$. Combining this with $n = \dfrac{c}{v}$ gives the relationships $\lambda_1 n_1 = \lambda_2 n_2$ and $v_1 n_1 = v_2 n_2$. Using these, the Snell's law can also be cast in the form

$$v_2 \sin \theta_i = v_1 \sin \theta_t. \tag{3.129}$$

In words, Snell's law states that light, in traveling from a medium of lower refractive index to a medium of higher refractive index, bends towards the normal to the interface.

An interesting consequence of Snell's law is the phenomenon of total internal reflection. Suppose that light is traveling from a medium of higher refractive index to a medium of lower refractive index (see Figure 3.20), thus, $n_1 > n_2$. By Snell's law, $\theta_t > \theta_i$. As θ_i increases, θ_t also increases and when $\theta_i = \theta_c$, the refracted ray is parallel to the interface with $\theta_t = \dfrac{\pi}{2}$. Any further increase in θ_i leads to the incident light being reflected back into medium 1 with no refraction. The angle θ_c is called the critical angle for internal reflection. Note that this angle is key to the operation of fibre optics.

3.6.3 Thin Lens Formula

Lenses are used for imaging objects. In a converging lens (Figure 3.21a), light rays parallel to the lens axis passing through the lens converge at a point called the focal point. In a diverging lens (Figure 3.21b), parallel rays incident upon the lens diverge after passing through the lens. The diverging rays appear to originate from a (virtual) point in front of the lens and this point is called the focal point for the diverging lens. Note that the lenses in Figure 3.21 are called double convex or concave lenses respectively, as they both have two

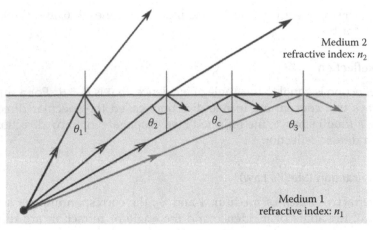

$$n_2 < n_1 \text{ and } \theta_1 < \theta_2 < \theta_c < \theta_3$$

FIGURE 3.20
Total internal reflection.

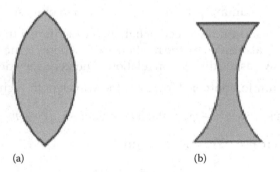

(a)   (b)

FIGURE 3.21
Schematic of a (a) converging lens and (b) diverging lens.

faces with equal but opposite form. Consider the case of a double convex lens shown in Figure 3.22. Let f denote the distance of the focal point of the lens from the vertical axis of symmetry. For a thin lens (as is the case here), f can also be taken to be the distance between the focal point of the lens to the lens surface. Denoting the distances of the object and the image from the vertical axis of the image by o and i respectively, it can easily be shown that f, o and i satisfy the so-called thin lens equation:

$$\frac{1}{o} + \frac{1}{i} = \frac{1}{f}. \tag{3.130}$$

Furthermore, it can easily be shown that the magnification M is given by

$$M = \frac{h_i}{h_o} = \frac{i}{o}. \tag{3.131}$$

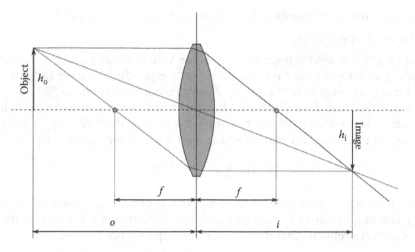

FIGURE 3.22
Image formation in a double convex lens.

3.6.4 Wave Representation of Light

The laws of reflection and refraction are presented in the previous section by using geometrical optics, where light is treated as a ray propagating in a rectilinear fashion. This theory is useful for studying optics involving objects that are significantly larger than the wavelength of light. However, this theory falls short when it comes to explaining phenomena such as interference, diffraction and polarization. Wave optics, where light is treated as a wave, is more suitable for studying these phenomena. In wave optics, light is treated as an electromagnetic wave where the electric field has 'vibrations' transverse to the direction of propagation of the wave.

A monochromatic light wave with wavelength λ and traveling with the wave speed v can be represented by the equation

$$E(\mathbf{r}, t) = \mathbf{E}_0 \cos[(k\mathbf{n} \cdot \mathbf{r} - \omega t) + \phi], \tag{3.132}$$

where $\mathbf{E}$ is the electric field vector with an amplitude of $\mathbf{E}_0$; $\mathbf{r}$ is the position vector of a point in space; k is the wave number given by $\dfrac{2\pi}{\lambda}$; ω is the circular frequency equal to $2\pi v$, where v is the frequency; and t is time. The quantity ϕ is the starting phase of the wave and $\mathbf{n}$ is the direction of propagation of the wave. It should be noted that the quantities $\mathbf{E}_0$ and ϕ may be functions of $\mathbf{r}$.

When the wave is planar and propagating in a specific direction (say, positive x), Equation 3.132 can be simplified to the scalar form

$$E(x, t) = E_0 \cos[(kx - \omega t) + \phi] \tag{3.133}$$

with the understanding that the electric field is oscillating transverse to the x-direction.

In the aforementioned expressions for light waves, the argument of the cosine function is known as the phase angle of the wave and often denoted by θ. The derivative of the phase angle with respect to time gives the circular frequency ω and the derivative with respect to the position gives the wave number. Furthermore, the phase speed of the wave defined by $\frac{\omega}{k}$ equals the wave speed of light.

The wave length λ of electromagnetic waves has values ranging from as low as 10^{-16} m (corresponding to γ-rays) to as large as 10^{10} m (corresponding to long radio waves). The visible light has wavelengths in the range of approximately 400 nm to 700 nm. The lower limit corresponds to the colour violet and the upper limit corresponds to the colour red.

A more convenient notation for representing the wave nature of a light wave uses the complex form. In this form, a monochromatic light wave is represented by

$$\mathbf{E}(\mathbf{r}, t) = \mathbf{E}_0 e^{-i[(\mathbf{kn \cdot r} - \omega t) + \phi]}. \tag{3.134}$$

Obviously, it is the real part of this expression that is of practical interest. The fact that it represents a rotating vector in complex space also makes it useful for modelling polarization and diffraction problems that are beyond the scope of this chapter.

3.6.5 Interference and Interferometry

Interferometry is a commonly used method for the accurate measurement of surface flatness and lengths (see Chapter 5). The technique relies on the phenomenon of interference that light exhibits. When two light waves of the same wavelength from coherent sources (that is the two waves maintain a constant phase relationship with each other) arrive at a point at the same time, the resulting field at the point is the vector sum of the fields of the two light waves. If the two waves interact to produce a resulting wave of amplitude higher than the individual amplitudes, then the interaction is said to be constructive interference. On the other hand, when the amplitude of the resulting wave is smaller than either of the two individual amplitudes, the interaction is said to be destructive interference.

The constructive and destructive interference phenomena are best illustrated by Thomas Young's famous double-slit experiment. A schematic of the experiment is shown in Figure 3.23. It consists of a monochromatic light source S and two screens A and B. Screen A contains two narrow slits S_1 and S_2 and screen B is the viewing screen. The two screens A and B are separated by a distance L. The two slits are spaced apart by a distance d. When light waves from S hit screen A, the two slits act as secondary light sources that are

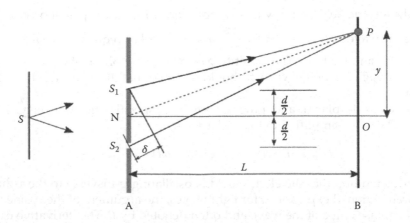

FIGURE 3.23
A schematic illustrating Thomas Young's double-slit experiment.

coherent. The waves emanating from these two slits interfere with viewing screen B producing a series of bright and dark bands called interference fringes. The bright bands result from constructive interference of the waves from S_1 and S_2, and the dark bands are due to destructive interference. The interference is due to the difference in the lengths of the optical paths from the two slits.

Consider a light wave travelling from slit S_1 and another from slit S_2, as shown in Figure 3.23. These waves meet at point P on screen B. Clearly, the path of S_2P is longer than S_1P with the difference being given by the quantity δ as shown in the figure. From geometry, it can be shown that

$$\delta = S_2P - S_1P = d \sin \theta, \tag{3.135}$$

where θ is the angle between PN and NO. Constructive interference occurs when δ is an integral multiple of the wavelength. Destructive interference occurs when δ is an odd multiple of half-wavelength, that is

$$\delta = d \sin \theta = m\lambda, m = 0, \pm 1, \pm 2, \ldots \tag{3.136}$$

for constructive interference and

$$\delta = d \sin \theta = \left(m + \frac{1}{2} \right) \lambda, m = 0, \pm 1, \pm 2, \ldots \tag{3.137}$$

for destructive interference.

From geometry, assuming that θ is small, it can also be shown that the bright fringes are located at distances

$$y_b = m \frac{\lambda L}{d}, m = 0, \pm 1, \pm 2, \ldots \tag{3.138}$$

and the dark fringes are located at distances

$$y_d = \left(m + \frac{1}{2} \right) \frac{\lambda L}{d}, m = 0, \pm 1, \pm 2, \ldots . \tag{3.139}$$

Clearly, the bright and dark fringes alternate. The separation distance of any two successive bright or dark fringes is $(\lambda L)/d$, which is independent of m.

The illumination intensity of the fringes is given by

$$I = 4I_0 \cos^2 \frac{\phi}{2}, \text{ with } \phi = \frac{2\pi d}{\lambda} \sin \theta. \tag{3.140}$$

In Equation 3.140, I_0 is the illumination intensity on the viewing screen due to one of the two light sources. From Equation 3.140, it is clear that the bright fringes have an intensity of $4I_0$ and the dark fringes have zero intensity (maxima and minima).

The Michelson interferometer uses the interference properties of light as described earlier for accurate measurements of lengths, surface flatness, mechanical stage motion control and so forth. A schematic of the setup is shown in Figure 3.24. It consists of a coherent light source, a beam splitter, two mirrors and a glass plate. The mirror M_2 is movable in the direction shown. The beam splitter is a partially silvered mirror that splits the incoming monochromatic light from the light source into two halves. One half is transmitted towards mirror M_1 and the other half is reflected towards the mirror M_2. The two beams reflected by the two mirrors, recombine at the beam splitter and eventually arrive at the light detector

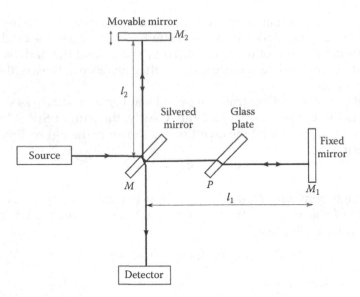

FIGURE 3.24
A schematic of the Michelson interferometer.

producing interference. The glass plate P has the same thickness as the mirror M and is also positioned at the same angle as M. This ensures that the two split beams pass through the same length of glass. The path difference between the two beams is equal to $2l_2 - 2l_1$, where l_1 and l_2 are the lengths of the paths of the transmitted and reflected waves from the beam splitter. The interference pattern observed in the detector depends on this difference. If this difference is an even multiple of $\lambda/2$, the interference will be constructive. On the other hand, if the difference is an odd multiple of $\lambda/2$, the interference will be destructive. If the movable mirror is moved by a distance of $\lambda/2$, the fringe patterns move by one on the reference point in the detector. If the fringe pattern shift is m, then the distance associated with this shift is simply $m\lambda/2$. Thus, by controlling the motion of the movable mirror, length measurements can be made accurately. Higher accuracy still can be achieved by using the knowledge that the fringe pattern is sinusoidal and electronically interpolating the signal.

3.6.6 Diffraction

Consider a plate with two slits. A set of parallel rays of lights are incident upon the plate from the left. According to Huygens' principle, when the rays reach the slits, the light waves bend and spread out as shown in Figure 3.25. Such bending of light around openings or obstacles is called diffraction. If a screen is placed sufficiently far to the right of the plate, an alternating pattern of bright and dark fringes (called diffraction patterns) will be observed.

The formation of the diffraction patterns can be easily understood by considering a single-slit plate receiving planar monochromatic light waves of wavelength λ from the left. When a wave arrives at the slit, by Huygens' principle, each point on the slit acts a second source of light waves. The intensity of the light at a point on the screen depends on the cumulative effect of the light waves emanating from the slit and reaching this particular point. The cumulative effect could be constructive or destructive interference. In fact, it can be shown that there will be a thick bright spot along the axis (i.e. $\theta = 0$) and on either side of it (vertically), there will be alternative dark and bright fringes. The dark fringes correspond to

Incident light
rays

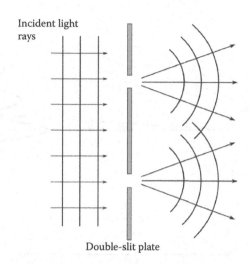

Double-slit plate

FIGURE 3.25
Diffraction of light as it passes through the slits of a double-slit plate.

destructive interference (or the minima). The locations of these minima can be shown to be given by

$$\sin \theta_m = \frac{m\lambda}{a} \text{ with } m = \pm 1, \pm 2, \dots .$$ (3.141)

In Equation 3.141, a is the slit width, and θ_m is the angle that the light rays that contribute to mth minimum make with the axis at the slit. Furthermore, for small angles, $\sin \theta_m \approx \frac{y_m}{L}$, assuming that $L \gg a$. Then, the location of the minima along the screen are given by

$$y_m = \frac{m\lambda L}{a} \text{ with } m = \pm 1, \pm 2, \dots .$$ (3.142)

The maxima (corresponding to the points of constructive interference) are approximately halfway between each pair of successive minima (see Figure 3.26).

If there are N number of long narrow slits parallel to each other with a slit spacing of d, then, it can be shown that the maxima occur at θ_m given by

$$d \sin \theta_m = m\lambda \text{ with } m = 0, \pm 1, \pm 2, \pm 3, \dots$$ (3.143)

when the incident light waves are parallel to the grating. On the other hand, when the angle of incidence is at angle θ_i, Equation 3.143 modifies to

$$d(\sin \theta_m - \sin \theta_i) = m\lambda \text{ with } m = 0, \pm 1, \pm 2, \pm 3, \dots .$$ (3.144)

This equation is known as the grating equation. It is interesting to note that the locations of the maxima are independent of N and depend on the ratio of the light wavelength and the slit spacing. For different values of m, different orders of maxima are obtained: zeroth-order corresponds to $m = 0$, first-order to $m = 1$ and so on. Knowing the values of θ_m and the slit spacing d, the wavelength of the incident light can be calculated from this equation.

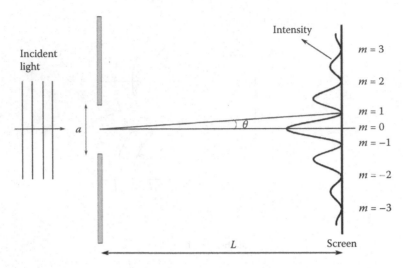

FIGURE 3.26
The locations of maxima and minima in the diffraction patterns of a single-slit experiment.

A diffraction grating is a device with an arrangement of a large number of long narrow slits parallel to each other. Depending on the nature of the slits and the material, the grating can be either a reflection grating or a transmission grating. For example, a transmission grating is obtained by placing parallel notches on a transparent glass plate. An example of a reflection grating is a glass plate coated with a metal thin-film with the notches placed on the metal side. In either case, the locations of the maxima are given by the grating equation.

For an N-slit grating, it can be shown that the angular spacing (also known as the angular line width) $\Delta\theta$ of an mth-order maximum is given by

$$\Delta\theta = \frac{2\lambda}{Nd\cos\theta_m} \tag{3.145}$$

assuming that θ_i is constant. Clearly, the angular spacing of the maxima decreases as N increases. Thus, the maxima become sharper with an increase in N. In addition, when the light source consists of multiple wavelengths, the angular dispersion (denoted by $\mathcal{D}$), defined as the variation of the angular line width with the wavelength, can be shown to be (Hecht 2002, Halliday et al. 2013)

$$\mathcal{D} = \frac{m}{d\cos\theta_m}. \tag{3.146}$$

Large values of $\mathcal{D}$ are desired since wavelengths that are close to each other can be easily separated on the screen. This can be achieved by using a small value for d and considering higher orders, that is larger m.

Another quantity of interest in diffraction gratings is the resolving power $\mathcal{R}$. When wavelengths that are nearly the same, the maxima of these wavelengths are difficult to differentiate. This problem can be overcome by using a diffraction grating with a high resolving power. The resolving power of a grating is defined as the ratio of λ_{mean} and $\Delta\lambda$, where λ_{mean} is the mean of the two nearly equal wavelengths and $\Delta\lambda$ is the difference in the two wavelengths. It can be shown (Hecht 2002, Halliday et al. 2013) that $\mathcal{R}$ is given by

$$\mathcal{R} = Nm. \qquad (3.147)$$

Thus, by increasing N, a high resolving power can be achieved for a diffraction grating.

Exercises

1. Strain rosettes: Strain rosettes are used to measure normal strains on the surfaces of a stressed body. Each rosette consists of three strain gages arranged at a point in specific directions to measure the normal strains at the point in those directions. Consider the 45° strain rosette as shown in Figure 3.27. Let the normal strains measured by the rosette in the three directions be a, b and c respectively. Find the components ε_{11}, ε_{12} and ε_{22} of the strain tensor with respect to the Cartesian coordinate frame shown in Figure 3.27.

2. Principal values and directions: Find the principal values and principal directions of the symmetric matrix with the components

$$\left[S_{ij} \right] = \begin{bmatrix} 7 & 2 & 0 \\ 2 & 6 & -2 \\ 0 & -2 & 5 \end{bmatrix}.$$

3. Stress vector: Consider a long, slender solid (Figure 3.28) which has length l and rectangular cross-section with dimensions of b by h. Introduce a coordinate system so that

$$0 \le x_1 \le l, \quad -\frac{h}{2} \le x_2 \le \frac{h}{2}, \quad -\frac{b}{2} \le x_3 \le \frac{b}{2}, \quad (b, h \ll l).$$

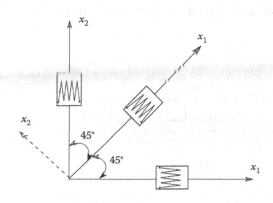

FIGURE 3.27
A 45° strain rosette.

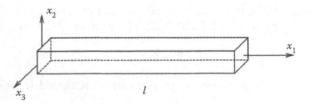

FIGURE 3.28
A schematic of a long-slender beam considered in Section 3.6.3.

Suppose that the stress distribution in the material is given by

$$S_{11} = \frac{P}{I_z}(l - x_1)x_2, S_{12} = -\frac{P}{2I_z}\left(\frac{h^2}{4} - x_2^2\right), S_{13} = S_{22} = S_{23} = S_{33} = 0,$$

where P is a point load applied at $x_1 = l$ and $I_z = \frac{bh^3}{12}$. Determine the traction vector and the resultant force on the cross-section $x_1 = 0$. Define the moment vector as the following:

$$\mathbf{M} = \int_A \mathbf{r} \times \mathbf{s}dA,$$

where $\mathbf{r}$ is the position vector of $\mathbf{x}$ with respect to a fixed point $\mathbf{X_0}$. Compute the moment due to traction on the cross-section $x_1 = 0$. Take the fixed point $\mathbf{X_0}$ to be the centre of the cross-section under consideration.

4. Mohr's circle: Consider a simply supported beam with a rectangular cross-section as shown in Figure 3.29. The length of the beam is 2.4 m and the cross-sectional dimensions are 4 cm × 8 cm. The beam is subjected to a distributed load of 5000 N·m^{-1}. The stress-state at any point in the beam is determined by the bending moment and shear force at the corresponding cross-section.

Suppose that the point of interest A is at 0.2 m from the left and at a height of 2 cm from the neutral surface. Determine the principal stresses at this point.

5. Hertz contact: Consider two steel spheres in contact under a compressive normal force of 6 N. Young's modulus is 200 GPa and Poisson's ratio is 0.3. Given that the radii of the

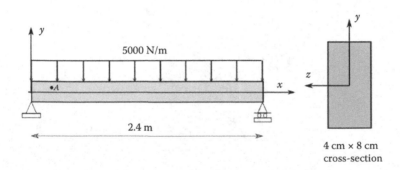

FIGURE 3.29
A simply supported beam subjected to a distributed load. Point A is at a distance of 0.2 m from the left and 2 cm above the neutral surface.

two spheres are 12 mm and 18 mm respectively, determine the contact radius, contact pressure, the compression δ and the contact stiffness.

6. Maximum shear stress due to hertz contact: Determine the depth at which maximum shear stress (along the z-axis) occurs in the normal contact between two spheres made of the same material. Consider materials with $v = 0.0, 0.1, 0.2, 0.3, 0.32, 0.4$, and 0.5 respectively.

7. Optics: Derive the thin-lens Equation 3.130 for a thin converging lens.

8. Diffraction: Light from a sodium vapour lamp illuminates a diffraction grating with 5000 lines per centimetre. Given that the light contains two components with wavelengths 589.0 nm and 589.59 nm, determine the separation between the first-order lines corresponding to these two components on a screen at a distance of 2 m.

9. Standard normal distribution: Show that the mean and variance of a standard normal distribution are zero and one respectively.

10. Fourier series representation of a unit square wave: Show that the unit square wave function has the Fourier expansion given by Equation 3.47.

11. Vibrations of an elastic string: Consider a taut elastic string of length L fixed at both the ends. Let the x-axis be along the length of the string. Let $u(x,t)$ denote the lateral displacement of the string. The initial displacement is $\bar{u}(x)$ and initial velocity is $\bar{v}(x)$. The partial differential equation that governs the motion of the string is given by

$$\frac{\partial^2 u}{\partial x^2} = \frac{1}{c^2} \frac{\partial^2 u}{\partial t^2},$$ (3.148)

where $u(x,t)$ is the lateral displacement of a point at x at time t. The quantity c is called the wave speed and is given by $\sqrt{F_0/\rho}$, where F_0 is the initial tension in the string and ρ is the linear density of the string. If the string is plucked in the middle and released with a zero initial velocity, show that the displacement field is given by

$$u(x,t) = \sum_{n=0}^{\infty} \frac{8L}{(2n+1)^3 \pi^3} \sin \frac{(2n+1)\pi x}{L} \cos \frac{(2n+1)\pi c t}{L}.$$ (3.149)

References

Abramowitz, M., Stegun, I. A. 1972. *Handbook of mathematical functions*. Dover Publications.

Aaslak, U. M., Greif, C. 2011. *A first course on numerical methods*. Society for Industrial and Applied Mathematics.

Cattaneo, C. 1938. Sul contatto di due corpi elastici. *Rendiconti dell'Accademia nazionale dei Lincei*, 6(27), 342–348, 434–436, 474–478.

Crandall, S., Dahl, N., Lardner, T. 1978. *An introduction to the mechanics of solids*. McGraw-Hill.

Hale, L. C. 1999. *Principles and techniques for designing precision machines*. PhD dissertation, Massachusetts Institute of Technology, Cambridge.

Halliday, D., Resnick, R., Walker, J. 2013. *Fundamentals of physics* (10th ed.). Wiley.

Hecht, E. 2002. *Optics* (4th ed.). Addison-Wesley.

Holický, M. 2013. *Introduction to probability and statistics for engineers*. Springer Science & Business Media.

Johnson, K. L. 1987. *Contact mechanics*. Cambridge University Press.

Kreyszig, E. 2007. *Advanced engineering mathematics*. John Wiley & Sons.

Mindlin, R. D. 1949. Compliance of elastic bodies in contact. *J. Appl. Mech.*, **16**, 259–268.

Mindlin, R. D., Deresiewicz, H. 1953. Elastic spheres in contact under varying oblique forces. *J. Appl. Mech.*, **20**, 327–344.

Puttock, M. J., Thwaite, E. G. 1969. *Elastic compression of spheres and cylinders at point and line contact.* Commonwealth Scientific and Industrial Research Organization.

Riley, K. F., Hobson, M. P., Bence, S. J. 2006. *Mathematical methods for physics and engineering: A comprehensive guide.* Cambridge University Press.

Sackfield, A., Hills, D. A., Nowell, D. 2013. *Mechanics of elastic contacts.* Elsevier.

Slaughter, W. S. 2002. *The linearized theory of elasticity.* Birkhäuser.

Tolstov, G. P. 2002. *Fourier series.* Translated into English by Richard A. Silverma, Dover Publications.

Vu-Quoc, L., Zhang, X., Lesburg, L. 2001. Normal and tangential force–displacement relations for frictional elasto-plastic contact of spheres. *Int. J. Solids Struct.*, **38**(36), 6455–6489.

Weisstein, E. W. F-Distribution. From MathWorld—A Wolfram Web Resource. http://mathworld.wolfram.com/F-Distribution.html (accessed March 21, 2017).

Appendix

Tables of Statistical Distributions

TABLE 3.A.1

Standard Normal Distribution Table

z_0	Δz									
	0.0000	0.0100	0.0200	0.0300	0.0400	0.0500	0.0600	0.0700	0.0800	0.0900
0.0000	0.5000	0.5040	0.5080	0.5120	0.5160	0.5199	0.5239	0.5279	0.5319	0.5359
0.1000	0.5398	0.5438	0.5478	0.5517	0.5557	0.5596	0.5636	0.5675	0.5714	0.5753
0.2000	0.5793	0.5832	0.5871	0.5910	0.5948	0.5987	0.6026	0.6064	0.6103	0.6141
0.3000	0.6179	0.6217	0.6255	0.6293	0.6331	0.6368	0.6406	0.6443	0.6480	0.6517
0.4000	0.6554	0.6591	0.6628	0.6664	0.6700	0.6736	0.6772	0.6808	0.6844	0.6879
0.5000	0.6915	0.6950	0.6985	0.7019	0.7054	0.7088	0.7123	0.7157	0.7190	0.7224
0.6000	0.7257	0.7291	0.7324	0.7357	0.7389	0.7422	0.7454	0.7486	0.7517	0.7549
0.7000	0.7580	0.7611	0.7642	0.7673	0.7704	0.7734	0.7764	0.7794	0.7823	0.7852
0.8000	0.7881	0.7910	0.7939	0.7967	0.7995	0.8023	0.8051	0.8078	0.8106	0.8133
0.9000	0.8159	0.8186	0.8212	0.8238	0.8264	0.8289	0.8315	0.8340	0.8365	0.8389
1.0000	0.8413	0.8438	0.8461	0.8485	0.8508	0.8531	0.8554	0.8577	0.8599	0.8621
1.1000	0.8643	0.8665	0.8686	0.8708	0.8729	0.8749	0.8770	0.8790	0.8810	0.8830
1.2000	0.8849	0.8869	0.8888	0.8907	0.8925	0.8944	0.8962	0.8980	0.8997	0.9015
1.3000	0.9032	0.9049	0.9066	0.9082	0.9099	0.9115	0.9131	0.9147	0.9162	0.9177
1.4000	0.9192	0.9207	0.9222	0.9236	0.9251	0.9265	0.9279	0.9292	0.9306	0.9319
1.5000	0.9332	0.9345	0.9357	0.9370	0.9382	0.9394	0.9406	0.9418	0.9429	0.9441
1.6000	0.9452	0.9463	0.9474	0.9484	0.9495	0.9505	0.9515	0.9525	0.9535	0.9545
1.7000	0.9554	0.9564	0.9573	0.9582	0.9591	0.9599	0.9608	0.9616	0.9625	0.9633
1.8000	0.9641	0.9649	0.9656	0.9664	0.9671	0.9678	0.9686	0.9693	0.9699	0.9706
1.9000	0.9713	0.9719	0.9726	0.9732	0.9738	0.9744	0.9750	0.9756	0.9761	0.9767
2.0000	0.9772	0.9778	0.9783	0.9788	0.9793	0.9798	0.9803	0.9808	0.9812	0.9817
2.1000	0.9821	0.9826	0.9830	0.9834	0.9838	0.9842	0.9846	0.9850	0.9854	0.9857
2.2000	0.9861	0.9864	0.9868	0.9871	0.9875	0.9878	0.9881	0.9884	0.9887	0.9890
2.3000	0.9893	0.9896	0.9898	0.9901	0.9904	0.9906	0.9909	0.9911	0.9913	0.9916
2.4000	0.9918	0.9920	0.9922	0.9925	0.9927	0.9929	0.9931	0.9932	0.9934	0.9936
2.5000	0.9938	0.9940	0.9941	0.9943	0.9945	0.9946	0.9948	0.9949	0.9951	0.9952
2.6000	0.9953	0.9955	0.9956	0.9957	0.9959	0.9960	0.9961	0.9962	0.9963	0.9964
2.7000	0.9965	0.9966	0.9967	0.9968	0.9969	0.9970	0.9971	0.9972	0.9973	0.9974
2.8000	0.9974	0.9975	0.9976	0.9977	0.9977	0.9978	0.9979	0.9979	0.9980	0.9981
2.9000	0.9981	0.9982	0.9982	0.9983	0.9984	0.9984	0.9985	0.9985	0.9986	0.9986
3.0000	0.9987	0.9987	0.9987	0.9988	0.9988	0.9989	0.9989	0.9989	0.9990	0.9990
3.1000	0.9990	0.9991	0.9991	0.9991	0.9992	0.9992	0.9992	0.9992	0.9993	0.9993
3.2000	0.9993	0.9993	0.9994	0.9994	0.9994	0.9994	0.9994	0.9995	0.9995	0.9995
3.3000	0.9995	0.9995	0.9995	0.9996	0.9996	0.9996	0.9996	0.9996	0.9996	0.9997
3.4000	0.9997	0.9997	0.9997	0.9997	0.9997	0.9997	0.9997	0.9997	0.9997	0.9998
3.5000	0.9998	0.9998	0.9998	0.9998	0.9998	0.9998	0.9998	0.9998	0.9998	0.9998

(Continued)

TABLE 3.A.1 (CONTINUED)

Standard Normal Distribution Table

					Δz					
	0.0000	0.0100	0.0200	0.0300	0.0400	0.0500	0.0600	0.0700	0.0800	0.0900
3.6000	0.9998	0.9998	0.9999	0.9999	0.9999	0.9999	0.9999	0.9999	0.9999	0.9999
3.7000	0.9999	0.9999	0.9999	0.9999	0.9999	0.9999	0.9999	0.9999	0.9999	0.9999
3.8000	0.9999	0.9999	0.9999	0.9999	0.9999	0.9999	0.9999	0.9999	0.9999	0.9999
3.9000	1.0000	1.0000	1.0000	1.0000	1.0000	1.0000	1.0000	1.0000	1.0000	1.0000

Note: The table lists the values of $F(z)$ for various values of z. The z value is $z_0 + \Delta z$. The value $F(z)$ is obtained from the element in the (i,j) cell, where i is the row of z_0 and j is the column of Δz. Note that $F(-z) = 1 - F(z)$.

TABLE 3.A.2

Standard Normal Distribution Table

F(z) (%)	z	F(z) (%)	z	F(z) (%)	z	F(z) (%)	z
1	−2.326	31	−0.496	61	0.279	91	1.341
2	−2.054	32	−0.468	62	0.305	92	1.405
3	−1.881	33	−0.44	63	0.332	93	1.476
4	−1.751	34	−0.412	64	0.358	94	1.555
5	−1.645	35	−0.385	65	0.385	95	1.645
6	−1.555	36	−0.358	66	0.412	96	1.751
7	−1.476	37	−0.332	67	0.44	97	1.881
8	−1.405	38	−0.305	68	0.468	97.5	1.960
9	−1.341	39	−0.279	69	0.496	98	2.054
10	−1.282	40	−0.253	70	0.524	99	2.326
11	−1.227	41	−0.228	71	0.553		
12	−1.175	42	−0.202	72	0.583	99.1	2.366
13	−1.126	43	−0.176	73	0.613	99.2	2.409
14	−1.08	44	−0.151	74	0.643	99.3	2.457
15	−1.036	45	−0.126	75	0.674	99.4	2.512
16	−0.994	46	−0.1	76	0.706	99.5	2.576
17	−0.954	47	−0.075	77	0.739	99.6	2.652
18	−0.915	48	−0.05	78	0.772	99.7	2.748
19	−0.878	49	−0.025	79	0.806	99.8	2.878
20	−0.842	50	0	80	0.842	99.9	3.09
21	−0.806	51	0.025	81	0.878		
22	−0.772	52	0.05	82	0.915	99.91	3.121
23	−0.739	53	0.075	83	0.954	99.92	3.156
24	−0.706	54	0.1	84	0.994	99.93	3.195
25	−0.674	55	0.126	85	1.036	99.94	3.239
26	−0.643	56	0.151	86	1.08	99.95	3.291
27	−0.613	57	0.176	87	1.126	99.96	3.353
28	−0.583	58	0.202	88	1.175	99.97	3.432
29	−0.553	59	0.228	89	1.227	99.98	3.54
30	−0.524	60	0.253	90	1.282	99.99	3.719

Note: The table lists the values of z given $F(z)$.

TABLE 3.A.3

t-Distribution Table

	$F(t) = P(T \le t)$						
	.900	.925	.950	.975	.99	.995	.999
v							
1	3.0777	4.1653	6.3138	12.706	31.821	63.657	318.31
2	1.8856	2.2819	2.9200	4.3027	6.9646	9.9248	22.327
3	1.6377	1.9243	2.3534	3.1824	4.5407	5.8409	10.215
4	1.5332	1.7782	2.1318	2.7763	3.7470	4.6041	7.1732
5	1.4759	1.6994	2.0150	2.5706	3.3648	4.0322	5.8934
6	1.4398	1.6502	1.9432	2.4469	3.1426	3.7074	5.2076
7	1.4149	1.6166	1.8946	2.3646	2.9979	3.4995	4.7851
8	1.3968	1.5922	1.8595	2.3060	2.8965	3.3554	4.5007
9	1.3830	1.5737	1.8331	2.2622	2.8214	3.2498	4.2968
10	1.3722	1.5592	1.8125	2.2281	2.7638	3.1693	4.1437
11	1.3634	1.5476	1.7959	2.2010	2.7181	3.1058	4.0247
12	1.3562	1.5380	1.7823	2.1788	2.6810	3.0545	3.9296
13	1.3502	1.5299	1.7709	2.1604	2.6503	3.0123	3.8520
14	1.3450	1.5231	1.7613	2.1448	2.6245	2.9768	3.7874
15	1.3406	1.5172	1.7531	2.1314	2.6025	2.9467	3.7328
16	1.3368	1.5121	1.7459	2.1199	2.5835	2.9208	3.6862
17	1.3334	1.5077	1.7396	2.1098	2.5669	2.8982	3.6458
18	1.3304	1.5037	1.7341	2.1009	2.5524	2.8784	3.6105
19	1.3277	1.5002	1.7291	2.0930	2.5395	2.8609	3.5794
20	1.3253	1.4970	1.7247	2.0860	2.5280	2.8453	3.5518
21	1.3232	1.4942	1.7207	2.0796	2.5176	2.8314	3.5272
22	1.3212	1.4916	1.7171	2.0739	2.5083	2.8188	3.5050
23	1.3195	1.4893	1.7139	2.0687	2.4999	2.8073	3.4850
24	1.3178	1.4871	1.7109	2.0639	2.4922	2.7969	3.4668
25	1.3163	1.4852	1.7081	2.0595	2.4851	2.7874	3.4502
26	1.3150	1.4834	1.7056	2.0555	2.4786	2.7787	3.4350
27	1.3137	1.4817	1.7033	2.0518	2.4727	2.7707	3.4210
28	1.3125	1.4801	1.7011	2.0484	2.4671	2.7633	3.4082
29	1.3114	1.4787	1.6991	2.0452	2.4620	2.7564	3.3962
30	1.3104	1.4774	1.6973	2.0423	2.4573	2.7500	3.3852

Note: The table lists the values of *t* for various values of $\Gamma(t)$ and the number of degrees of freedom *v*.

TABLE 3.A.4

χ^2-Distribution Table

					$F(y) = P(Y \leq y)$					
	0.005	0.010	0.025	0.050	0.100	0.900	0.950	0.975	0.990	0.995
v										
1	0.000	0.000	0.001	0.004	0.016	2.706	3.841	5.024	6.635	7.879
2	0.010	0.020	0.051	0.103	0.211	4.605	5.991	7.378	9.210	10.597
3	0.072	0.115	0.216	0.352	0.584	6.251	7.815	9.348	11.345	12.838
4	0.207	0.297	0.484	0.711	1.064	7.779	9.488	11.143	13.277	14.860
5	0.412	0.554	0.831	1.145	1.610	9.236	11.070	12.833	15.086	16.750
6	0.676	0.872	1.237	1.635	2.204	10.645	12.592	14.449	16.812	18.548
7	0.989	1.239	1.690	2.167	2.833	12.017	14.067	16.013	18.475	20.278
8	1.344	1.646	2.180	2.733	3.490	13.362	15.507	17.535	20.090	21.955
9	1.735	2.088	2.700	3.325	4.168	14.684	16.919	19.023	21.666	23.589
10	2.156	2.558	3.247	3.940	4.865	15.987	18.307	20.483	23.209	25.188
11	2.603	3.053	3.816	4.575	5.578	17.275	19.675	21.920	24.725	26.757
12	3.074	3.571	4.404	5.226	6.304	18.549	21.026	23.337	26.217	28.300
13	3.565	4.107	5.009	5.892	7.042	19.812	22.362	24.736	27.688	29.819
14	4.075	4.660	5.629	6.571	7.790	21.064	23.685	26.119	29.141	31.319
15	4.601	5.229	6.262	7.261	8.547	22.307	24.996	27.488	30.578	32.801
16	5.142	5.812	6.908	7.962	9.312	23.542	26.296	28.845	32.000	34.267
17	5.697	6.408	7.564	8.672	10.085	24.769	27.587	30.191	33.409	35.718
18	6.265	7.015	8.231	9.390	10.865	25.989	28.869	31.526	34.805	37.156
19	6.844	7.633	8.907	10.117	11.651	27.204	30.144	32.852	36.191	38.582
20	7.434	8.260	9.591	10.851	12.443	28.412	31.410	34.170	37.566	39.997
21	8.034	8.897	10.283	11.591	13.240	29.615	32.671	35.479	38.932	41.401
22	8.643	9.542	10.982	12.338	14.041	30.813	33.924	36.781	40.289	42.796
23	9.260	10.196	11.689	13.091	14.848	32.007	35.172	38.076	41.638	44.181
24	9.886	10.856	12.401	13.848	15.659	33.196	36.415	39.364	42.980	45.559
25	10.520	11.524	13.120	14.611	16.473	34.382	37.652	40.646	44.314	46.928
26	11.160	12.198	13.844	15.379	17.292	35.563	38.885	41.923	45.642	48.290
27	11.808	12.879	14.573	16.151	18.114	36.741	40.113	43.195	46.963	49.645
28	12.461	13.565	15.308	16.928	18.939	37.916	41.337	44.461	48.278	50.993
29	13.121	14.256	16.047	17.708	19.768	39.087	42.557	45.722	49.588	52.336
30	13.787	14.953	16.791	18.493	20.599	40.256	43.773	46.979	50.892	53.672

Note: The table lists the values of y for various values of $F(y)$ and the number of degrees of freedom v.

TABLE 3.A.5

F-Distribution Table

						m					
	1	2	3	4	5	6	7	8	9	10
n										
1	161.448	199.500	215.707	224.583	230.162	233.986	236.768	238.883	240.543	241.882
2	18.513	19.000	19.164	19.247	19.296	19.330	19.353	19.371	19.385	19.396
3	10.128	9.552	9.277	9.117	9.013	8.941	8.887	8.845	8.812	8.786
4	7.709	6.944	6.591	6.388	6.256	6.163	6.094	6.041	5.999	5.964
5	6.608	5.786	5.409	5.192	5.050	4.950	4.876	4.818	4.772	4.735
6	5.987	5.143	4.757	4.534	4.387	4.284	4.207	4.147	4.099	4.060
7	5.591	4.737	4.347	4.120	3.972	3.866	3.787	3.726	3.677	3.637
8	5.318	4.459	4.066	3.838	3.687	3.581	3.500	3.438	3.388	3.347
9	5.117	4.256	3.863	3.633	3.482	3.374	3.293	3.230	3.179	3.137
10	4.965	4.103	3.708	3.478	3.326	3.217	3.135	3.072	3.020	2.978
11	4.844	3.982	3.587	3.357	3.204	3.095	3.012	2.948	2.896	2.854
12	4.747	3.885	3.490	3.259	3.106	2.996	2.913	2.849	2.796	2.753
13	4.667	3.806	3.411	3.179	3.025	2.915	2.832	2.767	2.714	2.671
14	4.600	3.739	3.344	3.112	2.958	2.848	2.764	2.699	2.646	2.602
15	4.543	3.682	3.287	3.056	2.901	2.790	2.707	2.641	2.588	2.544
16	4.494	3.634	3.239	3.007	2.852	2.741	2.657	2.591	2.538	2.494
17	4.451	3.592	3.197	2.965	2.810	2.699	2.614	2.548	2.494	2.450
18	4.414	3.555	3.160	2.928	2.773	2.661	2.577	2.510	2.456	2.412
19	4.381	3.522	3.127	2.895	2.740	2.628	2.544	2.477	2.423	2.378
20	4.351	3.493	3.098	2.866	2.711	2.599	2.514	2.447	2.393	2.348
21	4.325	3.467	3.072	2.840	2.685	2.573	2.488	2.420	2.366	2.321
22	4.301	3.443	3.049	2.817	2.661	2.549	2.464	2.397	2.342	2.297
23	4.279	3.422	3.028	2.796	2.640	2.528	2.442	2.375	2.320	2.275
24	4.260	3.403	3.009	2.776	2.621	2.508	2.423	2.355	2.300	2.255
25	4.242	3.385	2.991	2.759	2.603	2.490	2.405	2.337	2.282	2.236
26	4.225	3.369	2.975	2.743	2.587	2.474	2.388	2.321	2.265	2.220
27	4.210	3.354	2.960	2.728	2.572	2.459	2.373	2.305	2.250	2.204
28	4.196	3.340	2.947	2.714	2.558	2.445	2.359	2.291	2.236	2.190
29	4.183	3.328	2.934	2.701	2.545	2.432	2.346	2.278	2.223	2.177
30	4.171	3.316	2.922	2.690	2.534	2.421	2.334	2.266	2.211	2.165

Note: The table lists the values of x for $F(x) = 0.95$ and the number of degrees of freedom n and m.

TABLE 3.A.6

F-Distribution Table

					m					
	1	2	3	4	5	6	7	8	9	10
n										
1	4052.18	4999.50	5403.35	5624.58	5763.65	5858.98	5928.35	5981.07	6022.47	6055.84
2	98.503	99.000	99.166	99.249	99.299	99.333	99.356	99.374	99.388	99.399
3	34.116	30.817	29.457	28.710	28.237	27.911	27.672	27.489	27.345	27.229
4	21.198	18.000	16.694	15.977	15.522	15.207	14.976	14.799	14.659	14.546
5	16.258	13.274	12.060	11.392	10.967	10.672	10.456	10.289	10.158	10.051
6	13.745	10.925	9.780	9.148	8.746	8.466	8.260	8.102	7.976	7.874
7	12.246	9.547	8.451	7.847	7.460	7.191	6.993	6.840	6.719	6.620
8	11.259	8.649	7.591	7.006	6.632	6.371	6.178	6.029	5.911	5.814
9	10.561	8.022	6.992	6.422	6.057	5.802	5.613	5.467	5.351	5.257
10	10.044	7.559	6.552	5.994	5.636	5.386	5.200	5.057	4.942	4.849
11	9.646	7.206	6.217	5.668	5.316	5.069	4.886	4.744	4.632	4.539
12	9.330	6.927	5.953	5.412	5.064	4.821	4.640	4.499	4.388	4.296
13	9.074	6.701	5.739	5.205	4.862	4.620	4.441	4.302	4.191	4.100
14	8.862	6.515	5.564	5.035	4.695	4.456	4.278	4.140	4.030	3.939
15	8.683	6.359	5.417	4.893	4.556	4.318	4.142	4.004	3.895	3.805
16	8.531	6.226	5.292	4.773	4.437	4.202	4.026	3.890	3.780	3.691
17	8.400	6.112	5.185	4.669	4.336	4.102	3.927	3.791	3.682	3.593
18	8.285	6.013	5.092	4.579	4.248	4.015	3.841	3.705	3.597	3.508
19	8.185	5.926	5.010	4.500	4.171	3.939	3.765	3.631	3.523	3.434
20	8.096	5.849	4.938	4.431	4.103	3.871	3.699	3.564	3.457	3.368
21	8.017	5.780	4.874	4.369	4.042	3.812	3.640	3.506	3.398	3.310
22	7.945	5.719	4.817	4.313	3.988	3.758	3.587	3.453	3.346	3.258
23	7.881	5.664	4.765	4.264	3.939	3.710	3.539	3.406	3.299	3.211
24	7.823	5.614	4.718	4.218	3.895	3.667	3.496	3.363	3.256	3.168
25	7.770	5.568	4.675	4.177	3.855	3.627	3.457	3.324	3.217	3.129
26	7.721	5.526	4.637	4.140	3.818	3.591	3.421	3.288	3.182	3.094
27	7.677	5.488	4.601	4.106	3.785	3.558	3.388	3.256	3.149	3.062
28	7.636	5.453	4.568	4.074	3.754	3.528	3.358	3.226	3.120	3.032
29	7.598	5.420	4.538	4.045	3.725	3.499	3.330	3.198	3.092	3.005
30	7.562	5.390	4.510	4.018	3.699	3.473	3.304	3.173	3.067	2.979

Note: The table lists the values of x for $F(x) = 0.99$ and the number of degrees of freedom n and m.

4

Introduction to Dynamics: Implications on the Design of Precision Machines

Patrick Baird and Stuart T. Smith

CONTENTS

ABSTRACT Newton's laws provide solutions to dynamic systems with unprecedented precision. However, the limits of application of these laws are often due to lack of precision in the manufacture and assembly of mechanical mechanisms that leads to complex behaviour such as hysteresis, friction and non-linearity. Much of the rest of the book is concerned with methods for optimising precision by eliminating these undesired features of machines and mechanisms.

All dynamic systems for small perturbations can be modelled as linear systems and understood in terms of frequency response functions and characteristic roots. Consequently, while dynamics is a large topic developed over many centuries, this current chapter has focused only on this aspect of the theory.

Ideal objects such as springs, rigid masses and frictionless pendulums are used to illustrate the basic principles of linear dynamics. These model objects can be used as ideal building blocks of a larger and more complex system of interacting bodies, such as may be encountered in a typical precision measurement system. An understanding of the basic concepts of motion in *simple* systems of one or two bodies illuminate foundational principles that apply to more complex, multiple-component systems. As examples of approximately linear, single-input single-output, dynamic processes, two commonly used contact measurement systems have been investigated: the properties and motion of a heavily damped contact probe from a coordinate measurement system, used widely in metrology; and the amplitude and frequency characteristics of a typical cantilever sensor from a scanning force microscope, which is used to analyse material surface properties on a much smaller scale. Such approximations are often adequate to model the response of the system.

Methods of calculating the motions of multi-body dynamical systems, and matrix calculations, with the use of eigenanalysis to obtain solutions of the system response to a variety of applied forces, are introduced towards the end of this chapter.

Notwithstanding the narrow limits of this chapter, the use of generalised coordinates and the Lagrange equation forms a very broad foundation of all of classical mechanics and justifies the effort required to appreciate the underlying principles of this approach.

4.1 Introduction: A World in Motion

Dynamic systems have evolved since the early days of industrial technology. There are many challenges presented by the design, manufacture and use of precision machines. In addition to all the usual aspects that related in the past to the efficiency of industrial machines—smoothness of surfaces and overcoming friction, the effects of weight, material properties such as stiffness, the build up of heat and so on—there are other challenges that result from the miniaturisation of components and the demands for high-speed precision devices. Wafer scanners, commonly used in the photolithographic manufacture of integrated circuits, are an example of lightweight, fast-moving components that must control motion of a wafer and mask with nanometre precision. The active components are to be kept at a precise temperature because small expansions or contractions can result in distorted wafer patterns on the substrate, sometimes with errors of less than 10 nm resulting in complete failure of the manufactured circuit. The wafer stage is moved very accurately and precisely by linear motors, which may be inductive or piezoelectric, with their positioning accurately controlled to determine the amount of exposure required in each part of the scanning process.

The analysis of dynamics makes use of scalars, vectors and coordinate systems, which were introduced in Chapter 3. For objects in motion, it is important to initially distinguish between distance and speed (both scalar quantities), and displacement and velocity (both vector quantities which take the direction of the path into account). Distance is the overall length of a path travelled by an object in any direction; displacement is the change in position in a particular direction. Speed is the distance travelled in unit time; velocity is the displacement per unit time. Vectors are typically displayed in bold type: for example if v is speed, **v** is velocity; however, for simplicity when describing derivatives of displacements, the parameter $\dot{x}$ may be used without bold type. In other cases, where convenient, normal type will be used rather than bold type to represent the magnitude of a vector, in which the actual direction is not contextually important or defined along a coordinate axis.

For simple systems, the mass of an object may be approximated as a point-like particle; the total path travelled by such an object is what is measured by scalars and the individual path(s) and direction(s) measured by vectors. The displacement and average velocity of a particle after travelling some distance may be zero if it returns to its initial position and all the vector quantities involved in its path cancel. The average velocity is the total distance over a defined path per unit time, that is $v = x/t$. The instantaneous velocity is given by the time derivative of distance

$$v = \frac{dx}{dt} = \dot{x}, \tag{4.1}$$

where the dot above x indicates differentiation with respect to time and is called Newton's fluxion notation. It is also necessary, for examples later in the chapter which involve angular motion (the simplest of which is circular), to introduce notation for rotational parameters.

If an angle is given by θ, the angular velocity measured in radians per second is given by

$$\omega = \frac{d\vec{\theta}}{dt}. \tag{4.2}$$

Angular velocity is another vector; the related scalar quantity in this case is angular frequency ω, which is defined as the number of rotations per unit time, without defining any given direction. In rotational motion, which is not necessarily circular or simple harmonic, the vector form, typically oriented along the axis of rotation, becomes useful. In simpler examples, the term ω will be used for convenience.

For small rotations, the sine of the angle is approximately equal to the angle and, therefore, the ratio of the arc s to the radius r is $\theta = s/r$, which provides the relations $v = r\omega = r\dot{\theta}$.

For classical mechanics in general, to begin with, the concept of a force, another vector quantity, is necessary (see Serway 2013 for a general introduction and basic concepts and examples; see Morin 2008 for a more detailed treatment). A force can be static or dynamic; it can be defined as a gradient of potential and something that, when insufficiently opposed, causes a change in motion or an acceleration. In the case of motion, it is the net force, once all balanced or opposing forces are taken into account, which results in the change in movement of an object. This acceleration can be a change in linear motion and/or direction, such as a rotation. Acceleration is also, typically by convention, a vector quantity. The term usually relates to a change in velocity rather than speed, although the term can also be loosely used to define a change in speed only.

Newtonian mechanics involving forces can be broadly classified into two subjects: statics and dynamics. A third branch of mechanics is kinematics, which considers motion in a geometrical sense without reference to the forces that may have caused or influenced it. Statics relates to systems not moving or in constant motion. Statics typically involves forces, but all forces acting are assumed balanced (in equilibrium). In reality everything is in motion, but large-scale influences such as the earth's rotation or gravitational field can often be ignored in a model that can be, considered as an isolated system. Dynamics (or kinetics) relates to systems in which there is a net imbalance of forces resulting in a change in motion, the direction of velocity being determined by the resultant vector sum of all the forces acting.

In a static system, the sum of all forces (applied and reaction) is zero, thus

$$\sum_{i=1}^{\text{All forces}} F_i = 0. \tag{4.3}$$

Applied forces are often designated as body forces (gravitation, electromagnetic) or tractions (applied externally to the body often transmitted through surface contact). More fundamentally, these forces themselves are derived from gradients of potential, such as those experienced between the atoms of a solid being pushed together or, indirectly, as electrons being driven through a conductor in a magnetic field. The microscopic details of an interaction can often be overlooked; however, there is a lot of balancing and cancelling out of forces acting in solids, as will be discussed later.

In a dynamic system comprising a body of fixed mass m, assuming no rotation effects, the net result of unbalanced forces is an acceleration

$$\sum \mathbf{F}_i = m\ddot{\mathbf{x}}. \tag{4.4}$$

The two conditions in Equations 4.3 and 4.4 relate to Newton's first and second laws (see Section 4.2). Because the force and acceleration are in the same direction and, often, the displacement is constrained to a linear path, Equation 4.4 can be written in the scalar form $F = m\ddot{x}$.

4.2 Newton's Laws of Motion for Particles and Rigid Bodies

Newton's laws (Newton 1726) form the basis of classical mechanics and are a sufficient method to use for many mechanical systems, especially those that can be approximated as a system of rigid bodies. The full dynamics of extended rigid bodies, including their rotational motion, and the mechanics of deformable bodies, requires a more generalised approach, to be dealt with later in this chapter.

Newton's first law of motion, sometimes called the law of inertia because it refers to inertial or nonaccelerating frames of reference, sets the background. Newton's first law of motion states that *all objects stay at rest or continue in motion with constant velocity (a constant speed in a straight line) unless acted on by an external force.* In this case, inertia can be described as the property of mass to resist a change in motion, but the term must be used more carefully, or avoided, in the context of moments. This nonaccelerating frame of reference is a starting point for the application of the other laws of motion: for an object not under the influence of forces, a reference frame can be specified in which the object has no acceleration. Newton's first law can be restated including the requirement of an inertial reference frame for the observer also, to avoid confusion from any apparent acceleration of an object as seen by an observer moving in a non-inertial or accelerating frame.

Newton's second law is the law of acceleration, in which the change in motion (acceleration) is proportional to the applied (or net) force and inversely proportional to the mass, thus

$$\mathbf{F} = m\mathbf{a} = m\frac{d\mathbf{v}}{dt} = \frac{d\mathbf{p}}{dt}. \tag{4.5}$$

Equation 4.5 also indicates a rate of change of linear momentum ($\mathbf{p} = m\mathbf{v}$), and the direction of the acceleration is the same as the direction of the net force on the object. For the same force applied to different objects, the mass is what dictates the difference in resulting acceleration. Since the mass is inversely proportional to the acceleration, small masses are accelerated more easily.

The angular equivalent of Newton's second law is the torque τ related to the polar second moment of mass I (often incorrectly called a first moment of inertia)

$$\tau = I\alpha = I\dot{\omega} = \frac{d\mathbf{L}}{dt}, \tag{4.6}$$

where the angular acceleration $\alpha = d\omega/dt$ and the torque is the rate of change of angular momentum $\mathbf{L}$ (not to be confused with the scalar Lagrangian $\mathcal{L}$ used later in this chapter). The angular momentum, which is conserved, is the moment of momentum about a rotation axis, expressed as

$$\mathbf{L} = \mathbf{r} \times \mathbf{p} = mr^2\omega = I\omega, \tag{4.7}$$

where $I = mr^2$ is the polar second moment of mass (in its simplest definition for particles), r being the distance from the axis of rotation (note that the × symbol in Equation 4.7 denotes a cross product, not a multiplication; see Chapter 3, Section 3.1.3). These parameters will be discussed in more detail in Sections 4.5 and 4.6, where rotation of rigid bodies is investigated.

The concept of work is also useful for discussions later in this chapter. The definition of the sum of the work done W by a force over the path or trajectory of a point is given by

$$W = \int_C \mathbf{F} \cdot d\mathbf{s} = \int_C F ds, \tag{4.8}$$

where s is the path taken and C represents the path travelled. The far right-hand definition in Equation 4.8 is only true when the force is always in the direction of the path of travel.

An example of a force causing an acceleration is that due to gravity. In many situations, gravity is relevant but can be considered as a constant acceleration acting in one direction only, especially if the objects in question are situated on or near to the surface of the earth, being the main influence. In the dynamics of moving bodies, gravity may be seen as a restoring force being worked against (Exercise 1). For two masses, the magnitude of the gravitational force, F_g, acting by one on the other, also formulated by Newton, is

$$F_g = \frac{\partial U}{\partial r} = -G\frac{Mm}{r^2}, \tag{4.9}$$

where U is the potential of the system; G is the universal gravitational constant, 6.67408 (31) $\times 10^{-11}$ N·m²·kg⁻²; M and m can be considered to be large and small masses respectively; and the numbers in the parentheses for the value of G represent the uncertainty in its value. The negative sign appears due to the direction of r being, by convention, positive in terms of separation distance. The force due to gravity is still strictly a vector, as it is directed towards the centre of mass, typically towards the ground; to display it as such, a unit vector $\hat{r}$ can be placed on the right-hand side of the equation. If the masses M and m are both small relative to, for example, the size of the earth, the gravitational force is extremely weak and would be very difficult to detect in the presence of much stronger forces that are typically present, such as electromagnetism, the origin of many external or driving forces that would be involved in engineering problems. The magnitude of G can be determined, for example, using a torsion balance (the experiment was first performed by Henry Cavendish in 1798). For the case of a mass situated near the earth's surface, Newton's second law requires the mass to accelerate at a constant value denoted by g:

$$F_g = \frac{GMm}{R^2} = mg, \tag{4.10}$$

where M and R are the mass and radius of the earth. Consequently, the gravitational acceleration at the earth's surface is

$$g = \frac{GM}{R^2}, \tag{4.11}$$

the value of which is approximately 9.8 m·s⁻². A centripetal acceleration combined with a perpendicular velocity results in the elliptic (near-circular) orbit of satellites around the earth, or the earth and other planets around the sun. On the earth, the acceleration g can be considered as a constant for many applications, although slight variations in the density and radius of the earth, relative to its centre of mass, result in a variation of about 0.05 m·s^{-2} across the surface at a similar distance from the centre (for example, sea level). As a consequence, g is often called the gravity acceleration rather than the gravitational acceleration.

Newton's third law, the law of reciprocal action, states that *every action has an equal and opposite reaction*. The resulting force pairs, equal in magnitude but opposite in direction, are called action and reaction forces. The third law applies to all forces, whether in static equilibrium or causing acceleration. It becomes more obvious in a static situation, such as an object on the ground; the object exerts the same force on the ground, equal and opposite to mg. If the object is falling towards the ground, it still exerts an equal and opposite force on the earth, but the resulting acceleration of the earth is negligible, due to its comparatively large mass. Once at rest, the object and ground are in static equilibrium. It is important to realise that the third law acts equally on different objects in opposing directions and, therefore, does not oppose motion as the forces in the contact region cancel. This is a useful concept for rigid body dynamics, as seen later in this chapter. Application of the third law leads to the principle of conservation of momentum (see, for example, Feynman et al. 1964). From the second law, the rate of change of linear momentum of two interacting particles can be dp_1/dt and dp_2/dt respectively. Due to the third law, the forces between them are equal and opposite and, therefore,

$$\frac{d(p_1 + p_2)}{dt} = 0. \tag{4.12}$$

The quantity $p_1 + p_2$ (or $mv_1 + mv_2$)—the total momentum of the system—does not change, provided the system is isolated from its surroundings (an ideal situation). The conservation of angular momentum (more fully described as the conservation of moment of momentum) follows by a similar argument. The conservation of energy is a consequence of the second law, which can be used to show that the rate of change of the total energy in an isolated system is zero and, therefore, the total energy remains constant (Exercise 2). This arises as a direct consequence of the forces due to motion and position being balanced.

Where, for straight-line motion in a single direction x, the kinetic energy is given by T, the potential energy by V and the total energy E is simply

$$E = T + V = \frac{mv^2}{2} + V(x). \tag{4.13}$$

Using the product and chain rules, the total time derivative of the energy can be shown as

$$\frac{dE}{dt} = Fv + v\frac{dV}{dx}, \tag{4.14}$$

and since $F = -dV/dt$, the total energy is also conserved in a closed system.

The concept of friction, a mechanism that dissipates energy in the form of heat, is now introduced (see also Chapter 7, Section 7.2.1.1 for a more involved discussion of friction between metal surfaces). A system of forces is shown in the example in Figure 4.1 for modelling a frictionless vehicle on a slope.

There is no net force in the y direction due to Newton's third law, where $\mathbf{n}$ is the normal reaction force of the slope on the vehicle. The acceleration in the x direction is, therefore,

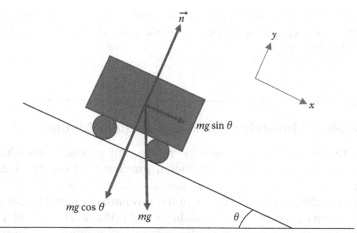

FIGURE 4.1
Forces for vehicle on a slope.

given by

$$\sum F_x = mg \sin \theta = ma_x. \tag{4.15}$$

A frictionless surface is an ideal concept. With friction taken into consideration, there is a resistance to motion, as shown in Figure 4.2. Static friction f_s between two surfaces is defined by the condition (Serway 2013)

$$f_s \leq \mu_s n, \tag{4.16}$$

where μ_s is the coefficient of static friction (a dimensionless constant, dependent on surface parameters) and n is the magnitude of the force in the normal direction. Once moving, the kinetic friction f_k is given by $\mu_k n$, typically less than the static friction μ_k and can vary with speed.

Whilst in equilibrium, net forces in both the x and y directions are zero. Static equilibrium occurs when θ is less than a critical angle. The frictional force f_s is $mg\sin \theta$ and using $n = mg\cos \theta$, it is also seen to be $n\tan \theta$. At a critical angle θ_c, the coefficient of static friction μ_s is equivalent to $\tan\theta_c$. Once the critical angle is exceeded, the mass will undergo acceleration

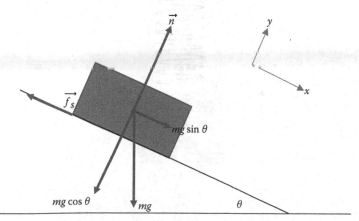

FIGURE 4.2
Object on a slope with friction.

that will depend on the dynamic friction that, as mentioned, can correspondingly vary with speed. This critical angle, whether it be for the static or dynamic coefficient of friction, is often called the friction angle.

4.3 A Simple Model Involving Linear and Angular Motion

In general, the forces that cause motion in engineering problems are electromagnetic in origin: motors, drives and so forth. Additional gravitational effects to consider are the constant pull of the earth on parts of the mechanism, always in one direction, and which can be significant depending on the geometry of the mechanism, simple examples being horizontal levers with masses on the end, pendulums and pulleys. The case of a rigid mass on a massless spring (within the limit of elastic behaviour) provides an introduction to the concept of stiffness, elastic behaviour, simple harmonic motion and the effect, and flow, of energy in a system.

Two important conditions for linear periodic motion are (1) elasticity, which for any spring is limited to within a deformation boundary, allowing for a spring to return to an equilibrium state; and (2) a source of energy which can be stored in the spring as potential energy when compressed or stretched and released where it transforms to kinetic energy when allowed to move towards equilibrium.

The forces at equilibrium are shown in Figure 4.3. When the spring is stretched or compressed in the x direction by an external force within its linear (non-deforming) range, by Hooke's law (see Section 3.5.3), the restoring force is

$$F = -kx, \tag{4.17}$$

where k is a spring constant, a measure of the stiffness of the spring; the spring concept can be applied to any elastic body. Not all springs are linear (follow the simple relationship in

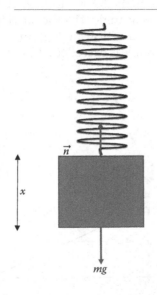

FIGURE 4.3
Mass on a spring.

Equation 4.17), but for non-linear springs the work done is still equal to the potential energy stored in the spring, as discussed later for the linear case. In any real system, there is a limit of elastic behaviour beyond which plastic deformation occurs that dissipates energy and the spring does not return to its equilibrium state. Note that the negative sign is by convention only. The applied force stretching or compressing the spring to give it potential energy is by convention positive; because the restoring force on release (the force pulling the spring back towards equilibrium) acts in the opposite direction, it is given a negative sign here.

Equating forces, ignoring the effects of gravity and friction in the ideal case,

$$m\ddot{x} = -kx,\tag{4.18}$$

so

$$\ddot{x} = -\frac{k}{m}x.\tag{4.19}$$

It can be seen that the units of k/m in Equation 4.19 need to be the square of angular frequency (s^{-2}).

Periodic motion suggests a harmonic solution in displacement of the form

$$x = A\cos(\omega_n t),\tag{4.20}$$

where ω_n represents a 'natural' frequency of the system. This can be shown in a more fundamental way through conservation of energy, but the presence of this type of function becomes clear through differentiation, as shown later. Note that $\omega_n t$ in Equation 4.20 is the phase term of the oscillation period, relating to the time evolution of the mass position in a vertical direction, as described by simple harmonic motion; it is not to be confused with a spatial angle between two directions, which will be the argument of sine or cosine functions, typically denoted by, for example, θ elsewhere in this chapter. The phase of an oscillation will be shown to be an important measurement parameter later in the chapter. Sometimes, a phase offset term such as ϕ may be added to the $\omega_n t$ argument, depending on the initial conditions (for example, the phase at specified at time $t = 0$). The second derivative of Equation 4.20 is

$$\ddot{x} = -\omega_n^2 x\tag{4.21}$$

and, therefore,

$$\omega_n^2 = \frac{k}{m},\tag{4.22}$$

so the natural frequency depends on the spring constant and the mass only. This makes physical sense as the frequency increases proportionally with stiffness (tendency to return to equilibrium) and decreases with mass (resistance to motion).

At the maximum displacement, the potential energy is

$$V_{\text{max}} = \frac{1}{2}kx_{\text{max}}^2.\tag{4.23}$$

The potential energy is entirely converted to kinetic energy at the equilibrium position (ignoring gravity for now), where the mass is moving fastest

$$T_{max} = \frac{1}{2} m v_{max}^2.$$ (4.24)

Due to conservation of energy, V_{max} and T_{max} must be equal, resulting in

$$\left(\frac{v_{max}}{x_{max}}\right)^2 = \frac{k}{m} = \omega_n^2.$$ (4.25)

This can be shown also by expressing $x(t)$ as a periodic function with amplitude x_{max} and taking the first derivative, which results in the expression in Equation 4.22 being equal to ω_n^2 (Exercise 3). In the case of simple harmonic motion, x can also be substituted for the radial parameter r, where $v = r\omega$, resulting in the same expression for ω_n as Equation 4.22. The dynamics of oscillating structures, including damping factors, will be investigated later in this chapter.

There is a variety of flexible elements, such as leaf, notch and coil springs used in precision systems (see Chapter 7). Springs are used in precision systems due to their restoring property. This is important for precise motion to and from a repeatable location. One example that will be investigated here is that of a touch trigger probe, a device for measuring surfaces of objects on a coordinate measuring system (CMS; see Chapter 5, Section 5.6 for a more thorough discussion of CMSs and touch trigger probes). A schema of a touch trigger probe is shown in Figure 4.4.

A probe such as that shown in Figure 4.4 is fixed to a multi-axis stage to trigger a measurement when the probe ball contacts a surface. A CMS can be programmed so that its motor-driven axes approach an object at various locations on the object's surface to obtain an adequate measurement of its geometry. The mechanism is spring-loaded to return the probe ball to its equilibrium position after contact (some have sub-micrometre repeatability). The tripod seat is a stable resting position under the spring load. Following contact, the deflection of the probe ball at the end of the stylus results in one of the cylinder-bearing electrical contacts in the probe head, which form part of an electric circuit, being broken. The three contacts are each in the form of a cylinder resting on two spheres. Strain gauges may be

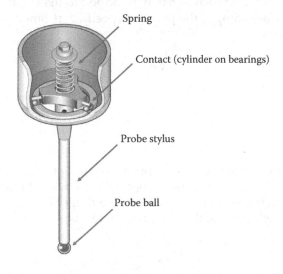

FIGURE 4.4
Touch trigger probe. (Adapted from Renishaw. Document: H-1000-8006-01-B (Touch-trigger_probing_technology) .ppt. Available from http://resources.renishaw.com/en/download/presentation-touch-trigger-probing-technology –70270.)

located above the contacts to register the movement of the internal mechanism. Because of the elasticity of the stylus, there is always a small deflection before the contact is broken; this is called pre-travel, a small pre-measurement distance of movement of the stylus (see Figure 4.5).

Because of the geometry of the contacts, however, the pre-travel distance varies with the angle of contact about the probe's longitudinal axis (stylus), and generates 'lobe'-shaped variations (deviations from a circle) observed when measuring, for example, a ring gauge oriented along the same axis as the probe stylus. These variations in the circle, depending on the probe approach direction, are largely repeatable errors and can be corrected for by measuring a ring gauge and subtracting the errors from subsequent measurements taken in the perpendicular plane. A best-fit through surface points is often good enough to estimate the dimensions of a simple geometric shape but is not ideal in all situations. Variation in pre-travel is also observed in the latitudinal direction whilst contacting a sphere towards its centre from a certain latitude (see Figure 4.6).

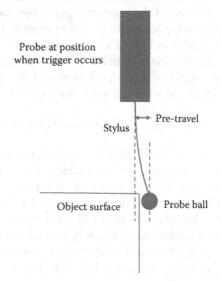

FIGURE 4.5
Probe pre-travel.

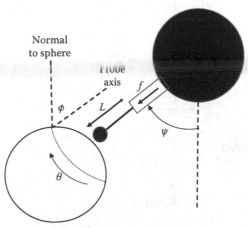

FIGURE 4.6
Offset angle probe and measurement parameters.

Spherical coordinates can be used as a reference for correcting the errors of pre-travel in all measurements, but there are other factors, such as the length of the stylus (more elastic bending with length), the size of the measuring ball (which affects friction and hence slippage), the angle of the stylus with the vertical direction (the probe often needs to be rotated to access some locations around an object which introduces the effect of gravity into the mechanism), the speed of contact and the set force on the spring mechanism.

The systematic errors of a CMS measurement system are a multivariate problem and can be corrected for approximately in an empirical way, for example, by numerical fitting of error functions or interpolation to gauge measurement errors (Baird 1996) or other numerical methods (Yang et al. 1996); however, it is an interesting exercise to examine the system of forces involved in the mechanism (see, for example, Shen and Zhang 1997). At the instant when the probe ball contacts the surface of an object prior to deflection, there are seven contact points in total: two for each of the three electrical contacts in the probe body and one for the physical contact with the object being measured.

The inset in Figure 4.7 shows one of the contact regions between the cylindrical parts of the contact mechanism and the support angle α. The system of forces at the seven contact regions is shown diagrammatically in Figure 4.8.

The coordinate system can be aligned with one of the cylindrical arms for convenience, for example, the x axis with the couple of contacts shown on the right, and forces equated at equilibrium just before the stylus is deflected, with, for example, in the x direction:

$$\sum F_x = (-F_{21} + F_{22} + F_{31} - F_{32}) \cos \alpha \cos 30 + F_p \sin \theta \cos \phi = 0. \qquad (4.26)$$

The cosine terms in the contact mechanism force expression in Equation 4.26 project the force first with the perimeter of the circle and then along the x axis. The terms in Equation 4.26, which must balance the contact, are aligned in the x direction by the measurement angles ϕ and θ. Similar equilibrium equations are obtained for forces and moments in all three axes, and the system can be solved to result in expressions for all six contact forces in the mechanism as functions of longitudinal angle ϕ, latitudinal angle θ, probing force F_p, spring force F_s, tripod-stylus weight W, stylus length L, tripod arm length R and kinematic support angle α. The probe is shown vertically aligned in Figure 4.8, but when the stylus is at an angle to the vertical, the weight term, distance from its centre of gravity, and another

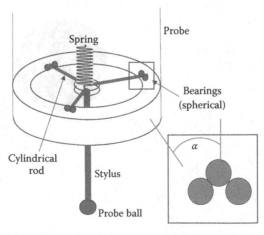

FIGURE 4.7
Trigger probe kinematic tripod mount.

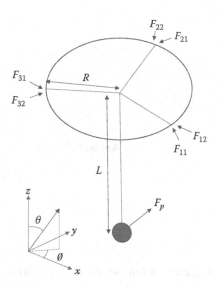

FIGURE 4.8
Trigger probe contact forces, geometric parameters and coordinates.

parameter, the angle of stylus orientation prior to deflection, need to be taken into account. Once this system of forces and geometric parameters is established, a trigger condition, such as a contact force threshold or electrical resistance at which triggering occurs, can be specified, and the pre-travel estimated depending on the probe and measurement configuration.

From cantilever dynamics, the deflection of the probe bending perpendicular to its axis can be described as

$$\delta_b = \frac{F_p L^3 \sin\theta}{3EI},$$ (4.27)

where θ is the angle of contact with the stylus as above, L is the stylus length, E is the elastic modulus and I is the second moment of area about the neutral axis, which can be calculated using the effective shaft radius (Exercise 4). There is also a deflection component along the shaft axis which is small enough compared with the bending deflection to be ignored. The force on the probe F_p is calculated from trigger contact force expressions such as Equation 4.26, depending on the orientation of measurement. There is an additional component to the bending δ_{bw} due to the offset weight of the probe when it is itself oriented at an angle to the vertical (direction of gravity, angle ψ in Figure 4.6), which depends on this probe orientation angle using a similar expression to Equation 4.27. The resulting pre-travel deflection for any measurement approach can then be estimated as

$$\delta_{pt} = (\delta_b + \delta_{bw}\cos\phi)\sin\theta.$$ (4.28)

To investigate angular motion, a simple pendulum is shown in Figure 4.9. For the dynamics of pendulums and other rigid systems, refer to Thornton and Marion (2008).

The mass m traces out a circular path s. The net force is given by

$$F_{net} = -mg\sin\theta.$$ (4.29)

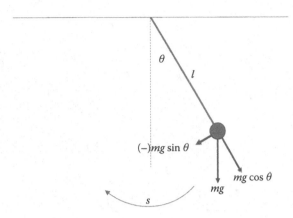

FIGURE 4.9
Simple pendulum.

For small amplitudes, this is another example of simple harmonic motion. The forces are balanced as

$$m\ddot{s} = -mg \sin \theta \qquad (4.30)$$

and since

$$\ddot{s} = l\ddot{\theta} \qquad (4.31)$$

then (in the small angle approximation $\sin\theta \approx \theta$)

$$\ddot{\theta} + \frac{g}{l}\theta = 0 \qquad (4.32)$$

is the equation of motion. For simple harmonic motion, the angular motion is of the form

$$\theta = A \sin \omega_n t \qquad (4.33)$$

so

$$\ddot{\theta} = -\omega_n^2 \theta \qquad (4.34)$$

giving the natural frequency of the pendulum as

$$\omega_n^2 = \frac{g}{l}, \qquad (4.35)$$

which is a similar result to the mass-spring oscillation (Exercises 5 and 10). It can be seen that g is the restoring parameter (analogous to k in the spring) and l provides the resistance to angular motion (analogous to the mass on the spring).

4.4 Effect of Damping in the System

Damping modifies the force equilibrium, seen in the simple example of an oscillating spring, with a drag factor (due to internal friction or external forces such as viscous drag or electrical eddy currents induced by magnetic fields) that is proportional to the velocity. If, as in this case, the magnitude of the damping force is small compared with the restoring force, the system is underdamped. (For good coverage of this subject and oscillations in general, see Morin 2008 or Thornton and Marion 2008). Viscous-type damping introduces a first derivative term in addition to the second derivative

$$m\ddot{x} = -kx - b\dot{x}, \tag{4.36}$$

where b is the damping coefficient (a mass flow parameter). For convenience, the natural angular frequency (or undamped resonant frequency) is donated as ω_n, and sets the parameter

$$\gamma = \frac{b}{2m} = \zeta\omega_n \tag{4.37}$$

(ζ is a damping ratio, see later), which simplifies the differential Equation 4.36 to

$$\ddot{x} + 2\gamma\dot{x} + \omega_n^2 x = 0. \tag{4.38}$$

Trying a solution of the form $x(t) = ae^{\lambda t}$ results in the quadratic characteristic equation

$$\lambda^2 + 2\gamma\lambda + \omega_n^2 = 0, \tag{4.39}$$

with characteristic roots

$$\lambda_{1,2} = -\gamma \pm \sqrt{\gamma^2 - \omega_n^2}, \tag{4.40}$$

so it is clear that γ is a parameter relating to the oscillation, and the square root term is complex if $\gamma^2 < \omega_n^2$. In Section 4.6, λ will be called an eigenvalue. Setting the damped oscillation frequency $\omega_d = \sqrt{\omega_n^2 - \gamma^2} = \omega_n\sqrt{1 - (\gamma/\omega_n)^2}$, gives a solution of the form

$$x(t) = Ae^{-(\gamma-i\omega_d)t} + Be^{-(\gamma+i\omega_d)t}. \tag{4.41}$$

Because both sides of this equation must be real, the constants A and B, called the eigenvectors (see Section 4.6), are a complex conjugate pair, as is the eigenvalue in Equation 4.40. Equation 4.41 can be reduced to the form

$$x(t) = ae^{-\gamma t}\cos(\omega_d t + \phi). \tag{4.42}$$

The damping parameter γ appears as an attenuation envelope of the oscillation, but also as a function that modifies the oscillation frequency (where it appears in the expression for the damped oscillation frequency ω_d).

An important dimensionless quantity, the Q-factor, is defined as the ratio of the energy stored and the energy dissipated in the system. The resonance and damping are related proportionally to the energy stored and dissipated in the oscillating system. It can be shown, through analysing the energy of the system, that the parameter γ is closely related to the bandwidth of the damped oscillation, which increases proportionally with damping, such that for γ significantly smaller than ω_n

$$\Delta\omega = 2\gamma\sqrt{3}, \tag{4.43}$$

where $\Delta\omega$ is defined as the full width at half maximum of the magnitude frequency response of the damped oscillation. The ratio λ/ω_n was seen earlier to modify the frequency. The Q-factor can now also be interpreted as the ratio of the resonant frequency to its bandwidth (aside from a multiplicative factor), which relates to the sharpness of an oscillating system (Exercise 6). An indicator of how close it is to true resonance (no damping being an ideal situation) is

$$Q = \frac{\omega_n}{2\gamma} = \sqrt{3}\frac{\omega_n}{\Delta\omega}. \tag{4.44}$$

Another dimensionless quantity, the damping ratio ζ, is defined as

$$\zeta = \frac{b}{2\sqrt{km}} \equiv \frac{\gamma}{\omega_n} \equiv \frac{1}{2Q}, \tag{4.45}$$

resulting in the alternative forms of the differential Equation 4.38 given by

$$\frac{d^2x}{dt^2} + 2\zeta\omega_n\frac{dx}{dt} + \omega_n^2 x = \frac{d^2x}{dt^2} + \frac{\omega_n}{Q}\frac{dx}{dt} + \omega_n^2 x = 0 \tag{4.46}$$

and a similar solution of the form

$$x = Ae^{-\zeta\omega_n t}\cos(\omega_d t + \phi). \tag{4.47}$$

The damped angular frequency is modified as

$$\omega_d = \omega_n\sqrt{1 - \zeta^2} \tag{4.48}$$

and it can be seen how convenient the parameter ζ is in describing the damping of the oscillation frequency in a simple way. The resonant frequency ω_{res} (the undamped natural frequency being an ideal case) occurs when the amplitude of oscillation for a given harmonic excitation is at a peak, and is a function of the natural frequency and damping ratio and can be calculated from

$$\omega_{res} = \omega_n\sqrt{1 - 2\zeta^2}. \tag{4.49}$$

In the example of the touch trigger probe of Figure 4.4, when the probe is moved away from contact with a workpiece, it will be pushed into the three bearing mounts with sufficient force to maintain contact and with substantial interface damping at these bearing mounts six contacts. Consequently, it will return to the equilibrium position and stop without observed oscillations. In this case, it is considered that $\gamma > \omega_n,(\zeta > 1)$ and the system is overdamped. The characteristic equation has real roots with a solution of the form

$$x(t) = e^{-\gamma t}[A\cosh(\omega_d t) + B\sinh(\omega_d t)] \tag{4.50}$$

and there is no oscillation. Oscillation only occurs in non-damped or underdamped systems (or, as will be shown later, in forced damped oscillations). In the special case where $\gamma = \omega_n$, ($\zeta = 1$), the system is sometimes called critically damped, for which the response is

$$x(t) = e^{-\gamma t}(A + Bt). \tag{4.51}$$

The touch trigger probe system may be close to this value, as is the case with many recoil systems (for example, constrained door hinges). In practice, for control of dynamic systems, a damping ratio of between 0.4 and 0.8 is typically optimal.

Many systems will be subject to harmonically varying forces of different frequencies; this can be harmful if they produce cyclic loads that could induce fatigue or can be useful for sensing and other diagnostic applications. This latter type of system is illustrated here by examining the scanning cantilever of an atomic force microscope (AFM) (Binnig et al. 1986, Sarid 1992; also see Chapter 5, Section 5.7.4). An AFM cantilever is used as a sensor to determine when a sharp tip attached to its free end is either in contact, or within close proximity to, a specimen surface, on a much smaller scale than is typical of the touch trigger probe. It is used to measure variations in surface height or profile down to molecular scales and, in very precise equipment, sometimes with atomic level resolution. The AFM cantilever is a small beam fixed at one end and free at the other. For a full treatment of the dynamics of cantilever beams of various types, see, for example, Gere (2004). The deflection dynamics can be described using a similar expression to that given earlier for the touch trigger probe shaft (see Equation 4.27). The typical material for AFM cantilevers that is made using semiconductor manufacturing technology is pure silicon or silicon nitride. The tip on the end of the cantilever is a conical or pyramidal structure on its underside. There are many variations in cantilever tip type and geometry, and some are also coated in metal for investigating electric charge and electromagnetic fields when the tip is moved to be close to a surface. A schematic side view is shown in Figure 4.10.

In the case of the forced damped oscillator, the coordinate x is replaced with z as the traditionally used linear coordinate corresponding to the direction normal to the plane of the specimen surface. With this replacement x and y correspond to the directions in the plane of a sample surface. Deflection of the cantilever relative to its base is typically measured using a laser beam directed at an angle onto its top surface and reflected onto a photodetector creating an optical lever. To obtain a surface profile, the cantilever is lowered toward the surface until the tip interaction results in a measurable change in the deflection of the cantilever. In the simplest implementation, the deflection of the cantilever as it contacts the surface can be used to determine the surface location and the net contact force between the tip and specimen. The specimen is then translated in the x direction while the

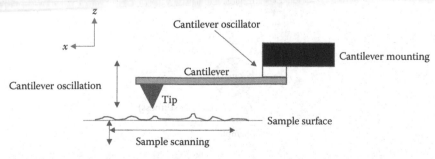

FIGURE 4.10
Atomic force microscope (AFM) cantilever assembly.

probe base is moved in the direction normal to the surface to maintain a constant interaction force. Effectively, this results in the cantilever being deflected by the same amount as the tip traverses over the surface at constant force. Consequently, the motion of the cantilever base to keep deflection and force constant is then used to represent the surface profile. In practice, the limiting precision with which the surface profile can be measured is related to the tip radius, which in turn is limited by the sensitivity with which the tip-to-surface interaction can be measured. Another, more sensitive method for measuring contact interactions is to attach a piezoelectric actuator to the cantilever to apply a harmonically oscillating force, so that it undergoes forced vibration at a constant frequency. This can be achieved by making the cantilever itself from a piezoelectric material, depositing a piezo-electric film onto its surface or exciting the support at the base of the cantilever as shown in Figure 4.10.

Both the magnitude and phase response of the cantilever is then measured by either the laser optical lever or the charge to the piezoelectric actuator. From frequency response measurement, both the amplitude and phase of the response can be collected with the use of frequency lock-in equipment. The variations of both amplitude and phase can also provide valuable, highly localised information about a surface's physical (electrostatic, van der Waals and meniscus forces) and material properties (elastic modulus, energy loss), and is essential when investigating electrical forces and surface charge distribution.

As a relatively simple and practically useful model, the AFM cantilever can be approximated as a lumped mass on a spring with a damper, as shown in Figure 4.11. The equation of motion for the forced damped oscillator is a non-linear second order differential equation of the form (modified with a tip-sample interaction)

$$m\ddot{z} + b(\dot{z} - \dot{u}) + (k + k_c)z - ku = 0,\qquad(4.52)$$

$$\ddot{z} + 2\zeta\omega_n(\dot{z} - \dot{u}) + \omega_n'^2 z - \omega_n^2 u = 0,\qquad(4.53)$$

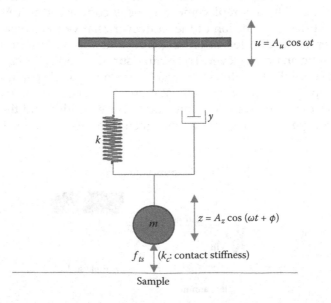

FIGURE 4.11
Point mass approximation schematic for driven oscillating cantilever.

where

$$u = A_u \cos(\omega t) = \text{Re}\left\{A_u e^{j\omega t}\right\} \tag{4.54}$$

is the driving oscillation where j is used to represent $\sqrt{-1}$. The stiffness term k_c represents the added stiffness due to contact of the tip with the surface being measured and ω'_n is defined in Equation 4.57 where it is discussed in detail. In practice, there will be a number of interactions at the tip requiring more advanced, and non-linear, models for predicting interaction forces, the most significant of which in AFM is the repulsion between the tip and surface due to overlap of electronic orbitals. It is, however, further complicated by the presence of van der Waals forces (polarization interaction), and other interaction modes when in contact, such as cantilever friction causing lateral bending in the xy-plane and water layer adhesion, and also cohesive forces; while not in contact, electrostatic (Baird et al. 1999) and magnetic forces may contribute due to localised charge variation, potential difference and the presence of dipoles (Baird and Stevens 2005). z is the tip position resulting from induced cantilever oscillation with angular frequency ω, and u (also in the z-direction) represents the piezo oscillation. When considering forced oscillations it is usual to ignore the transient effects that will decay after a period that will depend upon the damping of the system. To determine the steady-state frequency response $H(j\omega)$ of the tip, a solution is assumed to be of the form

$$z = H(j\omega)u = \text{Re}\left\{A_u H(j\omega) e^{j\omega t}\right\}$$
$$= A_u |H(j\omega)| \cos(\omega t + \arg(H(j\omega))) = A_z \cos(\omega t + \phi), \tag{4.55}$$

where ϕ is a phase offset between the driven oscillation and resulting tip oscillation and A_z is the amplitude of the tip oscillation, and is a linear function of the input amplitude and drive frequency. Substitution of Equation 4.55 into Equation 4.53 gives

$$H(j\omega) = \frac{\dfrac{\omega_n^2}{\omega_n'^2} + j2\zeta\omega\dfrac{\omega_n}{\omega_n'^2}}{\left(1 - \dfrac{\omega^2}{\omega_n'^2}\right) + j2\zeta\omega\dfrac{\omega_n}{\omega_n'^2}}, \tag{4.56}$$

where the primed frequency is the natural frequency of the system modified by the additional stiffness of the contact and can be written in the form

$$\omega_n'^2 = \frac{\lambda + \lambda_c}{m} = \omega_n^2 + (\Delta\omega)^2. \tag{4.57}$$

To determine the response of the tip it is only necessary to compute the magnitude and argument of the response at any given frequency. The magnitude of this frequency response represents the ratio of the amplitude at the tip divided by the amplitude of excitation at the cantilever base. A plot of the magnitude and phase of the frequency response as a function of the excitation frequency when the tip is free from the surface (i.e. $\omega_n = \omega_n'$) is shown in Figure 4.12 for six different damping ratios. As the tip comes into contact with a surface, the added stiffness will result in this response shifting to an increased frequency with corresponding changes in magnitude and phase. If the cantilever is driven at the natural frequency (i.e. $\omega = \omega_n$), then substitution of Equation 4.57 into Equation 4.56 yields the

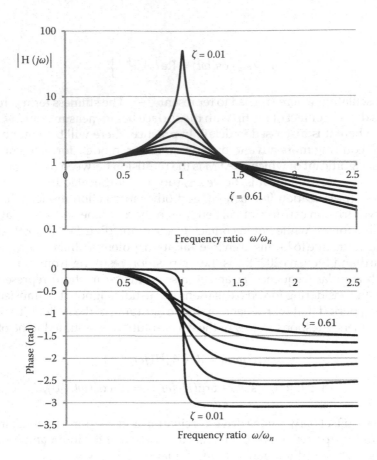

FIGURE 4.12
Amplitude plots for oscillating cantilever with different damping ratios when not in contact with a surface (i.e. $\omega_n = \omega'_n$).

response function

$$H(j\omega) = \frac{1 + j2\zeta}{\dfrac{(\Delta\omega)^2}{\omega_n^2} + j2\zeta}.$$

(4.58)

If the damping is small (i.e. $\zeta^2 \approx 0$), and using the relationship $\tan^{-1}(1/x) = \pi/2 - \tan^{-1}(x)$, then the amplitude and phase of the response can be approximated by

$$|H| \approx \frac{\omega_n^2}{(\Delta\omega)^2} = \frac{k}{k_c},$$

$$\arg(H) = \tan^{-1}(2\zeta) - \tan^{-1}\left(\frac{2\zeta\omega_n^2}{(\Delta\omega)^2}\right) \approx Q\frac{k_c}{k} - \frac{\pi}{2}.$$

(4.59)

Equation 4.59 shows that changes in amplitude of oscillation of the tip represent a measure of the stiffness of the contact, while the phase shift will combine a measure of both stiffness and energy dissipation in the contact region of the tip, with the sensitivity being proportional to the Q-factor of the probe system. More generally, as the AFM tip is brought into proximity with the surface, the interaction forces will modify each of the parameters in this frequency response by changing mass, stiffness and damping (typically all three will increase). Observed variations in these parameters as the tip is scanned over the surface are representative of variations in AFM tip–sample interactions. Depending on the experimental arrangement, these variations can provide information on surface topography and various material properties at the highly localised region of the tip.

4.5 Multi-Body Dynamics

The examples shown up to this point are from single rigid or flexible bodies, or 'particles'. Multi-body dynamics is the study of rigid bodies that may be connected to one another but with a certain number of degrees of freedom of movement. Solution by application of Newton's laws becomes more complicated with an increasing number of bodies due to the vector character of forces and momenta, typically represented in a rectilinear (Cartesian) coordinate system. The application of analytical mechanics, developed by D'Alembert, Euler, Lagrange, Hamilton and others becomes useful here as it takes advantage of energy-related scalar functions. Use is made of concepts, such as virtual work and least action, conservative forces and generalised coordinates, which make the best use of system constraints to simplify calculations on a multi-body system of particles or rigid bodies (Thornton and Marion 2008).

It is useful to distinguish between two types of force: conservative and non-conservative. A conservative force is a force in which the work done in moving a particle between two points is independent of the path taken, and consequently if the path is a closed loop, the work done is zero, hence

$$W = \oint F dr = 0, \tag{4.60}$$

where W is the work done and F is the force experienced by the particle as it moves an incremental distance dr. The potential energy is changed by an amount that is independent of the path taken to get from one location to another. Conservative forces are derivable from a potential function that is often a fixed distribution in space (gravity and electrostatics are examples). The Lagrangian formulation is an alternative to the Newtonian one for conservative forces and makes certain types of calculations more mathematically approachable. Lagrangian formulation is easier to use as it is independent of the choice of coordinates, whereas the Newtonian/Cartesian approach involves coordinate transformations, which become complicated and laborious.

Generalised coordinates are directions defined in the configuration space of a system, which depend on the constraints. The configuration space can consist of a combination of linear and rotational axes. Each set of values of the generalised coordinates defines a point in the configuration space and represents a state of the mechanical system. As the system changes in time, the point traces out a curve in configuration space, called the generalised trajectory. If a system has constraints that restrict its motion in certain directions, the number of degrees of freedom can be reduced. Coordinate axes can be defined based on the direction of constraints. For example, a freely moving particle (in the ideal sense) has three degrees of freedom, two connected particles have five and three connected particles have six, due to the additional rotations possible. Additional connected particles do not add further degrees of freedom so that a rigid body comprising many point masses will always have only six degrees of freedom. The same extends to rigid bodies of arbitrary shape, if considered to be a system of particles joined together in a constrained sense. If there are constraints on one or more of the directions of motion, the number of degrees of freedom (and hence generalised coordinates required) is less than that of a freely moving system. Examples of constraints are bodies confined to an axis or a surface, such as a fixed wheel which is constrained to move in one rotational direction only (see Chapter 6 for more on constraints and degrees of freedom).

There are differential and integral principles involved. Starting with the virtual work done on a system of particles, which is given by

$$\sum_{k=1}^{N}(m_k\mathbf{a}_k - \mathbf{R}_k - \mathbf{f}_k) \cdot \delta\mathbf{r}_k, \qquad (4.61)$$

where the f_k are all non-constrained forces (conservative or non-conservative), the $\mathbf{R}_k$ are the forces of constraints, the $\delta\mathbf{r}_k$ are virtual (infinitesimal) displacements and N represents the number of Cartesian coordinates for each mass (if each mass is free, the total system will have $3N$ degrees of freedom). As stated earlier, the constraints might, for example, be a cylinder that guides a piston so that it can only translate along its axis. In this case, the force of constraint will be normal to the axis of the cylinder so that no motion in this direction is possible. Hence, the work done by these forces is always zero and, therefore, the total virtual work done δW reduces to

$$\sum_{k=1}^{N}(m_k\mathbf{a}_k - \mathbf{f}_k) \cdot \delta\mathbf{r}_k. \qquad (4.62)$$

Equation 4.61 provides the framework of the principle of virtual work known as D'Alembert's principle. There is only one true path of motion but many virtual ones. These displacements may still overlap with the orientation of constraints. To eliminate the complexity that often occurs when trying to express the state of the system using only Cartesian coordinates, generalised coordinates are used relating to the configuration space of the constrained system. These generalised coordinates could be any path such as that traced out by a bead sliding along a wire of arbitrary shape. However, for the purposes of this book, these coordinates will be either linear or rotational, and all will be identified by the parameter q_k. Equation 4.62 can be further simplified by recognising that, with the addition of constraints, some of the coordinates will not be necessary. Additionally, the vector representation results in three terms in the summation of Equation 4.62. Consequently, the same equation can be written in scalar form with individual summations being required for the total number of degrees of freedom n (this would equal $3N$ for a system of free particles). With these considerations, Equation 4.62 can be written in a final form

$$\sum_{k=1}^{n}(m_k\ddot{x}_k - F_{x_k})\delta x_k = 0. \qquad (4.63)$$

While this form of D'Alembert's equation is a complete set of equations governing the dynamics of the system, it is frequently desired to express the equations of motion in terms of an equal number of generalised coordinates. The relationship between Cartesian and generalised coordinates can be expressed in functional form for each coordinate

$$x_k = f_k(q_1, q_2.., q_n).$$ (4.64)

Differentiating Equation 4.64 yields for any one degree of freedom k

$$dx_k = \frac{\partial f_k}{\partial q_1} dq_1 + \frac{\partial f_k}{\partial q_2} dq_2. = \sum_{j=1}^{n} \frac{\partial x_k}{\partial q_j} dq_j.$$ (4.65)

Further differentiating Equation 4.65 with respect to time and realising that the partial derivatives are functions of the coordinate geometry only provides the relationship between the velocities in each coordinate system

$$\dot{x}_k = \sum_{j=1}^{n} \frac{\partial x_k}{\partial q_j} \dot{q}_j.$$ (4.66)

Finally, because the displacements and velocities can all be treated as independent variables, it is possible to differentiate Equation 4.66 with respect to the velocity in a generalised coordinate giving the soon-to-be-useful relation

$$\frac{\partial \dot{x}_k}{\partial \dot{q}_j} = \frac{\partial x_k}{\partial q_j},$$ (4.67)

which is often referred to as the cancelling of the dots.

To transform from D'Alembert's equation to an equivalent form involving only generalised coordinates, consider the total kinetic energy T of the system given by

$$T = \sum_{k=1}^{n} \frac{m_k}{2} \dot{x}_k^2.$$ (4.68)

Differentiating Equation 4.68 with respect to a generalised velocity and making use of the cancelling of the dots yields

$$\frac{\partial T}{\partial \dot{q}_j} = \sum_{k=1}^{n} m_k \frac{\partial \dot{x}_k}{\partial \dot{q}_j} \dot{x}_k = \sum_{k=1}^{n} m_k \frac{\partial x_k}{\partial q_j} \dot{x}_k.$$ (4.69)

Using the product rule of calculus to take the time derivative of Equation 4.69 produces

$$\frac{d}{dt} \frac{\partial T}{\partial \dot{q}_j} = \sum_{k=1}^{n} m_k \frac{\partial x_k}{\partial q_j} \ddot{x}_k + \sum_{k=1}^{n} m_k \frac{\partial \dot{x}_k}{\partial q_j} \dot{x}_k.$$ (4.70)

Consider now the derivative of the kinetic energy of the system with respect to a generalised coordinate

$$\frac{\partial T}{\partial q_j} = \sum_{k=1}^{n} m_k \frac{\partial \dot{x}_k}{\partial q_j} \dot{x}_k.$$ (4.71)

Subtracting Equation 4.71 from 4.70 and summing over all degrees of freedom multiplied by a virtual displacement in each coordinate yields the relationship

$$\sum_{j=1}^{n}\left[\frac{d}{dt}\frac{\partial T}{\partial \dot{q}_j} - \frac{\partial T}{\partial q_j}\right]\delta q_j = \sum_{j=1}^{n}\sum_{k=1}^{n}m_k\frac{\partial x_k}{\partial q_j}\ddot{x}_k\delta q_j = \sum_{k=1}^{n}m_k\ddot{x}_k\delta x_k. \tag{4.72}$$

Consequently, it can be seen that the left-hand side of Equation 4.72 contains only generalised coordinates and the left term in D'Alembert's Equation 4.63 contains only Cartesian coordinates. To complete the transformation of D'Alembert's equation into a form containing generalised coordinates, it is only necessary to replace the term involving forces applied to the masses. These will generally be considered to originate from gradients of potential (conservative forces), viscous type dissipative forces and those being applied from outside of the system.

Conservative forces can generally be derived directly from the negative gradient of the potential. Consequently, the virtual work in either coordinate system can be equated to provide the relationships

$$\sum_{j=1}^{n}Q_j\delta q_j = \sum_{k=1}^{n}F_{x_k}\delta x_k = \sum_{j=1}^{n}\sum_{k=1}^{n}F_{x_k}\frac{\partial x_k}{\partial q_j}\delta q_j, \tag{4.73}$$

where Q_j is the generalised force in the direction of the corresponding coordinate q_j and the final term in Equation 4.73 is derived using Equation 4.65. From Equation 4.73 the generalised force can be related directly to forces in Cartesian coordinate directions from

$$Q_j = \sum_{k=1}^{n}F_{x_k}\frac{\partial x_k}{\partial q_j}. \tag{4.74}$$

Once again, the virtual work of viscous-type dissipation elements will be the same for both coordinate systems, giving the relation

$$\delta W = \sum_{j=1}^{n}Q_j\delta q_j = \sum_{k=1}^{n}b_k\dot{x}_k\delta x_k$$

$$= \sum_{j=1}^{n}\sum_{k=1}^{n}b_k\dot{x}_k\frac{\partial x_k}{\partial q_j}\delta q_j = \sum_{j=1}^{n}\sum_{k=1}^{n}b_k\dot{x}_k\frac{\partial \dot{x}_k}{\partial \dot{q}_j}\delta q_j, \tag{4.75}$$

where b_k represents damping forces in Cartesian directions and cancelling of the dots has been reversed in the last term of this equation. Equation 4.75 provides the relation between viscous dissipation forces in either coordinate system

$$Q_j = \sum_{k=1}^{n}b_k\dot{x}_k\frac{\partial x_k}{\partial q_j}. \tag{4.76}$$

Consider now the dissipation function D, suggested by Rayleigh (1873), to be the average power dissipated

$$D = \sum_{k=1}^{n}\frac{b_k}{2}\dot{x}_k^2. \tag{4.77}$$

Differentiating this dissipation function with respect to a generalised velocity and summing over all coordinates yields

$$\sum_{j=1}^{n} \frac{\partial D}{\partial \dot{q}_j} \delta q_j = \sum_{j=1}^{n} \sum_{k=1}^{n} b_k \dot{x}_k \frac{\partial \dot{x}_k}{\partial \dot{q}_j} \delta q_j = \sum_{j=1}^{n} Q_j \delta q_j. \tag{4.78}$$

From the left- and right-hand sides of Equation 4.78, it is clear that the viscous-type dissipation forces can also be determined in terms of the generalised coordinates. The relationship between viscous damping forces in both Cartesian and generalised coordinates follows from Equation 4.75:

$$Q_j = \sum_{k=1}^{n} b_k \dot{x}_k \frac{\partial x_k}{\partial q_j}. \tag{4.79}$$

Collecting together all of the preceding terms, D'Alembert's equation can be rewritten in terms only of generalised coordinates and forces resulting in the Lagrange equation given by

$$\frac{d}{dt}\left(\frac{\partial T}{\partial \dot{q}_k}\right) - \frac{\partial T}{\partial q_k} + \frac{\partial D}{\partial \dot{q}_k} + \frac{\partial V}{\partial q_k} = Q_k, k = 1, 2..., n. \tag{4.80}$$

Alternatively, it is possible to develop the equations governing motion for a conservative system using calculus of variations. As a first step, a quantity called the Lagrangian is defined. This is the 'difference' between the kinetic and potential energy of a system (assuming the force relating to the potential is a restoring force in the direction of equilibrium which, by convention, has a negative sign). The principle of least action, as an alternative to virtual work, is used to arrive at the Euler-Lagrange differential equations. The Lagrangian that is a function of the coordinates, their derivatives and time is defined as

$$\mathcal{L}(q, \dot{q}, t) = T - V, \tag{4.81}$$

where T is the kinetic energy and V is the potential energy. The Lagrangian is chosen so that the path of least action is the path along which Newton's laws are followed, and the negative sign for V is chosen by convention, where the force due to a potential is defined by the negative gradient of potential; this ensures that the application of the Lagrange equations becomes equivalent to Newton's second law (Exercise 7). The action is defined as the integral of the Lagrangian over time:

$$S = \int_{t_1}^{t_2} \mathcal{L}(q, \dot{q}, t) dt. \tag{4.82}$$

This action integral is an example of a functional; it depends on an entire function (a continuum of values between two limits), takes this function as its input and returns a scalar, in this case, the Lagrangian integrated between two endpoints. The action happens to have the same units as angular momentum. Applying the principle of least action (the more general formulation being known as Hamilton's principle of stationary action), the evolution of the system is described by a stationary point (no variation) of the action requiring the condition

$$\frac{\delta S}{\delta q(t)} = 0. \tag{4.83}$$

The extremum occurs where there is the least change from nearby paths. If S is a stationary value, any function very close to the path for stationary action yields the same action, up to first order in the deviation (the definition of a stationary value). It is important to specify here that, while not being absolutely necessary for a full development of the theory, a class of functions whose endpoints $x(t_1)$ and $x(t_2)$ are fixed is being considered, that is any small perturbation (virtual displacement) is zero at the endpoints: $\varepsilon(t_1) = \varepsilon(t_2) = 0$. Using calculus of variations, the perturbation to first order is derived, then integration by parts and making use of the aforementioned boundary condition, results in the Euler-Lagrange equation (see Exercise 8)

$$\frac{\partial \mathcal{L}}{\partial q} - \frac{d}{dt}\left(\frac{\partial \mathcal{L}}{\partial \dot{q}}\right) = 0, \tag{4.84}$$

which can be used to generate the equations of motion by calculating derivatives of kinetic and potential energy with respect to the generalised coordinates q and their time derivatives $\dot{q}$. (Derivation of the Euler-Lagrange is followed through in detail in Thornton and Marion, 2008, and other similar texts; Susskind and Hrabovsky, 2014, provides a simplified and clear account.) The Euler-Lagrange equations always take the same form and can be applied to each generalised coordinate separately. In terms of units and equality, Equation 4.84 is equivalent to equating the inertial forces to the rate of change in momentum, that is Newton's second law, or the angular form, torque being the rate of change of angular momentum, depending on the generalised coordinates chosen according to constraints on the system (the constraints must, however, be holonomic, that is not dependent on velocities or higher order derivatives of position). Starting the calculations using generalised coordinates and scalar quantities makes a big difference. For a large number of rigid bodies, this becomes much easier than using force and momentum vectors throughout the system. Lagrangian mechanics works well in this formalism for systems involving predominantly conservative forces; however, non-conservative forces (often dissipative forces such as friction) and externally driven forces can be accounted for with a set of modified equations. (In an extended formulation of the Lagrange equations, constraints can also be included as additional equations; these involve Lagrange multipliers.)

A useful but relatively simple example to compare the Newtonian and Lagrangian approach is that of a particle of mass constrained on the surface of a sphere of radius R under a conservative force, allowing two degrees of freedom, the latitudinal and longitudinal angles θ and ϕ, which are set as the generalised coordinates:

$$F = F_\theta \tilde{r}_\theta, \tag{4.85}$$

where $\tilde{r}_\theta$ is a unit vector. The kinetic energy is

$$T = \frac{1}{2}mv_\theta{}^2 + \frac{1}{2}mv_\phi{}^2 = \frac{1}{2}mR^2\dot{\theta}^2 + \frac{1}{2}mR^2\dot{\phi}^2\sin^2\theta \tag{4.86}$$

and the potential energy, set to zero when $\theta = \phi = 0$, is

$$V = -F_\theta R\theta, \tag{4.87}$$

giving rise to the Lagrangian

$$\mathcal{L} = T - V = \frac{1}{2}mR^2\dot{\theta}^2 + \frac{1}{2}mR^2\dot{\phi}^2\sin^2\theta + F_\theta R\theta. \tag{4.88}$$

The derivatives of the Lagrangian are calculated for the generalised coordinates q_i (in this case θ and ϕ)

$$\frac{\partial\mathcal{L}}{\partial\theta} = mR^2\dot{\theta}^2\sin\theta\cos\theta + F_\theta R, \tag{4.89}$$

$$\frac{\partial\mathcal{L}}{\partial\phi} = 0, \tag{4.90}$$

$$\frac{d}{dt}\left(\frac{\partial\mathcal{L}}{\partial\dot{\theta}}\right) = \frac{d}{dt}\left(mR^2\dot{\theta}\right) = mR^2\ddot{\theta}, \tag{4.91}$$

$$\frac{d}{dt}\left(\frac{\partial\mathcal{L}}{\partial\dot{\phi}}\right) = \frac{d}{dt}\left(mR^2\dot{\phi}\sin^2\theta\right) = mR^2\left(2\dot{\theta}\dot{\phi}\sin\theta\cos\theta + \ddot{\phi}\sin^2\theta\right). \tag{4.92}$$

Substituting Equation 4.89 to 4.92 into the Euler-Lagrange equation (Equation 4.84) results in the equations of motion

$$F_\theta = mR\left(\ddot{\theta} - \dot{\phi}^2\sin\theta\cos\theta\right), \tag{4.93}$$

$$0 = mR^2\sin\theta\left(\ddot{\phi}\sin\theta + 2\dot{\theta}\dot{\phi}\cos\theta\right). \tag{4.94}$$

The advantage of using this method here is that it is only required to write out the energy terms in the generalised coordinates and differentiate directly with respect to them (θ and ϕ). In the Newtonian approach, the Cartesian coordinates or their unit vectors in spherical polar form (differential operators) need to be derived, and their first and second time derivatives calculated. Even in this simple case, the Lagrangian method begins to become quicker to apply (see Exercises 9 and 10). The real advantage becomes clear with much more complex multi-body systems.

In the case of non-conservative forces, the virtual work can be divided into conservative and non-conservative types, resulting in the extended Lagrange equation (Equation 4.80).

4.6 Lumped Mass Models

A rigid body is a useful concept in engineering and is defined as a group of particles whose individual separations are constrained enough to be seen as fixed to one another. This is an ideal system, because there are atomic and molecular motions happening at a sub-microscopic level. These tiny motions can be ignored when investigating the macroscopic motion of an entire body. Some of the macroscopic motion, such as elastic deformations, can also be neglected for the purpose of obtaining equations of motion for a rigid body. A rigid body can be seen as a collection of discrete particles (requiring particle summation) or a continuous distribution of matter (requiring integration of mass density distribution). Either of these concepts can be used depending on convenience.

A rigid body, because of its extended shape, can exhibit modes of vibration that have directional characteristics beyond those of simple oscillations, such as those represented by ideal springs and pendulums. These modes of vibration can be solved as an eigensystem, which makes use of matrix representation and calculations. Eigensystems were originally used to investigate rigid body rotations, but now have many diverse applications, including stability and vibration analysis and, more generally, matrix algebra and differential operators. The whole subject of eigensystems was generalised into spectral theory by David Hilbert, using complex mathematical spaces for use in quantum theory (eigenstates and eigenfunctions).

The concepts of centre of mass and second moments of mass are very useful for calculating the motion of extended rigid bodies. In a system of particles bound together in a solid, the internal forces can be assumed to cancel out due to Newton's third law. The weak form of this law is all that is required, and for two particles 1 and 2 states that the magnitude of the forces acting between them are equal, as follows:

$$f_{12} = f_{21}.$$ (4.95)

If R is a vector describing the position of the centre of mass of a rigid body from a defined origin, then for n particles in a rigid body

$$\sum_{i=1}^{n} m_i(r_i - R) = 0$$ (4.96)

if the body is assumed to move as if it were a single particle (all the particles follow the weak version of Newton's third law and move together). This leads to the total moment of mass of the constituent particles being

$$\sum_{i=1}^{n} m_i r_i = MR,$$ (4.97)

where M is the total mass and, therefore,

$$R = \frac{1}{M} \sum_{i=1}^{n} m_i r_i,$$ (4.98)

or, for a continuous distribution,

$$R = \frac{1}{M} \int r \, dm.$$ (4.99)

This balance of internal forces, and their effect considered from the centre of mass, leads to a number of important concepts, such as the conservation of linear momentum for the system being equivalent to that for a particle of mass M at R, the total angular momentum (or moment of momentum) about the origin being equivalent to the sum of the angular momentum of the coordinate origin and the angular momentum of the body relative to the centre of mass. The total kinetic energy is equivalent to that of the total mass with a velocity of the centre of mass plus that of the constituent particles relative to the centre, and the total energy (in a conservative system) is constant. Equation 4.99 enables much simplification of the system in terms of calculation of rotational quantities, such as angular momentum and rotational kinetic energy.

To see the relation between the second moment of mass and useful parameters, such as angular momentum and kinetic energy, in simple (non-vector) notation:

$$\text{Angular momentum:} \, L = \sum_{i=1}^{n} r_i p_i = \sum_{i=1}^{n} r_i m_i \omega r_i = \omega \sum_{i=1}^{n} m_i r_i^2 = I\omega \quad (4.100)$$

$$\text{Rotational kinetic energy:} \, T_{rot} = \sum_{i=1}^{n} \left(\frac{1}{2} m_i \omega^2 r_i^2 \right) = \frac{1}{2} \omega^2 \sum_{i=1}^{n} m_i r_i^2 = \frac{1}{2} I\omega^2 \quad (4.101)$$

It can be seen that I is a constant of proportionality between rotational parameters here, and relates the mass distribution and shape of an object to how it moves when subject to a torsional moment. Therefore, if structural information is available for a rigid body, it is a simple matter to calculate its rotational motion using the second moment of mass that can be represented simply as a scalar as above, assuming the axis of rotation is known and used as a convenient axis of symmetry. In most instruments and machines, the axes of rotation are known. The rotational response of the body is represented in matrix or, equivalently, tensor form. This is called the inertia matrix or 'inertia tensor' (rank 2). If the axis of rotation is the axis of symmetry, the off-diagonal components cancel.

The advantage of the inertia tensor is that it can be pre-calculated for a body of known structure (there are some well-known forms for useful geometric shapes, for example for a rotating sphere this is $I = \frac{2}{5} mr^2$). The inertia tensor enables the rotational properties of these objects to be calculated using the earlier simple formulae.

In vector form,

$$L = \sum_{i=1}^{n} r_i \times p_i = \sum_{i=1}^{n} m_i r_i \times (\omega \times r_i) = \sum_{i=1}^{n} m_i \left[r_i^2 \omega - r_i (r_i \cdot \omega) \right]$$
$$= \mathbf{I}_I \cdot \boldsymbol{\omega}, \quad (4.102)$$

where $\mathbf{I}_I$ is the 3×3 inertia tensor. Shown in matrix form this is

$$\begin{pmatrix} L_x \\ L_y \\ L_z \end{pmatrix} = \begin{pmatrix} I_{xx} & I_{xy} & I_{xz} \\ I_{yx} & I_{yy} & I_{yz} \\ I_{zx} & I_{zy} & I_{zz} \end{pmatrix} \begin{pmatrix} \omega_x \\ \omega_y \\ \omega_z \end{pmatrix}. \quad (4.103)$$

The inertia matrix is a real symmetric matrix (equivalent to its transpose), which enables this problem to be solved as an eigenvalue problem, $\mathbf{I}_I \cdot \boldsymbol{\omega} = \lambda \boldsymbol{\omega}$, in which the three orthogonal eigenvectors $\boldsymbol{\omega}$ (analogous to the angular velocity) define the principal axes of rotation of the rigid body and the eigenvalues are second moments (and cross moments) of mass about these axes.

For eigensystems (for example, Stroud and Booth 2011), generally if A is a real symmetric matrix and λ is a scalar, and

$$Ax = \lambda x, \tag{4.104}$$

then any column vector x which satisfies the equation is an eigenvector of A and the associated scalar λ is an eigenvalue of A. In general, for physical problems the structure A is related to the vibration or stretching characteristics of the system. The eigenvalue problem can be represented as

$$(A - \lambda I)x = 0, \tag{4.105}$$

where I is the identity matrix and, therefore, the determinant is

$$|A - \lambda I| = 0. \tag{4.106}$$

If A is of dimension n, this is equivalent to a nth order polynomial equation for λ, and there are n roots (possibly complex for systems with viscous damping), n eigenvalues and n associated eigenvectors of A. It can be shown by taking the transpose and complex conjugate, that both λ and x are real for conservative systems. These come in complex conjugate pairs for all engineering systems that always have damping or other forms of energy dissipation.

This type of system is applicable to vibration analysis of mechanical structures, in which the eigenvalues are the natural frequencies of vibration and the eigenvectors indicate the directional characteristics or mode shapes. This is a generalised eigenvalue problem (there is a matrix on either side of the set of equations). To illustrate this, the simple case of masses and springs is reconsidered, where some coupled oscillations, as shown in Figure 4.13, are used to start the process (Exercise 9).

While this is a relatively simple two degree of freedom system for which the two generalised coordinates x_1 and x_2, and forces F_1 and F_2 are collinear, it is informative to use the Lagrange equation to obtain the two equations governing motion of this system. The three terms for the Lagrange equation can be readily derived

$$T = \frac{1}{2}m_1\dot{x}_1^2 + \frac{1}{2}m_2\dot{x}_2^2,$$

$$V = \frac{1}{2}k_1x_1^2 + \frac{1}{2}k_2(x_1 - x_2)^2 + \frac{1}{2}k_3x_2^2, \tag{4.107}$$

$$D = \frac{1}{2}b_1\dot{x}_1^2 + \frac{1}{2}b_2(\dot{x}_1 - \dot{x}_2)^2 + \frac{1}{2}b_3\dot{x}_2^2.$$

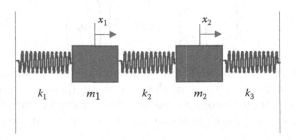

FIGURE 4.13
Two masses and three springs.

Substitution of Equation 4.107 into 4.80 yields the two equations

$$m_1 \ddot{x}_1 + (b_1 + b_2)\dot{x}_1 - b_2 \dot{x}_2 + (k_1 + k_2)x_1 - k_2 x_2 = F_1,$$
$$m_2 \ddot{x}_2 + (b_2 + b_3)\dot{x}_1 - b_2 \dot{x}_1 + (k_2 + k_3)x_1 - k_2 x_1 = F_2. \tag{4.108}$$

Equation 4.108 can be written in the matrix form (Exercise 10)

$$\begin{bmatrix} m_1 & 0 \\ 0 & m_2 \end{bmatrix} \begin{Bmatrix} \ddot{x}_1 \\ \ddot{x}_2 \end{Bmatrix} + \begin{bmatrix} b_1 + b_2 & -b_2 \\ -b_2 & b_2 + b_3 \end{bmatrix} \begin{Bmatrix} \dot{x}_1 \\ \dot{x}_2 \end{Bmatrix} + \begin{bmatrix} k_1 + k_2 & -k_2 \\ -k_2 & k_2 + k_3 \end{bmatrix} \begin{Bmatrix} x_1 \\ x_2 \end{Bmatrix} = \begin{Bmatrix} F_1 \\ F_2 \end{Bmatrix}$$

or

$$\mathbf{m}\ddot{\mathbf{x}} + \mathbf{b}\dot{\mathbf{x}} + \mathbf{k}\mathbf{x} = \mathbf{F}. \tag{4.109}$$

In practice, any system subject only to small perturbations from equilibrium will reduce to a linear equation of the form given by Equation 4.109. Additionally, and most important, the matrices are square and symmetric leading the Maxwell's reciprocity theorem to be used shortly. This theorem states that, given a force applied at one location resulting in a specific displacement at another location, if reversed, the same force applied at the location where the previous displacement was measured will produce the same displacement at the location where previously the force was applied. The reciprocity theorem can be applied to a large number of processes described by linear differential equations.

Because of the importance of Equation 4.109, there have been a number of methods for computing solutions. In the simplest case, consider a free system with no energy dissipation so that Equation 4.109 reduces to

$$\mathbf{m}\ddot{\mathbf{x}} + \mathbf{k}\mathbf{x} = 0. \tag{4.110}$$

One solution to this might be of the form

$$\mathbf{x} = \{X\}e^{\beta t}, \tag{4.111}$$

where the terms on the column matrix $\{X\}$ that has dimensions $n \times 1$ are constants corresponding to the characteristic shapes of this freely moving system and are the eigenvectors associated with a particular eigenvalue. Substitution of Equation 4.111 into 4.110 produces an equation of the form

$$(\beta^2 \mathbf{I} + \mathbf{m}^{-1}\mathbf{k})\{X\} = (\mathbf{A} - \lambda \mathbf{I})\{X\} = 0, \tag{4.112}$$

where $A = \mathbf{m}^{-1}\mathbf{k}$ is an $n \times n$ square matrix and the equation on the left-hand side is in the form of a standard eigen equation. In this form, there will be n eigenvalues each with a distinct set of eigenvectors that are associated with the mode shapes of the lumped mass system. For passive spring mass systems, each eigenvalue will be a real positive number and will correspond to the undamped natural frequencies $\beta = \sqrt{-\lambda} = j\sqrt{\lambda}$ of the system.

For any engineering system, damping will always be present, and it becomes necessary to directly solve for the eigenvalues and eigenvectors. In this case, the assumed solution is of similar form

$$\mathbf{x} = \{X\}e^{\lambda t}. \tag{4.113}$$

For a linear system with damping, using Equation 4.113 yields

$$\lambda^2\mathbf{I}\{X\} = \lambda\mathbf{I}\{W\} = -\mathbf{m}^{-1}\mathbf{b}\{W\} - \mathbf{m}^{-1}\mathbf{k}\{X\},$$

$$\{W\} = \lambda\mathbf{I}\{X\} \tag{4.114}$$

where $\{W\}$ is called the velocity eigenvector. Equation 4.114 can be expressed in the matrix form

$$\lambda\mathbf{I}\left\{\begin{array}{c}\{X\}\\\{W\}\end{array}\right\} = \left[\begin{array}{cc}\mathbf{0} & \mathbf{I}\\-\mathbf{m}^{-1}\mathbf{k} & -\mathbf{m}^{-1}\mathbf{b}\end{array}\right]\left\{\begin{array}{c}\{X\}\\\{W\}\end{array}\right\} = \mathbf{A}\{Z\}, \tag{4.115}$$

that can be readily seen as a standard eigen equation. However, in this case the eigenmatrix $\mathbf{A}$ is of dimension $2n \times 2n$, and the eigenvalue and eigenvector solutions will be in the form of complex conjugate pairs. For the example of Figure 4.13, there will be two complex conjugate pairs of eigenvalues representing the two roots of this system. As is the case with the roots of a single degree of freedom system given by Equation 4.40, the real and imaginary parts of each eigenvalue can be thought of as representing the decay and damped natural frequency for each mode of the system. For a two degree of freedom system, this can be expressed as

$$\lambda_1 = -\zeta_1\omega_{n1} \pm j\omega_{n1}\sqrt{1 - \zeta_1^2} = -\zeta_1\omega_{n1} \pm j\omega_{d1},$$

$$\lambda_2 = -\zeta_2\omega_{n2} \pm j\omega_{n2}\sqrt{1 - \zeta_2^2} = -\zeta_2\omega_{n2} \pm j\omega_{d2}. \tag{4.116}$$

In Equation 4.116, 1 and 2 represent the first and second modes of the two degree of freedom system with corresponding damping ratios and natural frequencies relating the decay and frequency of each mode. For further discussion of this analysis and interpretation of eigenvalues and eigenvectors, see Newland (1989).

For zero damping, these eigenvalues (or roots) will become purely imaginary, as was found for Equation 4.110. As the damping increases, so too will the real part (the decay term) of these eigenvalues, while the damped natural frequency (the imaginary part) decreases. Choosing specific values for the parameters of the model in Figure 4.13, a plot of the roots of this system with different damping values is plotted in Figure 4.14. Because the eigenvalues are in the form of complex conjugate pairs, only the upper portion of the plot is necessary. Also shown in Figure 4.14 is a grey-shaded region bounded by a damping ratio of $\zeta = 0.4$ at the upper boundary and $\zeta = 0.8$ at the lowest boundary. These damping values are often considered desirable for optimising the dynamic performance of mechanical systems either open loop or under closed loop control (see Chapters 8 and 14). Depending on the application, the lower value of the damping ratio is often considered rather low so that the system will tend to oscillate (too wobbly), while the higher damping ratio will not overshoot (too slow).

When the damping is considered to originate from distortion and motion of the elements of a mechanism, it is sometimes reasonable to assume that the dissipation matrix

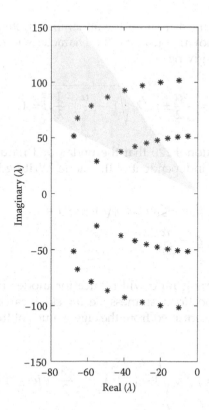

FIGURE 4.14
Roots locus for the model shown in Figure 4.13. Parameters for this model are $m_1 = 12$, $m_2 = 8$, $k_1 = 25000$, $k_2 = 35000$, $k_3 = 30000$, $b_1 = s180/2.5$, $b_2 = b_3 = s70/2.5$, $s = 1, 2, \ldots, 8$.

is proportional to the spring or mass matrix. A major reason for this is the ease of modelling and computation of eigenvalues and eigenvectors. An immediate consequence of this is that the mode shapes are not altered by the damping and these can be used to directly predict displacements of the elements of the mechanism. Because this preserves the mode shapes, which are mathematically orthogonal, this enables solutions in the form of these mode shapes, an approach called modal analysis, a topic that is beyond the scope of this book. Consider the case of damping proportional to mass (i.e. $b = \alpha m$), for which the equation governing free motion becomes

$$m\ddot{x} + \alpha m\dot{x} + kx = 0. \tag{4.117}$$

Again, assuming a solution of the form given by Equation 4.111

$$\beta^2 I\{X\} + \beta \alpha I\{X\} + m^{-1}k\{X\} = \left[(\beta^2 + \beta \alpha)I + A\right]\{X\} = 0, \tag{4.118}$$

which can immediately be written in the form of an eigen equation, where the eigenvalues are

$$\lambda = -\beta^2 - \beta \alpha \tag{4.119}$$

and the eigenmatrix is the same as that for the undamped system. Because the eigenvalues will be real positive and known constants, the characteristic roots of the system can be solved from Equation 4.109, giving

$$\beta_i = -\frac{\alpha}{2} \pm j\sqrt{\lambda_i}\sqrt{1 - \left(\frac{\alpha}{2}\right)^2 \frac{1}{\lambda_i}}, i = 1...n. \tag{4.120}$$

It can be seen from Equation 4.120 that the undamped natural frequency remains the same while the decay rate is independent of the mode. Writing Equation 4.120 in the form

$$\beta_i = -\zeta_i \omega_i \pm j\omega_i \sqrt{1 - \zeta_i^2}, i = 1...n$$

$$\zeta_i = \frac{\alpha}{2\omega_i}, \tag{4.121}$$

it is apparent that the damping ratio will reduce for modes having increasingly higher natural frequencies. For proportional stiffness (i.e. $b = \varepsilon k$), it can be similarly shown that the roots of the system can be determined from the eigenvalues of the undamped system using the equation

$$\beta_i = -\frac{\varepsilon \lambda_i}{2} \pm j\sqrt{\lambda_i}\sqrt{1 - \left(\frac{\varepsilon}{2}\right)^2 \lambda_i} = -\frac{\varepsilon \lambda_i}{2} \pm j\omega_i \sqrt{1 - \left(\frac{\varepsilon \omega_i}{2}\right)^2} \tag{4.122}$$

$$= -\zeta_i \omega_i \pm j\omega_i \sqrt{1 - \zeta_i^2}.$$

For proportionate damping, it is apparent that the decay rate will increase along with the damping ratio for higher modes; a trend commonly found in many mechanisms. Once the eigenvalues have been computed, the roots locus plot can be analytically derived and evaluated for varying parameters of the system. However, it has to be remembered that the proportionate mass and stiffness assumptions do not represent real systems. In practice, suitably chosen values may be used to develop reasonably precise models for at least two or more of the modal frequencies in a system that can be of most importance for a specific application.

While the root locus provides general information about the dynamics of a system, the common method for evaluating performance is to measure or model the steady-state frequency response of a system that is subject to harmonically varying input forces resulting in small perturbations about an equilibrium state. Although the term transfer function is often used instead of frequency response, a transfer function is typically used to describe processes that might also be non-linear and, therefore, is not used here. Such systems can again be fully represented by Equation 4.109. To derive the steady-state frequency response, the system of Figure 4.13 will again be considered. Because this is a two degree of freedom linear system, the most general equation of motion for a system subject to harmonic forces at some arbitrary frequency ω is given by

$$\begin{bmatrix} a_{11} & a_{12} \\ a_{21} & a_{22} \end{bmatrix} \begin{Bmatrix} \ddot{q}_1 \\ \ddot{q}_2 \end{Bmatrix} + \begin{bmatrix} b_{11} & b_{12} \\ b_{21} & b_{22} \end{bmatrix} \begin{Bmatrix} \dot{q}_1 \\ \dot{q}_2 \end{Bmatrix} + \begin{bmatrix} c_{11} & c_{12} \\ c_{21} & c_{22} \end{bmatrix} \begin{Bmatrix} q_1 \\ q_2 \end{Bmatrix} = \begin{Bmatrix} Q_1 \\ Q_2 \end{Bmatrix} = \begin{Bmatrix} F_1 e^{j\omega t} \\ F_2 e^{j\omega t} \end{Bmatrix}, \tag{4.123}$$

where $a_{rc} = a_{cr}$, $b_{rc} = b_{cr}$, $c_{rc} = c_{cr}$ are real constants. Because this is a linear differential equation, the response will also be linear. Being linear, for any harmonically varying steady-state input force, the output displacement response will be at the exact same frequency and the amplitude response will be proportional to the input force, often called the gain of the frequency response. Additionally, because the output is at the same frequency as the input, while there will be a time delay between input force and output response, the phase difference at any frequency, by definition, must be constant. By using the exponential representation of a harmonic function, the amplitude and phase are modelled as vectors in the complex plane and, therefore, phase shifts and gains are incorporated in the frequency response derivations typically defined by the function $H_{rc}(j\omega)$, where $j\omega$ is used to indicate that this is a complex number representation of the response and is a function of the input frequency only. The subscripts r and c indicate that the frequency response represents the ratio of input forces at coordinate r and the displacement response at coordinate c. The gain and phase of the response can be obtained from the magnitude and argument of this complex number at any given frequency. Based on these definitions, the responses at each coordinate can be written as

$$q_1(t) = \text{Re}\left\{ F_1 H_{11}(j\omega)e^{j\omega t} + F_2 H_{21}(j\omega)e^{j\omega t} \right\}$$

$$q_2(t) = \text{Re}\left\{ F_1 H_{12}(j\omega)e^{j\omega t} + F_2 H_{22}(j\omega)e^{j\omega t} \right\}. \tag{4.124}$$

Typically, the fact that only the real part of these equations are used is left out of the analysis and will also be dropped here. The main reason for this is that only the gain and phase effects are of interest in most analyses, although it will be seen that the complex representation also provides tools for visualisation of dynamics.

To determine the frequency responses of this linear system model, it is also noted that, if the output at one coordinate to any individual input is derived, then the effect of providing another input at another frequency anywhere in the system is to simply add the two individual responses in isolation of one another. This ability to add individual solutions to determine responses to a composite of inputs is known as the principle of superposition. Consequently, to determine the frequency response functions, it will only be necessary to consider an applied harmonic force at coordinate one for which the responses at coordinates one and two will be given by the first terms in the Equations 4.124. Substitution of these solutions into the general Equation 4.123, results in the simultaneous linear equations

$$\left(-a_{11}\omega^2 + j\omega b_{11} + c_{11}\right)H_{11}(j\omega) + \left(-a_{12}\omega^2 + j\omega b_{12} + c_{12}\right)H_{12}(j\omega) = 1$$

$$\left(-a_{11}\omega^2 + j\omega b_{11} + c_{11}\right)H_{11}(j\omega) + \left(-a_{22}\omega^2 + j\omega b_{22} + c_{22}\right)H_{12}(j\omega) = 0$$

or
$$\tag{4.125}$$

$$\begin{bmatrix} e_{11} & e_{12} \\ e_{21} & e_{22} \end{bmatrix} \left\{ \begin{array}{c} H_{11}(j\omega) \\ H_{12}(j\omega) \end{array} \right\} = \mathbf{e} \left\{ \begin{array}{c} H_{11}(j\omega) \\ H_{12}(j\omega) \end{array} \right\} = \left\{ \begin{array}{c} 1 \\ 0 \end{array} \right\},$$

where again $H_{rc} = H_{cr}$, $e_{rc} = e_{cr}$ and the second form of Equation 4.125 can be readily solved using Cramer's rule for determination of the response of any system of arbitrary degrees of freedom from

$$H_{rc}(j\omega) = \frac{1}{\Delta}\frac{\partial \Delta}{\partial e_{rc}}, \tag{4.126}$$

where the symbol Δ represents the determinant of the matrix $\mathbf{e}$. One important feature of this equation is that all responses share a common denominator. The roots of this common denominator correspond to the eigenvalues of the system for which, in the absence of damping, at frequencies equal to these roots, the denominator will go to zero, resulting in the responses going to infinity. Because of this, when the magnitude of the response is plotted against frequency, at frequencies corresponding to these roots, the plot will go to infinity and, therefore, these roots of the characteristic equation are often called poles. Sometimes, there will also be roots in the numerator. For the same reason these are also often called zeros. For the parameters given in Equation 4.109 corresponding to the model of Figure 4.13, the coefficients of e are

$$e_{11} = -m_1\omega^2 + j\omega(b_1 + b_2) + (k_1 + k_2),$$

$$e_{12} = e_{21} = -j\omega b_2 - k_2, \tag{4.127}$$

$$e_{22} = -m_2\omega^2 + j\omega(b_2 + b_3) + (k_2 + k_3).$$

Using Equations 4.127 and 4.126, the frequency responses of this system are given by

$$H_{11}(j\omega) = \frac{\begin{vmatrix} 1 & e_{12} \\ 0 & e_{22} \end{vmatrix}}{\begin{vmatrix} e_{11} & e_{12} \\ e_{21} & e_{22} \end{vmatrix}} = \frac{e_{22}}{e_{11}e_{22} - e_{12}^2},$$

$$H_{12}(j\omega) = H_{21}(j\omega) = \frac{\begin{vmatrix} e_{11} & 1 \\ e_{21} & 0 \end{vmatrix}}{\begin{vmatrix} e_{11} & e_{12} \\ e_{21} & e_{22} \end{vmatrix}} = \frac{-e_{21}}{e_{11}e_{22} - e_{12}^2} = \frac{j\omega b_2 + k_2}{e_{11}e_{22} - e_{12}^2}, \tag{4.128}$$

$$H_{22}(j\omega) = \frac{\begin{vmatrix} e_{11} & 0 \\ e_{21} & 1 \end{vmatrix}}{\begin{vmatrix} e_{11} & e_{12} \\ e_{21} & e_{22} \end{vmatrix}} = \frac{e_{11}}{e_{11}e_{22} - e_{12}^2} = \frac{\left(-m_1\omega^2 + j\omega(b_1 + b_2) + (k_1 + k_2)\right)}{e_{11}e_{22} - e_{12}^2}.$$

Expanding the first of Equation 4.128 yields the rather cumbersome equation

$$
\begin{aligned}
H_{11}(j\omega) &= \frac{-m_2\omega^2 + j\omega(b_2 + b_3) + (k_2 + k_3)}{(-m_1\omega^2 + j\omega(b_1 + b_2) + (k_1 + k_2))(-m_2\omega^2 + j\omega(b_2 + b_3) + (k_2 + k_3)) - (j\omega b_2 + k_2)^2} \\[2mm]
&= \frac{-m_2\omega^2 + (k_2 + k_3) + j\omega(b_2 + b_3)}{\begin{aligned}&m_1 m_2 \omega^4 + \omega^2[m_1(k_2 + k_3) + m_2(k_1 + k_2) - b_2(b_1 + b_3)] + k_1(k_2 + k_3) + k_2 k_3 \\ &- j\omega\{\omega^2[m_2(b_1 + b_2) + m_1(b_2 + b_3)] - [b_1(k_2 + k_3) + b_2(k_1 + k_3) + b_3(k_1 + k_2)]\}\end{aligned}} \\[2mm]
&= \frac{A(\omega) + jB(\omega)}{C(\omega) + jD(\omega)} = E(\omega) + jF(\omega).
\end{aligned}
$$

$$\tag{4.129}$$

While containing a number of terms, there are still some attributes of linear systems that can be readily determined. Because the last form of Equation 4.129 applies to all frequency responses of linear systems, the gain (or magnitude response) and phase can be readily computed from

$$|H_{11}(j\omega)| = \left(\frac{A^2(\omega) + B^2(\omega)}{C^2(\omega) + D^2(\omega)}\right)^{1/2},$$

$$\arg H_{11}(j\omega) = \tan^{-1}\left(\frac{B(\omega)}{A(\omega)}\right) - \tan^{-1}\left(\frac{D(\omega)}{C(\omega)}\right).$$

(4.130)

A second observation is that the imaginary terms in both numerator and denominator contain only products of damping parameters with other parameters of the system. Consequently, setting all of the damping parameters to zero, Equation 4.129 and the other responses reduce to

$$H_{11}(j\omega) = \frac{-m_2\omega^2 + (k_2 + k_3)}{m_1 m_2 \omega^4 + \omega^2[m_1(k_2 + k_3) + m_2(k_1 + k_2)] + k_1(k_2 + k_3) + k_2 k_3},$$

$$H_{12}(j\omega) = \frac{k_2}{m_1 m_2 \omega^4 + \omega^2[m_1(k_2 + k_3) + m_2(k_1 + k_2)] + k_1(k_2 + k_3) + k_2 k_3},$$

(4.131)

$$H_{22}(j\omega) = \frac{-m_1\omega^2 + (k_1 + k_2)}{m_1 m_2 \omega^4 + \omega^2[m_1(k_2 + k_3) + m_2(k_1 + k_2)] + k_1(k_2 + k_3) + k_2 k_3}.$$

All of Equation 4.131 contains a second-order polynomial in the denominator in the square of the frequency. Hence, there will be two roots of this polynomial corresponding to the squares of the natural frequencies that are equal to the eigenvalues of the characteristic equation of the undamped system.

It is noticed that the diagonals of the response matrix have a single root at which the responses will be zero. Because these responses correspond to the response to a force applied in the same coordinate direction, this means that the force does not produce any displacement and is, therefore, doing no work. In this particular model, and with the frequency corresponding to the zero in the numerator, the mass has effectively become infinitely rigid. This is possible because the transient response of the system has been ignored. In practice, there will be an initial displacement that transmits force to the second mass that will, in turn, oscillate producing forces that transmit back to the first mass. The amplitude of motion of the second mass will increase until it produces a force that matches that applied at the first mass. To see why the first mass ceases to move, consider the steady-state condition for which the first mass effectively becomes a rigid boundary. In this case, the second mass, m_2 in this example, is rigidly clamped between two springs of stiffness k_2 and k_3 that effectively add to produce a simple spring–mass system with infinite gain at the undamped natural frequency. Consequently, when the force applied at the first mass is at this frequency of $\sqrt{(k_2 + k_3)/m_2}$, the response at the second mass will always increase to be infinitely larger than that at the first mass. Effectively, the other mass will totally absorb any displacement of the first mass and it is often intentionally used to create an 'absorber'. The same effect and explanation can be applied to forces at the second mass being absorbed by the first mass at a frequency of $\sqrt{(k_1 + k_2)/m_1}$. This effect is also commonly used to create oscillators giving the convenient property that the mechanism is not moving at the location that the force is being applied, which often considerably simplifies the design.

The number of terms in the response equations means that for systems having two or more degrees of freedom, it is preferable to look at either root locus type plots or plots of the frequency response. Because the frequency response is a complex number and a function of frequency, three parameters must be known to determine the gain and phase from such a plot. Such a three-dimensional plot of the real and imaginary components of the frequency response for the same parameters used for the root locus plot of Figure 4.14 is shown in Figure 4.15a. It can be seen that the response plot is a string with distinct loops around each of the characteristic roots of the system. The magnitude, frequency and phase of the system are measured as the cylindrical coordinate values from the coordinate origin. Figure 4.15b and c show the real and imaginary parts of the frequency response. Of particular interest in the real and imaginary plots is the fact that the resonant peak is typically shown in the imaginary plot, while the undamped natural frequencies can be identified by zero crossings in the real component of the frequency response. Figure 4.15d is called a Nyquist or polar plot from which the magnitude and phase can be readily measured.

Another commonly viewed plot is the magnitude response for which all three responses are shown in Figure 4.16. It is noted that phase plots that can readily be computed from the frequency response functions can also be valuable, particularly for calculating the phase

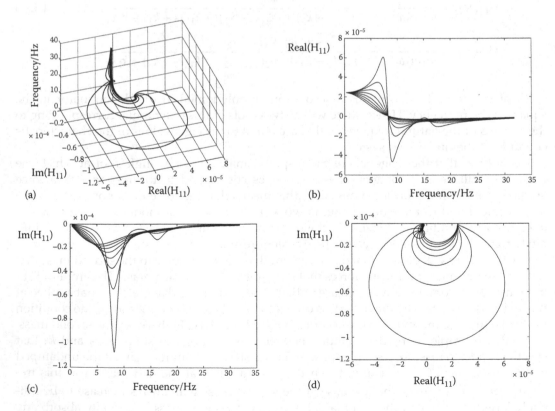

FIGURE 4.15
Different views of the frequency response functions corresponding to the parameters in Figure 4.14: (a) three-dimensional plot, (b) plot viewed in the real and frequency plane, (c) plot viewed in the imaginary and frequency plane, (d) plot viewed in the real and imaginary plane and called a Nyquist plot.

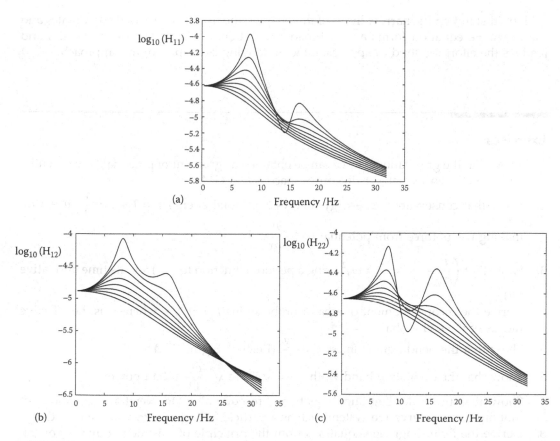

FIGURE 4.16
Magnitude responses plotted on a logarithmic scale as a function of frequency for the response functions: (a) H_{11} $(j\omega)$, (b) $H_{12}(j\omega) = H_{21}(j\omega)$, (c) $H_{22}(j\omega)$.

margins to assess stability of control systems. The poles and zeros are often the most easily identified features from this representation of the frequency response functions.

4.7 Summary

Newton's laws provide solutions to dynamic systems with unprecedented precision. However, the limits of application of these laws are often due to lack of precision in the manufacture and assembly of mechanical mechanisms that leads to complex behaviour such as hysteresis, friction and non-linearity. Much of the rest of the book is concerned with methods for optimising precision by eliminating these undesired features of machines and mechanisms.

All dynamic systems for small perturbations can be modelled as linear systems and understood in terms of frequency response functions and characteristic roots. Consequently, while dynamics is a large topic developed over many centuries, this current chapter has focused only on this narrow field of the theory.

Notwithstanding the narrow limits of this chapter, the use of generalised coordinates and the Lagrange equation forms a very broad foundation of all of classical mechanics and justifies the effort required to appreciate the underlying principles of this approach.

Exercises

1. Show that the gravitational force can be obtained as gradient of potential. The potential energy between two objects in empty space is $U = GMm/r$.

2. Show that conservation of energy follows from total energy $E = T + V = \frac{1}{2}mv^2 + V(x)$ making use of force from potential $F = -\frac{\partial V}{\partial x}$.

3. Show that $\left(\dfrac{v_{max}}{x_{max}}\right)^2 = \dfrac{k}{m} = \omega_n^2$ using a periodic function for $x(t)$ and its time derivative $v(t)$.

4. Derive the second moment of area of probe stylus $I = \pi r_c^4/4$, where r_c is the effective radius of the stylus shaft.

5. Show that the pendulum stiffness $k = \dfrac{mg}{l}$. (Refer to Figure 4.9.)

6. Show that the oscillation bandwidth $\Delta\omega = \sqrt{3}2\gamma \equiv \sqrt{3}\dfrac{\omega_n}{Q}$ at half power.

7. Show that the Euler-Lagrange equations are equivalent to Newton's equations of motion for a conservative system such as a particle falling in a gravitational field.

8. Derive the Euler-Lagrange equations from the principle of least action by way of calculus of variations making use of integration by parts.

9. For a rod undergoing longitudinal vibration derive the equations governing motion if the rod is cut into small sections of rigid mass connected by massless springs each being equal to the mass and stiffness of the elements (Figure 4.17). Show that, in the limit that the segments become infinitesimally small, these equations reduce to a wave equation.

10. Derive the equations of motion for a damped double pendulum as modelled in Figure 4.18, making use of the velocity diagram in Figure 4.19. Show that for small displacements, the equation governing motion becomes linear and can be represented in matrix form where all matrices are symmetric.

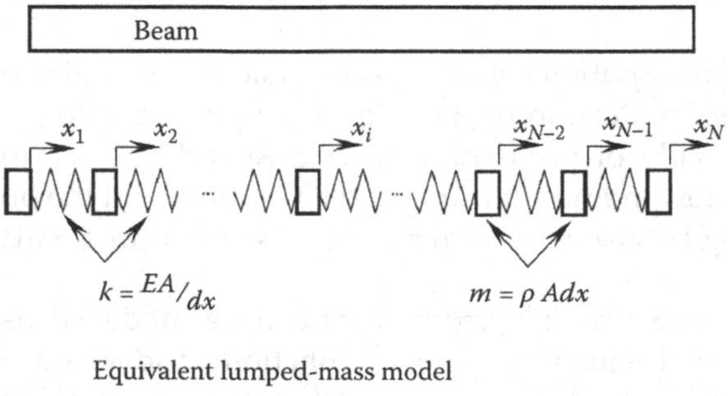

Equivalent lumped-mass model

FIGURE 4.17
Lumped mass model of a beam subject to axial motion.

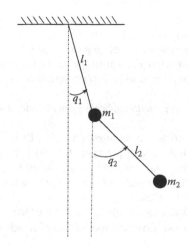

FIGURE 4.18
Parametric model for a double pendulum.

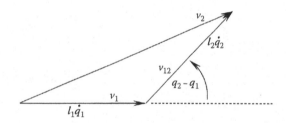

FIGURE 4.19
Velocity diagram for the lower mass of the double pendulum of Figure 4.18.

References

Baird, P. J. 1996. *Mathematical modelling of the parameters and errors of a contact probe system and its application to the computer simulation of coordinate measuring machines.* PhD thesis, Brunel University, London.

Baird, P. J., Bowler, J. R. and Stevens, G. C. 1999. Quantitative methods for non-contact electrostatic force microscopy. *Inst. Phys. Conf. Ser.* **163**: 381–386.

Baird, P. J. and Stevens, G. C. 2005. Nano- and meso-measurement methods in the study of dielectrics. *IEEE Trans. Dielec. Electr. Insul.* **12**: 979–992.

Binnig, G., Quate, C. F. and Gerber, C. 1986. Atomic force microscopy. *Phys. Rev. Lett.* **56**: 930–933.

Feynman, R. P., Leighton, R. B. and Sands, M. 1964 (revised and extended 2005). *Lectures on physics.* Addison-Wesley.

Gere, J. M. 2004. *Mechanics of materials* (6th ed.). Brooks/Cole.

Morin, D. 2008. *Introduction to classical mechanics.* Cambridge University Press.

Newland, D. R. 1989. *Mechanical vibration analysis and computation.* Longman Scientific and Technical publications.

Newton, I. 1726. *Philosophiae Naturalis Principia Mathematica* (3rd ed.). (A popular modern translation: Cohen, I. B., Whitman, A. and Budenz, J., 1999, *The principia*, University of California Press.)

Rayleigh, J. W. S. 1873. Some general theorems relating to vibrations. *Proc. Lond. Math. Soc.*, **4**: 357–368.

Renishaw. Document: H-1000-8006-01-B (Touch-trigger_probing_technology).ppt. Available from http://resources.renishaw.com/en/download/presentation-touch-trigger-probing-technology –70270.

Sarid, D. 1992. *Scanning force microscopy with applications to electric, magnetic and atomic forces.* Oxford University Press.

Serway, R. A. 2013. *Physics for scientists and engineers* (9th ed.). Brooks/Cole.

Shen, Y.-L. and Zhang, X. 1997. Modelling of pretravel for touch trigger probes on indexable probe heads on coordinate measuring machines. *Int. J. Adv. Manuf. Technol.* **13**: 206–213.

Stroud, K. A. and Booth, D. J. 2011. *Advanced engineering mathematics* (5th ed.). Industrial Press.

Susskind, L. and Hrabovsky, G. 2014. *Classical mechanics: The theoretical minimum.* Basic Books.

Thornton, S. T. and Marion, J. B. 2008. *Classical dynamics of particles and systems* (5th ed.). Brooks/Cole.

Yang, Q., Butler, C. and Baird, P. 1996. Error compensation of touch trigger probes. *Measurement* **18**: 47–57.

5

Dimensional Metrology

Massimiliano Ferrucci, Han Haitjema and Richard Leach

CONTENTS

ABSTRACT   Dimensional metrology is one of the cornerstones of mechanical engineering and is key to manufacturing. Dimensions have to be confirmed by measurement, and such measurements allow tolerances and uncertainties to be calculated. Following on from the metrology basics covered in Chapter 2, this chapter will present the multitude of instruments and techniques that are available for dimensional metrology, including the measurement of length, three-dimensional coordinates, and surface form and texture. Single and multi-dimensional measuring instruments will be discussed along with their capabilities and limitations. Reversal methods will also be covered.

5.1 Introduction

Dimensional metrology is the study of geometrical measurements, for example length, area, volume, flatness and roundness. Some examples of dimensional measurements made in the precision industry include the dimensions of an automotive component that are critical to later assembly and function, for example fuel injectors; or the size of a wafer support in a photolithography scanner. Confidence in the dimensions of these products is critical for their operation and, derived from this, for commerce, as it establishes an agreement between the vendor and the customer on the quantity being traded. In the absence of such confidence, the customer would have the arduous task of verifying the performance of a product before purchase—a time-consuming, expensive and unacceptable responsibility for the customer.

The introduction of interchangeable parts is an important milestone in achieving mass production. Products that used to be made by individual craftsmen in a linear manufacturing process were now made by several craftsmen, each of whom was responsible for making one part. The final product was then created by assembling the individual parts. To ensure that all parts would fit together, conformance was necessary in the way parts were made and dimensions were measured.

5.2 Length Standards

A measurement standard is a practical realisation of the definition of a measurement unit (see Chapter 2). The first internationally accepted standard of length was a platinum-iridium bar, named the International Prototype Metre (Quinn and Kovalevsky 2005, Quinn 2012).

The realisation of the metre was originally intended as the distance between two parallel ends of this bar, making the prototype an end-standard (see Section 5.2.1). In later versions of the metre bars, the length was given by the distance between two parallel lines engraved on their surfaces, making the metre bar a line standard (see Section 5.2.2).

Concerns over the long-term stability of a material standard (i.e. a physical object) led the international community to search for a stable and reproducible definition of the metre. In 1960, the metre was redefined in terms of the wavelength of light emitted in a vacuum by excitation of krypton-86 gas (CIPM 1960), which was shown to occupy a narrow and reproducible bandwidth of the electromagnetic spectrum so that the metre was essentially defined by a fundamental physical quantity. Following on from the invention of the laser, a further redefinition of the metre occurred in 1983, after which it was and still is, defined as the length of the path travelled by light in a vacuum in 1/299 792 458 seconds (CIPM 1984). As a result of this new definition, the speed of light is fixed at 299 792 458 $m \cdot s^{-1}$ and the metre is dependent on the definition and realisation of the unit of time, the second. The 1983 definition allows for the realisation to be achieved by the wavelength of any stable source of light, which removes the limitation of the previous definition to light emitted by krypton-86.

Length standards serve as references for disseminating the definition of the metre, that is transferring traceability, during a calibration. Standards can be based on fundamental physical principles on natural constants, such as the specified wavelength of light from a known source, or material standards (physical objects) that contain feature(s) with calibrated lengths. Fundamental principles are used by national metrology institutes (NMIs) to realise the definition of the metre in the laboratory and to then calibrate customer material standards (see Chapter 2 for more information on the types of standards and NMIs). The uncertainty in realising the metre with fundamental principles is much lower than the uncertainty associated with the calibrated value of material standards.

The traceability chain (see Chapter 2) for a length-measuring instrument, such as a micrometer, is illustrated in Figure 5.1. At the top of the traceability chain is the realisation of the metre by NMIs. With each calibration, the uncertainty of a reference standard increases.

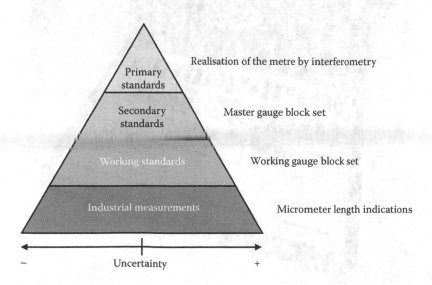

FIGURE 5.1
A typical traceability chain for measurements made with a micrometer. Traceability is transferred between each standard by way of calibration, which includes evaluation of uncertainty in the realised standard length.

At the bottom of the traceability chain are the length indications, that is the scale, of the measuring instrument used to perform measurements. It is important to note that traceability of the measured quantity can only be achieved if the uncertainty in its realisation is assessed and reported in compliance with the relevant standards (see Chapter 9).

5.2.1 End Standards

End standards are material standards, in which the calibrated feature is the distance between two parallel planes (Figure 5.2). The most common and well-known example of an end standard is the gauge block (Figure 5.3), which was a consequence of technological advancements during the industrial revolution (see Leach 2014a for more information on end standards, including a little history).

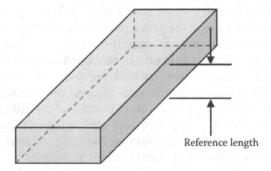

Reference length

FIGURE 5.2
The reference length in end standards is given by the distance between two parallel planes.

FIGURE 5.3
A set of 47 metric gauge blocks.

Gauge blocks are practical for industry due to their dimensional stability, ease of use and low cost. Gauge blocks are typically sold in sets of various sizes but can be obtained individually; Figure 5.3 shows a typical example set. The most common type of gauge block is rectangular, while square or 'Hoke' gauge blocks are still used in the United States (see Figure 5.4). The reference length is defined as the distance between two lapped surfaces—known as 'measuring faces'—on opposite ends of the gauge block (ISO 3650 1998). The probing points for which the reference length is specified are typically centred on each measuring face to avoid the discrepancies due to flatness deviations.

A useful property of gauge blocks is their ability to be wrung together. Wringing is a phenomenon by which two lapped surfaces adhere to each other when they are slid together under some applied force and in the presence of a small amount of fluid (see Leach et al. 1999 for more information on wringing). The attractive force between two or more well-wrung gauge blocks can reach 300 N, depending on the material (Doiron and Beers 2005). When multiple gauge blocks are wrung together, the resulting 'stack' can be used as a reference length equal to the cumulative lengths of each individual block. Although, by definition wringing should not add any length, the presence of a wringing film means that the actual length of the gauge block stack is not exactly the same as the cumulative lengths; the thickness of the film depends on the quality of wringing but is typically somewhere in the range of 5 nm to 20 nm. The contribution to the actual length by the wringing film must be taken into account when using a gauge block stack for calibration.

Gauge blocks are available in both metric and imperial (inch) denominations. Common sets of metric gauge blocks contain between 81 and 112 gauge blocks ranging in length from 0.5 mm to 100 mm. In Table 5.1, gauge block denominations in a 87-block metric set are shown.

Gauge blocks can be used as secondary standards and working standards. As secondary standards, gauge block lengths are typically calibrated by interferometry. These 'master' gauge blocks can then be used to calibrate the lengths of 'working' gauge blocks by comparison (see Section 5.3). Working gauge blocks are commonly used to calibrate length scales on measuring tools such as micrometers, calipers and dial indicators. Depending on the application, various gauge block grades are available. ISO specification standard 3650

FIGURE 5.4
Rectangular and Hoke gauge blocks. (Courtesy of NIST Digital Archives.)

TABLE 5.1

Gauge Block Denominations in Typical 87-Block Metric Set

Number of Blocks	Range/mm	Step/mm
9	1.001–1.009	0.001
49	1.01–1.49	0.01
19	0.5–9.5	0.5
10	10–100	10

TABLE 5.2

Limit Deviation of Length at Any Point from Nominal Length and Tolerance for the Variation in Length t_v for Metric Gauge Blocks of Nominal Lengths

Nominal Length, l_n/mm	Calibration Grade K		Grade 0		Grade 1		Grade 2	
	± t_e mm/ μm	t_v mm/ μm	± t_e mm/ μm	t_v mm/ μm	± t_e mm/ μm	t_v mm/ μm	± t_e mm/ μm	t_v mm/ μm
$0.5 \leq l_n \leq 10$	0.2	0.05	0.12	0.1	0.2	0.16	0.45	0.3
$10 < l_n \leq 25$	0.3	0.05	0.14	0.1	0.3	0.16	0.6	0.3
$25 < l_n \leq 50$	0.4	0.06	0.2	0.1	0.4	0.18	0.8	0.3
$50 < l_n \leq 75$	0.5	0.06	0.25	0.12	0.5	0.18	1	0.35
$75 < l_n \leq 100$	0.6	0.07	0.3	0.12	0.6	0.2	1.2	0.35

Source: Adapted from ISO 3650, 1998, Geometrical product specifications (GPS)—Length standards—Gauge block, International Organization on Standardization.

(1998) defines the maximum ('limit') deviation of length from nominal and the tolerance for variation in length for four gauge block grades: K, 0, 1 and 2. The values of these limits for gauge blocks up to a nominal length of 100 mm are shown in Table 5.2.

Other end standards include cylindrical gauges (pin/wire), ring gauges and ball gauges. The main difference between these standards and gauge blocks is that the measuring faces are not limited to two discrete planes. Instead, reference lengths are specified for any (180°) opposing surfaces and can be on the outside or the inside. Form deviations in the measuring faces should be considered when choosing the probing points as these can introduce variance in the measured results. More detailed information on gauge blocks can be found in Doiron and Beers (2005) and in Leach (2014a).

5.2.2 Line Standards

Line standards provide the reference length as the distance between two parallel lines on the surface of the object. Line standards are often used for calibrating the scales of optical (vision) systems, since the position of the lines can be visually discerned (Zhang and Fu 2000, Coveney 2014). The advantage of line standards is the ability to include multiple reference lengths onto the same gauge, eliminating the need for multiple/discrete gauges. These types of line standards are known as line scales (Figure 5.5) and can have a length of 1000 mm or longer. Longer line scales may be manufactured on thermally stable materials, such as Zerodur® (Schott AG n.d.), to mitigate the effects of thermal expansion due to temperature fluctuation during measurement. When used as displacement scales, line

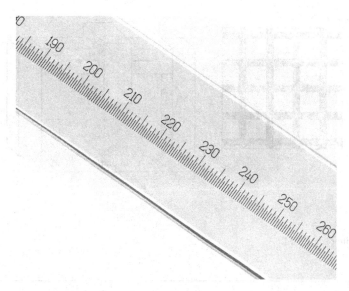

FIGURE 5.5
Line scale. (Courtesy of NPL, Crown Copyright.)

standards may also be designed to match the thermal expansion of the machines on which they will be attached. Typical resolutions in line scale indications are on the order of several hundreds of nanometres, with high-precision scales achieving resolutions of a few nanometres (de Groot et al. 2016).

The areal, or two-dimensional, equivalent of a line scale is the grid plate (Figure 5.6), which provides length indications along two dimensions. Crosshairs are used instead of parallel lines to denote reference points on the grid plate; reference lengths are defined between crosshair intersections. Grid plates are popular reference standards for calibrating the lateral axes of vision-based coordinate measuring systems (Swyt 2001; also see Section 5.6.6), as shown in Figure 5.7. Typical resolutions of two-dimensional grid plate indications are 0.5 µm (Doiron 1988).

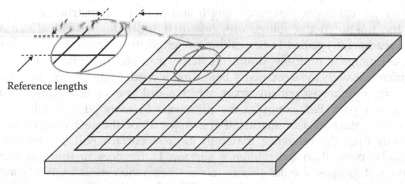

Reference lengths

FIGURE 5.6
Schema of a grid plate.

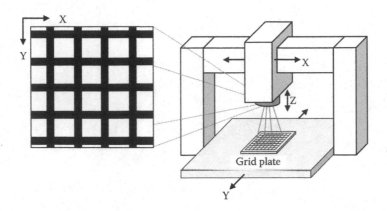

FIGURE 5.7
Schema of a grid plate on a CMM.

5.3 Length Comparators

5.3.1 Comparator Principle

The principle of comparing the measurement of a calibrated reference to the measurement of an unknown test part is known as the 'comparator principle'. Usually, the calibrated reference value is chosen to be as close to the value of the part being measured, therefore, reducing the effects of any systematic error (bias) in the measurement (this process is often referred to as 'nulling'). Repeatability of measurement can be determined from repeat measurements of the reference object. The test object is then measured and differences between the two measurements are applied to the calibrated values to determine the test quantities. Similarly to the comparison of standard gauges, the differences between the reference and test objects must be considered when assigning a measurement value.

5.3.2 Length Comparators

Length comparators are measuring instruments that compare the measurement of a calibrated reference gauge to the measurement of an unknown test gauge. A fundamental aspect of comparator measurements is that the major scale used to measure the test gauge is provided by the calibrated reference length and not by the instrument. The reference gauge and the unknown test gauge are measured individually. The difference in the instrument's readout is applied to the calibrated length to evaluate the length of the test gauge. The test gauge is typically similar in shape, size and material composition as the calibrated reference with measurements being made under similar conditions. Any dissimilarity can result in measurement errors, which should be considered in calculating the length of the test part. For example, the difference in material will affect mechanical deformation in contact measurements. If the material of the test object has a smaller elastic modulus than the reference material, the test surface will be deformed by the contacting probe more than the reference surface. Differences in thermal expansion of reference and test gauges should also be considered when measurements are not performed at standard reference temperature (20 °C). Materials with different coefficients of thermal expansion (CTEs) will contract and expand differently. Given the length l of a

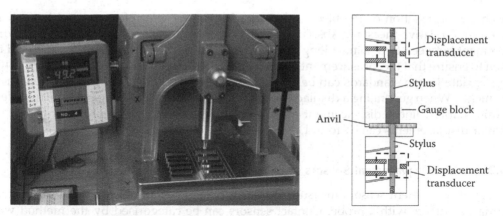

FIGURE 5.8
Mechanical gauge block comparator. (Courtesy of NIST.)

gauge block at reference temperature T_{ref}, the change in length Δl at a new temperature T can be calculated as

$$\Delta l = \alpha \left(T - T_{ref} \right) l, \tag{5.1}$$

where the units for CTE (α) are per kelvin.

The length measurement of the test object can be considered traceable if the calibrated value of the reference object is taken into account; the response of the instrument is shown to be linear and all sources of uncertainty, including the uncertainty in the reference standard, are combined in the calculation of the final length uncertainty.

Mechanical gauge block comparators (see Figure 5.8) consist of two opposing anvils with contact styli (Doiron and Beers 2005). During measurement, the gauge block is placed so that its measuring faces are contacted by the anvils. The measurement procedure typically involves making multiple measurements of both reference and test blocks. Diameter standards can also be measured by comparators. For example, at the National Institute of Standards and Technology (NIST; United States), a wire micrometer is used to perform comparator measurements of pin gauges and cylindrical standards (Doiron and Stoup 1997). The measurement design of the wire micrometer is similar to that used for the gauge block comparator, except that the measurement axis is horizontal. The contact points consist of a fixed cylindrical anvil and a moveable flat anvil. Displacement of the moveable anvil is tracked interferometrically. The calibrated diameter of a reference diameter gauge is used as the scale for assigning a diameter measurement to the unknown test gauge. Differences in deformation due to gauge material should also be considered for measurement of diameter standards.

5.4 Displacement Measurement

5.4.1 Introduction to Displacement Measurement

The length measurements discussed so far provide an estimate for the absolute distance between two fixed points, as embodied in physical length standards. Displacement is the change in position of an object or feature with respect to a reference position and is, therefore, a relative measure. Absolute lengths can be determined by measuring the displacement between two points along the desired length segment; however, care should be taken to ensure that the measurement axis is parallel to the length segment (see Chapter 10). Appropriate length standards can be applied to check the reliability of displacement measurements. When generating a displacement, the change in length or angle can be recorded. In this chapter, linear displacement measurement is covered, and more information on angular displacements can be found elsewhere (Morris 2001).

5.4.2 Contact Displacement Sensors

Contact displacement sensors measure the linear position of an object by physically contacting its surface with a probe. Contact sensors can be categorised by the method with which the position of the probe tip is measured, namely mechanical and electromagnetic.

5.4.2.1 Indicators

Indicators are some of the earliest displacement measuring devices and are mechanical systems. The contact probe is located at the end of a spring-loaded plunger that can be retracted or extended, and the movement of the plunger is tracked by way of a rack-and-pinion mechanism (Figure 5.9). The reading display on the indicators can be an analogue dial, in which case a needle indicates the measurement position along a circular array of graduations. More recently, digital displays (Figure 5.10) have replaced the more traditional dials. Dial test indicators (Figure 5.11) are mechanical indicators with a different geometrical construction. The appropriateness of a particular type of indicator depends on the measurement task and space limitations in mounting the indicator. Sub-micrometre displacements can be resolved with some of the more precise indicators.

5.4.2.2 Linear Variable Differential Transformers

Linear variable differential transformers (LVDTs) measure the position of a contact plunger by transforming the mechanical displacement of its ferromagnetic core into an electrical signal. Inside the housing of the LVDT, three solenoidal coils are coaxial to the

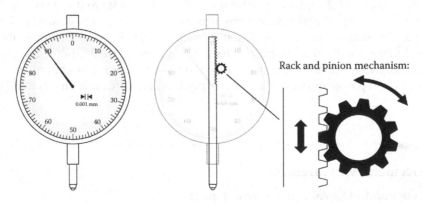

FIGURE 5.9
(Left) Mechanical indicator with analogue (dial) display. (Right) A rack and pinion mechanism tracks the displacement of the plunger.

FIGURE 5.10
Indicator with digital readout. (Courtesy of The L.S. Starrett Company Ltd.)

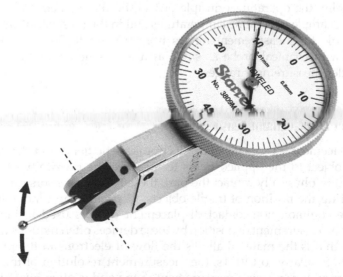

FIGURE 5.11
Dial test indicator. (Courtesy of The L.S. Starrett Company Ltd.)

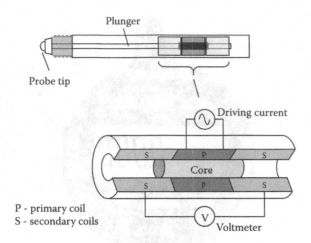

FIGURE 5.12
Linear variable differential transducer (LVDT). (Courtesy TE Connectivity.)

ferromagnetic core (Figure 5.12). The central coil is known as the primary coil, while the external coils are known as secondary coils. An alternating current is driven through the primary coil, which induces a voltage in each secondary coil. The secondary coils are connected in series to the same circuit, generating a voltage output equal to the difference between the two voltages (known as a voltage opposition circuit). The value of the induced voltage in each secondary coil is proportional to the length of the core within the respective secondary coil. An output voltage of zero occurs when the core is centred and occupies equal lengths in the secondary coils. The magnitude of the differential output provides the magnitude of the displacement, while the phase of the output signal provides direction. Given that there is no friction between the core and the solenoidal coils, LVDTs are known for their relatively long life cycle and the core does not require electrical connection. Additionally, the operating principle of LVDTs allows them to be used in harsh environments, for example in extreme temperatures and in the presence of strong vibrations. LVDTs are available for measurement ranges from 0.5 μm to 500 mm (Fleming 2013). Displacements of a mechanical probe as small as a few nanometres can be resolved for LVDTs with smaller measurement ranges.

5.4.3 Non-Contact Displacement Sensors

Non-contact displacement sensors generate electromagnetic fields and detect the response of the measured object to the applied fields to measure the proximity of its surface. By removing the need to physically contact the part, these types of sensors can be particularly useful for measuring the position of fragile objects. Capacitive and inductive sensors are some of the more common non-contact displacement devices used in industry and are discussed here. The measurement of position by these devices often requires the test material to be conductive; that is the material allows the flow of electrons in the presence of electromagnetic fields. Similarly to LVDTs, the measurement resolution of these instruments generally increases with decreasing measurement range and is often quoted in the specifications as a percentage of the maximum measurable distance.

5.4.3.1 Capacitive Sensors

Capacitive sensors operate on the principle of mutual capacitance between two conductive surfaces. Mutual capacitance is the transfer of electric charge between the surfaces in the presence of a potential difference. The concept of mutual capacitance is illustrated in Figure 5.13a. In the simple case of two parallel conductive plates of surface area A separated by a gap of distance d, the capacitance C is given by Equation 5.2:

$$C = \frac{\varepsilon_0 \varepsilon_r A}{d},$$

(5.2)

where ε_0 is the permittivity of free space (vacuum) and ε_r is the permittivity of the material in the gap between the two conductive plates. If ε_0, ε_r and A are kept constant, the mutual capacitance of the parallel plates is inversely proportional to the distance of the gap between them. The probing surface of the capacitive sensor represents one of the two conductive surfaces (Figure 5.13b) of a parallel plate capacitor, while the target object is the other. An alternating current is driven through the device, thereby generating a potential difference between the two surfaces. This potential difference is measured by the sensor and translated to a distance measurement (Wilson 2005).

The specific material is of no consequence to position measurements. Capacitive displacement measurements are, however, sensitive to the gap material and are typically most reliable when only air is present between sensor and target. The measurement range for capacitive sensors is from 10 µm up to 10 mm. Capacitive displacement sensors are known for high sampling frequencies (20 kHz or greater) and nanometre resolution (Wilson 2005). Sub-nanometre resolution can be achieved for instruments with smaller measurement range (Fleming 2013).

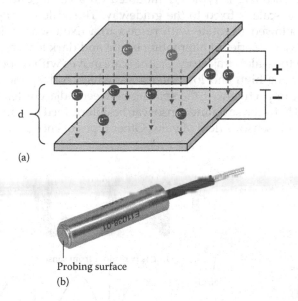

(a)

Probing surface

(b)

FIGURE 5.13
(a) In the presence of a voltage difference, two conductive surfaces in proximity to one another will exhibit a transfer of charge. Mutual capacitance is a measure of the amount of charge transferred between the two surfaces in proximity. (b) Capacitive sensors operate on the principle of capacitance between two conductive surfaces. The probing surface of the sensor acts as one of the two conductive surfaces. (Courtesy of Lion Precision.)

5.4.3.2 Inductive (Eddy-Current) Sensors

Inductive sensors operate on the principle of inductance, that is the capacity of electrical currents to be induced in the presence of a changing magnetic field. Inductive displacement measuring devices generate alternating magnetic fields by driving an alternating current through a coil located at the end of the probe. The magnetic fields penetrate the measured object and within it induce small looping electric currents, known as eddy currents. These currents produce magnetic fields that oppose the original magnetic field and effect a change in the impedance of the probe coil. By measuring this change in the coil's inductance, the distance of the object from the probe can be determined. The response of the target to the original magnetic field varies with its material. Therefore, measurements from inductive sensors should be specifically calibrated for the material of the measured object.

Correct operation of inductive sensors demands that the target object satisfy certain size requirements. The surface area of the target object should be at least three times the surface area of the sensor's probe to provide the necessary interaction with the original magnetic field. The thickness of the target object should also be large enough to ensure the generation of eddy currents by the penetrating magnetic field. Inductive measurements are not sensitive to the presence of contaminants or liquids in the gap area between sensor and target, thereby making inductive sensors ideal for measuring in harsh environments. Measurement resolutions on the order of several nanometres are possible. Inductive sensors can measure distances from several hundred micrometres up to 80 mm.

5.4.4 Optical Encoders

Optical encoders are used to measure linear or angular position by employing optical sensors, such as photodetectors, that read their position along a dedicated scale (Figure 5.14). The sensor on linear encoders is typically mounted on a carriage that can be translated along a guideway; the scale is fixed to the guideway. The scale for angular encoders is a circular disk that is allowed to rotate with respect to a fixed sensor. Position or angular information is codified as a series of alternating bright and dark features, for example lines, equally spaced along the scale's trajectory. As the scale moves with respect to the sensor, the sensor detects alternating high and low intensities, which correspond to bright and dark features on the scale, respectively. If the spacing between adjacent features is known, the relative displacement between scale and sensor can be determined by counting the number of features that traverse the sensor's field of view. Other types of encoders employ magnetic or

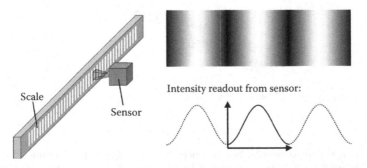

Intensity readout from sensor:

FIGURE 5.14
Linear optical encoder.

inductive scales coupled with appropriate sensors; more information on these and other types of encoders can be found elsewhere (Webster and Eren 2014).

The spacing between features can be very small, for example in the order of a few micrometres for linear encoder scales. By discriminating the phase between bright and dark lines, the encoder can interpolate between lines and achieve higher resolutions in its displacement. The light source in optical encoders can be located on the same side of the reference scale as the sensor, in which case the intensity peaks correspond to reflected light. When the light source is on the opposite side of the scale from the sensor, the intensity peaks correspond to transmitted light.

Encoders are typically incorporated into larger instruments to track the position of movable axes, for example the kinematic axes of coordinate measuring systems (see Section 5.6), and motion tables for precision manufacturing. Accuracies of linear optical encoders are length dependent and are typically 5 µm per metre (Fleming 2013). For displacements below 270 mm, the accuracy can be as high as 0.5 µm per metre. The measurement range of linear optical encoders can be up to several metres. Accuracy for angular optical encoders can be as low as 0.5 arc seconds (Wilson 2005).

5.4.5 Interferometry

Interferometry is a measuring technique that exploits the interfering nature of super-imposed waves. One of the simplest types of optical interferometers for displacement measurement is the Michelson-type interferometer. The construction of a Michelson-type interferometer with retroreflectors instead of flat mirrors is shown in Figure 5.15. Retro-reflectors consist of three plane mirrors arranged as the three sides of a cube's corner. Irrespective of the incidence angle, the return beam is always parallel to the incident beam. For all points of reflection except the corner, that is the intersection point of the three mirrors, the return beam is shifted in space from the incident beam.

A light source emits an ideally narrow spectrum of light and directs it as a spatially coherent beam towards a beam splitter. Here the signal is split into two separate beams

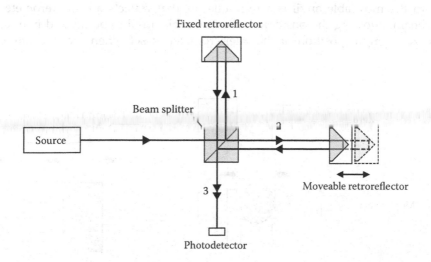

FIGURE 5.15
Interferometer design based on the Michelson type. (1) Reference beam reflected by a fixed retroreflector. (2) Test beam reflected by a movable retroreflector. (3) Superimposed reference and test beams directed to a photodetector.

travelling in different directions. Beam 1 is directed to a fixed retroreflector and is known as the 'reference' beam. Beam 2 is directed towards a retroreflector that can be moved longitudinally with respect to the incident light wave, henceforth the 'measurement' beam. Both beams are reflected back to the beam splitter where they are combined into beam 3 and directed towards a fixed photodetector. The signal contained within beam 3 is known as the interference signal and is characterised by the superposition of beams 1 and 2. Assuming a perfectly aligned interferometer (and very small beam cross-sections), the intensity of beam 3 is uniform across its cross-sectional profile. The intensity of beam 3 as measured by the photodetector is a function of the difference in path length between beams 1 and 2.

The intensity of the interference signal is a function of the phase difference between the superimposed beams. A phase difference occurs as a result of a difference in the length of the paths travelled by the two beams. When the phase difference between the beams is 0°, interference is constructive and the intensity of the interference signal is at a maximum. If the beams are 180° out-of-phase, interference is destructive and the intensity of the interference signal is at a minimum. The phase difference, hence the intensity of the interference signal, repeats with displacements of the retroreflector equal to one half wavelength of light. The sequence of constructive and destructive interference signals is known as a fringe pattern, given that constructive interference results in bright fringes, while destructive interference results in dark fringes. In the case of a monochromatic light source, displacement measurements for lengths larger than one wavelength of light can only be made by counting how many fringes pass the photodetector as the reference mirror is moved. Sub-nanometre resolutions can be achieved by interpolating the phase within a fringe period. Interferometers can be used to measure displacements of several metres. In some cases, displacements up to tens of metres can be determined with light signals of high spatial coherence.

Helium-neon gas (He-Ne) lasers are commonly used as light sources for interferometry due to the ability to achieve relatively high spatial coherence in the beam and due to the stability of the generated wavelength (temporal coherence). An example of displacement interferometry being applied to perform length measurements is the NIST wire micrometre (Figure 5.16), which is used for calibrating diameter standards, for example pin gauges and wire gauges. The instrument consists of a fixed semi-cylindrical anvil and a moveable anvil. Attached to the moveable anvil is a retroreflector that reflects an interferometer's measurement beam, allowing the position of the moveable anvil to be tracked interferometrically. The zero length position of the interferometer is set when the two anvils are in

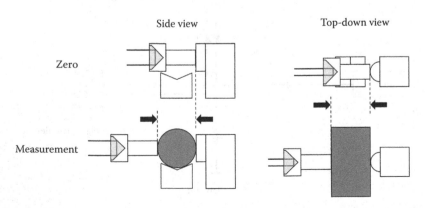

FIGURE 5.16
The NIST wire micrometre is used to calibrate diameter standards. The position of a probing anvil is tracked interferometrically.

contact. The number of interference fringes that traverse the photodetector's field of view are counted as the cylinder is placed between the two anvils. More information about displacement interferometers, including practical guidelines for setting up and aligning them, can be found elsewhere (Ellis 2014).

5.5 Form Measurement

The geometrical form of a part can be critical for its effective function. Form is typically described as the deviation of the measured surface or path from the fundamental definition of the corresponding geometrical primitive, also known as the reference geometry feature. The following form parameters are common to all geometries presented here.

Peak-to-reference deviation—Maximum deviation of measured points in the positive direction with respect to the reference geometry

Reference-to-valley deviation—Maximum deviation of measured points in the negative direction with respect to the reference geometry

Peak-to-valley deviation—A measure of the maximum variation in measured points, distance between peak-to-reference deviation and reference-to-valley deviation

Root-mean-square deviation—A measure of the distribution of measured points with respect to the reference geometry

Form deviations are often quoted only as positive values since form parameters are defined for a specific direction from the reference geometry. In some cases, reference-to-valley deviations are quoted as negative values, in which case its absolute value is used in determining peak-to-valley deviation. In practice, parallelism is the only feature for which the measurement is a comparison to a physical reference instead of a fundamental geometry.

5.5.1 Straightness Measurement

The reference geometry for straightness is a line, defined as the shortest path between two points in space. Straightness is typically used in the context of an instrument's measurement axis. For example, perfect straightness of a micrometer spindle means that the contact point does not exhibit lateral displacements as it traverses its path. The same applies to kinematic guideways, for example the Cartesian axes on a coordinate measuring system (see section 5.6). Two lateral displacement planes—straightness profiles—are defined along the measurement axis (Figure 5.17). Straightness deviations can be evaluated for each straightness profile and are defined normal to a defined reference line, for example according to the minimum zone or least-squares principle. Within each straightness profile, the straightness deviation parameters given in Table 5.3 can be evaluated. More information on straightness deviation can be found in ISO 12780-1 (2011).

Straightness of machine axes is most commonly measured optically. Early methods employed an optical telescope and a visual target mounted on the moving carriage. At the starting position of the carriage, the optical telescope is aligned to the visual target, that is the centre of the target is coincident with the centre of the telescope's field of view. Lateral movements of the carriage result in decentralisation of the target centre from the field of

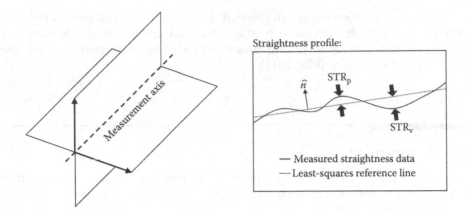

FIGURE 5.17
Straightness profile and straightness deviation parameters with respect to least-squares reference line.

TABLE 5.3

Straightness Deviation Parameters for Each Straightness Profile

Parameter	Description
STR_p	Peak-to-reference straightness deviation
STR_v	Reference-to-valley straightness deviation
STR_t	Peak-to-valley straightness deviation
STR_q	Root-mean-square straightness deviation

view. Any angular discrepancies between the optical path of the telescope and the axis of motion can be determined by fitting a line to the start and end positions and subtracting the line from the measured lateral movements. Modern techniques for straightness measurement typically use interferometry.

5.5.2 Roundness and Cylindricity Measurement

Roundness and cylindricity are conditions for surfaces of revolution (ASME Y14.5 2009). The fundamental geometries for roundness and cylindricity are the circle and the cylinder, respectively. A circle is defined by the set of points that are equidistant to a common centre, whereas a cylinder is defined by the set of points that are equidistant to a common axis. It follows that roundness is the degree to which measured surface points deviate from equidistance to a common centre point, while cylindricity is the degree to which measured surface points deviate from equidistance to a common axis. Both measurements are defined for bodies of revolution, hence the results are presented as a function of rotation angle (Figure 5.18). Roundness measurements can be applied to objects with circular cross-sections, for example spheres, cones and cylinders, whereas cylindricity measurements are strictly for cylindrical objects. It is important to note that roundness and cylindricity are not measures of the object's size, but instead are measures of variations in the object's surface from the ideal form.

Roundness parameters are defined with respect to a reference circle, for example a circle that is least-squares fit to the measured data (see Muralikrishnan and Raja 2010).

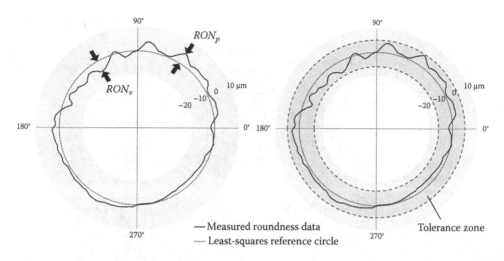

FIGURE 5.18
A roundness chart. (Courtesy of Taylor-Hobson Ltd.)

Peak-to-reference roundness deviation (*RONp*) corresponds to the maximum deviation of the measured surface outside the reference circle, while reference-to-valley roundness deviation (*RONv*) corresponds to the maximum deviation inside the reference circle (Figure 5.18, left). A tolerance zone can be defined by two concentric circles whose radii are specified with respect to the reference circle (Figure 5.18, right). A complete list of roundness parameters can be found in the ISO 12181-1 (2011).

Parameters for cylindricity can be considered three-dimensional extensions of roundness parameters (Figure 5.19). Other parameters such as taper angle provide a measure for the vertical profile of the cylinder. A complete list of cylindricity parameters can be found in the ISO 12180-1 (2011).

A roundness tester is a dedicated instrument for measuring roundness and cylindricity (Figure 5.20). A precision rotary stage is used to rotate the object and a stylus probe provides displacement measurement of the object's surface. The probe position can be adjusted horizontally and vertically to allow measurement of various object sizes. Cylindricity of an object can be measured by probing the surface at various vertical positions of the probe during a measurement run. For this purpose, the instrument's vertical column is ideally straight and parallel to the rotary stage axis.

5.5.3 Flatness and Parallelism Measurement

Flat surfaces are desirable for mating parts, for example wringing of a gauge block to a platen or to another gauge block. The reference geometry for flatness is the plane, hence flatness is the degree to which the measured surface deviates from a plane. Commonly used parameters for flatness are as follows. Peak-to-reference flatness deviation (*FLTp*) and reference-to-valley flatness deviation (*FLTv*), both of which are often stated with respect to a plane that is least-squares fit from the measured data. Peak-to-valley flatness deviation (*FLTt*) is given by the addition of *FLTp* and *FLTv*. Flatness deviation measurement from surface data is illustrated in Figure 5.21. The tolerance zone is defined by the distance of two planes parallel to the reference plane. *FLTq* is the root-mean-square flatness deviation

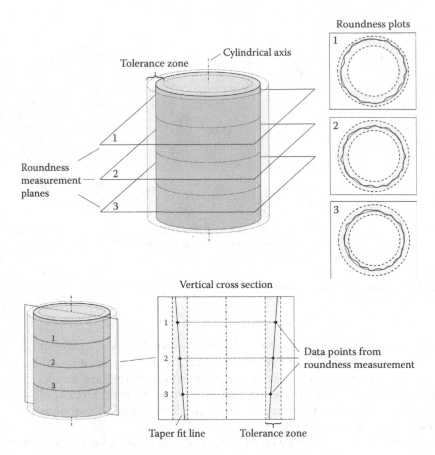

FIGURE 5.19
(Top) A common procedure for testing cylindricity is to perform roundness measurements at various heights along the cylindrical axis. (Bottom) Taper of the cylinder surface with respect to its axis can be determined from the collection of best-fit reference circles.

and provides a measure of flatness over the entire measured surface. ISO 12781-1 (2011) provides more details on flatness specifications.

One method to measure flatness is by probing the surface with a coordinate measuring system (CMS; see Section 5.6). However, measurement by CMS is a time-consuming task as surface points are acquired sequentially. An alternative to individually probing the test surface is a method based on principles similar to those in interferometry. An optical flat—a high grade glass disk with extremely parallel circular surfaces (often wedged to avoid internal interference effects)—is placed on top of the test surface and illuminated with monochromatic light (Figure 5.22). Differences in the path of light between the bottom surface of the optical flat and the test surface result in interference fringes reflected back through the flat. Uneven spacing and curvature in the interference pattern are indications of flatness deviations in the test surface. Parallel fringes denote a perfectly flat surface. The surface of the test object must be reflective for its flatness to be accurately measured by the optical flat method.

Parallelism is the degree to which two surfaces are parallel to each other. Unlike flatness, parallelism is evaluated with respect to a datum—a physical reference plane. In testing parallelism, a tolerance zone is created by establishing two planes parallel to the datum plane and a specified distance from each other (Figure 5.23). If the measured surface points

FIGURE 5.20
A roundness tester. (Courtesy of Taylor-Hobson Ltd.)

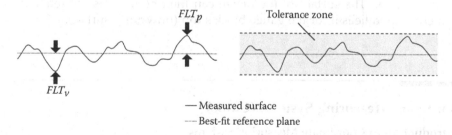

— Measured surface
---- Best-fit reference plane

FIGURE 5.21
Flatness deviation and tolerance zone.

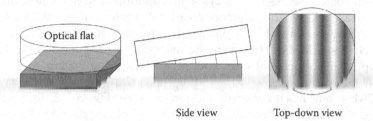

Side view   Top-down view

FIGURE 5.22
Optical flat method for measuring flatness.

are within this tolerance zone, then parallelism is satisfied. Parallelism between a gauge block's two gauging surfaces is a critical requirement for its correct function. One method to measure the parallelism of a gauge block's two surfaces is interferometrically. The gauge block is first wrung to a platen. If the wringing is performed correctly, that is the platen and gauge block are cleaned from dirt and the wringing force is strong, then the two wrung

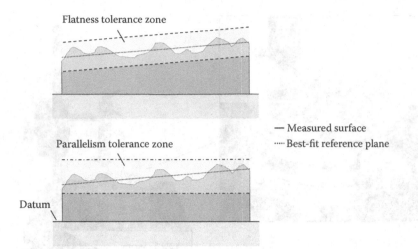

FIGURE 5.23
Parallelism is specified with respect to a datum. A tolerance zone is defined by two planes parallel to the datum plane a given distance apart. The testing of flatness and parallelism is illustrated here for the same measured surface.

surfaces are parallel. The surface of the platen can then be used as the reference for the measurement of parallelism for the gauge block's top (unwrung) surface.

5.6 Coordinate Measuring Systems (CMSs)

5.6.1 Introduction to Coordinate Measuring Systems

A coordinate measuring system (CMS) is an instrument for measuring the physical geometry of an object. Typically a CMS has the ability to move a probing system that detects the location of the surface on an object (see Section 5.6.2) and the capability to determine spatial coordinate values on the surface relative to a reference coordinate system. The first CMSs were contact types, similar to that shown in Figure 5.24, and were previously referred to as coordinate measuring machines (CMMs). However, the relevant ISO technical committee (ISO 213 Working Group 10) introduced the term 'coordinate measuring system' to cover all types of instruments for measuring geometry, including those using optical and x-ray techniques. Contact-type CMSs come in a number of configurations (see Figure 5.25 for a non-exhaustive list) and a range of sizes, from those able to measure something the size of a large aerospace component to the micro-scale versions described in Section 5.6.9. However, the majority of industrial CMSs have working volumes of cubes of sides of approximately 0.5 m to 2 m. By far the most common configuration of CMS is the moving-bridge type (Figure 5.25b). Conventional contact-type CMSs generally incorporate three linear axes and use Cartesian coordinates, but CMSs are available with four (and more) axes, where the fourth axis is generally a rotary axis (referred to as the C axis). CMSs are often housed in temperature-controlled rooms held close to 20 °C. The first CMSs became available in the late 1950s and early 1960s (see Hocken and Periera 2011 for a thorough description of CMSs and some history, and Flack and Hannaford 2005 for an overview of CMS use).

FIGURE 5.24
Bridge-type CMM. (Courtesy of Hexagon, Global EVO.)

These days there are many different types of CMSs (see Section 5.6.6). However, to illustrate the basic precision engineering principles, only the contact-type CMS will be considered in detail in the following sections.

5.6.2 CMS Probing Systems

CMSs measure either by discrete probing, where data from single points on the surface are collected, or by scanning, where data are collected continuously as the mechanical stylus tip is moved across the surface. The stylus tip in contact with the surface is usually a synthetic ruby ball, although other geometries and materials are possible, for example, cylindrical or ceramic stylus tips.

The majority of CMSs use a touch trigger probe (see Hocken and Pereira 2011 and Flack 2001 for further details about all mechanical probe types, and Chapter 4 for a detailed description of the workings of a touch trigger probe). It is, therefore, helpful to have some basic understanding of how a touch trigger probe works. The problem faced by the designer of touch trigger probes is that the probes must operate to a higher degree of accuracy than the accuracy required for the manufacture of the measured object. Only by the application of kinematic principles can the design of the probe be such that its accuracy in operation does not depend entirely on its accuracy of manufacture (see Chapter 6). The touch trigger probe, therefore, employs a form of kinematic location to retain a stylus in a highly repeatable manner. A typical mechanism (see Figure 5.26 for a photograph and Chapter 4, Figure 4.8 for

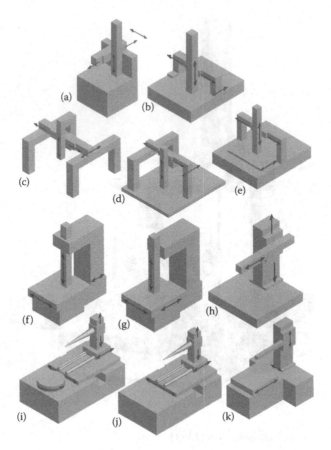

FIGURE 5.25
Typical CMS configurations. (a) Fixed table cantilever, (b) moving bridge, (c) gantry, (d) L-shaped bridge, (e) fixed bridge, (f) moving table cantilever, (g) column, (h) moving ram horizontal-arm, (i and j) fixed table horizontal-arm, (k) moving table horizontal-arm.

a solid model) consists of three cylindrical rods, each pressed in the groove formed between a pair of spheres. This action constrains all six degrees of freedom of the stylus so that it returns to the same position after deflection has taken place. An electrical circuit is made through all six contacts such that a trigger signal is generated whenever the stylus is deflected in any direction, in this case by the continuity of the circuit being broken. At the moment of contact, this trigger signal notifies the computer to record the machine position. Because the stylus can relatively freely move following contact, this can accommodate over-travel during the necessary time for the CMS to decelerate. When the probe is retracted, the stylus moves off the surface and the spring force causes the stylus mounting to reseat. The mechanism allows the probe to detect surface contact with sub-micrometre repeatability.

For higher accuracy applications, analogue probes tend to be the preferred option. On a typical analogue probe, the probing system consists of three orthogonal spring parallelograms (or flexures, see Chapter 7) and sensors to measure their deflection. Each parallelogram is clamped in its neutral position; the zero points of the sensors are adjusted to this position. Typically, a moving coil system generates the measuring force when contact is made with the object being measured. When the probing system has adjusted into a near-zero position, the machine co-ordinates and the digitised residual deflections of the probe

FIGURE 5.26
A touch trigger probe.

head are transferred to the computer. Often, the probe head is pre-deflected in the probing direction to ensure that the probe head can be stopped within its deflecting range in case of a contact or collision. Measurement with an analogue probe is static, which results in a considerable increase in accuracy when compared to a touch trigger probe. A typical touch trigger probe has a repeatability of around 0.5 μm and a form error of around 1 μm; a typical analogue probe can be fives times better. In some cases, contact can be continuously measured so that a plot of deflection as a function of contact force can be determined. From this it is also possible to extrapolate to identify more precisely the point of first contact. This can be particularly important when measuring materials of low elastic modulus or compliant mechanisms.

The data collected by the CMS are essentially ball centre data. Therefore, the stylus tip in contact with the surface needs to be qualified to determine the effective stylus radius and the position of the centre of the tip relative to the coordinate reference in the CMS. Stylus qualification is carried out by measuring a known artefact, usually a high-precision ceramic reference sphere (Flack 2001).

5.6.3 CMS Software

The measured data from a CMS forms what is referred to as a point cloud, which is a set of data points with (x,y,z) coordinates (note that some instruments produce polar coordinates and many optical instruments produce a map of z height data at (x,y) positions, often called range data). The point cloud needs to be aligned with either the component engineering drawing or, more usually these days, a computer-aided design (CAD) model. This alignment is often carried out with reference to defined datum features on the drawing or model (see Section 5.6.4). However, for complex components, such as those with freeform geometry (i.e. geometry not conforming to a regular shape), alignment is usually carried out by a mathematical best-fit operation, most commonly a least-squares fit (Forbes and Minh 2012). Once data are collected from the probe and machine axes, they are analysed by a software package. Along with the geometry data, the software will also collect data from various environmental sensors on or around the CMS, for example, temperature, barometric pressure and humidity sensors.

The CMS software mathematically fits associated features (circles, planes, etc.) to the collected data that can then be used to calculate intersection points, distances between features, locations of features in the workpiece coordinate frame, distances between features and form errors such as roundness, cylindricity and so on.

In addition to the aforementioned functions, the CMS software will create alignments relating to the part in question (see Section 5.6.4), report the data and compare against CAD data where necessary. Modern CMS software is capable of being programmed directly from a CAD model. Furthermore, once data are collected, the actual points can be compared to the nominal points and pictorial representations of the errors created. Point clouds can also be best-fitted to the CAD model for alignment purposes.

CMS software needs to be tested, and this is covered in ISO 10360 part 6 (2001). These tests use reference data sets and reference software to check the ability of the software to calculate the parameters of basic geometric elements (Forbes et al. 2015).

5.6.4 CMS Alignment

To measure an object on a CMS, its alignment relative to the coordinate system of the machine needs to be determined. The alignment process confirms to the software the physical part location on the CMS; more advanced alignments use the CMS to provide more accuracy in the process. Alignment is usually made using datum features on the object; datum features are essentially locations used as guides to tell the machine where it is or as directions on how to get to a particular position (note that these physical datum features may or may not be the same as the datums on the associated CAD drawing). The alignment needs to control the following:

- The part spatial rotation (two degrees of freedom). For a plane, the spatial rotation defines the direction perpendicular to it (levelling), for example, the top plane on a part; for a cylinder or cone, the spatial rotation defines the direction parallel to the main axis; and for a 3D line, the spatial rotation defines the direction three dimensionally parallel to it, for example, the axis of a shaft.

- The part planar rotation (one degree of freedom). For a plane, this rotation defines an axis to be perpendicular to the plane, for example, one of the square sides of a block; for a cylinder or cone this rotation defines an axis to be parallel to its axis; and for a 2D or 3D line this rotation defines an axis to be parallel to its 2D projection.

- The part origin (three degrees of freedom).

As an example, for a rectangular block, the alignment process would typically be:

1. Measure a plane on the top surface (defines rotation axis and z zero)
2. Measure a line on the side face (defines planar rotation about z axis and y zero)
3. Measure a point on a face orthogonal to the side face (x zero)

Other alignments are possible, for example, best-fit alignments and reference point alignments are used for freeform geometries.

5.6.5 Prismatic against Freeform

CMSs are used to measure highly complex objects, but it is useful to split the geometry types into the following:

- Prismatic components, examples of which include cylindrical engine block cavities, brake components and flat pad break bearings
- Freeform components, examples of which include car door panels, aircraft wing sections and mobile phone covers

Prismatic components can be broken down into elements that are readily defined mathematically, for example, planes, circles, cylinders, cones and spheres. A measurement will consist of breaking down the component into these geometries and then looking at their interrelationships, for example the distance between two holes or the diameter of a circle. Freeform components cannot be broken down as with prismatic components. Generally, the surface is contacted at a large number of points and a surface approximated (fit) to the data. Many real-world components are a mixture of freeform surfaces and geometric features; for example a mobile phone cover may have location pins that need to be measured.

If a CAD model exists, then the point cloud of data can be compared directly against the CAD model. Having a CAD model is an advantage for freeform surfaces, as the nominal local slope at the contact point is known in advance. The local slope is needed to determine the point of contact on the surface based on the centre location of the probe sphere at contact and the surface normal vector at that point.

5.6.6 Non-Standard Types of CMSs

Conventional contact CMSs are large and bulky machines, often with masses of several hundreds of kilograms. Where a smaller footprint is needed, for example for use on an assembly station, low mass portable CMSs with articulated arms are available. Articulated CMSs have six or seven axes that are equipped with rotary encoders, instead of linear axes (see Figure 5.27a, and Hocken and Pereira 2011 for more details).

The contact-type CMS has been used extensively in advanced manufacturing and engineering, but it has two major limitations. First, there is a need to physically contact the object being measured. The finite contact force always causes some deflection of the surface and may cause damage in the case of plastic deformation, especially with low modulus or low hardness objects, for example polymer parts. Second, the need to physically contact the surface at each measurement point means that contact CMSs are fundamentally serial measurement devices, which results in potentially long measurement times and/or only a small number of data points. These two limitations can be overcome by the use of optical and x-ray CMSs.

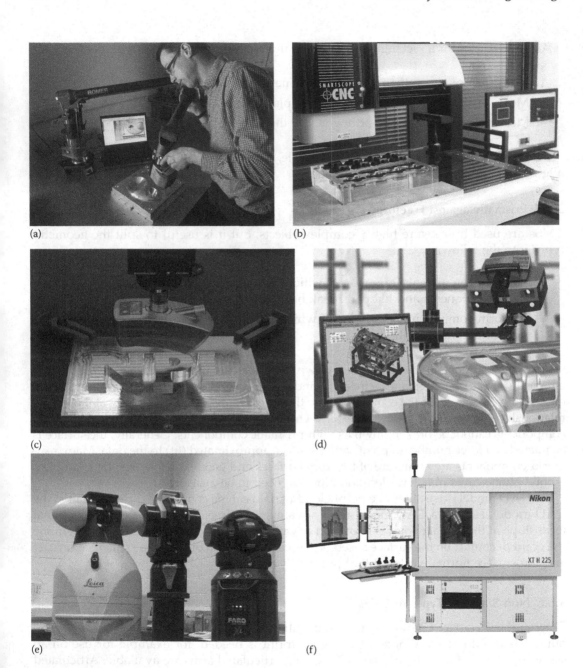

FIGURE 5.27
Non-conventional CMS systems. (a) Articulated arm system. (Courtesy of NPL, Crown Copyright.) (b) SmartScope CNC 500 vision system, a large travel bridge-design system which also allows additional sensors, including touch trigger probes, micro-probes and laser scanners. (Courtesy of OGP UK, ogpuk.com.) (c) Optical scanning system. (Courtesy of NPL, Crown Copyright.) (d) ATOS structured light scanning system completing inspection on automotive sheet metal part. (Courtesy of GOM.) (e) Laser tracker. (Courtesy of NPL, Crown Copyright.) (f) X-ray computed tomography system. (Courtesy of Nikon.)

There are a number of commercially available optical probes that can be used on some specialised CMSs. Optical probes can be either:

- 2D vision systems (see Figure 5.27b) (Coveney 2014)—Vision systems usually comprise a two- or three-axis motion system with a long stand-off microscope-type objective as the probe. They can also incorporate rotary axes and a rotatable probe head. Typical uncertainties with vision CMSs are of the order of 0.05 mm over distances of up to a metre.

- Point or line scanning triangulation or chromatic confocal systems (see Figure 5.27c) (Leach 2011)—Such probes are usually used in place of the contact probe with conventional CMS motion systems. Typical uncertainties with these probes are of the order of 5 µm over distances less than a metre.

- Structured light, otherwise known as fringe projection, systems (see Figure 5.27d) (Harding 2014)—Fringe projection systems illuminate the part being measured a with structured light pattern, for example a sinusoidally modulated intensity distribution, and calculate the geometry by analysing the distortion of the structured light pattern due to the geometry. Such systems can be very fast as they illuminate a large area of the part in a parallel fashion, but they do not currently have the same accuracy levels as contact and point/line scanning optical techniques. Typical uncertainties with fringe projection systems are of the order of 0.01 mm over distances of up to a metre.

- Laser trackers (see Figure 5.27e) are used to measure relatively large objects by determining the positions of optical targets held against the object (Schmitt et al. 2016). Typical uncertainties with laser trackers are of the order of 0.025 mm over distances of several metres.

There are also many commercial CMSs that include multiples of the above probes, for example, a vision system with a contact probe. An important consideration with such multi-sensor systems is the referencing of the different probes to the same coordinate system.

There is also an increasing use of x-ray computed tomography (XCT), a method of forming 3D representations of an object by taking many x-ray images through a component while it is rotated around an axis and using algorithms to reconstruct a 3D model (see Figure 5.27f) (De Chiffre et al. 2014, Carmignato et al. 2017). Whilst XCT shows promise as a much faster method than contact CMSs (although high-resolution 3D scans can take hours) and the benefit of being able to measure internal geometries, there are still no standardised ways to calibrate and performance-verify them; this is an active field of research (Ferrucci et al. 2015).

5.6.1 CMS Errors

A typical CMS (i.e. those depicted in Figure 5.25) has twenty-one sources of geometric error. These can be accounted for by noting that each axis has one linear error, three rotation errors and two straightness errors (six per axis gives eighteen). In addition, there are orthogonality errors between any two pairs of axes. The twenty-one geometrical errors are minimised during manufacture of the CMS and can also be error-mapped (volumetric error compensation) with corrections to geometric errors made in software (Hocken and Periera 2011, Schwenke et al. 2008).

CMS geometric errors are determined in one of the four following manners:

- Using artefacts and instruments such as straight edges, autocollimators and levels
- Using a laser interferometer system and associated optics
- Using a calibrated-hole plate (Lee and Budekin 2001)
- Using a tracking laser interferometer (Schwenke et al. 2008)

5.6.8 Standards, Traceability, Calibration and Performance Verification

CMS calibration is the measurement of the twenty-one geometrical error sources to enable mechanical correction or error mapping of the CMS. Performance verification is a series of tests that allows the manufacturer of the CMS to demonstrate that an individual machine meets the manufacturer's specification (see Flack 2001 for a thorough description of the various methods). Note that calibration can be part of the performance verification.

The ISO 10360 series of specification standards defines the procedure for performance verification of CMSs. The series is broken down into seven parts, which are listed next and described in detail in Hocken and Periera (2011).

- Part 1: Vocabulary (2000)
- Part 2: CMMs used for measuring linear dimensions (2009)
- Part 3: CMMs with the axis of a rotary table as the fourth axis (2000)
- Part 4: CMMs used in scanning measuring mode (2000)
- Part 5: CMMs using single and multiple-stylus contacting probing systems (2010)
- Part 6: Estimation of errors in computing Gaussian-associated features (2001)
- Part 7: CMMs equipped with imaging probing systems (2011)
- Part 8: CMMs with optical distance sensors (2013)
- Part 9: CMMs with multiple probing systems (2013)
- Part 10: Laser trackers for measuring point-to-point distances (2016)

Traceability of measurements carried out by CMSs is difficult to demonstrate. The formulation of an uncertainty budget is impracticable for the majority of the measurement tasks for CMSs due to the complexity of the measuring process. It used to be the case that the only way to demonstrate traceability was to carry out ISO 10360-type performance verification tests on the CMS. However, if a CMS is performance-verified, this does not automatically mean that measurements carried out with this CMS are calibrated and/or traceable. A performance verification only demonstrates that the machine meets its specification for measuring simple lengths, that is it is not task specific.

A better method to achieve at least a degree of traceability is described in ISO 15530 part 3 (2011). This specification standard makes use of calibrated artefacts to essentially use the CMS as a comparator. The uncertainty evaluation is based on a sequence of measurements on a calibrated object or objects, performed in the same way and under the same conditions as the actual measurements. The differences between the results obtained from the measurement of the objects and the known calibration values of these calibrated objects are used to estimate the uncertainty of the measurements.

The criterion mainly used to quantify the performance of a CMS is the maximum permissible error (MPE), which is defined (in the ISO 10360 standards) as the largest error or

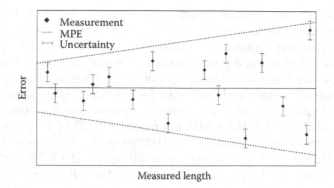

FIGURE 5.28
Plot of MPE for a CMS, also showing how uncertainty can be expressed.

deviation of a measurement from a reference quantity value. The MPE of a CMS is often specified as a constant value plus a length-dependent term, for example MPE = 2.4 + 2L/1000 µm, where L is the measured length and is in millimetres. A CMS's MPE serves as a benchmark for its performance when compared to other similar instruments. Often, prior to purchasing a measuring instrument, several tests with dedicated reference objects are performed to determine if the particular instrument satisfies its specified MPE. In this regard, the MPE of a measuring instrument can have contractual significance. If the proven performance of an instrument does not satisfy the stated MPE (and assuming the user satisfied the measurement conditions under which the MPE is specified), then the instrument vendor has an obligation to rectify the deficiencies.

An instrument's MPE is often mistaken for the uncertainty of measurements performed on the instrument. While MPE is a maximum expected deviation of measurements by a particular instrument from a reference value, uncertainty is a statistical dispersion of values that can be attributed to the result of a measurement (see Chapter 9). A measurement result expressed as a single value corresponding to the estimate of the measurand is incomplete without a statement of uncertainty surrounding the estimate. Figure 5.28 illustrates the concepts of MPE and uncertainty as they pertain to measurements on features with various lengths. In this plot, the error of each measurement result from the corresponding reference value is plotted as a function of the measured length. Some of the measurement points consist of uncertainty intervals that exceed the MPE (dotted line). The use of these measurement results in deciding whether the instrument satisfies the MPE would prove inconclusive. Finally, unlike the MPE of a measuring instrument, measurement uncertainty is a task-specific quantity.

Alternative methods that are consistent with the *Guide to the Expression of Uncertainty in Measurement* (GUM) (see Chapter 9) can be used to determine the task-specific uncertainty of coordinate measurements. One such method that evaluates the uncertainty by Monte Carlo methods (see Chapter 9) is described in ISO/TS 15530 part 4 (2008). To allow CMS users to easily create uncertainty statements, CMS suppliers and other third-party companies have developed uncertainty-evaluating software, also known as 'virtual CMMs' (Balsamo et al. 1999, Flack 2013).

5.6.9 Micro-CMMs

Modern manufacturing has seen an increase in the reduction in size of many components and this has resulted in the need for metrology systems that can measure on ever-smaller

scales. The increased demand on the manufacture of micro-scale products requires CMSs to
be able to measure micro-scale parts accurately. So-called micro-CMMs have been developed to meet this demand. Generally, a micro-CMM can be designed by miniaturisation
of the traditional CMS or using a probe that employs optical technology. State-of-the-art
micro-CMMs typically have working ranges of tens of millimetres with hundreds of
nanometre volumetric accuracy and can be used to measure features with milli- to micro-
scale dimensions. Typical examples of commercial systems are the Zeiss F25 micro-CMM
(Vermeulen et al. 1998), the IBS Isara 400 Ultra precision CMM (Widdershoven et al. 2011)
and the SIOS Nanomeasuring Machine (NMM) (Jäger et al. 2016). The Zeiss F25 CMM has
a measurement volume of 100 mm × 100 mm × 100 mm, and a MPE of 0.25 + $L/666$ µm,
where L is the measurement length in millimetres. The Isara 400 minimises the Abbe error
(see Chapter 10) by aligning three linear interferometers to the centre of the stylus tip. The
measurement volume is 400 mm × 400 mm × 100 mm, and the stated 3D measurement
uncertainty is 109 nm (at $k = 2$), although it is not clear from this statement what exactly is
being measured (this is an example of bad practice when specifying an instrument). The
NMM is a laser interferometer-based micro-CMM developed by the Ilmenau University of
Technology. The measurement range is 200 mm × 200 mm × 25 mm with a sub-nanometre
resolution of motion. In addition, several other micro-CMMs exist and several reviews of
existing micro-CMMs can be found elsewhere (Leach 2014a, Thalmann et al. 2016).

5.7 Surface Texture Metrology

5.7.1 Introduction to Surface Texture Metrology

The surface topography of a component part can have a profound effect on the function of
the part. In tribology, it is the surface interactions that influence such quantities as friction,
wear and the lifetime of a component (see Chapter 7). In fluid dynamics, it is the surface
that determines how fluids flow and it affects such properties as aerodynamic lift, therefore influencing efficiency and fuel consumption of aircraft. Examples of surface–function
relationships can be found in almost every manufacturing sector, both traditional and high-
tech (Bruzzone et al. 2008). To control surface topography, and hence the function of a
component, it must be measured and useful parameters extracted from the measurement
data. Figure 5.29 shows the typical range of surface texture values expected from a variety
of common machining operations; the parameter Ra is described in Section 5.7.5.

The following definitions are used in this book:

Surface topography—Local deviations of a surface from a perfectly flat plane or all the
surface features treated as a continuum of spatial wavelengths (Leach 2014a).

Surface texture—The geometrical irregularities present at a surface. Surface texture
does not include those geometrical irregularities contributing to the form or shape
of the surface (Leach 2015).

Surface form—Underlying shape of a part (Leach 2014a) or fit to a measured surface
(ISO 10110-8 2010).

Surface profile measurement is the measurement of a line across the surface that can be
represented mathematically as a height function with lateral displacement $z(x)$. Areal surface
texture measurement is the measurement of an area on the surface that can be represented

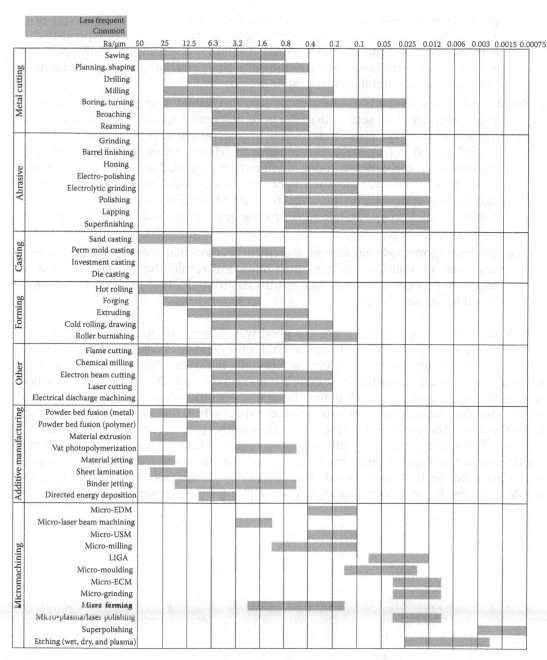

FIGURE 5.29
Surface textures for various machining operations.

mathematically as a height function with displacement across a plane $z(x,y)$. Surface texture characterisation, both profile and areal, is discussed in Section 5.7.5 and presented in detail elsewhere (Whitehouse 2010, Leach 2013, Leach 2014a, Muralikrishnan and Raja 2008).

There is a range of instrumentation for measuring surface texture, but this book will only consider the techniques that have been standardised. ISO 25178 part 6 (2010) defines three classes of methods for surface texture measuring instruments:

Line profiling method—Surface topography method that produces a 2D graph or profile of the surface irregularities as measurement data that may be represented mathematically as a height function $z(x)$. Examples given in ISO 25178 part 6 include: stylus instruments, phase shifting interferometry, circular interferometric profiling and optical differential profiling.

Areal topography method—Surface measurement method that produces a topographical image of the surface that may be represented mathematically as a height function $z(x,y)$. Often, $z(x,y)$ is developed by juxtaposing a set of parallel profiles. Examples cited in ISO 25178 part 6 include: stylus instruments, phase shifting interferometry, coherence scanning interferometry, confocal microscopy, confocal chromatic microscopy, structured light projection, focus variation microscopy, digital holography microscopy, angle resolved SEM, SEM stereoscopy, scanning tunnelling microscopy, atomic force microscopy, optical differential and point autofocus profiling.

Area-integrating method—Surface measurement method that measures a representative area of a surface and produces numerical results that depend on area-integrated properties of the surface texture. An example given in ISO 25178 part 6 is total integrated scatter.

Amplitude–wavelength (AW) space is a good way to map the specifications of instruments (Stedman 1987, Jones and Leach 2008). Each operational constraint (for example, range, resolution, tip geometry, lateral wavelength limit) can be modelled and parameterised, and relationships between these parameters derived. The relationships are best represented as inequalities, which define the area of operation of the instrument. A useful way to visualise these inequalities is to construct a space where the constraining parameters form the axes. The constraint relationships (inequalities) can be plotted to construct a polygon. This shape defines the viable operating region of the instrument. Constraints that are linear in a given space must form a flat plane across that space, solutions on one side of which are valid. Such a plane can only form a side of a convex polyhedron containing the viable solutions. Figure 5.30 shows an AW space plot for three common instruments.

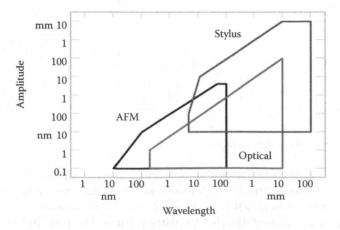

FIGURE 5.30
Amplitude–wavelength plot for the most common instrument types.

5.7.2 Contact Stylus Instruments

The contact stylus instrument has been used to measure surface texture for over one hundred years and is still the most utilised method in manufacturing industry (Whitehouse 2010, Leach 2014b). A stylus instrument consists of a sharp tip placed in contact with the surface being measured. By scanning the tip across the surface and monitoring its response to surface heights, it is possible to measure surface topography. A schema of a stylus instrument is shown in Figure 5.31. Stylus instruments can have vertical resolutions below 1 nm, although values of tens of nanometres are more common in industry.

As the stylus tip is scanned across the surface, its vertical displacement is recorded and converted into an electrical signal using an electromechanical transducer (ISO 25178-601 2010, Whitehouse 2010, Leach 2014b). The tip plays a critical role in the performance of a stylus instrument, as it is in physical contact with the surface. The stylus tip is usually made of diamond but other materials, such as aluminium oxide, are often employed depending of the material of the surface being measured. Other parameters that should be taken into account are the shape and size of the stylus tip, which directly affect the spatial frequency response of the instrument. Depending on the application, the stylus tip can have different geometries; the most frequently used has a conical shape with a rounded contacting edge and radius of curvature ranging from 2 μm to 10 μm, and a 60° or 90° slope angle (Leach 2014b).

The main sources of error associated with a stylus instrument are

- Surface deformation
- Amplifier distortion
- Finite stylus dimensions
- Lateral deflection

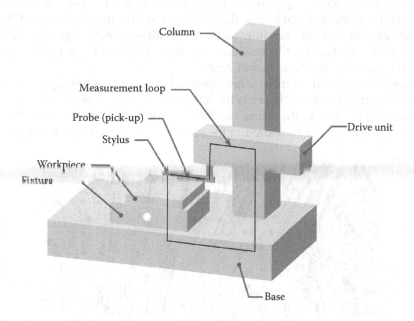

FIGURE 5.31

Schema of a typical stylus instrument for measuring surface topography. (From Leach, R. K., 2014, *Fundamental principles of engineering nanometrology*, 2nd edn., Elsevier.)

- Effect of skid or other datum
- Relocation upon repeated measurements
- Effect of filters (electrical or mechanical)
- Quantisation and sampling effects
- Dynamic effects
- Environmental effects
- Effect of incorrect data-processing algorithms

Methods to quantify these effects and their contribution to the measurement uncertainty are given by Haitjema (2015a). One effect that should be taken into account is the systematic error associated with the shape of the stylus tip that will distort the measured topography (McCool 1984). This issue has been intensively studied, and mathematical models have been proposed to compensate the distortion effect and improve the accuracy of measurements (Mendeleyev 1997, Lee et al. 2005). ISO 25178-601 (2010) states that the stylus force should be 0.75 mN, but this is rarely checked by the user. The value of 0.75 mN was chosen so as not to cause significant damage in steel with a 2 µm radius stylus, but it does cause damage in many other materials. Small radius asperities (particularly those of significantly smaller radius than the stylus) may be plastically deformed and the stylus tip will faithfully traverse over the longer wavelength features. This 'burnishing' of high-frequency asperities and removal of surface films often result in a visible line where the stylus has traversed the surface, a little like the path in snow where a person has walked over a hill. However, for many measurements, such as tribological performance, the measured texture is still relevant to functional performance. Figure 5.32 shows the direction of stylus measurement on an aluminium surface manufactured using selective laser melting (an additive manufacturing method). Smaller forces limit the measurement speed due to the risk of 'stylus flight'. Modern stylus instruments regularly obtain measurements of surface texture with subnanometre height resolution, but traceability of these measurements in each of their axes is relatively new and has not yet been fully taken up in industry. Although stylus instruments are widely used for surface profile measurement, if used with a large radius spherical tip, they can also be used to measure surface form.

A drawback of a stylus instrument when operated in an areal scanning mode is the time to take a measurement. It may be perfectly acceptable to take several minutes to make a profile measurement, but if the same number of points are required in the direction orthogonal to the scan as are measured in the scan direction, then measurement times can

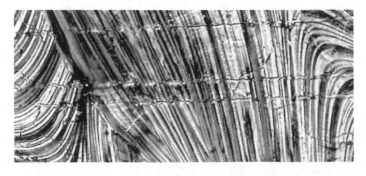

FIGURE 5.32
Scratch in a surface due to a stylus instrument trajectory. The scratch marks are the linear features from left to right.

be up to several hours. For example, if the drive mechanism can scan at 0.1 mm·s^{-1} and 1000 points are required for a profile of 1 mm, then the measurement will take 10 s. If a square grid of points is required for an areal measurement, then the measurement time will be approximately 2.7 hours. This often precludes the use of a stylus instrument in a production or in-line application.

5.7.3 Optical Techniques

There are many different types of optical instruments that can measure surface topography, both surface texture and surface form (only texture instruments are discussed in this section, but some form instruments are discussed in Section 5.6.6). The main instruments that are available commercially and have been standardised are simply listed here and some general principles are given. Each method is discussed in detail in Leach (2011) and the associated specification standard:

- Chromatic confocal microscopy (ISO 25178-602 2010)
- Phase shifting interferometry (ISO 25178-603 2013)
- Coherence scanning interferometry (ISO 25178-604 2013)
- Point autofocus instrument (ISO 25178-605 2014)
- Imaging confocal microscopy (ISO 25178-606 2012)
- Focus variation microscopy (ISO 25178-607 2016)

There are many more optical instruments, or variations on the aforementioned instruments listed, most of which are listed in ISO 25178 part 6 (2010). Optical instruments have a number of advantages over stylus instruments. They do not physically contact the surface being measured and hence do not present a risk of damaging the surface (although relatively high-powered lasers can be used with confocal instruments, therefore, presenting a risk of burning). This non-contact nature can also lead to much faster measurement times for the optical scanning instruments. The area-integrating methods can be faster still, sometimes only taking some seconds to measure a relatively large area. However, care must be taken when interpreting the data from an optical instrument (compared to that from a stylus instrument). Whereas it is relatively simple to predict the output of a stylus instrument by modelling it as a ball of finite diameter moving across the surface, it is not such a trivial matter to model the interaction of an electromagnetic field with the surface. Often many assumptions are made about the nature of the incident beam or the surface being measured that can be difficult to justify in practice. Optical instruments have a number of limitations, some of which are generic and some that are specific to instrument types. This section briefly discusses some of the generic limitations and specific limitations can be found in Leach (2011).

Many optical instruments use a microscope objective to magnify the features on the surface being measured. It is worth noting that the magnification of the objective is not the value assigned to the objective, but the combination of the objective and the microscope's tube length. The tube length may vary between 160 mm and 210 mm, and thus, if the nominal magnification of the objective assigned by the manufacturer is based on a 160 mm tube length, then the magnification of this objective on a system with 210 mm tube length will be about 30% greater, as magnification equals tube length divided by the focal length of the objective. Magnifications vary from 2.5× to 200× depending on the application and the type of surface being measured.

Instruments employing a microscope objective will have two fundamental limitations. First, the numerical (or angular) aperture (NA) determines the largest slope angle on the surface that can be measured and affects the optical resolution. The NA of an objective is given by

$$A_N = n \sin \alpha,$$ (5.3)

where n is the refractive index of the medium between the objective and the surface (usually air, so n can be approximated by unity), and α is the acceptance angle of the aperture (see Figure 5.33, where the objective is approximated by a single lens). The acceptance angle will determine the slopes on the surface that can specularly reflect light back into the objective lens and hence be measured. Note that, if there is some degree of diffuse reflectance (scattering) from a rough surface, some light can still reflect back into the aperture, allowing larger angles than those dictated by Equation 5.3 to be detected (see Figure 5.33).

The second limitation is the optical resolution of the objective. The resolution determines the minimum distance between two lateral features on a surface that can be measured. The resolution is given by

$$r = \frac{k\lambda}{A_N},$$ (5.4)

where λ is the wavelength of the incident radiation. For a theoretically perfect optical system with a filled objective pupil, the optical resolution is given by the Rayleigh criterion, where k in Equation 5.4 is replaced by 0.61. Yet another measure of the optical resolution is the Sparrow criterion, or the spatial wavelength where the instrument response drops to zero and where the k in Equation 5.4 is replaced by 0.47. Equation 5.4 and the Rayleigh and Sparrow criteria are often used almost indiscriminately, so the instrument user should always check which expression has been used where optical resolution is a limiting factor. Also, Equation 5.4 sets a minimum value (although the Sparrow criterion will give a smaller numerical value; this is down to the manner in which 'resolved' is defined). If the objective is not optically perfect (i.e. aberration free) or if a part of the beam is blocked (for example when a steep edge is measured) the value becomes higher (worse).

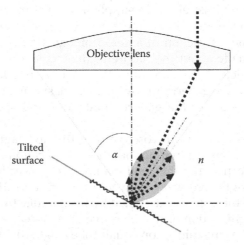

FIGURE 5.33
Microscope aperture imaging a tilted surface.

For some instruments, it may be the distance between the pixels (determined by the image size and the number of pixels in the camera array) in the microscope camera array that determines the lateral resolution.

The optical resolution of the objective is an important characteristic of an optical instrument but its value can be misleading. When measuring surface texture, the spacing of points in an image must be considered, along with the ability to accurately determine the heights of features. One definition used in the ISO 25178 series of specification standards is the lateral period limit. This is defined as the spatial period of a sinusoidal profile for which the instrument response (measured feature height compared to actual feature height) falls to 50%. Methods to determine the lateral period limit are described elsewhere (Giusca and Leach 2013).

As well as lateral resolution, there is a similar effect due to diffraction in the direction of propagation, that is the diffraction disk has a finite thickness. The depth of field in the object plane is the thickness of the optical section along the principle axis of the objective lens within which the object is in focus. The depth of field Z is given by

$$Z = \frac{n\lambda}{A_N^2},\qquad(5.5)$$

where n is the refractive index of the medium between the lens and the object. Depth of field is affected by the optics used, lens aberrations and the magnification. Note that as depth of field increases, the lateral resolution proportionally decreases.

Many optical instruments, especially those utilising interference, can be affected by the surface having areas that are made from different materials (Harasaki et al. 2001). For a dielectric surface there is a π phase change on reflection (at normal incidence), that is a π phase difference between the incident and reflected beams. The phase change on reflection δ is given by

$$\tan\delta = \frac{2n_1 k_2}{1 - n_2^2 - k_2^2},\qquad(5.6)$$

where n and k are the refractive and absorption indexes of the surrounding air (medium 1) and the surface being measured (medium 2) respectively. For dielectrics, k will be equal to zero but for materials with free electrons at their surfaces (i.e. metals and semiconductors), a finite k will lead to a $(\pi + \delta)$ phase change on reflection. For the example of a chrome step on a glass substrate, the difference in phase change on reflection gives rise to an error in the measured height of approximately 16 nm (at a wavelength of approximately 633 nm) when measured using an optical interferometer (see Exercise 9). A stylus instrument would not be subject to this error in height (although there may be comparable errors due to the different indentation properties of the materials).

Finally, it is important to note that surface roughness plays a significant role in measurement quality when using optical instrumentation. Many researchers have found that estimates of surface roughness derived from optical measurements differ significantly from other measurement techniques. The surface roughness is generally over-estimated by optical instrumentation (this is not necessarily true when considering area-integrating instruments) and this may be attributed to multiple scattering. Although it may be argued that the local gradients of rough surfaces exceed the limit dictated by the NA of the objective and, therefore, would be classified as beyond the capability of optical instrumentation, measured values with high signal-to-noise ratio are often returned in practice.

There are many more limitations of optical instruments that are out of the scope of this book. The user is encouraged to investigate these limitations for their specific instrument type; see Leach (2011) and the specific part of the ISO 25178 series of specification standards.

5.7.4 Scanning Probe Microscopes

Scanning probe microscopes (SPMs) are a family of instruments that are used for measuring surface topography, usually on a smaller scale than that of conventional stylus instruments and optical instruments. Along with the electron microscope (not discussed in this book as there are many textbooks on the subject, see for example Egerton 2008), they are the instruments of choice when surface structure needs to be measured with spatial wavelengths smaller than the diffraction limit of an optical instrument (typically around 500 nm). The theory, principles, operation and limitations of SPMs are discussed thoroughly elsewhere (Meyer et al. 2003, Leach 2014a). The SPM is a serial measurement device that uses a nanometre-scale probe to trace the surface of the sample based on local physical interactions (in a similar manner to a stylus instrument). While the probe scans the sample with a predefined pattern, the signal of the interaction is recorded and is usually used to control the distance between the probe and the sample surface. This feedback mechanism and the scanning of a nanometre-scale probe form the basis of all SPMs.

There are many different types of SPMs but the most common type is the atomic force microscope (AFM) (Maganov 2008). In a conventional AFM, the sample is scanned continuously in two axes underneath a force-sensing probe consisting of a tip that is attached to, or part of, a cantilever. A scanner is also attached to the z axis (height) and compensates for changes in sample height, or forces between the tip and the sample. The presence of attractive or repulsive forces between the tip and the sample will cause the cantilever to bend and this deflection can be monitored in a number of ways. The most common system to detect the bend of the cantilever is the optical beam deflection system, wherein a laser beam reflects off the back of the cantilever onto a photodiode detector. Such an optical beam deflection system is sensitive to sub-nanometre deflections of the cantilever. In this way, an areal map of the surface is measured, usually over a few tens to hundreds of micrometres squared, although longer-range instruments are available.

5.7.5 Surface Texture Parameters

A surface texture parameter, be it profile or areal, is used to give the surface texture of a part a quantitative value. Such a value may be used to simplify the description of the surface texture, to allow comparisons with other parts (or areas of a part) and to form a suitable measure for a quality system. Surface texture parameters are also used on engineering drawings to formally specify a required surface texture for a manufactured part. Some parameters give purely statistical information about the surface texture and some can describe how the surface may perform in use, that is to say, its functionality.

All the surface texture parameters are calculated once the form has been removed from the measurement data. Form removal is not discussed in detail here, but the most common methods use the least-squares technique. Most instruments and surface characterisation software packages will have built-in form removal routines, and background information can be found in Forbes (2013). Following form removal, the surface texture data is filtered to remove unwanted detail and to define the spatial bandwidth over which the parameters are calculated. Again, the process of filtering is not discussed here but is described in detail elsewhere (Seewig 2011, Leach 2014a).

In profile characterisation, there are three types of parameters which depend on how the surface has been filtered. The primary profile is defined as the total profile after application of the short wavelength (low-pass) filter and an example primary profile parameter is *Pa*. The roughness profile is defined as the profile derived from the primary profile by suppressing the long-wave component using a long-wavelength (high-pass) filter, and an example roughness parameter is *Ra*. Roughness is usually the process marks produced by the action of the cutting tool or machining process, but may include other factors such as the structure of the material. The waviness profile is derived by the application of a band-pass filter to select the surface structure at longer wavelengths than the roughness and an example waviness parameter is *Wa*. Waviness is usually produced by instabilities in the machining process, such as an imbalance in a grinding wheel, or by deliberate actions in the machining process (waviness is an important consideration in the optics industry). Figure 5.34 shows examples of the same measured profile after the various filtering operations.

Once the form has been removed and the various filtering operations have been completed, the parameters can be calculated. For historical reasons, the *Ra* parameter is the most common of the all the surface texture parameters and is dominant on most engineering drawings when specifying surface texture. The *Ra* parameter will be used as an example and is the arithmetic mean of the absolute ordinate values $z(x)$ within the sampling length l:

$$Ra = \frac{1}{l} \int_0^l |z(x)| dx. \tag{5.7}$$

The derivation of *Ra* can be illustrated graphically as shown in Figure 5.35. The areas of the graph below the centre line within the sampling length are placed above the centre line. The *Ra* parameter value is the mean height of the resulting profile. Note that, due to the nature of the absolute value operation in Equation 5.7, the *Ra* value does not provide any information as to the shape of the irregularities on a surface. It is possible to obtain similar *Ra* values for surfaces having very different structures.

There are a number of profile surface texture parameters, of which *Ra* is an example, that can be used for a given engineering application, a number of which are utilised in the tribology discussions in Chapter 7, Section 7.2.1. These parameters are given in ISO 4287 (2000), described in detail in Leach (2014b) and simply presented in Table 5.4 (only the roughness parameters are displayed in the table, but there are also equivalent parameters for waviness and primary profile).

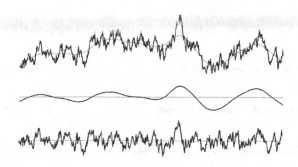

FIGURE 5.34
Primary (top), waviness (middle) and roughness (bottom) profiles. (From Leach, R. K., 2014, *Fundamental principles of engineering nanometrology*, 2nd edn., Elsevier.)

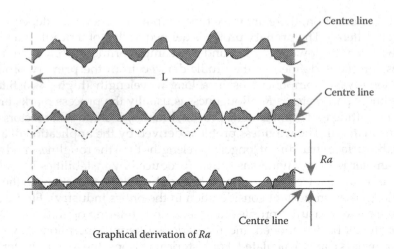

Graphical derivation of *Ra*

FIGURE 5.35
The derivation of the *Ra* parameter. (From Leach, R. K., 2014, *Fundamental principles of engineering nanometrology*, 2nd edn., Elsevier.)

TABLE 5.4

Surface Profile Parameters

Symbol	Description	Equation
Rp	Maximum profile peak height	
Rv	Maximum profile valley depth	
Rz	Maximum height of the profile	$Rp + Rv$
Rc	Mean height of the profile elements	$Rc = \dfrac{1}{l} \displaystyle\int_0^l z(x)dx$
Rt	Total height of the surface	
Ra	Arithmetical mean deviation of the assessed profile	$Ra = \dfrac{1}{l} \displaystyle\int_0^l \lvert z(x)\rvert dx$
Rq	The root mean square deviation of the assessed profile	$Rq = \sqrt{\dfrac{1}{l} \displaystyle\int_0^l z^2(x)dx}$
Rsk	Skewness of the assessed profile	$Rsk = \dfrac{1}{Rq^3}\left[\dfrac{1}{l}\displaystyle\int_0^l z^3(x)dx\right]$
Rku	Kurtosis of the assessed profile	$Rku = \dfrac{1}{Rq^4}\left[\dfrac{1}{l}\displaystyle\int_0^l z^4(x)dx\right]$
RSm	Mean width of the profile elements	$RSm = \dfrac{1}{m}\displaystyle\sum_{i=1}^{m} X_i$
R\Delta q	Root mean square slope of the assessed profile	$R\Delta q = \sqrt{\dfrac{1}{l}\displaystyle\int_0^l \left(\dfrac{dz(x)}{dx}\right)^2 dx}$

Source: ISO 4287, 2000, Geometrical product specification (GPS)—Surface texture: Profile method—Terms, definitions and surface texture parameters, International Organization of Standardization.

There are inherent limitations with profile surface measurement and characterisation. A fundamental problem is that a profile does not necessarily indicate functional aspects of the surface. With profile measurement and characterisation it is also often difficult to determine the exact nature of a topographic feature (many surfaces are inherently 3D). To overcome some of these limitations with profile characterisation, a set of areal (often called 3D, but they are mathematically 2D) surface texture parameters has been developed over the last twenty years. Distinct from the profile system, areal surface characterisation does not require three different groups (profile, waviness and roughness) of surface texture. For example, in areal parameters only Sq is defined for the root mean square parameter rather than the primary surface (Pq), waviness (Wq) and roughness (Rq) as in the profile case. The meaning of the Sq parameter depends on the type of scale-limited surface used, and the scale limitation is set with 3D filters. Areal characterisation is still a hot topic of research and will not be covered in this book. All aspects of areal characterisation, including a range of case studies, can be found in Leach (2013).

5.8 Reversal and Error Separation Techniques in Dimensional Metrology

In Section 5.5, various techniques and considerations in form measurement are given. Contrary to length measurements, there is no 'primary' standard in form measurements. This means that when performing a form measurement, the user must ask where/what is the reference for the ideal form. Physical references are available in several realisations: For flatness there are optical reference flats and surface plates. For squareness there may be a reference squareness standard that is calibrated by an NMI. For roundness and cylindricity, the manufacturer of the roundness/cylindricity measurement may deliver a roundness standard and a reference cylinder with a certificate that states the deviation of the ideal form and an uncertainty. But these realisations do not necessarily address the principal issue: That even NMIs and accredited laboratories have to deal with the fact that for form, no primary standards are defined. The reason for this is that there is a multitude of methods to realise form measurements 'error-free'; this means that the reference form is eliminated from the measurement and the measurement can be assumed to be relative to a 'perfect' reference form. The most straightforward methods of doing this are called 'reversal methods'. The term 'reversal' means that by reversing an orientation or measurement direction in a measurement, any deviation of the reference form from the ideal form can be eliminated. When two or more directions or measurements are needed, the more general term 'error separation' is used.

A number of reversal and error separation methods come from rather different fields, ranging from woodworking to geodesy, and are sometimes considered more as 'tricks' rather than fundamental dimensional methods. A systematic inventory, classification of such methods and more elaborate examples are given elsewhere (Evans et al. 1996). It should be emphasised that these methods only work for repeatable measurements.

5.8.1 Level Reversal

The simplest everyday reversal method is the level reversal. Assuming that a surface is inclined at some angle α from the true horizontal and the null reading of the level indicator

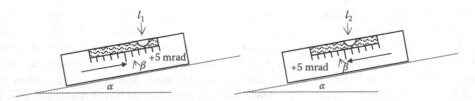

FIGURE 5.36
Illustration of the level reversal: (left) the normal measurement, (right) the measurement with the level 180°
reversed.

has an offset of an angle β to the reference surface of the level, the level outputs (l_1 in the
normal orientation and l_2 with the level rotated 180° around the vertical axis) are

$$l_1 = \alpha + \beta$$
$$l_2 = \alpha - \beta.$$

(5.8)

This gives directly

$$\alpha = \frac{l_1 + l_2}{2}$$
$$\beta = \frac{l_1 - l_2}{2}.$$

(5.9)

Level reversal is illustrated in Figure 5.36. The level indications in Figure 5.36 are $l_1 = 3$ mrad and $l_2 = 1$ mrad, giving $\alpha = 2$ mrad and $\beta = 1$ mrad (1 mrad is 1 milliradian; this is a common notation for small angles). After this reversal measurement, both the surface tilt α and the level deviation β are determined bias-free, or 'separated', hence the term error separation.

This process does not need to be repeated for every subsequent measurement of the level; once β is established, or it is established to be negligible, the level can be used in one orientation, saving measurement time. This example uses single numbers; further examples will use profiles/functions.

5.8.2 Straightedge Reversal

A straightedge is measured relative to a reference guideway that can be a machine bed or, in a vertical setup, a cylindrical square or a height meter. This is illustrated in Figure 5.37. An indicator I, typically an LVDT-type displacement sensor, measures the straightness $S(z)$ of an object, relative to its reference $R(z)$. The indicator outputs for the two positions are

$$I_1(z) = R(z) + S(z)$$
$$I_2(z) = S(z) - R(z),$$

(5.10)

so that

$$R(z) = \frac{I_1(z) - I_2(z)}{2}$$
$$S(z) = \frac{I_1(z) + I_2(z)}{2}.$$

(5.11)

After using Equation 5.11, the reference line must be removed, as any tilt can be added while reversing the object. After such a measurement, an estimation of $S(z)$ can be made that

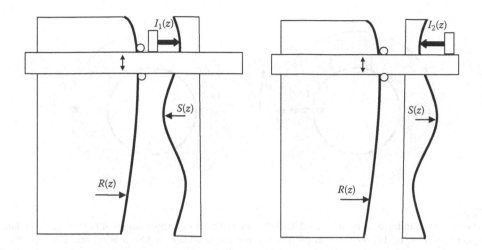

FIGURE 5.37
Illustration of straightedge reversal: (left) the normal straightness measurement, (right) the measurement in reversal mode.

is more accurate than the measurement $I_1(z)$ alone. From the measurement of the reference $R(z)$, it can be established whether it is within its tolerance, that is whether it is small enough to be neglected in subsequent use. When using a CMS for performing such measurements, the individual measurement points must be available to make the calculation according to Equation 5.11. When a measuring instrument can only output the straightness as its final value, the reversal measurements can still be useful to estimate upper limits of the deviations.

5.8.3 Spindle Roundness Calibration

A reversal method can be used to calibrate the spindle of a machine tool or a roundness tester (see Figure 5.20). For its application on a roundness tester, it is assumed that the spindle deviation in the direction of the probe P is given by $S(\theta)$, where θ is the angle of the object table of the roundness tester. On this roundness tester, a sphere is measured, where the roundness deviation of the sphere at its equatorial plane is given by $B(\theta)$. This measurement is given by $I_1(\theta)$. As the reversal measurement, a second measurement is taken with the sphere rotated over 180°, and the probe touching the sphere from the other side, so at $\theta = 0$ it measures the sphere at the same position as in the first measurement. The difference is that if the probe is configured such that it measures the ball deviations as positive, the spindle deviations changes sign in the second measurement. The recorded profile is given by $I_2(\theta)$. This situation is illustrated in Figure 5.38. For the measured signals it can be written that

$$I_1(\theta) = B(\theta) - S(\theta)$$
$$I_2(\theta) = B(\theta) + S(\theta),$$

(5.12)

so that

$$B(\theta) = \frac{I_2(\theta) + I_1(\theta)}{2}$$

$$S(\theta) = \frac{I_2(\theta) - I_1(\theta)}{2}.$$

(5.13)

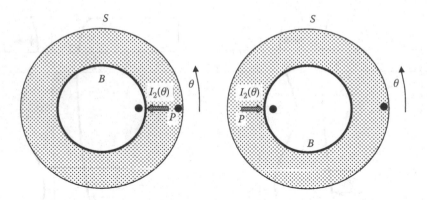

FIGURE 5.38

Illustration of the roundness reversal method: (left) the normal roundness measurement $I_1(\theta)$ of a ball B relative to a precision spindle S, (right) the measurement in reversal mode where the ball B is rotated 180° and the measurement $I_2(\theta)$ is taken from the opposite side.

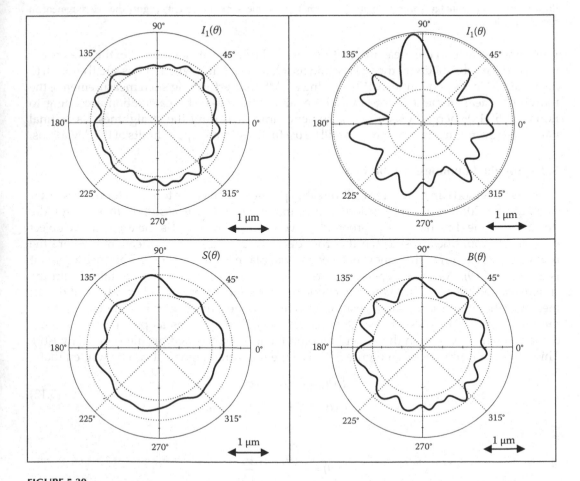

FIGURE 5.39

Example of the measurements and evaluation results of roundness reversal measurements. $I_1(\theta)$ and $I_2(\theta)$ indicate the measurements in normal and reversal mode respectively. $B(\theta)$ and $S(\theta)$ are the roundness diagrams of the ball and spindle deviation respectively.

An example of measurements $I_1(\theta)$ and $I_2(\theta)$ and the resulting roundness deviation of the spindle $S(\theta)$ and ball $B(\theta)$ are given in Figure 5.39. As with straightness reversal, the hardware and software of the roundness tester must enable a measurement in the reverse configuration in the first place, and the measurement software must subsequently enable the manipulation of the data according to Equation 5.13.

For the cases where the reversal is physically impossible, so-called multi-step methods are used. In multi-step methods, the sphere is rotated in a number of steps, and the measurement is taken in the standard configuration every time. The effect of the sphere roundness deviation is averaged if the mean value of all these measurements is taken, and the spindle effect is reduced when the roundness measurement results are rotated to keep the orientation of the sphere constant. A more robust analysis shows that some harmonic components of the sphere and spindle deviations will be separated, and some are not (see Haitjema 2015b).

Such reversal measurements take much more time and effort than a standard roundness measurement, so a cost-benefit analysis needs to be undertaken to determine whether the extra effort and smaller uncertainty is really needed for the measurement. Common practice is that roundness testers are supplied with a reference sphere or hemisphere that is used to check the roundness tester in its normal configuration. Usually, such a reference sphere has a small roundness deviation with a small uncertainty so that the required check consists of a single measurement of the reference, and when the resulting roundness is small enough, it is concluded that the spindle is within specification. The calibration of such reference spheres takes place using the mentioned reversal methods or variations of these (Haitjema et al. 1996).

5.8.4 Square Reversal

Reversal methods can be used to determine the squareness of metal or granite squares. In its simplest form, the same setup and methods as described in Section 5.8.2 are used, and the same Equations 5.10 and 5.11 apply. However, with square reversal, it is essential that the reference lines are not removed from the equations, and that the reversal is carried out carefully so that the reference horizontal line is not changed. A somewhat more elaborate method is shown in Figure 5.40, where a cylindrical square is used on a surface plate. A cylindrical square is a round bar whose sides are at right angles to its ends. The sides must be parallel and the ends must be square to the sides. The indicator can be attached and moved to two sides of the cylindrical square A, whose two sides deviate by angles of β and γ from true squareness from the base. The square to be measured, square B, deviates with an angle α from true squareness. The square's angle α is measured relative to the angles β and γ of the cylinder, giving measurements $I_1(z)$ and $I_2(z)$. The surface plate that acts as a base for these measurements is assumed to have no significant straightness deviations in the area where it is used for those measurements. The reversal is obtained by reversing the cylinder A, which means in this case that it is put upside-down. In this position, the square angle α is measured relative to the angles δ and ε, giving measurements $I_3(z)$ and $I_4(z)$. The measurements $I_j(z)$ can be transformed to angles I_j from the slopes of a least-squares fit to the outputs from a displacement indicator when moved along the cylinder. In this situation, the following equations apply:

$$I_1 = \alpha + \gamma$$

$$I_2 = \alpha + \beta$$

$$I_3 = \alpha + \delta$$

$$I_4 = \alpha + \varepsilon = \alpha - \gamma - \beta - \gamma,$$

(5.14)

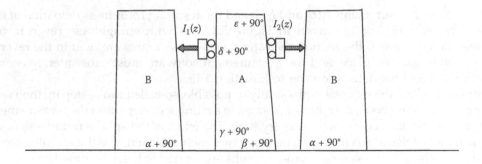

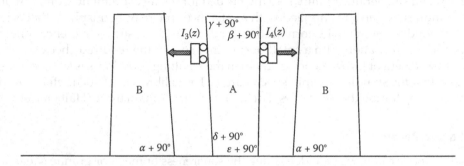

FIGURE 5.40
Schematic of a reversal squareness measurements. As a reference, a squareness cylinder A is used that measures the object B squareness deviation angle α relative to its two reference angles β and γ. After reversing the cylinder A, putting it upside-down, the measurement is repeated.

where the fundamental property

$$\beta + \gamma + \delta + \varepsilon = 0 \tag{5.15}$$

has been used. The sum of the angles of a four-sided square is defined as 360°. This is essentially stated in Equation 5.15 and it is used to solve Equation 5.14. In general, an operation such as that represented by Equation 5.15, is called 'closing the circle'; this is an important principle that is often used in angle metrology.

Solving Equation 5.14 gives

$$\alpha = \frac{I_1 + I_2 + I_3 + I_4}{4}$$

$$\beta = \frac{3I_2 - I_1 - I_3 - I_4}{4} . \tag{5.16}$$

5.8.5 Further Aspects of Reversal Methods

Section 5.8.1 to Section 5.8.4 present some elementary error separation techniques. These can be further extended to three or more surfaces for straightness measurements, additional rotations for flatness measurements and to more measurements for angle. Other techniques can result in a reduction of form deviations rather than an elimination, for example when a sphere is measured in random orientations on a roundness tester, the average profile will converge to the spindle deviation. This principle can also be used in straightness and flatness measurement. In addition to the question, What is the reference?, a further question is, What reversal or error separation technique can be used to eliminate form deviations from the measurement? It is also noted that similar approaches can be used to manufacture specific geometries such as square and flat surfaces for bearings and other applications, and is briefly discussed in Chapter 7, Section 7.2.1.2.

Exercises

1. Briefly describe the role of length standards in the traceability chain. Draw a diagram that links the definition of the metre to a length measurement on the production floor.
2. A steel gauge block (CTE = $(11.5 \pm 0.1) \times 10^{-6}$ K^{-1}) is used as the reference for comparative measurement of a similar length test block. The length of the steel gauge block was calibrated by interferometry at standard reference temperature (20 °C) and its calibrated length is l_{cal} = 35.0004 mm ± 0.000020 mm. What is the actual length l_{actual} of the steel gauge block if the comparative measurement is performed in a laboratory with a temperature of (21.5 ± 0.1) °C? Consider elastic compression from the probe to be negligible.
3. Why is fringe counting necessary for interferometric displacement measurements larger than half the wavelength of light?
4. A stabilised helium-neon laser (stable wavelength λ = 632.8 nm) is used in Michelson-type interferometric displacement measurements. The fringe-counter can discriminate the phase of the measured signal to $\dfrac{\pi}{50}$ radians, that is $\delta\varphi = \dfrac{\pi}{50}$. What is the smallest displacement δx in the test retroreflector that the interferometer can measure?
5. Briefly describe the roughness parameters Ra and Rz. Specify a limitation of each parameter.
6. Show that the difference between Ra and Rq for a sinusoidal surface is approximately 11%. Hint: Integrate between 0 and $\pi/2$.
7. Briefly describe the differences between stylus and optical surface texture measuring instruments. Explore the different types of instruments in terms of theory of operation and/or performance parameters of commercial instruments.
8. A stylus instrument has a tip radius of r. What is the minimum wavelength that it can faithfully measure, i.e. recover all the amplitude? Assume a sinusoidal surface.
9. A stylus instrument and an interferometer (λ = 633 nm) measure the height of a chrome step on a glass substrate. The stylus instrument measures the height as 150 nm. Given that the refractive index of air is unity, and the refractive index and absorption index of the chrome are 3.212 and 3.300 respectively, calculate the difference between the heights measured by the two instruments.
10. A sphere with a slightly elliptical form that gives a roundness deviation of 1 μm is measured on a roundness tester, where the spindle has an elliptical deviation in the same orientation, giving also 1 μm deviation.

   a. What roundness deviation will a roundness tester measure when operated normally?

b. The sphere is rotated 180° on the roundness tester. The probe is unchanged. What roundness deviation will the roundness tester measure?

c. The probe is moved over the sphere and measures the sphere from the other side. A measurement is taken. What deviation will the roundness tester measure?

d. Draw the roundness diagrams of the normal and reversal measurement, and conclude from there the roundness diagrams of the spindle and sphere.

References

ASME Y14.5. 2009. Dimensioning and tolerancing. American Society of Mechanical Engineers, New York.

Balsamo, A., Di Ciommo, M., Mugno, R., Rebaglia, B. I., Ricci, E. and Grella, R. 1999. Evaluation of CMM uncertainty through Monte Carlo simulations. *Ann. CIRP* **48**:425–428.

Bruzzone, A. A. G., Costa, H. L., Lonardo, P. M. and Lucca, D. A. 2008. Advances in engineering surfaces for functional performance. *Ann. CIRP* **47**:750–769.

Carmignato, S., Dewulf, W. and Leach, R. K. 2017. *Industrial X-ray computed topography*. Springer.

CIPM. 1960. New definition of the meter: The wavelength of krypton-86. Proc. 11th General Council of Weights and Measures, Paris, France.

CIPM. 1984. Documents concerning the new definition of the metre. *Metrologia* **19**:165–166.

Coveney, T. 2014. *Dimensional measurement using vision systems*. NPL Good Practice Guide No. 39, National Physical Laboratory.

De Chiffre, L., Carmignato, S., Kruth, J.-P., Schmitt, R. and Weckenmann., A. 2014. Industrial applications of computed tomography. *Ann. CIRP* **63**:655–677.

De Groot, P., Badami, V. and Liesener, J. 2016. Concepts and geometries for the next generation of heterodyne optical encoders. Proc. ASPE, Portland, Oregon, October, pp. 146–149.

Doiron, T. 1988. Grid plate calibration at the National Bureau of Standards. *J. Res. NIST* **93**:41–51.

Doiron, T. and Beers, J. 2005. *The gauge block handbook*. NIST Monograph 180.

Doiron, T. and Stoup, J. 1997. Uncertainty and dimensional calibrations. *J. Res. NIST* **102**:647–676.

Egerton, R. F. 2008. *Physical principles of electron microscopy: An introduction to TEM, SEM and AEM*, 2nd edn. Springer, Heidelberg.

Ellis, J. 2014. *Field guide to displacement measuring interferometry*. SPIE Press.

Evans, C. J., Hocken, R. J. and Estler, T. W. 1996. Self-calibration: Reversal, redundancy, error separation and 'absolute testing'. *Ann. CIRP* **45**:617–635.

Ferrucci, M., Leach, R. K., Giusca, C. L., Dewulf, W. and Carmignato, S. 2015. Towards geometrical calibration of X-ray computed tomography systems—A review. *Meas. Sci. Technol.* **26**: 092003.

Flack, D. R. 2001. *CMM probing*. NPL Good Practice Guide No. 43, National Physical Laboratory.

Flack, D. R. 2013. *Co-ordinate measuring machines task specific measurement uncertainties*. NPL Good Practice Guide No. 130, National Physical Laboratory.

Flack, D. R. and Hannaford, J. 2005. *Fundamental good practice in dimensional metrology*. NPL Good Practice Guide No. 80, National Physical Laboratory.

Fleming, J. 2013. A review of nanometer resolution position sensors: Operation and performance. *Sensors & Actuators A: Phys.* **190**:106-126.

Forbes, A. B. 2013. Areal form removal. In: Leach, R. K. *Characterisation of areal surface texture*, chap. 5. Springer, Berlin.

Forbes, A. B. and Minh, H. D. 2012. Generation of numerical artefacts for geometric form and tolerance assessment. *Int. J. Metrol. Qual. Eng.* **3**:145–150.

Forbes, A. B., Smith, I. M., Härtig, F. and Wendt, K. 2015. Overview of EMRP joint research project NEW06 "Traceability for computationally-intensive metrology." In: *Advanced mathematical and computational tools in metrology and testing X*, pp. 164–170. World Scientific.

Giusca, C. L. and Leach, R. K. 2013. Calibration of the scales of areal surface topography measuring instruments: Part 3—Resolution. *Meas. Sci. Technol.* **24**:105010.

Haitjema, H. 2015a. Uncertainty in measurement of surface topography. *Surf. Topog.: Metr. Prop.* **3**:035004.

Haitjema, H. 2015b. Revisiting the multi-step method: Enhanced error separation and reduced amount of measurements. *Ann. CIRP* **64**:491–494.

Haitjema, H., Bosse, H., Frennberg, M., Sacconi, A. and Thalmann, R. 1996. International comparison of roundness profiles with nanometric accuracy. *Metrologia* **33**:67–73.

Harasaki, A., Schmit, J. and Wyant, J. C. 2001. Offset of coherent envelope position due to phase change on reflection. *Appl. Opt.* **40**:2102–2106.

Harding, K. 2014. *Handbook of optical dimensional metrology*. CRC Press.

Hocken, R. J. and Pereira, P. 2011. *Coordinate measuring machines and systems*, 2nd edn. CRC Press.

ISO 3650. 1998. *Geometrical product specifications (GPS)—Length standards—Gauge block*. International Organization on Standardization.

ISO 4287. 2000. *Geometrical product specification (GPS)—Surface texture: Profile method—Terms, definitions and surface texture parameters*. International Organization of Standardization.

ISO 10110 part 8. 2010. *Optics and photonics—Preparation of drawings for optical elements and systems— Part 8: Surface texture; roughness and waviness*. International Organization for Standardization.

ISO 10360 part 1. 2000. *Geometrical product specifications (GPS)—Acceptance and reverification tests for coordinate measuring machines (CMM)—Part 1: Vocabulary*. International Organization for Standardization.

ISO 10360 part 2. 2009. *Geometrical product specifications (GPS)—Acceptance and reverification tests for coordinate measuring machines (CMM)—Part 2: CMMs used for measuring size*. International Organization for Standardization.

ISO 10360 part 3. 2000. *Geometrical product specifications (GPS)—Acceptance and reverification tests for coordinate measuring machines (CMM)—Part 3: CMMs with the axis of a rotary table as the fourth axis*. International Organization for Standardization.

ISO 10360 part 4. 2000. *Geometrical product specifications (GPS)—Acceptance and reverification tests for coordinate measuring machines (CMM)—Part 4: CMMs used in scanning measuring mode*. International Organization for Standardization.

ISO 10360 part 5. 2010. *Geometrical product specifications (GPS)—Acceptance and reverification tests for coordinate measuring machines (CMM)—Part 5: CMMs using single and multiple-stylus contacting probing systems*. International Organization for Standardization.

ISO 10360 part 6. 2001. *Geometrical product specifications (GPS)—Acceptance and reverification tests for coordinate measuring machines (CMM)—Part 6: Estimation of errors in computing Gaussian associated features*. International Organization for Standardization.

ISO 10360 part 7. 2011. *Geometrical product specifications (GPS)—Acceptance and reverification tests for coordinate measuring machines (CMM)—Part 7: CMMs equipped with imaging probing systems*. International Organization for Standardization.

ISO 10360 part 8. 2013. *Geometrical product specifications (GPS)—Acceptance and reverification tests for coordinate measuring machines (CMM)—Part 8: CMMs with optical distance sensors*. International Organization for Standardization.

ISO 10360 part 9. 2013. *Geometrical product specifications (GPS)—Acceptance and reverification tests for coordinate measuring machines (CMM)—Part 9: CMMs with multiple probing systems*. International Organization for Standardization.

ISO 10360 part 10. 2016. *Geometrical product specifications (GPS)—Acceptance and reverification tests for coordinate measuring machines (CMM)—Part 10: Laser trackers for measuring point-to-point distances*. International Organization for Standardization.

ISO 12180 part 1. 2011. *Geometrical product specifications (GPS)—Cylindricity Part 1: Vocabulary and parameters of cylindrical form.* International Organization on Standardization.

ISO 12181 part 1. 2011. *Geometrical product specifications (GPS)—Roundness Part 1: Vocabulary and parameters of roundness.* International Organization on Standardization.

ISO 12780 part 1. 2011. *Geometrical product specifications (GPS)—Straightness Part 1: Vocabulary and parameters of straightness.* International Organization on Standardization.

ISO 12781 part 1. 2011. *Geometrical product specifications (GPS)—Flatness Part 1: Vocabulary and parameters of flatness,* International Organization on Standardization.

ISO 15530 part 3. 2011. *Geometrical product specifications (GPS)—Coordinate measuring machines (CMM): Technique for determining the uncertainty of measurement—Part 3: Use of calibrated workpieces or measurement standards.* International Organization for Standardization.

ISO/TS 15530 part 4. 2008. *Geometrical product specifications (GPS)—Coordinate measuring machines (CMM): Technique for determining the uncertainty of measurement—Part 4: Evaluating CMM uncertainty using task specific simulation.* International Organization for Standardization.

ISO 25178 part 6. 2010. *Geometrical product specification (GPS)—Surface texture: Areal—Part 6: Classification of methods for measuring surface texture.* International Organization for Standardization.

ISO 25178 part 601. 2010. *Geometrical product specifications (GPS)—Surface texture: Areal—Part 601: Nominal characteristics of contact (stylus) instruments.* International Organization for Standardization.

ISO 25178 part 602. 2010. *Geometrical product specification (GPS)—Surface texture: Areal—Part 602: Nominal characteristics of non-contact (confocal chromatic probe) instruments.* International Organization for Standardization.

ISO 25178 part 603. 2013. *Geometrical product specification (GPS)—Surface texture: Areal—Part 603: Nominal characteristics of non-contact (phase shifting interferometric microscopy) instruments.* International Organization for Standardization.

ISO 25178 part 604. 2013. *Geometrical product specification (GPS)—Surface texture: Areal—Part 604: Nominal characteristics of non-contact (coherence scanning interferometry) instruments.* International Organization for Standardization.

ISO 25178 part 605. 2014. *Geometrical product specification (GPS)—Surface texture: Areal—Part 605: Nominal characteristics of non-contact (point autofocusing) instruments.* International Organization for Standardization.

ISO 25178 part 606. 2012. *Geometrical product specification (GPS)—Surface texture: Areal—Part 606: Nominal characteristics of non-contact (imaging confocal) instruments.* International Organization for Standardization.

ISO 25178 part 607. 2016. *Geometrical product specification (GPS)—Surface texture: Areal—Part 607: Nominal characteristics of non-contact (focus variation) instruments.* International Organization for Standardization.

Jäger, G., Manske, E., Hausotte, T., Müller, A. and Balzer F. 2016. Nanopositioning and nano-measuring machine NPMM-200—A new powerful tool for long-range micro- and nanotechnology. *Surf. Topog.: Met. Prop.* **4**:034004.

Jones, C. W. and Leach, R. K. 2008. Adding a dynamic aspect to amplitude-wavelength space. *Meas. Sci. Technol.* **19**:055105.

Leach, R. K. 2011. *Optical measurement of surface topography.* Springer.

Leach, R. K. 2013. *Characterisation of areal surface texture.* Springer.

Leach, R. K. 2014a. *Fundamental principles of engineering nanometrology,* 2nd edn. Elsevier.

Leach, R. K. 2014b. *The measurement of surface texture using stylus instruments.* NPL Good Practice Guide. No. 37, National Physical Laboratory.

Leach, R. K. 2015. Surface texture. In: Laperrière, L. and Reinhart, G. *CIRP Encyclopaedia of production engineering.* Springer-Verlag, Berlin.

Leach, R. K., Hart, A. and Jackson, K. 1999. *Measurement of gauge blocks by interferometry: An investigation into the variability in wringing film thickness.* NPL Report CLM 3.

Lee, C.-O., Park, K., Park, B. C. and Lee, Y. W. 2005. An algorithm for stylus instruments to measure aspheric surfaces. *Meas. Sci. Technol.* **16**:1215.

Lee, E. S. and Burdekin, M. 2001. A hole plate artifact design for volumetric error calibration of a CMM. *Int. J. Adv. Manuf. Technol.* **17**:508–515.

Magonov, S. 2008. *Atomic force microscopy*. John Wiley & Sons.

McCool, J. I. 1984. Assessing the effect of stylus tip radius and flight on surface topography measurements. *ASME J. Tribol.* **106**:202–209.

Mendeleyev, V. 1997. Dependence of measuring errors of rms roughness on stylus tip size for mechanical profilers. *Appl. Opt.* **36**:9005–9009.

Meyer, E., Hug, H. J. and Bennewitz, R. 2003. *Scanning probe microscopy: The lab on a tip*. Springer, Berlin.

Morris, A. S. 2001. *Measurement and instrumentation principles*, 2nd edn. Butterworth-Heinemann, Oxford.

Muralikrishnan, B. and Raja, J. 2008. *Computational surface and roundness metrology*. Springer.

Muralikrishnan, B. and Raja, J. 2010. *Computational surface and roundness metrology*. Springer.

Quinn, T. 2012. *From artefacts to atoms*. Oxford University Press.

Quinn, T. and Kovalevsky, J. 2005. The development of modern metrology and its role today *Philos. Trans. R. Soc. A* **363**:2307–2327.

Schmitt, R., Peterek, M., Morse, E., Knapp, W., Galetto, M., Härtig, F., Goch, G., Hughes, E. B., Forbes, A. and Estler, W. T. 2016. Advances in large-volume metrology—Review and future trends. *Ann. CIRP* **65**:643–666.

Schott AG. n.d. ZERODUR® Extremely Low Expansion Glass Ceramic, http://www.schott.com /advanced_optics/zerodur.

Schwenke, H., Knapp, W., Haitjema, H., Weckenmann, A., Schmitt, R. and Delbressine, F. 2008. Geometric error measurement and compensation for machines—An update. *Ann. CIRP* **57**:660–675.

Seewig, J. 2011. Areal filtering methods. In: Leach, R. K., *Characterisation of areal surface texture*, chap. 4. Springer.

Stedman, M. 1987. Basis for comparing the performance of surface-measuring machines. *Precision Engineering* **9**:149–152.

Swyt, D. A. 2001. Length and dimensional measurements at NIST. *J. Res. NIST* **106**:1–23.

Thalmann, R., Meli, F. and Küng, A. 2016. State of the art of tactile micro coordinate metrology. *Appl. Sci.* **6**(5):150.

Vermeulen, M. M. P. A., Rosielle, P. C. J. N. and Schellekens, P. H. J. 1998. Design of a high-precision 3D-coordinate measuring machine. *Ann. CIRP* **47**:447–450.

Webster, J. G. and Eren, H. 2014. *Measurement, instrumentation, and sensors handbook: Spatial, mechanical, thermal, and radiation measurement*. CRC Press, Boca Raton, FL.

Whitehouse, D. J. 2010. *Handbook of surface and nanometrology*, 2nd ed. CRC Press.

Widdershoven, I., Donker, R. L. and Spaan, H. A. M. 2011. Realization and calibration of the "Isara 400" ultra-precision CMM. *J. Phys.: Conf. Ser.* **311**:012002.

Wilson, J. S. 2005. *Sensor technology handbook*. Newnes, Oxford, UK.

Zhang, G. X. and Fu, J. Y. 2000. A method for optical CMM calibration using a grid plate. *Ann CIRP*. **49**:399–402.

6

Kinematic Design

Shah Karim and Ulrich Weber

CONTENTS

ABSTRACT Kinematics is the study of motion, without regard for the cause of the motion, of a rigid body in space and its relation to other bodies. In designing precision mechanics, kinematic considerations are always important when parts of a mechanism move relative to each other both in operating conditions and during assembly of the mechanism. The accuracy and stability of precision mechanics can be improved significantly by following fundamental kinematic principles. Since the role of kinematics is to ensure the functionality of the mechanism, to be effective, a kinematic approach to design will include: (1) adherence to kinematic principles, (2) evaluation of kinematic design options based on practical considerations and (3) design, such as position, velocity and acceleration analysis, of the mechanism or machine. Hence, this chapter is organised focusing on the first two criteria, while the third is introduced in Chapter 8 (system modelling). In this chapter, the basic kinematic principles are explained, followed by a discussion of some practical considerations that demonstrate the motivation for the development of pseudo-kinematic design as opposed to pure kinematic design. Readers will also be familiarised with fundamental kinematic structures most commonly used in precision machines and mechanisms. Illustrated examples are used to demonstrate how kinematic principles are realised in design.

6.1 Introduction

The treatment of kinematics in the design of a machine, mechanism or any moving mechanical system is to establish (1) the functional relationships between the parts of a mechanism, machine or other mechanical system; (2) how these parts are interconnected; and (3) how these parts move relative to each other. These three factors are the fundamental and primary considerations of a design process about which designers must make their choices before other factors such as materials, manufacturing methods and costs are considered. Failure to give proper attention to the kinematics of a mechanism, machine or other mechanical system may lead to wrong choices for other design factors, such as bearing types and manufacturing tolerances, which are involved at the later stage of the design process, resulting in high design cost and/or substandard performance and/or unsatisfactory reliability (Eckhardt 1998, Mason 2001, Vinogradov 2000).

The following terms will be used throughout the chapter. A mechanism is a mechanical device that transfers motion and/or force from a source to an output. The term machine is usually used to indicate a complete product. As such, a car is a machine, but a windshield wiper is a mechanism. Both machines and mechanisms may denote assemblies and thus are comprised of parts. However, the identification of something as machine or mechanism has no bearing in kinematics discussion (Rider 2015).

A linkage consists of links generally considered as rigid bodies which are connected by joints, such as pivot joints or ball joints. Rigid body means that the body resists deformation so that deformation is insignificant and does not need to be considered as a contributor to the motion of components of the mechanism. Joints serve to constrain the motion of the links or rigid bodies so that they are not free to move with what would otherwise be six degrees of freedom for each rigid body. Constraints are also considered to be ideal so that, for example, a single degree of freedom hinge provides only a relative rotation between two links to which it connects, and no work is required by the joint. A kinematic chain with at least one fixed link is regarded as a mechanism, if at least two other links can move. Linkages can form simple mechanisms and can be designed to perform complex tasks (Rider 2015).

A large majority of mechanisms exhibit motion in such a way that all the links move in parallel planes, referred to as two-dimensional planar motion. Other mechanisms exhibit motion in three-dimensional space, referred to as three-dimensional spatial motion.

The displacement of a point is the difference between its position after motion (final position) and its position before that motion (initial position). It can be represented by a three-dimensional vector drawn from the initial position of the point to its final position (components of the displacement vector are measured in the reference coordinate frame). Note that many different paths and times can be used to produce the same displacement.

Rigid body displacements are more complicated than point displacements, as more than three parameters are needed to describe them. Rigid body displacements are the difference between the final position of the body and the initial position, all measured relative to a reference coordinate frame. Note that while a point in space is represented by its three coordinates relative to a reference coordinate frame, a rigid body requires a matrix to define its position (a position vector is described by three coordinate values in the matrix) and orientation (three directional vectors are described by nine coordinate values in the matrix) in space; all measurements are with respect to the reference coordinate frame (discussed in detail in Chapter 8).

6.2 Degrees of Freedom

In kinematics, the number of degrees of freedom indicates the number of different and independent modes of motion of a body. A mode of motion can be described as a rigid body translation, rigid body rotation or a combination of both (called a screw motion).

In a rigid body translation, every point of the rigid body performs the same linear displacement concerning magnitude and orientation, as illustrated in Figure 6.1, for the corner points A, B and C. The displacement vector from the start to the end position of every point is of equal length and parallel to the displacement vectors of other points.

In a rigid body rotation, all points of the body a move on concentric circular arcs, with the same rotation angle ϕ_{rot} around the rotation axis a_{rot}, as shown in Figure 6.2. Points on the rotation axis a_{rot} are not shifted.

In Figure 6.3, the rotation axis a_{rot} is perpendicular to the plane of the paper, so that the rotation axis a_{rot} is shown as a point in B and the perpendicular bisector planes m_A, m_D and m_E of the displacement vectors $\vec{u}_A$, $\vec{u}_D$ and $\vec{u}_E$ as lines. Figure 6.3 shows that the perpendicular bisector plane of a displacement vector always intersects with the rotation axis a_{rot}. The displacement vectors of every point on a connecting line from one point of the rigid body to the rotation axis are parallel to each other and the magnitude increases linearly with the distance to the rotation axis as the displacement vectors $\vec{u}_A$ and $\vec{u}_D$ in Figure 6.3 show. Note that the rotation axis a_{rot} is always oriented perpendicular to the displacement vectors and defines the orientation of the rigid body rotation.

In a planar mechanism, all displacement vectors at all points of all parts have to be parallel to a single plane at the same time (planar kinematics). For the purposes of the following discussions, the plane will be considered flat and, therefore, easily specified with Cartesian coordinates (in fact this is often called 2D Cartesian, or Euclidean, space). Rigid body rotation is possible only about axes that are perpendicular to this plane. The trajectory of the mechanism can be different from the displacement vector, which connects the end position to the start position with a straight line.

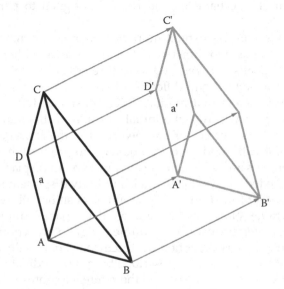

FIGURE 6.1
Translation of a rigid body with parallel and equidistant displacement vectors.

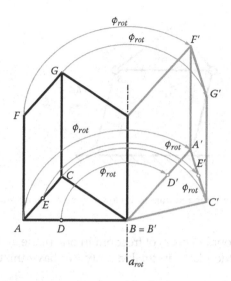

FIGURE 6.2
Rotation of a rigid body with rotation axis a_{rot} in perspective view.

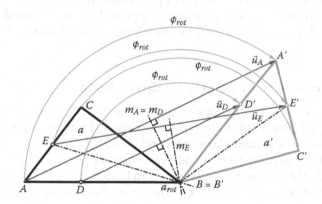

FIGURE 6.3
Rotation of a rigid body a with the points A, B and C, with the rotation axis a_{rot}; displacement vectors $\overrightarrow{u_A}$, $\overrightarrow{u_B}$, $\overrightarrow{u_C}$ and perpendicular bisectors m_A, m_B, m_C intersecting the rotation axis a_{rot}.

In Figure 6.4, the triangle a with the corner points A, B and C is shifted to the triangle a' with the corner points A', B' and C' by a parallelogram guiding mechanism. The corner points A, B and C move along the trajectories s_A, s_B and s_C, which are circular arcs, whereas the displacement vectors $\vec{u}_A$, $\vec{u}_B$ and $\vec{u}_C$ are straight, parallel and of equal lengths. The triangle a is shifted in a translation in spite of the curved trajectories s_A, s_B and s_C.

6.2.1 Degrees of Freedom of a Body in a Plane (Planar Motion)

A body can move in a plane with rigid body translations parallel to this plane and with a rigid body rotation perpendicular to this plane. For the displacement in every arbitrary position in the plane, two rigid body translations along independent directions are necessary.

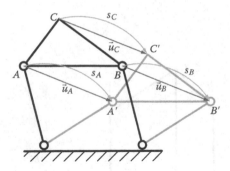

FIGURE 6.4
Difference between displacement vector and trajectory shown by means of a parallelogram.

The body has two translational degrees of freedom in one plane and one rotational degree of freedom perpendicular to the plane. In total, a body can have three degrees of freedom in a plane.

6.2.2 Degrees of Freedom of a Body in Spherical Kinematics

A spherical linkage is characterised by pivot joints with rotation axes intersecting at one point. Instead of two translational degrees of freedom and one rotational degree of freedom, a body in a spherical linkage has three rotational degrees of freedom with axes intersecting at one point.

6.2.3 Degrees of Freedom of a Body in Space (Spatial Motion)

There are three rigid body translations in independent directions necessary for the displacement of a body in every arbitrary position in space. In addition, there are three rotational degrees of freedom around three independent axes necessary for rotating the body around every arbitrary axis. In total, a body can have six degrees of freedom in space, as shown in Figure 6.5. The six degrees of freedom can be described by pure translations $\vec{u}_x$, $\vec{u}_y$ and $\vec{u}_z$ and pure rotations $\overrightarrow{\varphi_x}$, $\overrightarrow{\varphi_y}$ and $\overrightarrow{\varphi_z}$. However, they can also be represented by six independent modes of motion consisting of combinations of translations and rotations, where the translation and rotation are coupled.

A screw is a clear example of a mode of motion consisting of a combination of rotation and coupled translation. A screw motion is the most common mode of motion in space. Screw motion is discussed in Section 6.6 under spatial kinematics.

6.3 Constraints

In kinematics, the addition of constraints will reduce the number of degrees of freedom of a body and force it to move along specific trajectories. Constraints are formed usually by bearings typically configured to provide rotating or sliding joints (see Chapter 7). By transmitting reaction forces and reaction torques between two parts of a mechanism, bearings prevent locally defined modes of relative motion between two parts of a

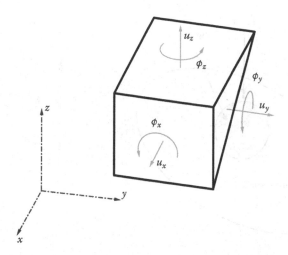

FIGURE 6.5
Six degrees of freedom for a body in space.

mechanism, such as translation or rotation, while constraining the parts to move along trajectories in the remaining degrees of freedom. The forces' torques are transmitted by sliding contacts or via intermediate rolling elements. The examination of prevented (con-strained) and possible (free) relative motion in a bearing, expressed in terms of its degrees of freedom, is the basis for the kinematical analysis of a mechanism. Other forms of constraints often either comprise or can be effectively modelled by surface contacts between different parts of mechanism.

6.3.1 Surface Contacts As Joints and Their Constraints

Consider a model of a simple mechanism in which a single degree of constraint can be attained by an infinitely rigid sphere pressed against an infinitely rigid, flat and frictionless surface (Figure 6.6). In this case, motion is restricted only in the vertical direction, say z, leaving the sphere to rotate about any of the three axes and to slide in the x and y directions. Thus, in the absence of the plane, the ball has six degrees of freedom, and introducing a point contact to the surface reduces that to five degrees of freedom, which means the contact imposes a single constraint (and hence, single degree of constraint).

Consider another model of a mechanism in which two spheres are joined together to make a rigid body and placed onto a surface (see Figure 6.7). In this model, the constraints are in the z axis and either the α or β axis, providing four degrees of freedom. Note that the first constraint comes from the point contact between the rigid sphere and the plane, while the second constraint comes from the fact that adding each link to a mechanism removes a rotary degree of freedom. An alternative way of achieving two degrees of constraint (in this case, in the z axis and either in the x or y axis) is by placing a single rigid sphere in a groove, producing four degrees of freedom (three rotations and a translation along the groove). Other models, all representing three degree of constraints, are shown in Figure 6.7. The examples demonstrate a very important principle of kinematic design that uses the number of contact points to construct useful models for predicting the motion of mechanisms in practice: *the number of contact points between any two perfectly rigid bodies is equal to the number*

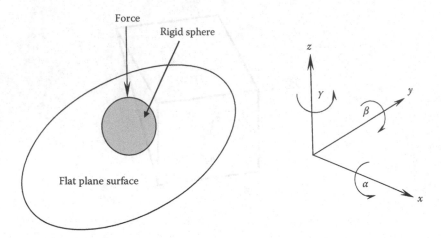

FIGURE 6.6
(Left) A single degree of freedom provided by a sphere on a flat surface. (Right) The degrees of freedom of the Cartesian coordinate system.

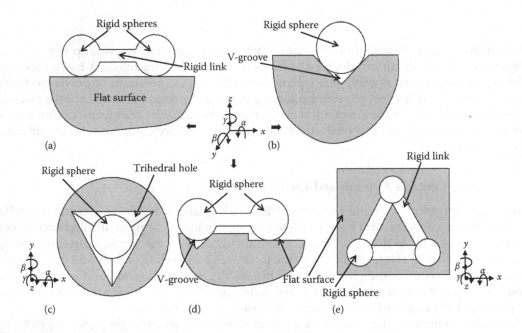

FIGURE 6.7
Models of idealized kinematic constraints: (a) two spheres, rigidly connected, resting on a flat surface, (b) a sphere in a V-groove, (c) a single sphere in a trihedral hole, (d) two spheres rigidly connected, one of which is resting in a V-groove and the other one is on a flat surface, (e) three rigidly attached spheres on a flat surface.

of their mutual constraints (i.e. degree of constraints) (Smith and Chetwynd 2003). Furthermore, the models shown in Figures 6.6 and 6.7 are often used in design to approximate the pure kinematic or semi-kinematic (pseudo-kinematic) behaviours of mechanisms (see Section 6.8).

Surface contacts can be characterised by the nominal contact points that produce the required degree of constraint to provide the mechanism with the desired degree of freedom. The nominal contact points are often determined by the intersection of both contact surfaces, for example, a plane-to-plane contact, where three points define a plane or a cone to sphere contact with an intersection circle defined by three points. Where the contact surfaces have high surface texture and form tolerances, surfaces can be in contact only at a small number of contact points.

However, characterising surface contacts requires careful considerations since there are instances for which contact might not be possible. One example is the contact of four points on a flat plane. Mathematically this is only possible for the contact points to be located on a flat plane. An example of this is the four-legged chair for which, because of the tolerances in manufacture of the chair and floor, contact of all legs with a floor is often only achievable if the legs elastically distort. This is common for adults who impose sufficient force for this to occur but less so for young children who happily rock back and forth between two three-point supports. Another instance is the contact between a link comprised of three spheres and two parallel V's. Again, unless the contacts exactly conform to the geometry of the V's, only five contacts will be possible. Mathematically, these additional requirements will present an indeterminate contact condition and result in an over-constrained mechanism. In practice, such over-constrained mechanisms rely upon elastic or plastic deformation to provide the desired function (see Chapter 7).

6.3.2 Surface Contacts in Planar Kinematics

The subject of planar kinematics considers contacts as either points or lines. In practice, any load applied to these 'ideal' contacts would produce unacceptably high stresses. As a consequence, such constraints are provided by contacts between bodies of spherical, cylindrical and other geometric shapes that can be considered approximations to this ideal. To understand the effect of these contacts on the subsequent mobility of the assembled mechanism (see Section 6.4), it is necessary to be able to specify the degree of freedom f_j of a joint j. For any joint within a planar mechanism, its freedom can be computed from the difference between the three unconstrained degrees of freedom in a plane and the lowest number of contact points, n_{pj}, which can be used to approximate the ideal contacts between the bodies of a joint:

$$f_j = 3 - n_{pj}. \qquad (6.1)$$

6.3.2.1 Point Contacts in Planar Kinematics

In a point contact, two bodies a and b are in contact at one point P, as shown in Figure 6.8. The bodies a and b cannot move relative to each other along the surface normal n_y of the contact point P without intersecting each other or lifting off.

If a lift off of a and b is not possible at P, the point contact prevents a displacement parallel to the surface normal n_y at P. The motion at the contact point is constrained to a translational displacement u_x perpendicular to n_y and a rotation ϕ_z around an axis perpendicular to the plane of the paper. The point contact is a joint with two degrees of freedom. In this case, Figure 6.8 represents a two-freedom joint with constraint in the y coordinate direction.

6.3.2.2 Straight Line Contacts in Planar Kinematics

Based on two point contacts, straight or arcuate line (in this case, the line of the arc comprises the chord of a circle) contacts can be formed. In the straight line contact, all contact points are situated on a straight line having parallel surface normals relative to each other.

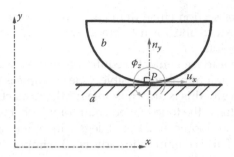

FIGURE 6.8
Point contact in planar kinematics.

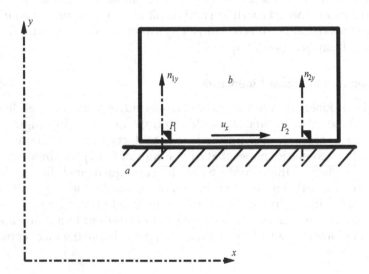

FIGURE 6.9
Straight-line contact in planar kinematics.

The straight line contact prevents a translational displacement along the surface normal n_{1y} and n_{2y} and rotation around an axis perpendicular to the plane of the paper, as shown in Figure 6.9. The motion between the two parts a and b is constrained to a translational displacement u_x perpendicular to n_{1y} and n_{2y}. The straight line contact is a joint with one degree of freedom, in the direction of coordinate x as illustrated in Figure 6.9.

6.3.2.3 Arcuate Line Contacts in Planar Kinematics

The arcuate line contact is based on two point contacts. The surface normals of all points of the contact line intersect at a common intersection point O, which is the centre of curvature of the contact line. All points of the contact line have the same distance to O. The arcuate line contact prevents all translations in the plane spanned by the surface normal n_{1y} and n_{2y} of the contact points and constrains the motion between the parts a and b to a rotation ϕ_z around an axis perpendicular to the plane spanned by n_{1y} and n_{2y}, as shown in Figure 6.10. The arcuate line contact is a joint with one degree of freedom, in this case a rotation about an axis normal to the plane and located at intersection point O.

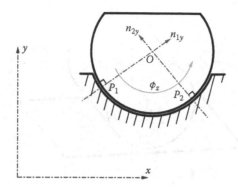

FIGURE 6.10
Arcuate line contact in planar kinematics.

6.3.3 Surface Contacts in Spatial Kinematics

There are point contacts, line contacts and area contacts in spatial kinematics. For any joint within a spatial mechanism, its freedom can be computed from the difference between the six unconstrained degrees of freedom in a space and the lowest number of contact points, n_{pj}, which can be used to approximate the ideal contacts between the bodies of a joint:

$$f_j = 6 - n_{Pj}. \qquad (6.2)$$

6.3.3.1 Point Contacts in Spatial Kinematics

In a point contact, two bodies a and b are in contact at one point P, as shown in Figure 6.11. The bodies a and b cannot move translationally along the surface normal n_z relative to each other without intersecting or lifting off. If a lift off of the two bodies a and b at P is not possible, the point contact prevents a translation parallel to the surface normal n_z. The motion at the contact point is constrained to three rotations φ_x, φ_y and φ_z and to two translations u_x and u_y perpendicular to n_z. The point contact is the basis for all other surface contacts. All other surface contacts can be formed by a combination of several point contacts.

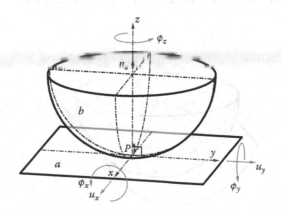

FIGURE 6.11
Point contact in spatial kinematics.

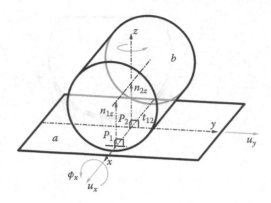

FIGURE 6.12
Straight-line contact in spatial kinematics.

6.3.3.2 Straight Line Contacts in Spatial Kinematics

With two point contacts, a straight line or an arcuate line (in this case, the line of the arc comprises the chord of a circle) contact can be formed. In the straight line contact, the surface normals of all points on the contact line t_{12} are parallel and in plane. A straight line contact can be realised with a plane a and a cylinder b, as shown in Figure 6.12. The straight line contact prevents a translation parallel to the surface normal n_{1z} and n_{2z} and a rotation around an axis perpendicular to the plane, spanned by n_{1z} and n_{2z} and the contact line t_{12}.

The straight line contact constrains the motion between a and b to a translation u_x parallel to t_{12}, to a translation u_y perpendicular to a plane spanned by n_{1z} and n_{2z} and t_{12}, to a rotation ϕ_x around an axis parallel to t_{12} and to a rotation ϕ_z around an axis parallel to n_{1z} and n_{2z}. The straight line contact is a joint with four degrees of freedom.

6.3.3.3 Arcuate Line Contacts in Spatial Kinematics

The arcuate line contact is based on two point contacts. The surface normals of all points of the contact line t_{12} intersect in the centre of curvature O of t_{12} and are in plane. The points on t_{12} have the same distance to O. An arcuate line contact can be realised with a cylinder a and a sphere b with the same radius, as shown in Figure 6.13 and resembles the interaction of a

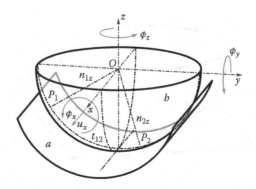

FIGURE 6.13
Arcuate line contact in spatial kinematics.

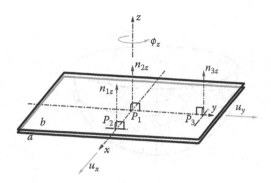

FIGURE 6.14
Planar contact in spatial kinematics.

ball and race in a ball bearing. The arcuate line contact prevents two translations in the plane spanned by the surface normal n_{1z} and n_{2z}. In this type of contact, the motion between a and b is constrained to a translation u_x perpendicular to the plane spanned by n_{1z} and n_{2z} and to the three rotations φ_x, φ_y and φ_z around all spatial axes. The arcuate line contact is a joint with four degrees of freedom and two translational constraints.

6.3.3.4 Planar Contacts in Spatial Kinematics

Based on three point contacts, a planar contact or a spherical contact can be formed. In the planar contact the surface normals of all points in the contact area are parallel. A planar area contact can be realised with two parallel planes a and b, as shown in Figure 6.14. The planar contact prevents a translation parallel to the surface normal n_{1z}, n_{2z} and n_{3z} and two rotations around independent axes, perpendicular to n_{1z}, n_{2z} and n_{3z}.

The planar contact constrains the motion between a and b to two translations u_x and u_y perpendicular to n_{1z}, n_{2z} and n_{3z} and to a rotation φ_z around an axis parallel to n_{1z}, n_{2z} and n_{3z} and is, therefore, a joint with three degrees of freedom.

6.3.3.5 Spherical Contacts in Spatial Kinematics

The spherical contact is based on three point contacts. The surface normals of all points in the contact area intersect in the centre of curvature O of the sphere. All points of the contact area have the same distance to O. A spherical contact can be realised with a hollow sphere a and a sphere b with nearly the same radii, as shown in Figure 6.15. Since three points on a sphere are always situated on a circle t_{123}, it is also possible to replace the hollow sphere a by a cone c, as shown in Figure 6.16, having only a line contact t_{123} between the cone c and sphere b. The spherical contact prevents all three spatial translations and constrains the motion between the sphere a or cone c and the sphere b to three independent rotations φ_x, φ_y, φ_z around axes through O. The spherical contact is a joint with three rotational degrees of freedom.

6.3.3.6 Cylindrical Contacts in Spatial Kinematics

The cylindrical contact is based on four point contacts. It can be realised by a pin with an outer cylinder surface b in a hole with an inner cylinder surface a having nearly the same

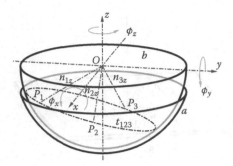

FIGURE 6.15
Spherical contact in spatial kinematics with two spheres.

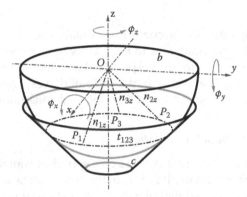

FIGURE 6.16
Spherical contact in spatial kinematics with a sphere and a cone.

radii, as shown in Figure 6.17. The surface normals of all points in the contact area intersect perpendicular to the common central axis a_o of both cylindrical surfaces a and b. The cylindrical contact prevents two independent translations perpendicular to a_o and two rotations around axes perpendicular to a_o. Note that the motion between a and b is constrained to a translation u_x along a_o and to a rotation φ_x around a_o. The cylindrical contact has two degrees of freedom. A rotary shaft in a journal can be approximated by a cylindrical contacts (also see Section 6.8.1).

6.3.3.7 Conical Contacts in Spatial Kinematics

The conical contact is based on five point contacts, and consists of an inner cone b and an outer cone a having the same central axis a_o and the same cone angle, as shown in Figure 6.18. The surface normals of all points in the contact area intersect the central axis a_o. The conical contact prevents all three spatial translations and two rotations around the axes perpendicular to a_o. The motion between the two cones a and b is constrained to a rotation φ_z around the central axis a_o. The conical contact is a joint with one degree of freedom. A rotary shaft using a tapered roller bearing can be approximated by conical contacts (also see Section 6.8.1).

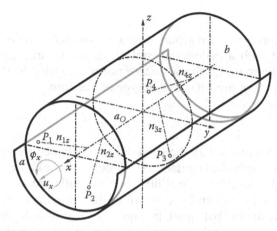

FIGURE 6.17
Cylindrical contact in spatial kinematics.

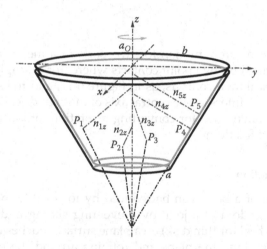

FIGURE 6.18
Conical contact in spatial kinematics.

6.3.4 Standard Joints and Their Constraints

Standard joints usually used in machinery consist of one or several surface contacts allowing only specific motion modes between two parts of a mechanism and are used to create the desired constraints represented in the pure kinematic models of Section 6.3.3.

6.3.4.1 Linear Guiding Systems

Linear guiding systems allow only a translation along one direction between two parts of a mechanism. A linear guiding system can be realised by combining a planar contact, as shown in Figure 6.14, and a straight line contact, as shown in Figure 6.12, with surface normals perpendicular to the surface normals of the planar contact. A linear guiding system has one degree of freedom.

6.3.4.2 Pivot Joint

The pivot joint allows only a rotation around one axis between two parts of a mechanism. A pivot joint can be built with a cylindrical contact, as shown in Figure 6.17, and a point contact, as shown in Figure 6.11, with a surface normal parallel to the central axis of the cylindrical contact. The pivot joint has one degree of freedom.

6.3.4.3 Ball Joint

A ball joint allows rotation around all three spatial axes between two parts of a mechanism and can be realised with a spherical contact, as in Figure 6.15 or Figure 6.16. The ball joint in Figure 6.15 offers a higher load bearing ability than the ball joint in Figure 6.16 because of the area contact between the ball and socket rather than the line contact between ball and cone. In comparison with the ball joint in Figure 6.15, the ball joint in Figure 6.16 is characterised by higher positional accuracy and repeatability, making the ball joint in Figure 6.16 more suitable for kinematic couplings, in which a precise relative position between two parts is required. The ball joint has three degrees of freedom.

6.3.4.4 Screw Thread

The screw thread allows a screw motion for two parts of a mechanism to each other. A screw thread can be formed by five point contacts where, for example, four point contacts are arranged as cylindrical area contacts, as in Figure 6.17, and the surface normal of the fifth point contact must not intersect the central axis of the cylindrical area contact of Figure 6.17. A screw thread is often used for converting a rotation into a translation. The screw thread has one degree of freedom.

6.3.5 Constraints by Friction

The degrees of freedom of a body can be reduced by forces due to friction. Friction can reduce the degrees of freedom of a joint by preventing sliding and transmitting friction forces. For example, a wheel (or thin disk) on a plane surface touches the plane with a point contact allowing sliding along the plane and rotating around the contact point (consult Figure 6.11 to have an idea of the contact between the plane and the wheel). Without friction, the wheel slides along the plane under an external force acting on the wheel parallel to the plane and rotates under an external torque acting on the wheel. With a possible friction force larger than the external force, the friction force prevents the contact point on the wheel from sliding. However, due to the torque created by the external force and friction force, the contact point acts as the instantaneous centre of rotation, allowing the wheel to roll with an instantaneous centre about the contact point, which represents one degree of freedom.

6.4 Mobility of Mechanisms

A mechanism consisting of several parts connected with hinges can be immovable, such as in a bridge frame, or movable in one or more degrees of freedom, depending on the number of links and joints.

The mobility M of a mechanism can be calculated by the formula for mobility of Tchebytchev, Gruebler and Kutzbach, using the number n of links (also called parts), the number j of connecting joints and the sum of all degrees of freedom Σf_j of all connecting joints (Liu and Wang 2014, Rider 2015).

In a mechanism consisting of n parts connected with j joints, every part is regarded separately. Thus, the mobility of a mechanism of n parts, one of which is connected to the ground and acts as an immobile frame relative to which the other parts move, is $6(n-1)$. However, in general, every rigid 'joint' between two parts reduces the number of moving parts, thereby reducing the total mobility of the mechanism by $6j$. Since, every joint also provides a specific number of freedoms to the mechanism, totalling $\Sigma_j f_j$, the net mobility M of the mechanism is

$$M = 6(n-1-j) + \sum_j f_j \tag{6.3}$$

or, alternatively,

$$M = 6(n-1) - \sum_j n_{Pj}, \tag{6.4}$$

where $\Sigma_j n_{Pj}$ is the total number of constraints of the joints of the mechanism (see Section 6.3.3).

In a planar mechanism, every part, regarded separately, has only three degrees of freedom, so the formula of Tchebytchev, Gruebler and Kutzbach has to be adapted; thus

$$M = 3(n-1-j) + \sum_j f_j. \tag{6.5}$$

6.4.1 Under-Constrained Mechanisms

With a mobility M larger than zero ($M > 0$), the mechanism is movable in M different degrees of freedom. In this case, the mechanism is under-constrained and will move when subject to an external load if the external load is not in balance with the internal reaction forces and moments of the mechanism.

The planar mechanism shown in Figure 6.19 has five parts a, b, c, d and e, which are connected with four pivot joints A, B, C and D and a linear guide E, guiding part e in the part d. The pivot joints and the linear guide each have one degree of freedom in planar kinematics. The mechanism has two degrees of freedom according to the formula of Tchebytchev, Gruebler and Kutzbach calculated as

$$M = 3(n-1-j) + \sum_j f_j = 3(5-1-5) + 1+1+1+1+1 = -2. \tag{6.6}$$

The independent freedoms of this mechanism corresponds to a four-bar link attached to e, where one of the links of the four-bar mechanism can be fixed. This mechanism has a linear translational motion of e and a rotational motion about one of the four bar mechanism links.

6.4.2 Exact Constraint Mechanisms

With a mobility M equivalent to zero ($M = 0$), the mechanism is exactly constrained and it can support all external loads with its internal reaction forces (and torques) (Blanding 1999). Thus, the mechanism is in balance with external loads and internal reaction forces.

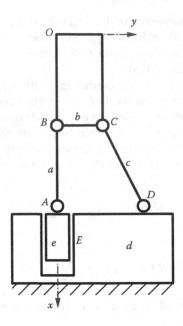

FIGURE 6.19
Under-constrained mechanism with two degrees of freedom.

In Figure 6.20, an exactly constrained mechanism is shown which has zero degrees of freedom, from

$$M = 3(n - 1 - j) + \sum_j f_j = 3(5 - 1 - 6) + 1 + 1 + 1 + 1 + 1 + 1 = 0. \tag{6.7}$$

The calculated value of zero degrees of freedom means that theoretically the mechanism would be 'locked up' or immovable. However, there are many practical cases in which

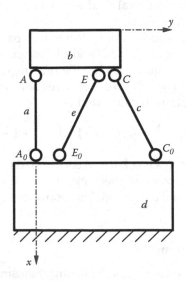

FIGURE 6.20
Exact constraint mechanism.

special dimensional or angular relationships of the parts in the mechanisms provide degrees of freedom where the computations indicate that degrees of freedom do not exist. Consideration of the geometric conditions for which this occurs will often provide insight into tolerance or assembly requirements. This is because the formula of Tchebytchev, Gruebler and Kutzbach cannot predict such 'practical' degrees of freedom, as it contains no considerations of dimensions (also see Section 6.4.4). In fact, the concept of exact constraint is important in kinematic design and needs to be understood from practical design considerations (see Section 6.8) (Eckhardt 1998).

Note that assemblies that function by geometric compatibility and force equilibrium alone are properly constrained; they can be 'just put together'. Therefore, a properly constrained mechanism can be assembled often without requiring forces during assembly to make things fit and/or the need for tight manufacturing tolerances, because the mechanism will have at least one degree of freedom left before adding the final part to the assembly (Hale 1999).

6.4.3 Over-Constrained Mechanism

With a mobility M less than zero ($M < 0$), the mechanism is over-constrained. Normally such mechanisms are not movable and can only be assembled with accurate manufacturing tolerances as opposed to properly constrained assemblies. The over-constrained mechanism shown in Figure 6.21 consists of six parts (a, b, c, d, e and g) and eight pivot joints ($A, A_0, C, C_0, E, E_0, G$ and G_0) and has, according to the formula of Tchebytchev, Gruebler and Kutzbach, a mobility M of minus one:

$$M = 3(n - 1 - j) + \sum_j f_j = 3(6 - 1 - 8) + 1 + 1 + 1 + 1 + 1 + 1 + 1 + 1 = -1. \quad (6.8)$$

Note that removing the unnecessary link g will give the mechanism five links and six single degree of freedom joints, giving mobility M of zero.

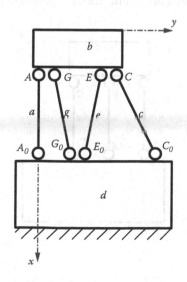

FIGURE 6.21
Over-constrained mechanism.

6.4.4 Singularities

With geometric singularities, it is possible to create movable mechanisms, which should be over-constrained or exactly constrained according to the formula of Tchebytchev, Gruebler and Kutzbach. The mechanism shown in Figure 6.22 consists of five parts (the parts b and d and the rod-shaped parts a, c and e) and six pivot joints (A, A_0, C, C_0, E and E_0), each having one degree of freedom. According to the formula of Tchebytchev, Gruebler and Kutzbach, the mechanism is immovable

$$M = 3(n - 1 - j) + \sum_j f_j = 3(5 - 1 - 6) + 1 + 1 + 1 + 1 + 1 + 1 = 0. \qquad (6.9)$$

But motion is still possible because of the parallelism of the three rods a, c and e. The rods a, c and e can only transmit forces along their rod axis between the parts b and d. Thus, the mechanism cannot support a force F_b, acting on part b perpendicular to the three rods a, c and e, so that the part b moves. Since the rods a, c and e have, indeed must have, the same length, the mechanism can be moved over a large range. Also, as part b is supported by three rods, the mechanism shown in Figure 6.22 has a higher stiffness and a higher load bearing ability along the rod axis than a parallelogram supporting part b only with two rods. Geometric singularities can be applied for stiffening or increasing the load bearing ability of mechanisms. In some cases, using many singularities can have the beneficial effect of elastically averaging effects of manufacturing tolerances. In such a case, the net motion of the parallel mechanism with many slightly variable length legs might more precisely move in the desired direction than would be the case for a parallel mechanism having two variable length legs. However, the stresses in individual legs would be unknown to within the required elastic deformation to accommodate the manufacturing tolerances.

In the mechanism shown in Figure 6.23, rod e is elongated. This mechanism moves under a force F_b, acting on part b perpendicular to the three rods a, c and e, because of the parallelism of the rods a, c and e. Since the pivot joint in point E moves along a circular arc with a larger radius than the radii of the circular arcs of the points A and C, rod e is compressed, whereas rods a and c are stretched. The mechanism is displaced until the external force F_b is in balance with the tensile forces in the rod a and c and the compression force in the rod e.

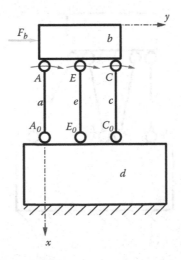

FIGURE 6.22
Exact constraint mechanism with one degree of freedom.

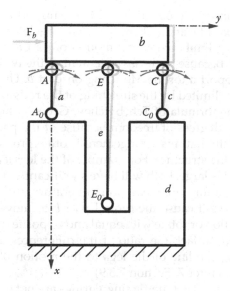

FIGURE 6.23
Exact constraint mechanism with one degree of freedom with short range.

The mechanism shown in Figure 6.24 consists of the parts b and d, the rod-shaped parts a and c and the pivot joints in the points A_0, A, C_0 and C. According to the formula of Tchebytchev, Gruebler and Kutzbach, the mechanism has one degree of freedom:

$$M = 3(n - 1 - j) + \sum_j f_j = 3(4 - 1 - 4) + 1 + 1 + 1 + 1 = 1. \qquad (6.10)$$

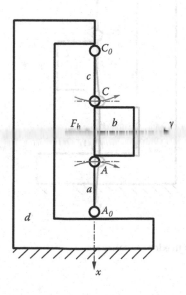

FIGURE 6.24
Four-bar linkage with two degrees of freedom (translation).

The mechanism shown in Figure 6.24 is not able to support a force F_b acting on part b perpendicular to the rod axis of a and c. Thus, part b moves laterally with stretching of the rods a and c. This stretching limits the lateral motion of part b, so that part b can only be moved over a small range. Because of the arrangement of the two rods a and c in one line, the mechanism cannot support a torque M_b acting on part b. Under a torque M_b, a small rotation of part b is possible, limited by the stretching of the rods a and c, as shown in Figure 6.25. In contradiction to the formula of Tchebytchev, Gruebler and Kutzbach, the mechanism can be moved in two degrees of freedom because of the parallel arrangement of the rods a and c in one line. Note that this arrangement is often problematic in the presence of temperature differences in the structure. For example, if the legs a and c are relatively small and the supporting frame d is large, both will have significantly different response times to temperature changes. Hence for an increase in temperature, the legs a and c will expand more than the frame. This will cause the central part b to move arbitrarily in either the positive or negative y direction or rotate with equal and opposite displacements of the joints at the top and bottom of b. In fact, the ratio of the differences in the axial displacement caused by expansion along the line of the legs to the motion of part b perpendicular is theoretically infinite (see Chapter 7, Section 7.3.9).

In conclusion, since sometimes geometric singularities are not easy to detect, especially in spatial mechanisms, the formula of Tchebytchev, Gruebler and Kutzbach has to be applied with caution.

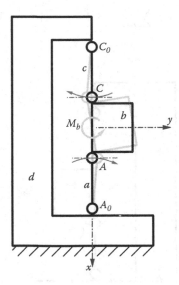

FIGURE 6.25
Four-bar linkage with two degrees of freedom (rotation).

6.5 Motion in Planar Kinematics

The formula of Tchebytchev, Gruebler and Kutzbach gives only a statement about the number of degrees of freedom of a mechanism, but not how the parts of a mechanism move relative to each other. In planar mechanisms, the instantaneous centres of rotation describe the motion of the parts relative to each other for small displacements.

6.5.1 Instantaneous Centre of Rotation in Planar Kinematics

The underlying principle of finding the centre of rotation for a combined motion is that a combination of translation and rotation around a defined axis can be described as a pure rotation around another axis. In Figure 6.26, the triangular part b with the corner points A, B and C is displaced translationally in the position b' with the corner points A', B' and C'. It is then rotated from the position b' to position b'' around the point A', which places the corner points at A'', B'' and C''. The centre of rotation $P_{bb''}$ for a pure rotation of part b from the initial position into the position b'' is situated in the intersection of the perpendicular bisector $m_{AA''}$ of the displacement vector from A to A'' and the perpendicular bisector $m_{BB''}$ of the displacement vector from B to B''. The perpendicular bisector $m_{CC''}$ of the displacement vector from C to C'' intersects also in point $P_{bb''}$. Note that a pure translation can be regarded as a rotation, whose centre of rotation is situated perpendicular to the direction of translation at a radius of infinite length (see Chasles' theorem in Section 6.6).

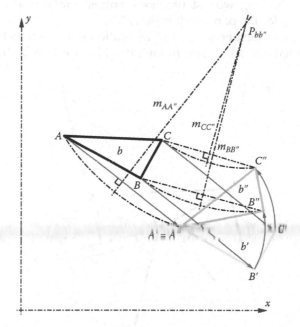

FIGURE 6.26
Superposition of translation and rotation with centre of rotation for the combined motion.

Regarding only the initial and final position, it is unimportant along which trajectories the points of a body reach their final position. The body can be always transferred into the final position by a rotation, at which the centre of rotation is not necessarily identical with the centres of curvature of the trajectories of the points along which the body is moving. Figure 6.27 shows a four-bar linkage with the parts a, b, c and d, with d as the solid frame part (and forms the global reference coordinate), and the pivot joints in the points A_0, A, C_0 and C. Part b with the corner points A, B and C can be transferred into the position b'' with the corner points A'', B'' and C'' with a pure rotation around point $P_{bb''}$, situated in the intersection of the perpendicular bisectors $m_{AA''}$, $m_{BB''}$ and $m_{CC''}$ of the displacement vectors $\overrightarrow{AA''}$, $\overrightarrow{BB''}$ and $\overrightarrow{CC''}$. In reality, the point A moves to A'' along a circular arc with centre of curvature in point A_0 and point B moves to B'' along a circular arc with centre of curvature in point B_0. The centre of rotation $P_{bb''}$ for part b from the initial position into position b'' is not identical with the centres of curvature of the trajectories from A to A'' and from B to B''.

Small motions around the initial position assume the trajectories of the points along which the body is moving and can be approximated by tangents. As shown in Figure 6.28, the instantaneous centre of rotation P_{bd} of part b is situated in the intersection of the perpendiculars to the tangents at the trajectories in the points A and B. The instantaneous centre of rotation can be determined if the directions of the instantaneous displacement of two points of a body are known, because the centre of rotation is situated in the intersection of the perpendiculars to the displacement directions.

The instantaneous centre of rotation is not a material point; it is a geometric point which can change its absolute position and its relative position with respect to the body as as can be seen in Figure 6.28, showing two instantaneous centres of rotation P_{bd} for the part b in the initial position and $P_{b''d}$ for the part b in position b''.

It is possible to derive instantaneous centres of rotation not only for a body relative to the frame part (global reference coordinate frame), but also for two bodies relative to each

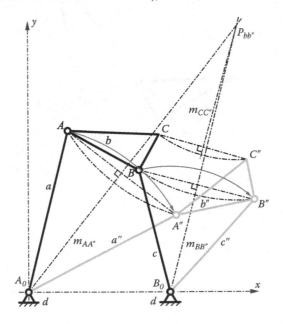

FIGURE 6.27
Four-bar linkage with centre of rotation.

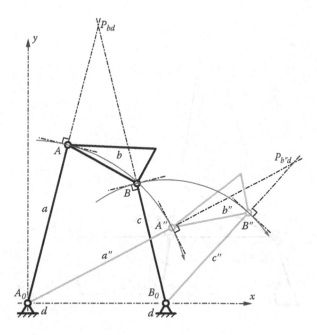

FIGURE 6.28
Four-bar linkage with instantaneous centre of rotation.

other. In Figure 6.29, part d is the frame part (global reference coordinate frame) and is connected with the rod-shaped part a by the pivot joint in point A_0. Thus point A_0 is the instantaneous centre of rotation P_{ad} of part a relative to part d. The point A_0 is also the instantaneous centre of rotation P_{da} of part d relative to part a. The pivot joint in point A connects the part b with the part a. Point A is the centre of rotation P_{ab} of part a relative to part b and also the centre of rotation P_{ba} of part b relative to part a. The pivot joint in point B represents the instantaneous centre of rotation P_{bc} of part b relative to part c and the pivot joint in point B_0 represents the instantaneous centre of rotation P_{cd} of part c relative to part d. The instantaneous centre of rotation P_{bd} of part b relative to part d is situated in the intersection of the perpendicular to the instantaneous moving direction in point A, which is also the connecting line of the centres of rotation P_{ad} and P_{ba}, with the perpendicular to the instantaneous moving direction of point B, which is the connecting line of the instantaneous centres of rotation P_{cd} and P_{bc}.

Thus, the instantaneous centre of rotation P_{bd} of part b relative to part d is the intersection point of the connecting line from P_{ad} to P_{ba} with the connecting line from P_{cd} and P_{bc}:

$$P_{bd} = P_{db} = \overline{P_{ad}\ P_{ba}} \times \overline{P_{cd}\ P_{bc}}. \qquad (6.11)$$

Similar to P_{bd}, the instantaneous centre of rotation P_{ac} is situated in the intersection of the connecting line from P_{ba} to P_{bc} and with the connecting line from P_{ad} and P_{cd}:

$$P_{ac} = P_{ca} = \overline{P_{ad}\ P_{cd}} \times \overline{P_{ba}\ P_{bc}}. \qquad (6.12)$$

Since the connecting lines $\overline{P_{ad}\ P_{cd}}$ and $\overline{P_{ba}\ P_{bc}}$ are parallel to each other, the instantaneous centre of rotation P_{ac} of part a relative to part c is situated at infinity. Part a moves, with

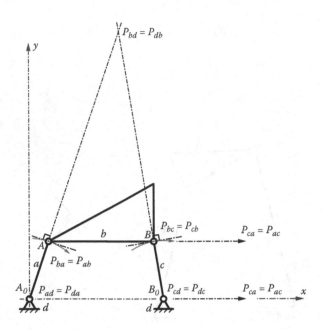

FIGURE 6.29
Four-bar linkage with relative instantaneous centres of rotation.

respect to part c, translationally, at least for small displacements as per the geometry shown in Figure 6.29.

6.5.2 Displacements due to Rotation

As shown in Figure 6.26, every combination of translation and rotation can be described as pure rotation in planar kinematics, considering only the displacement of the points of a body.

In Figure 6.30, the point A is displaced to A' and the point B to B' by a rotation around point P with the angle φ_z.

The displacement vector $\overrightarrow{u_A}$ from A to A' can be described by the coordinates x_A and y_A of point A, the coordinates x_p and y_p of the rotation point P, the angle α between the x-axis and the connecting line from P to A and the rotation angle φ_z:

$$\overrightarrow{u_A} = \begin{pmatrix} u_{Ax} \\ u_{Ay} \end{pmatrix} = \sqrt{(x_A - x_P)^2 + (y_A - y_P)^2} \begin{pmatrix} \cos(\alpha + \varphi_Z) - \cos\alpha \\ \sin(\alpha + \varphi_Z) - \sin\alpha \end{pmatrix}. \tag{6.13}$$

The addition theorem for cosines and sines can be used to create the following transformation for the displacement vector $\overrightarrow{u_A}$:

$$\overrightarrow{u_A} = \begin{pmatrix} u_{Ax} \\ u_{Ay} \end{pmatrix} = \sqrt{(x_A - x_P)^2 + (y_A - y_P)^2} \begin{pmatrix} \cos\alpha \cos\varphi_z - \sin\alpha \sin\varphi_z - \cos\alpha \\ \sin\alpha \cos\varphi_z + \cos\alpha \sin\varphi_z - \sin\alpha \end{pmatrix},$$

$$\overrightarrow{u_A} = \sqrt{(x_A - x_P)^2 + (y_A - y_P)^2} \begin{pmatrix} -\cos\alpha\,(1 - \cos\varphi_z) - \sin\alpha \sin\varphi_z \\ -\sin\alpha\,(1 - \cos\varphi_z) + \cos\alpha \sin\varphi_z \end{pmatrix}. \tag{6.14}$$

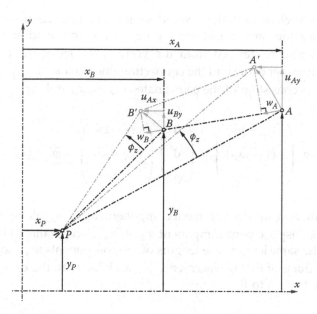

FIGURE 6.30
Displacements due to rotation.

The term $(1 - \cos \varphi_z)$ can be transformed with the addition theorem a second time, and the sine and cosine of angle α can be expressed with the coordinates x_A and y_A of point A and the coordinates x_p and y_p of point P. Hence,

$$\cos \alpha = \frac{(x_A - x_P)}{\sqrt{(x_A - x_P)^2 + (y_A - y_P)^2}}, \tag{6.15}$$

$$\sin \alpha = \frac{(y_A - y_P)}{\sqrt{(x_A - x_P)^2 + (y_A - y_P)^2}}, \tag{6.16}$$

$$\overrightarrow{u_A} = \begin{pmatrix} -(x_A - x_P)(1 - \cos \varphi_z) - (y_A - y_P) \sin \varphi_z \\ -(y_A - y_P)(1 - \cos \varphi_z) + (x_A - x_P) \sin \varphi_z \end{pmatrix}, \tag{6.17}$$

$$\overrightarrow{u_A} = 2\sin^2 \frac{\varphi_z}{2} \begin{pmatrix} -(x_A - x_P) \\ -(y_A - y_P) \end{pmatrix} + \sin \varphi_z \begin{pmatrix} -(y_A - y_P) \\ -(x_A - x_P) \end{pmatrix}.$$

For small angles, the sine of φ_z can be approximated by the angle φ_z itself. Equally $\sin \frac{\varphi_z}{2}$ can be expressed by $\frac{\varphi_z}{2}$, so

$$\overrightarrow{u_A} \approx \frac{\varphi_z^2}{2} \begin{pmatrix} -(x_A - x_P) \\ -(y_A - y_P) \end{pmatrix} + \varphi_z \begin{pmatrix} -(y_A - y_P) \\ -(x_A - x_P) \end{pmatrix}. \tag{6.18}$$

For a linearisation such as a rotation around an instantaneous centre of rotation, only the tangents to the trajectory are considered by neglecting the quadratic term of φ_z. This approximation can also be derived from the vector product of the rotation vector $\overrightarrow{\varphi_z}$ perpendicular to the paper plane and the connecting line vector $(\overrightarrow{r_A} - \overrightarrow{r_p})$ from the centre of rotation P to point A consisting of the coordinates x_A, y_A, x_P and y_P:

$$\overrightarrow{u_A} \approx \varphi_z \begin{pmatrix} -(y_A - y_P) \\ -(x_A - x_P) \\ 0 \end{pmatrix} = \begin{pmatrix} 0 \\ 0 \\ \varphi_z \end{pmatrix} \begin{pmatrix} (x_A - x_P) \\ (y_A - y_P) \\ 0 \end{pmatrix} = \overrightarrow{\varphi_z} \times (\overrightarrow{r_A} - \overrightarrow{r_p}). \quad (6.19)$$

For this approximation, the displacement component w_A of $\overrightarrow{u_A}$ along the connecting line from A to B and the displacement component w_B of $\overrightarrow{u_B}$ also parallel to the connecting line from A to B have the same length. The lengths of the components w_A and w_B are calculated with the scalar products of the displacements $\overrightarrow{u_A}$ and $\overrightarrow{u_B}$ with the direction vector $\overrightarrow{n_{AB}}$ of the connecting line from A to B:

$$w_A = \overrightarrow{u_A} \cdot \overrightarrow{n_{AB}}$$

$$= \varphi_z \cdot \begin{pmatrix} -(y_A - y_P) \\ -(x_A - x_P) \end{pmatrix} \cdot \frac{1}{\sqrt{(x_B - x_A)^2 + (y_B - y_A)^2}} \begin{pmatrix} (x_B - x_A) \\ (y_B - y_A) \end{pmatrix}$$

$$= \frac{\varphi_z}{\sqrt{(x_B - x_A)^2 + (y_B - y_A)^2}} [x_A \cdot y_B - x_B \cdot y_A + y_P \cdot (x_B - x_A) - x_P \cdot (y_B - y_A)] \quad (6.20)$$

$$= \varphi_z \cdot \begin{pmatrix} -(y_B - y_P) \\ -(x_B - x_P) \end{pmatrix} \cdot \frac{1}{\sqrt{(x_B - x_A)^2 + (y_B - y_A)^2}} \begin{pmatrix} (x_B - x_A) \\ (y_B - y_A) \end{pmatrix} = \overrightarrow{u_B} \cdot \overrightarrow{n_{AB}} = w_B.$$

In Equation 6.17, $\sin \varphi_z$ can be transformed into $2 \sin \frac{\varphi_z}{2} \cos \frac{\varphi_z}{2}$, so

$$\overrightarrow{u_A} = 2\sin^2 \frac{\varphi_z}{2} \begin{pmatrix} -(x_A - x_P) \\ -(y_A - y_P) \end{pmatrix} + 2 \sin \frac{\varphi_z}{2} \cos \frac{\varphi_z}{2} \begin{pmatrix} -(y_A - y_P) \\ -(x_A - x_P) \end{pmatrix},$$

and

$$\overrightarrow{u_B} = 2\sin^2 \frac{\varphi_z}{2} \begin{pmatrix} -(x_B - x_P) \\ -(y_B - y_P) \end{pmatrix} + 2 \sin \frac{\varphi_z}{2} \cos \frac{\varphi_z}{2} \begin{pmatrix} -(y_B - y_P) \\ (x_B - x_P) \end{pmatrix}. \quad (6.21)$$

Elimination of x_P and y_P by subtraction of displacements $\overrightarrow{u_A}$ and $\overrightarrow{u_B}$ and calculation of rotation angle φ_z yields

$$\overrightarrow{u_B} - \overrightarrow{u_A} = \begin{pmatrix} u_{Bx} - u_{Ax} \\ u_{By} - u_{Ay} \end{pmatrix}$$

$$= 2\sin^2 \frac{\varphi_z}{2} \begin{pmatrix} (x_A - x_B) \\ (y_A - y_B) \end{pmatrix} + 2\sin\frac{\varphi_z}{2}\cos\frac{\varphi_z}{2} \begin{pmatrix} (y_A - y_B) \\ -(x_A - x_B) \end{pmatrix}. \tag{6.22}$$

From linear equations for $2\sin^2\frac{\varphi_z}{2}$ and $2\sin\frac{\varphi_z}{2}\cos\frac{\varphi_z}{2}$:

$$2\sin^2\frac{\varphi_z}{2} = \frac{(x_A - x_B)(u_{Bx} - u_{Ax}) + (y_A - y_B)(u_{By} - u_{Ay})}{(x_A - x_B)^2 + (y_A - y_B)^2},$$

$$2\sin\frac{\varphi_z}{2}\cos\frac{\varphi_z}{2} = \frac{(y_A - y_B)(u_{Bx} - u_{Ax}) - (x_A - x_B)(u_{By} - u_{Ay})}{(x_A - x_B)^2 + (y_A - y_B)^2},$$

$$\tan\frac{\varphi_z}{2} = \frac{(x_A - x_B)(u_{Bx} - u_{Ax}) + (y_A - y_B)(u_{By} - u_{Ay})}{(y_A - y_B)(u_{Bx} - u_{Ax}) - (x_A - x_B)(u_{By} - u_{Ay})} \tag{6.23}$$

$$= \frac{(u_{Bx} - u_{Ax})(x_A - x_B) + (u_{By} - u_{Ay})(y_A - y_B)}{(u_{Bx} - u_{Ax})(y_A - y_B) - (u_{By} - u_{Ay})(x_A - x_B)},$$

$$\varphi_z = 2\arctan\left(\frac{(u_{Bx} - u_{Ax})(x_A - x_B) + (u_{By} - u_{Ay})(y_A - y_B)}{(u_{Bx} - u_{Ax})(y_A - y_B) - (u_{By} - u_{Ay})(x_A - x_B)}\right).$$

With known $2\sin^2\frac{\varphi_z}{2}$ and $2\sin\frac{\varphi_z}{2}\cos\frac{\varphi_z}{2}$, the linear equations for x_P and y_P in Equation 6.21 become

$$\overrightarrow{u_A} = \begin{pmatrix} u_{Ax} \\ u_{Ay} \end{pmatrix} = 2\sin^2\frac{\varphi_z}{2}\begin{pmatrix} -(x_A - x_P) \\ -(y_A - y_P) \end{pmatrix} + 2\sin\frac{\varphi_z}{2}\cos\frac{\varphi_z}{2}\begin{pmatrix} -(y_A - y_P) \\ -(x_A - x_P) \end{pmatrix},$$

$$\tag{6.24}$$

$$x_P = x_A + \frac{1}{2}\left[u_{Ax} - \frac{(u_{Ax} - u_{Bx})(y_A - y_B) - (u_{Ay} - u_{By})(x_A - x_B)}{(u_{Ax} - u_{Bx})(x_A + x_B) + (u_{Ay} - u_{By})(y_A - y_B)}u_{Ay}\right],$$

$$y_P = y_A + \frac{1}{2}\left[\frac{(u_{Ax} - u_{Bx})(y_A - y_B) - (u_{Ay} - u_{By})(x_A - x_B)}{(u_{Ax} - u_{Bx})(x_A + x_B) + (u_{Ay} - u_{By})(y_A - y_B)}u_{Ax} + u_{Ay}\right]. \tag{6.25}$$

Thus, it is possible to determine the displacement vectors of two points A and B, that is $\overrightarrow{u_A}$ and $\overrightarrow{u_B}$, with known coordinates of A and B, that is x_A, y_A, x_B and y_B, known coordinates of the centre of rotation P, that is x_P and y_P, and known rotation angle, that is φ_z (Equations 6.23–6.25).

6.6 Motion in Spatial Kinematics

In the previous sections, planar displacements were discussed. The nature of the displacements in three-dimensional space is briefly examined in this section. It was noted before that in the case of a plane, a displacement could be described as a rotation followed by a translation (see Equation 6.19). In fact, this combination of rotation and translation can be used to describe spatial displacement, although a choice of the reference point is needed. A convenient way to understand this idea is by imagining the case when describing a rigid body in space with respect to a reference coordinate frame indicated by coordinates x, y and z, and an origin, say O. Start by describing the position of a point, say A, in the rigid body with coordinates x_A, x_B and x_C being the distances in the x, y and z directions from the origin O to point A. Note that the rigid body could be rotated about the point A without changing any of the coordinates of point A. Now consider that a local coordinate frame is attached to the rigid body at point A. The axes of the local coordinate may be perfectly parallel to the axes of the reference coordinate or they may be at different angles with respect to a reference coordinate, say α, β and γ are the angles between the x, y and z axes of the reference coordinate and local coordinate. The parameters α, β and γ provide information about the orientation of the rigid body in space. Now consider the definition of translational and rotational displacement for this case. An example of translational displacement (or pure translation) will be a displacement of the rigid body A that will require the changes in the values of the coordinates x_A, x_B and x_C, but no change in the values of α, β and γ. An example of a rotational displacement (or pure rotation) will be a displacement of the body A that will involve no change of x_A, x_B and x_C, but α, β and γ will vary. In this case, displacement of point A does not take place and all points along a particular line through the non-displaced point A also show zero displacement. This line is called the axis of rotational displacement or simply axis of rotation. Note carefully that if a translational displacement can change the three parameters x_A, x_B and x_C by any desired amount and, if a rotational displacement can change the three parameters α, β and γ by any desired amount, then it is possible that a combination of a translation and rotation can attain the six parameters by any desired amount and thus can achieve any desired displacement. Hence, any displacement in three dimensions can be equivalent to a translation plus a rotation or vice versa (Eckhardt 1998).

However, there is a problem in describing spatial displacement using translation and rotation. This is because not all spatial displacements are rotations. For example, consider the screw displacement shown in Figure 6.31. In this motion, the rotation takes place about some line in space and the translation takes place simultaneously along the same line. For a screw motion, a point on the screw axis moves along the axis and a point off the axis moves along a helix. As there are no fixed points, it is not a rotation.

In fact, screw displacement provides a powerful description of spatial displacement and forms the basis of 'screw theory' or Chasles' theorem (Jazar 2007). The theorem states that *every spatial displacement is the composition of a rotation about some axis and a translation along the same axis*. A few terms are important to describe this theorem. A screw is a line or axis with an associated pitch, which is a ratio of linear to angular quantities. A twist is a screw

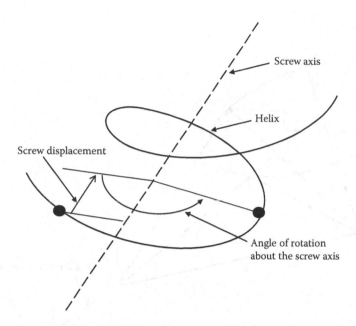

Screw axis

Helix

Screw displacement

Angle of rotation
about the screw axis

FIGURE 6.31
Screw displacement is a combination of a translation and a rotation about a common axis.

plus a scalar magnitude, providing a rotation about the screw axis and a translation along
the screw axis. The rotation angle is the twist magnitude and the translation distance is the
twist magnitude multiplied by the pitch (hence, pitch is the ratio of translation to rotation).
Thus Chasles' theorem can be re-stated as *every spatial displacement is a twist about some screw*
(Mason 2001).

A screw motion can be shown as a four-variable function š (h, $\emptyset$, a, $\vec{s}$), where $\hat{a}$ is a unit
vector or twist axis, $\vec{s}$ is a position vector, $\emptyset$ is the twist angle and h (or pitch p, which is $h/\emptyset$)
is translation (Figure 6.32). The position vector $\vec{s}$ indicates the position of a point on the
screw axis with respect to global reference coordinates. The twist angle $\emptyset$, the twist axis $\hat{a}$
and the pitch p (or translation h) are called screw parameters.

Chasles' theorem suggests that:

- Screw motion has only one degree of freedom because of the coupling of the
  translation to the rotation by the screw pitch.

- All modes of motion can be considered as screw motions: a rigid body rotation is a
  screw motion with a screw pitch of zero; a rigid body translation is a screw motion
  with an infinite screw pitch.

- The screw motion can be used to define infinitesimal motions in an ideal way. As
  such, any motion at a given time will have a screw and associated twist magnitude,
  describing the instantaneous velocity and angular velocity of the motion. This
  defines the instantaneous screw axis.

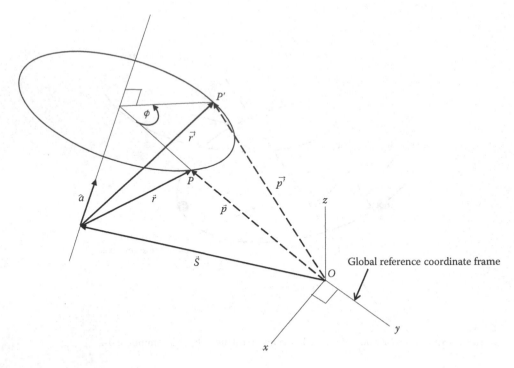

FIGURE 6.32
Screw motion of a rigid body.

6.7 A Short Case Study

To understand how the kinematic behaviour of a mechanism can affect design considerations, take the case of the spatial motion of some kind of 'end effector' within a machine.
A pick-and-place machine (this important class of machines are employed to move objects
from some initial positions to some final positions and are very common in automatic
manufacturing and automatic assembly systems) is required to change the location (position and orientation) of an object or objects. Based on what has been discussed, a body can
be displaced from any position to another position by a translation plus a rotation;
simplistically, this may suggest that a two degree of freedom machine can be used. This
translation and rotation can be achieved simultaneously, as suggested by Chasles' theorem,
meaning that a one degree of freedom machine consisting of a nut and screw (or any other
combination of parts that can also give a screw displacement) can be used to get the desired
displacement. However, there are other practical factors that may influence the choice of the
degrees of freedom to be used. For instance, consider the use of a pick-and-place machine in
a system where the initial position, in which the object is to be found, may vary (it is also
very much possible that the final position, in which object is to be placed, may vary). For
example, if an object is presented in its initial position on a table top, always right-side up
and always facing north, but such that it can be anywhere on the table top, then its initial
position should have two degrees of freedom, because it will be necessary to specify values
for two parameters (such as x and y) to describe its position. Therefore, to accommodate
these position variations in the initial position only, two degrees of freedom would be

required in the machine in addition to the theoretical minimum of one degree of freedom as suggested by Chasles' theorem, giving a total of three degrees of freedom. It is noteworthy that in cases in which variations must be accommodated in both the initial and final positions, the situation may become even more complicated because some degrees of freedom provided to accommodate one position variation may also accommodate the variations in the other position. Therefore, the degrees of freedom cannot simply be added (Eckhardt 1998).

This case demonstrates that when considering the kinematic behaviour of a mechanism during the design process, designers and engineers should carefully justify the predictions of the theoretical models (typically based on fundamental design principles) in light of the practical requirements of the application to which the mechanism will be used.

6.8 Kinematic Design and Realisation in Practice

6.8.1 Introduction

From the previous discussion, the position and orientation of one part (or link) with respect to another within a mechanism can be defined by six degrees of freedom: three translations and three rotations (see Sections 6.2.3 and 6.6). Thus, the requirements of precision location of a part in a mechanism are to constrain N of these degrees of freedom. The design parameters are the means, usually through connecting joints, which maintain position and orientation by providing resistance to motion in the N degrees of freedom. In other words, when the parts of a mechanism are where they are supposed to be, then the mechanism will deliver the function that it is designed for (this is called nominal design). When constraints are applied, degrees of freedom are taken away so that a part gets to where it supposed to be (i.e. the part can be located deterministically) (Slocum 1992, Whitehead 1954).

Avoidance of the unpredictable effects of over- or under-constrained designs is the major motivation for exact constraint or kinematic design in precision machines, that is, to isolate sensitive parts or systems such as a metrology frame from the influence of dimensionally changing supports and/or manufacturing tolerances (Hale 1999). In an exact constraint mechanism, each degree of freedom in a part is individually considered and constrained as per the design requirement. As such, an exact constraint device uses the minimum number of constraints to restrict the motion into the desired direction, while all the other degrees of freedom are free. In analysis, an exact constraint mechanism can be defined using vector loops for displacement and force where there is the same number of equations and unknowns. These equations can be used to determine the location of the parts for any manufacturing deviations of the part features. Therefore, the accuracy of exact constraint mechanisms can be directly correlated to the manufacturing process and easily corrected. Repeatability, on the other hand, often depends on the assembly process factors such as friction, tolerance effects in joint assemblies, environmental contamination and thermal errors (Slocum 1992, Blanding 1999, Hale 1999).

A purely kinematic design may be difficult (or expensive) to achieve in practice or the drawbacks of kinematic design, such as high stress at the contacts, limiting load capacity and mechanism stiffness, may limit its use for certain applications. The term semi-kinematic or pseudo-kinematic has been used in precision engineering literature to describe designs that are 'impure' to some extent. Impure does not mean that something is wrong; rather, there are trade-offs to make in design. In fact, it is important to understand the advantages and limitations of various kinematic design types so that these trade-offs can be made to find the best possible practical solution for a constraint device to be used in a mechanism (Hale 1999).

To illustrate the concepts of pure kinematic and pseudo-kinematic designs, consider the case of designing a rotary shaft, an example given by Smith and Chetwynd (2003) and Chapter 7, Section 7.2.2. Based on the kinematic principles discussed in Section 6.3, one solution can be to have a rigid link between two spheres, one constrained in a trihedral hole and the other in a groove, as shown in Figure 6.33. This five-contact system represents a correctly constrained single-axis rotation. But, what about the design option in which the two trihedral holes are used to locate two spheres? This is an over-constrained mechanism which will rotate but only with poor repeatability. This is because the shaft sits properly in the holes only if the length of the separating link of the spheres is exactly correct. This is possible when manufacturing is done to the exact dimensions, resulting in increased costs due to tight tolerancing. This indicates a general rule in design that *a divergence from pure kinematic design results in increased manufacturing costs* (Smith and Chetwynd 2003).

Now, for the same example, consider that the shaft needs to be designed to support a load. Since point contacts are generally unsuitable for load-bearing applications, as will be seen in later discussion, a possible solution to locate the shaft while maintaining its axial rotation can be the use of suitable bearings. The principles of kinematic constraint may suggest using a bearing with three degrees of freedom at one end and using a bearing with four degrees of freedom at another end of the shaft. What are the available choices of bearings for this case? Figure 6.33 shows a range of typical rolling element bearings and

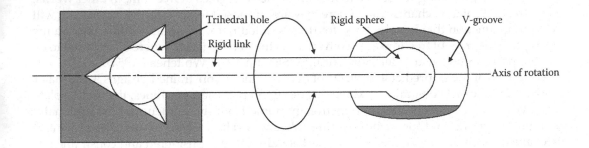

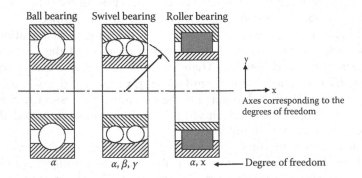

FIGURE 6.33
(Top) Kinematic model to represent the rotary shaft having five constraints. (Bottom) Some rolling element bearings and their respective axes corresponding to their degrees of freedom.

their respective axes of rotation or translation. The simple ball bearing is effectively a five-constraint mounting. For swivel bearings, the inner race can rotate about all three axes relative to the outer surface and, thus, can be kinematically considered equivalent to a sphere in a trihedral hole. For roller bearings, the inner race is free to rotate about the x-axis and translate along the same axis; this is equivalent to the sphere in a straight V-groove. When the axial load becomes significant, a tapered roller bearing can be used (also see Section 6.3.3.7). However, rolling element bearings become impractical for much higher load capacity of the shaft, and plain hydrodynamic or hydrostatic bearings may need to be considered (see Figure 6.34). Also, consult Chapter 7 for further discussion of different bearing types.

When employing bearings for the shaft of this example, one solution may be using two ball bearings at the two ends to support the shaft; but, this is a highly over-constrained mechanism (ten constraints). Alternatively, a combination of a swivel ball (also called self-aligning) bearing at one end and a roller bearing can be used at the other end (see Figure 6.34). This is still an over-constrained mechanism with seven constraints and one degree of freedom. Note that using the first solution may seem economically attractive and, in fact, it is common practice to use paired ball bearings because of their lower manufacturing cost as compared to the other rolling element types, for example, swivel and roller bearings. However, the requirement of maintaining close tolerance of the coincidence of the axes of the two bearings may incur significant manufacturing costs.

In fact, designs using rolling element bearings may actually be much more expensive than a simple kinematic solution. Using hydrostatic or hydrodynamic bearings also does not usually offer a cheaper alternative to a simple kinematic solution. Ideally, designers should

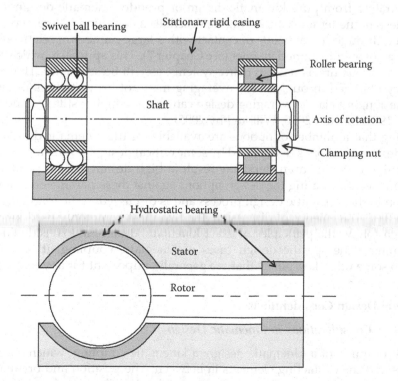

FIGURE 6.34
Rotary shaft: (top) using rolling element bearings, (bottom) using journal or hydrostatic bearings.

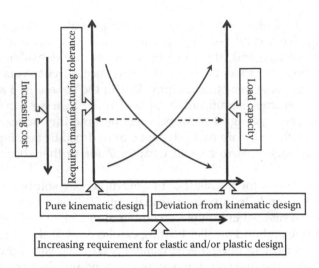

FIGURE 6.35
A simple diagram showing the drivers for the deviations of the kinematic design.

aim to find design solutions that closely approximate the kinematic or pseudo-kinematic (or semi-kinematic) design principles. Perhaps, the drivers for the kinematic or pseudo-kinematic design can conveniently be presented by a simple diagram, as shown in Figure 6.35, the theme of which will be made clearer in the following sections.

There is yet another design type: elastic design or elastic averaging, which can be considered separately from pure kinematic design or pseudo-kinematic design. Contrary to kinematic design, the term elastic averaging describes a condition where two solid bodies are connected through many points of contact with a large number of relatively compliant members in an over-constrained manner (see Chapter 7). This approach, which relies on the elasticity of the structure, is used in many systems, as will be seen later. The repeatability and accuracy obtained through elastic averaging may not be as high as in deterministic kinematic design, but elastic averaging design can provide higher stiffness and lower local stress (Hale 1999, Slocum 2010, Willoughby 2005).

Considering that a number of options are available for the kinematic design of a mechanism, the design selection is determined by some critical design parameters. Failure to pay proper attention to these parameters may result in high design and/or product cost, and poor performance. Evaluating the design options against these parameters is an important consideration in the kinematic design process and is discussed in the following sections in some detail. For convenience of discussion, the two most commonly used kinematic couplings, which follow the principles of exact kinematic design, are considered. However, before examining the specific design cases of kinematic coupling, it is worth paying attentions to some other key points that are generally important for any kinematic design.

6.8.2 General Design Considerations

6.8.2.1 System Considerations in Kinematic Design

Consider an example of a kinematic design: a kinematic coupling, which consists of two components and the contacting elements in between. The position and orientation of one component with respect to another component can be described by six relative degrees of freedom; thus, the requirements of precision location of one component with respect to

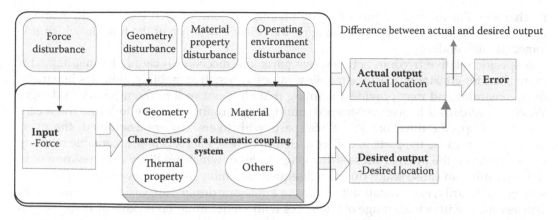

FIGURE 6.36
A simplistic model to represent a kinematic coupling as a system.

another are to constrain N of the degrees of freedom of the coupling system. This can be achieved by the contacting elements, which provide resistance to motion in the N degrees of freedom. The repeatability of the coupling to locate one component of the coupling with respect to the other will depend on the errors that arise from the sources related to the characteristic features of the coupling system. As such, if the kinematic coupling can be considered as a system, the coupling's characteristic features can be described by the system inputs, for example, the applied load in the coupling and the friction force between contact elements and coupling components; by the system's own characteristics, for example, geometry, material and thermal properties of the coupling; and by the system's output, which is the location of one component of the coupling with respect the other component in this case. This system concept of a kinematic coupling can be shown by drawing a simple diagram as can be seen in Figure 6.36. If, for instance, the applied load (input) in the coupling becomes large, it may change the contact geometry between the contacting element and the components of the coupling; this variation in input incurs error, affecting the repeatability of the coupling. Similarly, if, for instance, the dimension of the contacting element (for example, the diameters of the balls) changes, representing variation in system characteristics, it will affect the repeatability of the coupling. Thus, variations in the inputs and system characteristics of a kinematic system result in outputs that differ from the expected outputs, the differences of which are described as the errors or repeatability (Hale and Slocum 2001, Schouten et al. 1997).

6.8.2.2 Mobility Considerations in Kinematic Design

As seen in Section 6.4, the mobility M of a mechanism can be calculated by the formula of Tchebytchev, Gruebler and Kutzbach using the number n of parts, the number j of connecting joints and the sum of all degrees of freedom Σf_j of all connecting joints (see Equation 6.3). Designing mechanisms with mobilities of zero, one or two should, ideally, correspond to a kinematic clamp, a single- and a two degree of freedom mechanism, respectively. If the mobility is not the correct value for the required motion, identification of the reason why this has occurred will often provide insight about issues that can arise during assembly and during the application of the mechanism. However, it is acknowledged that there are sometimes instances where an additional freedom might not impact the function of the mechanism. For example, a push rod comprising ball joint at each end would be able to rotate about its own axis without changing the relative distance of the links to which it is joined at each end.

A specific relative freedom between two parts in a mechanism cannot be maintained if there is under-constraint (positive mobility value). A negative mobility indicates kinematic over-constraint and may provide a warning that all is not well in the mechanism design. When a mechanism is over-constrained, multiple constraints compete to locate the mechanism in a specific direction. When real parts with tolerances are assembled, the over-constraint can cause the parts to deflect or deform. These effects can be acceptable in many cases, however, the internal stresses (the magnitudes of which are likely to be unknown) in the assembly can cause issues during subsequent assembly stages or even cause failure in service. Contrarily, exact constraint designs or kinematic designs are easier to assemble and will assemble with a wide range of tolerances without deformation (Eckhardt 1998, Slocum 1992, Smith and Chetwynd 2003).

Just because a mechanism has the expected value for mobility, it may not be concluded that the design is fit for purpose. Recall the example of the chair. For a three-legged chair, leg length and compliance are nominally not critical. Three legs will always contact the ground under an applied load and the chair is properly constrained in three degrees of freedom, while being free in the other three. However, such a chair is more prone to tipping, as the load vector must pass through bounds of a triangle defined by the contacts. On the other hand, consider a five-legged chair where each leg has modest compliance, such that when a person sits on it, all the legs deform a little and so all legs make contact with the ground (elastic design). This over-constrained chair is more expensive to design and manufacture, but it has an advantage: loads can generally be applied anywhere within the polygon that bounds the contact points. Here, mobility alone may not be used as a means of confirming a design. If the design is not cost driven, a five-legged chair instead of a three-legged chair may be preferred. Thus, an improved design can be achieved when the calculation of the mobility is reconciled with an understanding of how the mechanism may actually behave physically and what manufacturing cost will be incurred to realise the design (Slocum 2010, Smith and Chetwynd 2003).

6.8.3 Kinematic Couplings

Figure 6.37 shows two of the most commonly used kinematic clamp designs. When a coupling is designed in such a way that three balls on one part of the coupling rest in a trihedral hole, in a V-groove (pointing towards the tetrahedron) and on a flat surface of another part, the coupling is referred to as a type I Kelvin clamp. The other common form of standard kinematic coupling is the type II Kelvin clamp that interfaces three balls on one part of the coupling to three V-grooves on the opposing part of the coupling (Dornfeld and Lee 2008, Slocum 1992). The geometry of a type I clamp offers a well-defined translational location based on the position of the trihedral hole, but it is asymmetric. For the type II Kelvin clamp, grooves at 120° meet the point which coincides with the centroid of the three spheres and this will be independent of thermal expansion effects. The geometry of type II clamps is also easier to manufacture. Both types of clamp are commonly used for mounting specimens so that they can be removed from an instrument and accurately relocated (Williamson and Hunt 1968). Gravity, magnets or springs are typically used to provide the force to ensure the contacts. However, type II is generally considered as a popular choice for applications such as theodolite mounts and other instruments that operate in a variable environment.

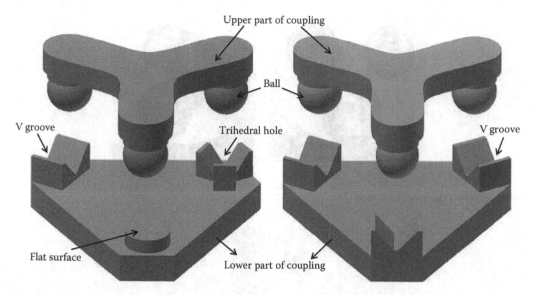

FIGURE 6.37
Common types of kinematic couplings: (left) type I Kelvin clamp, (right) type II Kelvin clamp.

A trihedral hole, as shown in Figure 6.37 and Figure 6.7, is difficult to make using conventional manufacturing methods. For many practical cases, it is acceptable to approximate the trihedral hole by other structures. One such structure is shown in Figure 6.38 in which three spheres are placed into three flat-bottomed holes. A fourth sphere placed on these three spheres will contact each at their point of common tangent. Similarly, V-grooves are very often used in practice using parallel cylinders, as shown in Figure 6.39. Figure 6.39 also shows the use of a type II clamp to build a reference gauge, which facilitates the measurement of the effective diameter of the probe stylus of a coordinate measuring system (CMS; see Chapter 5) in a way that is insensitive to the temperature of the gauge and which also minimises the uncertainty due to the positioning error of the CMS.

6.8.4 Some Specific Design Considerations of Kinematic Couplings

The performance of kinematic couplings will depend on some critical design parameters which must be carefully considered during the design process. While these design considerations are discussed in relation to the purely kinematic coupling, the general concepts can be applied to any kinematic coupling.

6.8.4.1 Load Capacity

The load capacity of a kinematic coupling is a direct function of the Hertz contact stress (see Chapter 3). In fact, the term Hertz contact symbolises the high stresses that arise between bodies in point or line contact (see Section 3.5.6). As Hertz stresses locally can be very high, they often act as initiation sites for spalling, crack growth and other failure mechanisms (Johnson and Johnson 1987, Slocum 2010).

Since point contacts are made in kinematic couplings, exact constraints are theoretically achieved, but they are also valid for all practical purposes as long as the loads are low. However, when heavier loads are applied, which may be due to the weight of the object or

FIGURE 6.38
Method for achieving trihedral contact.

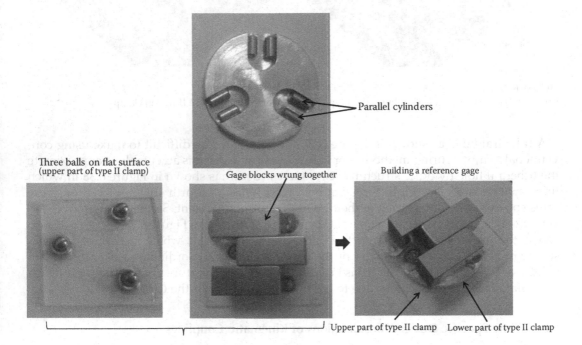

Parallel cylinders

Three balls on flat surface
(upper part of type II clamp)

Gage blocks wrung together

Building a reference gage

Upper part of type II clamp   Lower part of type II clamp

FIGURE 6.39
(Top) Use of parallel cylinders for achieving V-groove, (bottom) use of the approximate V-groove in a type II clamp to build a reference gauge.

preload, high Hertz contact stress at the contact points cause local deformations (point contacts become Hertzian contact ellipses) which act like additional orthogonal constraints. Surface friction also plays an important role during surface contact. Friction significantly alters repeatability during initial assembly of the coupling. When two coupling surfaces touch each other, friction between the surfaces builds up and creates forces that impede the motion of the entire coupling from settling into its lowest energy state. Each subsequent replacement of the coupling will settle into a different position based on a complex relationship between the initial position of each contact point and the exact direction of the applied force. Thus, the main limit to the repeatability of a kinematic coupling is the surface texture of the contact regions, while load capacity and stiffness are limited by the Hertz contact stress (Culpepper 2000, Slocum 2010, Willoughby 2005).

Since friction tends to limit the repeatability of kinematic locations, the use of hard surfaces with low interface shear strength is ideal for coupling design. This can be achieved by introducing a surface film of fatty acids within oils (lubrication) or by coating the contacts with thin polymeric or low strength metallic or lamellar films (Smith and Chetwynd 2003).

Controlling deformation and friction at the contact interface of the couplings is crucial for achieving a high level of repeatability. Hard ground steel surfaces can perform to the micrometre to sub-micrometre level. Coating with hard steel, for example with titanium nitride, can help prevent corrosion, but the difference in elastic modulus can lead to failure at high stress levels. Hard polished ceramic or tungsten carbide surfaces are preferable for couplings. For high-cycle applications (i.e. the number of interface engagement cycles of a coupling), use of corrosion-resistant materials, such as stainless steel, carbides or ceramic materials, for designing coupling interfaces can be advantageous. Because steels (other than stainless steel) are susceptible to fretting wear, they should be used only for low-cycle applications (Slocum 2010).

6.8.4.2 Geometric Stability

Geometric stability refers to the state of the coupling that will remain exactly constrained under design loads. The stability, and overall stiffness, of a kinematic coupling is maximised when the coupling ball and groove centerlines, and the normals to the planes containing pairs of contact force vectors, intersect at the centroid of the coupling triangle, as shown in Figure 6.40. In other words, the centrelines bisect the angles of the coupling triangle and intersect at a point called the coupling centroid. For static stability, the planes containing the pairs of contact force vectors must form a triangle. Beyond this triangle, stability is ensured by following careful design strategies, for example, by making sure that none of the contact forces reverse from a compressive state beyond this triangle and by applying a necessary preload to meet this condition (preload is the force applied to the coupling to hold it together). To get good stiffness for the coupling, preload needs to be high, repeatable and must not deform the rest of the structure (Culpepper 2000, Slocum 1992).

The stiffness of a kinematic coupling is also related to the coupling layout. Stiffness is equal in all directions when all the contact force vectors intersect the coupling plane at 45° angles. Coupling stiffness can be adjusted by changing the interior angles of the coupling triangle; elongating the triangle in one direction will increase the stiffness about an axis normal to the coupling plane and normal to the direction of elongation, and decrease the stiffness about an axis in the coupling plane and normal to the direction of elongation (Hart 2002, Slocum 1992).

6.8.4.3 Materials

Materials play a major role in the performance of a coupling. First, materials determine the maximum stress and, hence, the load capacity of the couplings. Second, they control the friction behaviour between coupling interfaces. Third, they also control the corrosion property of the coupling interface. As for the material selection, for high precision and high load applications, silicon nitride, silicon carbide or tungsten carbide are the best materials to provide high load capacity, to minimise the coefficient of friction and to prevent corrosion at the Hertz contact interface (Willoughby 2005).

Materials also affect the variations on surface geometry: surface texture, debris and fretting. Among the three, surface texture is a less difficult parameter to control as it can be specified and measured during manufacturing (see Chapter 5). Moreover, surfaces tend to

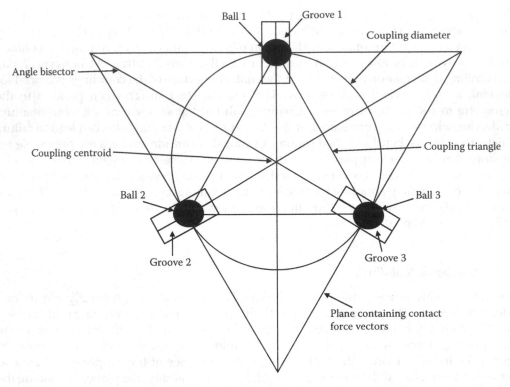

FIGURE 6.40
Geometric parameters for a ball-groove kinematic coupling. (From Hart, A. J., 2002, Design and analysis of kinematic coupling for modular machine and instrumentation structures, Master of science dissertation, Massachusetts Institute of Technology; and Slocum, A., 2010, Kinematic couplings: A review of design principles and applications, *Int. J. Mach. Tools Manufac.* 50:310–327.)

burnish or polish each other over time and with increased load. Debris and fretting corrosion are more difficult parameters to control, as these effects tend to develop over the lifetime of the components. In many cases, repeatability variations due to debris can be reduced by cleaning procedures and by placing a small layer of grease between the contact surfaces. The effect of fretting occurs between two surfaces of similar materials, particularly steel, when they are pressed together under large forces, causing small surface asperities to crush together and atomically bond. When the surfaces are separated, the bonds are ripped apart and the small pieces of the bonded materials remain as debris on the surface. The separated surfaces also cause the new material to become exposed to the environment; oxidation changes the surface hardness and surface texture. Thus, fretting causes repeatability problems by changing surface properties and topography. To avoid fretting, stainless steel, ceramics or a combination of dissimilar materials are recommended (Hart 2002, Slocum 2010, Willoughby 2005).

6.8.4.4 Manufacturing and Assembly

Figure 6.40 shows the primary geometric design parameters of ball-groove coupling; they are the effective ball radii, the coupling diameter, which is defined as the diameter of the circle on which the ball centres lie, and the groove orientations. The performance of coupling will depend on the ability of the manufacturing and assembling processes to conform to the design parameters. Thus, the sensitivity to these parameters on performance should be studied during the design process. Some tools, for example, the spreadsheets developed by Slocum, are readily available (Slocum 2010).

Generally, the accuracy of kinematic couplings depends on the tolerancing of the components and their assembly. In general, some averaging effects can be achieved during assembly; it has been observed that it is not unusual for the accuracy of the coupling to be two to three times better than the accuracy of the components used. Assembly factors, such as environmental contamination and thermal errors, also affect repeatability (Slocum 2010).

Another important design consideration is manufacturing and assembly cost. Two kinematic couplings—the Kelvin clamp and ball-groove coupling—can demonstrate component relocation at the micrometre-level of precision. However, the cost and time required to grind/polish the grooves are high for moderate to high volume applications. Also, these couplings are required to retain a gap between aligned components, thus sealing the assembly is a problem. This issue can be solved for some applications by using flexural kinematic couplings, but these are not suitable for many assembly processes due to the cost of integrated flexures (Culpepper et al. 2004).

6.8.5 Pseudo-Kinematic Design

A pure kinematic design for couplings, as was discussed in the previous section, can be difficult to achieve, if not impossible. Research shows that many practical couplings, which are developed following kinematic design principles for certain industrial applications, perform relatively well compared with the performance of pure kinematic couplings. In the following sections, a number of such design concepts will be examined.

6.8.5.1 Canoe Ball Couplings

The load capacity of a traditional ball-groove coupling, where the sphere diameters are approximately the widths of the V-grooves, is limited to its six point contacts. To build greater load capacity, yet maintain performance, Slocum developed the 'canoe ball' shape, which replaces the traditional hemispherical contacting element of a ball-groove coupling with a trapezoidal block (see Figure 6.41). This design accommodates the contact region of a ball as large as 1 m in diameter in an element as small as 25 mm across. A canoe ball mount sits on a standard V-groove, resulting in a much larger contact area than a ball of equivalent diameter that contacts the groove at the same points. This arrangement significantly increases the load capacity of a coupling while maintaining repeatability (Hart 2002, Slocum 2010, Willoughby 2005).

6.8.5.2 Quasi-Kinematic Couplings

Quasi-kinematic couplings (QKCs) are a passive means for precision location which combine elastic averaging (see Section 6.8.5.3) and kinematic design principles. As compared to the exact constraint provided by ball-groove couplings (to be more precise, they provide near-exact constraint in practice), QKCs employ slightly over-constrained attachments using simple, rotationally symmetric mating units (Culpepper 2000). Figure 6.42 shows a typical QKC, which has two parts: the upper platform with spherical surfaces is called a contactor, and the base is called a target. Each contactor engages in line contact with the

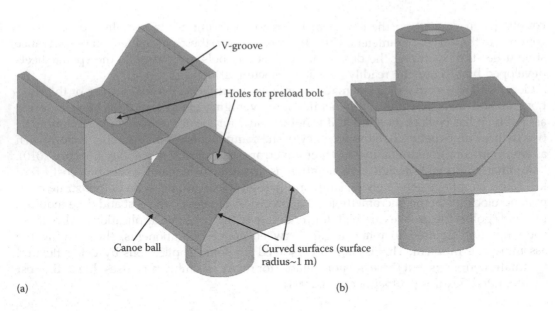

FIGURE 6.41
Canoe ball and groove elements: (a) ball-groove exploded assembly, (b) ball-groove assembly.

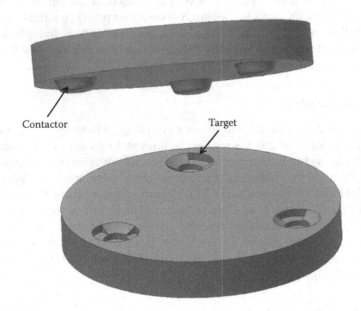

FIGURE 6.42
Components of a typical quasi-kinematic coupling.

corresponding target. Quasi-kinematic interfaces, by their design, leave a gap between the normal contact surfaces of the interface halves when the contactors and targets first touch. Then a preload is applied to seat the interface and close the gap. The preload serves to ensure the proper interface seating position by overcoming contact friction and by levelling surface inconsistencies at the contact areas. The deformation of the contactor and target, when the preload is applied, may be fully elastic or it may be partially elastic and partially plastic. When the gap is closed, the large mating horizontal surfaces, not the quasi-kinematic line contacts, dictate the normal stiffness. This high normal stiffness is desirable for high-load bearing machine applications (Culpepper 2000, Culpepper et al. 2004, Hart 2002).

It is interesting to note that this elastic and plastic deformation of this QKC can produce a burnishing effect on the contact surfaces. Surface texture, as was seen before, affects the repeatability and, hence, this burnishing effect acts as a built-in polishing operation, which otherwise is required to be achieved by expensive and time-consuming machining operations, such as polishing, during the manufacturing of the coupling. Thus, this deliberate introduction of plastic deformation into the design is an excellent example of plastic design.

QKCs are less repeatable compared to ball-groove couplings and contain unknown assembly stresses. However, the simple geometry reduces cost and enables direct machining of the coupling halves into mating components. Some high-volume precision manufacturing applications have used QKCs for this cost-to-performance trade-off, for example, in the assembly of a six-cylinder automotive engine (Culpepper 2000).

6.8.5.3 Elastic Averaging

Elastic averaging allows the performance of an interface to be improved by averaging errors using compliance between precision surfaces. The key to elastic averaging is to have a large number of features spread over a broad region that elastically deform when separate parts are forced into geometric compliance with each other. Although the repeatability and accuracy obtained through elastic averaging may not be as high as in deterministic systems, elastic averaging design can provide higher stiffness, lower local stress, and improved load distribution when compared to exact constraint designs. In a well-designed and preloaded elastically averaged coupling, the repeatability is approximately inversely proportional to the square root of the number of contact points (Hart 2002, Willoughby 2005).

One common example of elastic averaging is the concept of using multiple compliant wiffle tree supports to create a large number of support points relatively uniformly distributed over an area. One advanced wiffle tree application is that used for support and alignment of multi-segment telescopes (see Prochaska et al. 2016). A more familiar compliant wiffle tree design is incorporated into the structure of automobile windshield wipers to ensure that the wiper blade remains in contact with the windshield and distributes the point force uniformly across the blade. Another example of elastic averaging in machine design is commonly found in precision rotatory indexing tables (Figure 6.43), in which two face gears are mated and lapped together. Contacts are made over many teeth so that individual manufacturing errors are averaged out. This type of mechanism is often referred to as a Hirth coupling (Figure 6.43). A more detailed discussion of elastic averaging is covered in Chapter 7.

6.8.6 Comparison of Different Types of Coupling

Table 6.1 shows a comparison of the relative abilities of different coupling design options in precision engineering. Furthermore, Figure 6.44 shows the comparison of typical performance limits, in terms of repeatability, of the coupling designs (Culpepper 2000).

Considering that all kinematic design options (exact, quasi or elastic averaging) have their own advantages and limitations, designers are left with the available information, design tools and their own intuition to decide when choosing the most appropriate and cost-effective design for a given application. Specific coupling designs are also discussed in Chapter 7, Section 7.3.8.

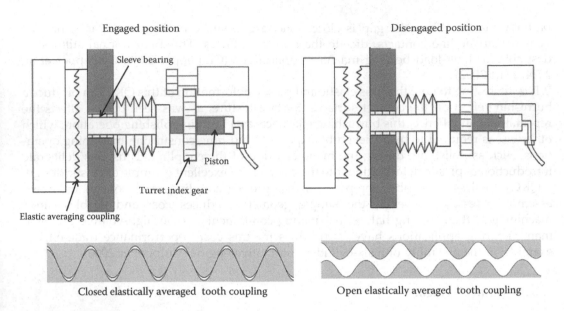

FIGURE 6.43
Application of elastic averaging: (top) schematic representation of a typical turret indexing and locking mechanism in engaged and disengaged positions, (bottom) Hirth coupling.

TABLE 6.1

Comparative Performance Criteria of Different Couplings

Coupling Type	Contact Type	Repeatability	Stiffness	Load Capacity
Pin joint	Surface	Poor	High	High
Elastic averaging	Mixed	Good	High	High
Quasi kinematic	Line	Good	Medium to high	High
Kinematic	Point	Excellent	Low	Varies

Source: Willoughby, P., 2005, Elastically averaged precision alignment, Doctoral dissertation, Massachusetts Institute of Technology.

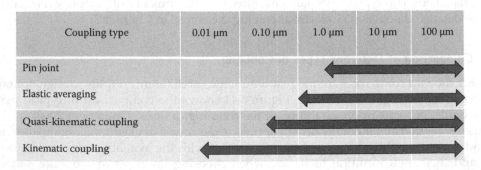

FIGURE 6.44
Performance limits of various types of couplings.

6.9 Kinematic Structure

6.9.1 Introduction to Structures

All machines or mechanisms have structures and the structures are often made up of parts. One of the first considerations in the design of a structure is choosing the manner by which the parts of the structure move in order to perform the functions for which the structure is designed. In kinematics, the motion of the part can be described by the position and direction in three-dimensional space in terms of a three-dimensional coordinate system (Cartesian, cylindrical and spherical coordinate systems can be identified, and transformations can be applied to transform from one system to other) (Niku 2010, Vinogradov 2000). Thus, a kinematic structure defines the layout of the parts and their respective axes. Essentially, a kinematic structure represents the kinematic chain without considering the detailed geometric, kinematic and functional aspects of the mechanisms used. A kinematic chain can be described as a set of rigid bodies attached to each other. For a metal cutting machine, a kinematic chain shows the flow of the motion in a kinematic structure using all axes, the work piece, the tool and the bed of the machine (Schwenke et al. 2008, Zhang 2009). Figure 6.45 shows the kinematic structure of a typical five-axis precision machine.

Kinematic structures can be arranged in serial, parallel and hybrid (combination of serial and parallel) kinematic topologies, with different numbers of links and connections to satisfy the functional requirements. The mobility of the kinematic structures will depend on the number of joints and links, and the joints' degrees of freedom, as can be determined by Equation 6.3.

6.9.2 Serial Kinematic Structures

6.9.2.1 Characteristic Features

Traditional precision machines and most industrial robots are based on serial kinematic structures (only having one open kinematic chain), in which multiple parts (considering the parts as rigid bodies, ignoring elasticity and any deformation caused by large load conditions) are connected in series (see Figures 6.46 and 6.47). Thus, in serial kinematic structures, the individual axes of the parts are built one on top of the other, with the lower axis carrying the one above it (Niku 2010, Warnecke et al. 1998). Each part involved in this structure is called a link and the combination of links is referred to as a linkage. A pair of links in a linkage is connected by a joint. The two most commonly used engineering joints are (1) prismatic joints (allows two links to produce relative displacement along the common axis), and (2) revolute joints, pivot joints or pin joints (allows two links to produce relative rotation about the joint axis) (see Figure 6.48). Combining these two types of joints, many useful mechanisms are made for precision machines and robots. Of course, a machine structure includes the supporting frame and the functional elements of the machine, for example, the drives and guides, which are added to this frame.

One of the fundamental requirements of a kinematic structure is to locate (position and orientation) its end effector, for example a tool or any part of the structure performing the task, in three-dimensional space. This requirement is satisfied when the kinematic structure is considered as a coordinate system and every part of the structure can be described, in terms of part's position and orientation, in the coordinate system. Thus, depending upon the geometry, a kinematic structure can be described by a Cartesian coordinate system, a cylindrical coordinate system or a spherical coordinate system. For example, a Cartesian

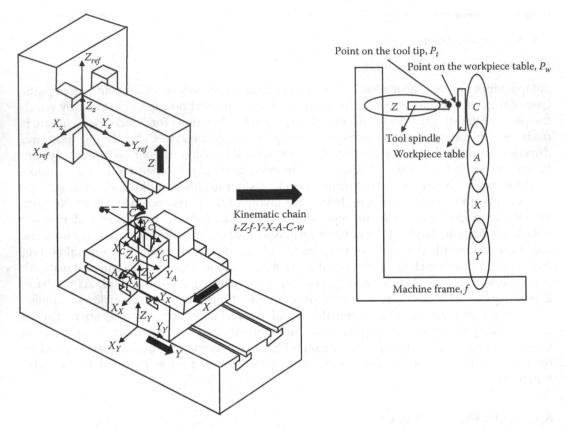

FIGURE 6.45
Schematic configuration and kinematic chain (t: tool spindle, Z: Z axis, b: frame, Y: Y axis, C: C axis, w: work piece table) of a typical five-axis precision machine.

robotic arm structure has three prismatic joints, a cylindrical robotic hand consists of one revolute joint and two prismatic joints, and a spherical robotic hand has two revolute joints and one prismatic joint (Eckhardt 1998, Niku 2010). The use of coordinate systems to represent kinematic structures is considered as the foundation of kinematic analysis, such as forward kinematic analysis and backward kinematic analysis (discussed in Chapter 8).

6.9.2.2 Benefits and Limitations of Serial Kinematic Structures

Serial kinematic machines usually have the advantages of a well-defined structure and configuration, and have comparatively large working volumes of simple shape. In serial structures, the direct correlation between positioning actuators and machine axes means the machine and workpiece coordinate systems can be the same, resulting in straightforward control of the structure (Allen et al. 2011). On the other hand, serial machines suffer from large bending moments and second moments of mass, especially at the extremes of axis travel if large masses are to be moved and have limited dynamics of the feed axes. Furthermore, these machine structures are highly dependent on the precision of each positioning element in the kinematic chain and are prone to compound error. Errors in the lower linkages of a kinematic chain will compromise the accuracy of the linkages stacked upon it. For this reason, as more positioning elements are added to a serial machine, the

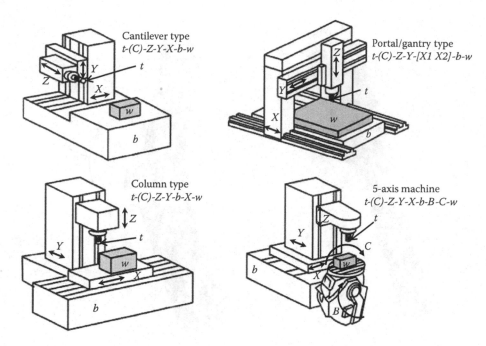

FIGURE 6.46
Serial kinematic machines abstracted to kinematic chain (*t*: tool, *b*: bed, *w*: work piece). (From Schwenke, H., Knapp, W., Haitjema, H., Weckenmann, A., Schmitt, R. and Delbressine, F., 2008, Geometric error measurement and compensation of machines—An update, *Ann. CIRP* 57:660–675.)

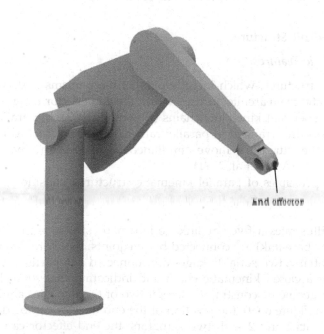

FIGURE 6.47
Six degree of freedom Puma robot (serial kinematic structure).

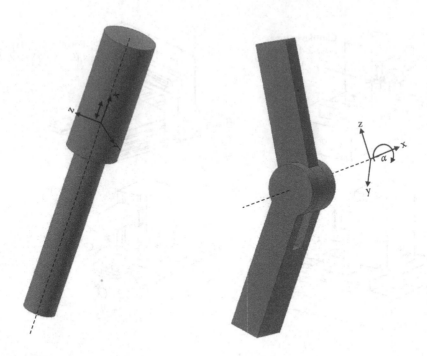

FIGURE 6.48
CAD models of (left) a prismatic joint and (right) a revolute joint.

greater the overall error is likely to become, causing greater deviations at the final position of the end effector (Allen et al. 2011, Warnecke et al. 1998).

6.9.3 Parallel Kinematic Structures

6.9.3.1 Characteristic Features

Parallel kinematic structures, which include parallel mechanisms and parallel machines, have closed kinematic chain architectures in which the end effector is connected to the base by at least two independent kinematic chains (Gao et al. 2002). In parallel structures, the positioning actuators are arranged in parallel rather than being stacked one on top of the other, requiring all the actuators to move simultaneously to effect any change in end effector position or orientation (Allen et al. 2011).

To illustrate the principles of parallel kinematic structures, consider the following two examples:

1. Figure 6.49 illustrates a five-bar linkage kinematic structure in which five links (including the base link) are connected by five joints, therefore, having a mobility of two. Essentially, two serial linkages are connected at a particular joint (joint 4), thus forming a closed kinematic chain and indicating that serial linkages must conform to a geometric constraint. Here, if two of the five joint angles are known ($\varnothing_1$ and $\varnothing_2$ in Figure 6.49), the position of the end effector can be determined. By controlling joints 1 and 2 with two actuators, the end effector can be positioned within the vertical plane of the structure. Thus, only two joints are active joints, driven by independent actuators, while the other three joints are passive joints, which are free to rotate (Tsai 1999).

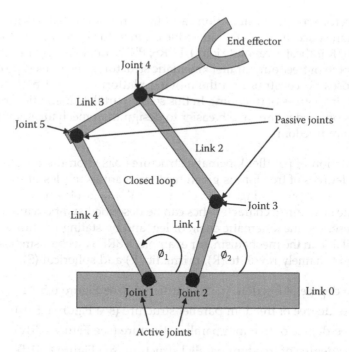

FIGURE 6.49
Five-bar link parallel kinematic structure.

2. Figure 6.50 shows a form of parallel kinematic structure known as a Stewart-Gough mechanism (or often just Stewart platform). This structure has six legs or kinematic chains. These legs, each of which consists of an actuator-controlled prismatic joint and two spherical joints at the ends, connect the moving platform to the base. Note that this structure uses spherical joints (joints that allow one link to rotate freely in three dimensions with respect to the other link about the centre of a sphere), which are used in many designs of parallel structures in addition to the other two commonly used joints, that is prismatic and revolute joints (Tsai 1999).

The parallel kinematic structure shown in Figure 6.50 uses SPS legs, which means that joints are laid out in each leg or kinematic chain in this way: spherical joint (S), prismatic joint (P), spherical joint (S). Each SPS leg has six degrees of freedom since

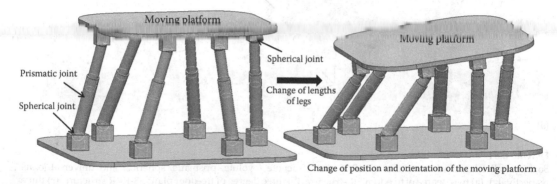

Change of position and orientation of the moving platform

FIGURE 6.50
Stewart-Gough type parallel kinematic structure.

its end effector has three translations and three rotations that are independent in the Cartesian space. In fact, any kinematic chain with six degrees of freedom, such as SPS, RSS (R indicates revolute joint), PPRS, PRPS and PPSR chains, can be the leg of a six degree of freedom parallel kinematic structure, since this type of kinematic chain imposes no constraint on the moving platform (a rigid body in space has maximum six degrees of freedom). In this sense, parallel kinematic structures with six degrees of freedom are much easier to design than mechanisms with less than six degrees of freedom.

A brief introduction of parallel kinematic structures based on the classification of space dimension and degrees of freedom is given next; relevant examples of different classes of parallel structure are shown in Figure 6.51 (Liu and Wang 2014; Zhang 2009). Note that a parallel kinematic structure's characteristics can be described symbolically by showing the types of joint present in the kinematic chain or leg, and by stating the number of kinematic chains/legs available in the mechanism, for example, a 3RPS mechanism has three legs and three types of joint, namely revolute (R), prismatic (P) and spherical (S).

1. Planar two degree of freedom parallel structure (see Figure 6.51a)
2. Planar three degree of freedom parallel structure (see Figure 6.51b)
3. Spatial three degree of freedom parallel structure (see Figure 6.51c)
4. Spatial four degree of freedom parallel structure (see Figure 6.51d)
5. Spatial five degree of freedom parallel structure (see Figure 6.51e)
6. Spatial six degree of freedom parallel structure (see Figure 6.51f)

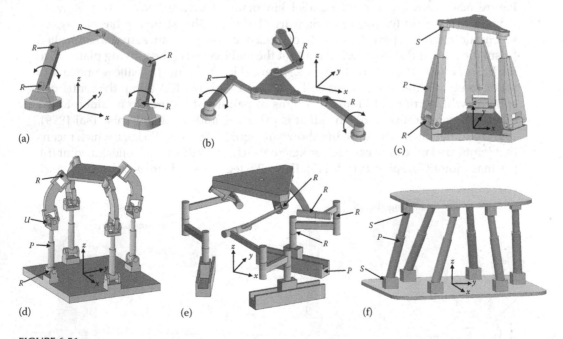

FIGURE 6.51
Some parallel kinematic structures (R, P, S and U denote revolute, prismatic, spherical and universal joints respectively): (a) two degree of freedom 5R structure, (b) three degree of freedom planar 3-RRR structure, (c) three degree of freedom 3-RPS structure, (d) four degree of freedom structure with four RPUR chains, (e) five degree of freedom structure with three PRRRR chains, (f) six degree of freedom general Stewart platform.

6.9.3.2 Benefits and Limitations of Parallel Kinematic Structures

Parallel kinematic structures offer higher stiffness, lower moving mass, higher acceleration, reduced installation requirements and mechanical simplicity as compared to existing conventional machines with serial kinematic structures. Parallel structures can potentially be more accurate than serial structures. This is because serial machines accumulate errors for every axis, since the parts are serially added in the kinematic chain. While calculating the errors of a serial machine, geometric errors of each axis, usually expressed in the forms of matrices, are successively multiplied, resulting in deviations of the expected positions of the tool point (see Chapter 8, Sections 8.3 and 8.7). In contrast, errors are averaged among the kinematic chains of the parallel machine structures. The parallel structures have the potential to be highly modular, highly reconfigurable and high-precision machines. Other potential advantages include high dexterity, the requirement for simpler and fewer fixtures, multi-mode manufacturing capability and a small footprint (Allen et al. 2011, Gao et al. 2002, Zhang 2009).

Some attributes of machine tools for serial and parallel kinematic structures can qualitatively be compared as presented in Table 6.2. Parallel kinematic structures have received considerable attention in industry. They have found their applications mainly in machine tools (for example, see Figure 6.52), medical applications for ophthalmic surgeries, manipulators (Delta robot for very fast pick-and-place tasks of light load, micromanipulator), flight/automobile/tank/earthquake simulators, and precision positioning for very large telescopes, receiving antennas and as satellite platforms for manoeuvring in space (Gao et al. 2002, Mekid 2008, Weck and Staimer 2002).

However, parallel kinematic structures have some limitations that pose challenges to their widespread adoption (Rugbani and Schreve 2012, Soons 1999):

- The main disadvantage of parallel structures is the limited workspace and the difficulty of their motion control due to singularity problems. Singularity is a point in the workspace where a kinematic machine/mechanism cannot move its end effector due to certain geometric configuration(s) of the structure, for example, collinear alignment of two or more axes of a robotic structure.

TABLE 6.2

Comparison of the Properties between Serial and Parallel Kinematic Structure

Attribute	Serial Kinematic Structure	Parallel Kinematic Structure
Working volume/total size of machine	Moderately suitable	Least suitable
Error accumulation	Least suitable	Most suitable
Overall accuracy	Most suitable	Moderately suitable
Static stiffness	Most suitable	Most suitable
Axes acceleration	Least suitable	Most suitable
Cutting forces	Least suitable	Most suitable
Machining of ≤5 faces in single setup	Least suitable	Moderately/most suitable (depending on machine structure)
Range of angular motion (reaching to 90°) dexterity	Most suitable	Moderately suitable

Source: Mekid, S., 2008, Introduction to parallel kinematic machines, In: Mekid, S., *Introduction to precision machine design and error assessment*, CRC Press.

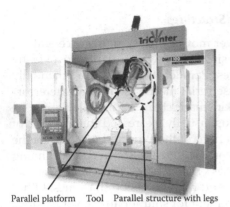
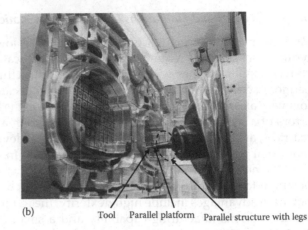

Parallel platform   Tool   Parallel structure with legs
(a)

(b)   Tool   Parallel platform   Parallel structure with legs

FIGURE 6.52
Machines with parallel kinematic structures for machining: (a) TriCenter as universal five-axis machine, (b) Ecospeed for machining complex monolithic aerospace parts. (From Weck, M. and Staimer, D., 2002, Parallel kinematic machine tools-current state and future potentials, *Ann. CIRP* 51:671–683.)

- Parallel structures have more complex working envelopes than traditional serial machines. These working envelopes often take the form of an onion or mushroom and depend on the exact layout of the machine structure.

- Many parallel structures have limited ranges of motion, particularly rotational motion.

- The positioning and orientation inaccuracies of parallel kinematic structures can stem from a number of sources: manufacturing errors of parallel structure elements, assembly errors, error resulting from distortion by force and heat, control system errors and actuator errors, errors due to overhanging structures, calibration and even the accuracy of the mathematical models used.

- The design-specific cost of parallel kinematic structures in comparison with conventional machines can be high.

6.9.4 Hybrid Kinematic Structures

6.9.4.1 Characteristic Features

In recent years, many new types of parallel structures have been developed to enhance the capabilities and performance of parallel kinematic machines, such as maximising workspace, and increasing rotational capability (El-Khasawneh 2012). However, considering that some limitations are inherent to parallel kinematic structures, a new shift in addressing the issues with parallel structures has occurred, motivating researchers to look at hybrid structures, which consist of parallel and serial kinematic structures. This shift, in fact, has created new research and development needs with promises of interesting innovations in mechanism design (Harib et al. 2012).

A mechanism with a hybrid structure is a combination of serial and parallel kinematic structures to exploit the advantageous characteristics of both. This is generally implemented by either connecting two parallel structures in series (one of the parallel structures is the upper stage and the other is the lower stage, where the moving platform of the lower stage is the base platform of the upper stage) or connecting the series and the parallel structures in series. Mechanisms with hybrid structures, as compared to the mechanisms with parallel structures, can improve the ratio of workspace to architecture size and accuracy.

To have an understanding of how a hybrid structure can be designed for a mechanism to achieve five degrees of freedom of motion, consider the following design options (Liu and Wang 2014):

1. A structure can be arranged such that the upper stage provides the motions of three degrees of freedom (rotations about the x and y axes and a translation along the z axis), while the lower stage provides the motions of two degrees of freedom (translations along the x and y axes). The hybrid motions (five degrees of freedom) of the structure can be realised in two ways:

   • 3SPR (spherical-prismatic-revolute) as upper stage and 'liner guiding system' as the lower stage.

   • 3SPR as upper stage and 3RRR (three revolute joints), which is a planar parallel structure, as the lower stage.

2. The hybrid structure can also be arranged such that the upper stage provides the motions of three degrees of freedom (translations along x, y and z axes), while the lower stage provides the motions of two degrees of freedom (rotations about the x and y axes). The motions of the structures can be achieved by a combination of a 3SRR parallel structure as the upper stage and a two degree of freedom spherical parallel structure as the lower stage. However, spherical parallel structures are complex to manufacture and have low stiffness, low precision and small workspace; hence, they are not currently popular in practice.

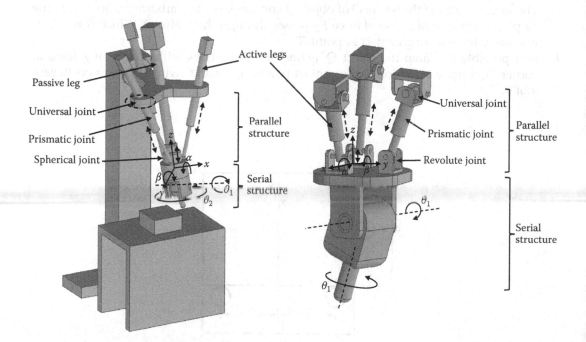

FIGURE 6.53
Schematic representation of two types of hybrid mechanism: (left) Tricept, (right) Exechon.

6.9.4.2 Some Applications of Hybrid Structures

Among the early hybrid kinematic designs, the Tricept was considered as the first commercially successful hybrid machine tool. Tricept has a three degree of freedom parallel kinematic structure and a standard two degree of freedom wrist end effector (Figure 6.53, left). Another hybrid machine, the Exechon, was introduced later as an improvement over the Tricept design (Figure 6.53, right). The Exechon has an over-constrained structure with eight links and a total of nine joints. However, regardless of the promising prospect of hybrid kinematic structures, detailed understanding of the involved kinematics, dynamics and design of these structures is still lacking (Harib et al. 2013, Milutinovic et al. 2013).

Exercises

An object Q of rectangular shape with edge lengths ($l = 80$ mm and $h = 60$ mm) has to be clamped against three bolts with the contact points A, B and C, as shown in Figure 6.54. The bolts B and C are situated in the distance $a = 30$ mm from the vertical centre line.

1. How many degrees of freedom has the system of object and bolts, if the object is in contact at the points A, B and C with three bolts?
2. A force F_D for pushing the object Q against the three bolts is applied at corner D. Which conditions has the direction of force F_D to meet for pushing the object Q against all three contact points at the same time without friction at the contact points A, B and C?
3. The force F_O acts in the centre O of object Q and can have any arbitrary direction in the xy-plane. The line of action of force F_D passes through the centre O. Which force F_D is necessary for ensuring contact at point B?
4. Is it possible to clamp the object Q against the three bolts with a clamping force at corner D, if there is friction at the contact points A, B and C with a friction coefficient μ of 0.3?

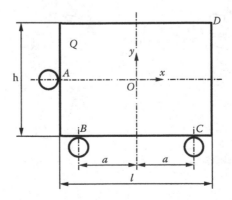

FIGURE 6.54
Geometry.

5. How does the centre O of object Q move, if the object Q is expanding relative to the
    environment with a very small scaling factor K?
6. How can the three contact points A, B and C be arranged, so that the line of action of
    force F_D does not pass between the instantaneous centres of rotation P_{AB} and P_{AC}?
7. How does the centre O of object Q in the suggested arrangement in the solution of
    Exercise 6 move if the object Q is expanding relative to the environment with a very
    small scaling factor K?
8. How can the three bolts in the solution of Exercise 7 be arranged by changing the shape
    of object Q to avoid motion of centre O due to an expansion of Q relative to the
    environment?
9. A four-bar linkage, as shown in Figure 6.55, has a driver link MN, a follower link OP
    and a coupler or connecting link NO. The points M, N, O and P represent revolute or pin
    joints. The fourth linkage is the base MP and can be referred to as the frame link or
    reference. Consider Figure 6.56, which shows a kinematic diagram of a planar four-bar

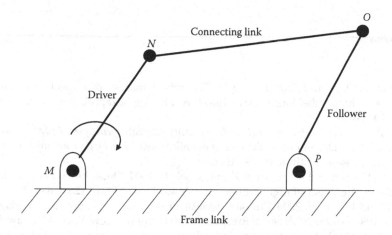

FIGURE 6.55
A four-bar linkage.

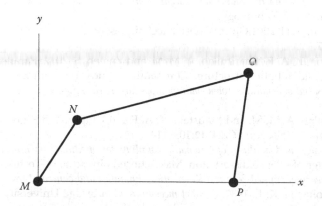

FIGURE 6.56
Geometry of a four-bar linkage.

linkage, representing the geometry of the linkage in xy coordinate system. Given that $MN = 4$ cm, $OP = 10$ cm, $NO = 16$ cm and $MP = 14$ cm, determine:

    a. The degrees of freedom of the mechanism.

    b. The angle θ between OP and MP when OP is in its extreme right position.

    c. The angle $\varnothing$ between OP and MP when OP is in its extreme left position.

    d. The total angle of rotation of OP for one revolution of link MN.

10. Consider the example of four-bar linkage, as shown in Figure 6.56. Instead of analysing the linkage already designed, as demonstrated in Exercise 9 where the lengths of the link were known, a linkage needs to be designed where follower link PO is to rotate through an angle of 45° while MN rotates through one revolution. If the suitable lengths of MP and OP are decided to be 14 cm and 10 cm respectively, what must be the lengths of the links MN and NO?

References

Allen, J. M., Axinte, D. A. and Pringle, T. 2011. Theoretical analysis of a special purpose miniature machine tool with parallel kinematic architecture: Free leg hexapod. *Proc. IMechE Part B: J. Eng. Manu.* **226**:412–430.

Blanding, D. L. 1999. *Exact constraint: Machine design using kinematic principles.* ASME Press: New York.

Culpepper, M. L. 2000. *Design and application of compliant quasi-kinematic couplings.* Doctoral dissertation, Massachusetts Institute of Technology.

Culpepper, M. L., Slocum, A. H., Shaikh, F. Z. and Vrsek, G. 2004. Quasi-kinematic couplings for low-cost precision alignment of high-volume assemblies. *Tran. ASME.* **126**:456–463.

Dornfeld, D. A. and Lee, D. E. 2008. *Precision manufacturing.* New York: Springer-Verlag.

Eckhardt, H. D. 1998. *Kinematic design of machines and mechanisms.* New York: McGraw-Hill.

El-Khasawneh, B. S. 2012. The kinematics and calibration of a 5 degree of freedom hybrid serial-parallel kinematic manipulator. *8th International Symposium on Mechanics and Its Applications,* 1–7.

Gao, F., Li, W., Zhao, X., Jin, Z. and Zhao, H. 2002. New kinematic structure for 2-, 3-, 4- and 5-DOF parallel manipulator designs. *Mech. Mach. Theory* **37**:1395–1411.

Hale, L. C. 1999. *Principles and techniques for designing precision machines.* Doctoral dissertation, Massachusetts Institute of Technology.

Hale, L. C. and Slocum, A. H. 2001 Optimal design techniques for kinematic couplings. *J. Int. Soc. Prec. Eng.* **25**:114–127.

Harib, K. H., Moustafa, K. A. F., Sharifullah, A. M. M. and Zenieh, S. 2012. Parallel, serial and hybrid machine tools and robotics structure: Comparative study on optimum kinematic designs. In: Kucuk, S., *Serial and parallel robot manipulators-kinematics, dynamics, control and optimization.* InTech.

Harib, K. H., Sharifullah, A. M. M. and Moustafa, K. A. F. 2013. Optimal design for improved hybrid kinematic machine tools. *Ann. CIRP* **12**:109–114.

Hart, A. J. 2002. *Design and analysis of kinematic coupling for modular machine and instrumentation structures.* Master of science dissertation, Massachusetts Institute of Technology.

Jazar, R. N. 2007. *Theory of applied robotics: Kinematics, dynamics and control.* New York: Springer.

Johnson, K. L. and Johnson, K. L. 1987. *Contact mechanics.* Cambridge University Press.

Liu, X. J. and Wang, J. 2014. *Parallel kinematics.* Springer-Verlag.

Mason, M. T. 2001. *Mechanics of robotic manipulation.* Cambridge: The MIT Press.

Mekid, S. 2008. Introduction to parallel kinematic machines. In: Mekid, S., *Introduction to precision machine design and error assessment*. CRC Press.

Milutinovic, M., Slavkovic, N. and Milutinovic, D. 2013. Kinematic modelling of hybrid parallel serial five axis machine tool. *FME Tran.* **41**:1–10.

Niku, S. B. 2010. *Introduction to robotics: Analysis, control, applications*. John Wiley & Sons.

Prochaska, J. X., Ratliff, C., Cabak, J., Tripsas, A., Adkins, S., Bolte, M., Cowley, D., Dahler, M., Deich, W., Lewis, H., Nelson, J., Park, S., Peck, M., Phillips, D., Pollard, M., Randolph, B., Sanford, D., Ward, J. and Wold, T. 2016. Detailed design of a deployable tertiary mirror for the Keck I microscope. *Proc. SPIE 99102Q-99102Q*.

Rider, M. J. 2015. *Design and analysis of mechanisms a planar approach*. John Wiley & Sons.

Rugbani, A. and Schreve, K. 2012. Modelling and analysis of the geometrical errors of a parallel manipulator micro-CMM. *Int. Precision Assembly Semin.* **371**:105–117.

Schouten, C. H., Rosielle, P. C. J. N. and Schellekens P. H. J. 1997. Design of a kinematic coupling for precision applications. *Prec. Eng.* **20**:46–52.

Schwenke, H., Knapp, W., Haitjema, H., Weckenmann, A., Schmitt, R. and Delbressine, F. 2008. Geometric error measurement and compensation of machines—An update. *Ann. CIRP* **57**:660–675.

Slocum, A. 2010. Kinematic couplings: A review of design principles and applications. *Int. J. Mach. Tools Manufac.* **50**:310–327.

Slocum, A. H. 1992. *Precision machine design*. Society of Manufacturing Engineers.

Smith, S. T. and Chetwynd, D. G. 2003. *Foundations of ultra-precision mechanism design* (Vol. 2). CRC Press.

Soons, J. A. 1999. Measuring the geometric errors of a hexapod Proceedings of the 4th LAMDAMAP Conference, New Castle Upon-Tyne, United Kingdom.

Tsai, L.-W. 1999. *Robot analysis: The mechanics of serial and parallel manipulators*. John Wiley & Sons.

Vinogradov, O. 2000. *Fundamentals of kinematics and dynamics of machines and mechanisms*. CRC Press.

Warnecke, H. J., Neugebauer, R. and Wieland, F. 1998. Development of Hexapod based machine tool. *Ann. CIRP* **47**:337–440.

Weck, M. and Staimer, D. 2002. Parallel kinematic machine tools-current state and future potentials. *Ann. CIRP* **51**:671–683.

Whitehead, T. N. 1954. *The design and use of instruments and accurate mechanism; underlying principles*. New York: Dover Press.

Williamson, J. B. P. and Hunt, R. T. 1968. Relocation profilometry. *J. Sci. Inst. (J. Phy. E).* **1**:749–752.

Willoughby, P. 2005. *Elastically averaged precision alignment*. Doctoral dissertation, Massachusetts Institute of Technology.

Zhang, D. 2009. *Parallel robotic machine tools*. Springer Science and Business Media.

7

Precision Machine Principles and Elements

Stuart T. Smith

CONTENTS

ABSTRACT   Mechanical elements and mechanisms form the foundational components that can be assembled to provide the manufacturing instruments and machines for producing anything from the large-scale (for example, ships, planes, trains and automobiles) to atomic-level fabrication (for example, lithography tools, and micro- and nano-electro-mechanical systems). Important principles that must be applied when designing, fabricating and using these machines and processes are outlined in other chapters in this book. Adherence to these principles is essential when precision is a major objective. This chapter considers those elements that are commonly utilised in precision machines not only in terms of performance, but also emphasises the limits of achievable precision and principles of installation so that these limits can be realised. While this chapter is mainly focused on motion guidance mechanisms, specific principles associated with stiffness and symmetry considerations are discussed, with a focus on the performance of integrated mechanisms. Operation and performance of specific machine elements are then reviewed, again focusing on limiting precision and the optimal conditions for this to be realised. These machine elements are split into two major categories: motion guidance mechanisms or 'bearings', and actuators that provide the motive power. The chapter concludes with a discussion of methods and mechanisms for coupling actuators to the motion systems.

7.1 Overview: Basic Principles of Machines

For many applications, commonly used machine components can be commercially purchased and assembled into processes with performance that adequately satisfies the functional requirements. For these cases, such mechanisms can be assembled and analysed using standard machine element design equations and principles, see for example Budynas and Nisbet (2014), Spotts and Shoup (2003) or Juvinall and Marshek (2011). In this chapter, it will be assumed that the reader has completed, at least at an introductory level, a course of machine component design.

For precision applications, particularly when performance requirements are close or beyond the limits of existing technologies, it becomes necessary to more carefully select and implement solutions or, when solutions are not available, to create new designs. The purpose of this chapter is to review machine components in terms of their implementation and limitations for precision applications. The chapter is split into three sections: basic principles, bearings, and actuators. Sub-sections contain reviews of the types of bearings and actuators commonly used in instruments and machines. Selection of these topics and the structure of this chapter might be appreciated by considering a generic measuring instrument or machine as shown in Figure 7.1. The machine contains an end effector (see also Chapter 11, Section 11.5.2) that might be a measuring probe or cutting tool that can be rotated about a vertical axis and translated in the vertical direction. A specimen that is subject to the end effector (possibly being measured or machined) is mounted onto a high resolution, short-range, translation stage that is, in turn, mounted onto a long-range slideway. Finally all of the components are fabricated into a stiff frame. Measurement of the various components is represented throughout the book with specific focus on metrology implementation being the subject of Section 11.5.

Mechanical machinery spans all mechanisms, processes and devices. Ultimately, the goal of a mechanical machine is to deliver work with specifically determined motion. Whether

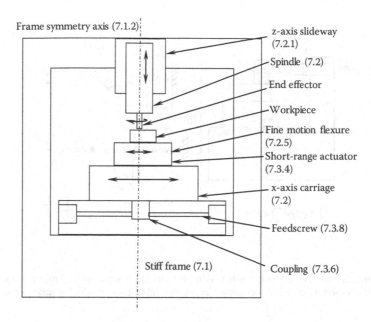

Frame symmetry axis (7.1.2)

z-axis slideway (7.2.1)

Spindle (7.2)

End effector

Workpiece

Fine motion flexure (7.2.5)

Short-range actuator (7.3.4)

x-axis carriage (7.2)

Feedscrew (7.3.8)

Stiff frame (7.1)

Coupling (7.3.6)

FIGURE 7.1
Schematic diagram of a generic measuring instrument or mechanical machine. Numbers in parentheses indicate sections in this chapter where these individual components are discussed.

this be aligning a telescope or pushing a cutting tool into a workpiece, for precision applications the desire is to deliver this work with precisely known, or controlled, motions and, often, to do this as fast as possible. The latter involves not only high forces and large displacements but also high power. Newton's laws of motion (see Chapter 4) imply that to accelerate a component to high speeds will require large forces, while Hooke's law (see Chapter 3, Section 3.5.3) indicates that large forces induce distortions. Consequently, before discussing specific machine components, it is helpful to consider machines as generic mechanisms comprised from elastic bodies connected by bearings.

7.1.1 Stiffness

Inside most workshops or manufacturing facilities there will be a variety of machine tools. Interestingly, the lower cost machine tools will be removing large amounts of material, while the expensive precision tools (particularly diamond turning machines) will be making lighter finishing cuts, often operating at slower feed rates. Clearly, the performance limiting aspect of the tools is not the strength of the materials from which it is made. The major difference is that distortions due to the cutting forces will be considerably lower for the precision machine. Hence, for the controlled delivery of power, limitations on motion control are primarily influenced by the stiffness of the components through which forces are transmitted (the force loop) and, in particular, those components where the force and metrology loops coincide (see Chapter 11 for a detailed analysis of force loops).

Figure 7.2 is a block diagram representation of a lathe showing key components (base, main carriage, cross-slide, tool-post, tool, workpiece, spindle and head-stock). The aim of the lathe is to cut material from the workpiece by applying a force to the tool at reference point P and to control the position of the tool in directions aligned with the axis of the spindle. In practice, the motion of the tool uses line-scales that are placed at convenient

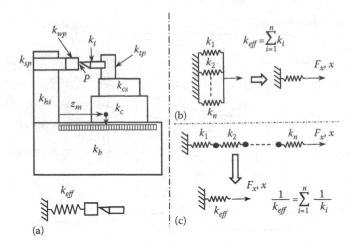

FIGURE 7.2
Block diagram representation of the components making up a machine structure: (a) major components of a standard lathe, (b) stiffness model for springs connected in parallel, (c) stiffness model for springs connected in series.

locations and will be mounted with some finite misalignment between the measurement axis and the desired axes at the workpiece. The effects of these offsets and misalignments result in significant cosine and Abbe errors (discussed in Chapter 10).

From Newton's laws of motion, any force on the tool must induce a reaction at the workpiece, that will, in turn, be transmitted to the spindle and so on. All components are made from elastic materials for which any force will introduce a stress. Stresses introduce strains that result in deformation. Precision necessitates that stresses be in the elastic range of the materials and, therefore, a linear relationship is expected between force and displacement for any of the components of the lathe assembly. In the absence of dynamic effects, the force passing through all components around the structure will have the same value.

Simplistically, each of the components of the process can be abstracted to a linear spring of stiffness k_x (where the subscript indicates the direction of deflection for the component of force in the same direction). Tracing some of the components around the lathe, as shown in Figure 7.2, the workpiece and spindle will be subject to direct forces in the axial direction. Therefore, these two elements can be modelled as two springs in series having values given by the axial stiffness. Other components (the head-stock and tool-post in particular) will be subject to forces that will produce bending moments. In this case, the total stiffness will comprise components of shear, direct stress and bending, each of which adds to the compliance (the reciprocal of stiffness). Generally, all of the components will distort and these distortions will accumulate around the loop. Consequently, motion of the workpiece due to the applied force at the tool tip can be computed from an effective stiffness:

$$\frac{1}{k_{eff}} = \frac{1}{k_t} + \frac{1}{k_{wp}} + \frac{1}{k_{sp}} + \frac{1}{k_{hs}} + \frac{1}{k_b} + \frac{1}{k_c} + \frac{1}{k_{cs}} + \frac{1}{k_{tp}} \tag{7.1}$$

or

$$c_{eff} = c_t + c_{wp} + c_{sp} + c_{hs} + c_b + c_c + c_{cs} + c_{tp}, \tag{7.2}$$

where the stiffness k terms are illustrated in Figure 7.2.

Equation 7.2 illustrates an important principle that springs connected in series can be combined by adding compliances while for springs connected in parallel, stiffness adds, see Figure 7.2b and c. This leads to a general conclusion that, as structural complexity increases, so too does compliance. Directly from this, the concept of 'reduction' in design will not only lead to a less expensive process, but reduction also contributes to optimising stiffness.

Another useful concept can be derived by considering the effect of multiple springs attached to a rigid, straight bar as shown in Figure 7.3. Figure 7.3a shows a pair of springs connected to a bar of length L. The forces, and subsequent displacements, at the two springs can be obtained by taking moments at each end of the bar. The displacement x and rotation θ at the point of applied force are given respectively by

$$x = \frac{F}{L}\left[\frac{L-a}{k_2} + \frac{a}{L}\left(\frac{a}{k_1} - \frac{L-a}{k_2}\right)\right] = F/k_x = Fc_x,$$

$$\theta = \frac{F}{L^2}\left[\frac{a}{k_1} - \frac{L-a}{k_2}\right] = F/k_\theta.$$

(7.3)

Dividing the first part of Equation 7.3 by the force and taking the first two derivatives with respect to a yields

$$\frac{dc_x}{da} = \frac{1}{L}\left[\frac{2a}{Lk_1} - \frac{L-2a}{Lk_2} - \frac{1}{k_2}\right],$$

$$\frac{d^2c_x}{da^2} = \frac{1}{L}\left[\frac{2}{Lk_1} + \frac{2}{Lk_2}\right].$$

(7.4)

To determine an expression for a, setting the derivative in the first part of equation (7.4) to zero and noting that the second derivative in the second part of Equation 7.4b is always positive yields

$$a = \frac{k_1 L}{k_1 + k_2}.$$

(7.5)

Substituting a in Equation 7.5 into the second part of Equation 7.3 gives the simple result $\theta = 0$. Hence, this demonstrates the principle that for parallel linear springs configured as shown in Figure 7.3 there is a location for the applied force for which the displacement of

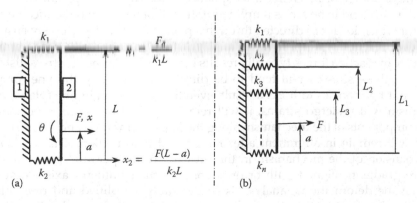

FIGURE 7.3
Two rigid links connected by co-linear springs: (a) two springs, (b) multiple spring model.

the connecting arm is a rotation-free rectilinear motion and this location corresponds to the maximum linear stiffness of the combined springs. Extending the preceding analysis it is possible to explore the effect of multiple springs. Here it will be assumed that the force is being applied at the location for pure rectilinear motion. Under this condition, all of the springs will experience the same linear displacement x and the moment M about the point of applied force will be zero, leading to the condition

$$M = [k_1(L_1 - a) + k_2(L_2 - a)..k_n(L_n - a)]x = x\sum_{i=1}^{n}k_i(L_i - a) = 0. \tag{7.6}$$

Rearranging Equation 7.6 yields the general equation

$$a = \frac{\sum_{i=1}^{n}k_i L_i}{\sum_{i=1}^{n}k_i}. \tag{7.7}$$

Equation 7.7 can be used to determine the location of a that is often called the 'centre of stiffness' of a rigid body that is connected by multiple flexible elements. From a precision engineering perspective, forces applied at this location will result in linear motion without rotations, a condition that is important for the reduction of Abbe errors (see Chapter 10, Section 10.1.1). For dynamic systems, it is necessary that the axis of motion is coincident with both the centre of mass and the centre of stiffness.

On a final note for this section, it is always important to keep in mind that structural stiffness is three dimensional, comprising both linear and angular (sometimes thought of as torsional) stiffness components. For designs subject to multiple forces, all components must be considered.

7.1.2 Symmetry

Symmetry is discussed in this overview section because of the important role that it plays in precision design. Not only does it provide numerous benefits that are demonstrated throughout this chapter, but it is also easy to recognise by the designer. For many structures, the major source of mechanical compliance is the distortion produced by bending moments that also tend to be stress amplifiers. Often, moments can be reduced by creating structures for which loads are directed through a plane of symmetry, thereby ensuring that the subsequent reaction is supported by equal structural forces on either side of the plane. The symmetry of reaction forces typically results in similarly symmetric stress distributions. Consequently, these stresses will tend to be directed either along the line (or plane) of symmetry or at right angles to it. As a result, even though a straight line (plane) drawn on the mechanism will undergo strains, it will remain straight or, more important, objects placed on components of the mechanism along the line of symmetry and aligned relative to each other will remain in alignment after mechanical deformation. This is also true for thermal expansion of the mechanism in the presence of temperature changes as well as temperature gradients aligned with or orthogonal to the symmetry axis. Because of the symmetry of the deformations, analysis is considerably simplified and often eliminated altogether. For the aforementioned example, if it was desired to know the distortion from a plane for a symmetric mechanism loaded along the line of symmetry, the answer is zero.

Similarly, an arbitrarily complex dynamic mechanism can often be perfectly balanced (in a specific direction) by simply duplicating the mechanism, fastening the two together and having each perform the same operation.

Numerous examples of symmetric structures can be found merely by browsing through this book, in particular in the chapters on kinematic design, loops, and alignment, and throughout this chapter. Consequently, they will not be expanded any further here, but the reader is encouraged to look at all designs with an eye for symmetry and, when spotted, to consider the benefits that this imparts.

7.1.3 Fasteners

Following this section, the remaining discussions in this chapter explore mechanisms for accurate and precise motion control and means of actuation. However, it is a common requirement that an object, once positioned, be fastened in place. When the body has been adjusted to the desired position, the mechanism for fastening should, desirably by itself, not cause the component to move, at least in one or more desired directions. While the ideas of kinematic clamps for relocation and kinematic design principles are discussed in Chapter 6, this brief section takes a look at what might be considered more permanent fastening methods. In all of the figures, bolts and nuts should have spring washers to provide a more gradual transition from loose to fully tight (see Chapter 10). Additionally, if fastening is required to be permanent, 'thread locking' glues (see Chapter 10, Section 10.1.6.2) would also be used to prevent loosening in the presence of vibrations and thermal loads.

For adjustment of apparatus or alignment of components where the goal is that, after adjustment, they be rigidly clamped, design approaches can be split into two categories: positioning of a component using a screw mechanism, or positioning using an external means followed by clamping. In the following two sections these are discussed as positioning and clamping respectively.

7.1.3.1 Positioning

Most instrumentation and machine slides will have a locking mechanism to secure the moving platform after positioning. For many machine tools, this is achieved by simply clamping the gib-strip of a V-slide (or dovetail) using a screw clamp, as shown in Figure 7.4a. While this makes an effective and stiff clamp, the action of the braking force will distort the components and dramatically change the load experienced at the bearing surfaces. As a result of this force, clamping will produce some relative motion of the slide, typically of the order of 20 μm to 70 μm, or, even more, depending on the state of the machine and adjustments of the gibs. Many slideways (all of those shown in Figure 7.4) are translated using a feedscrew and nut arrangement. Figure 7.4b shows the clamping method for the moving carriage in a jig boring machine. In this case, instead of directly clamping the slideway, a locking nut is provided to prevent the screw from rotating. Clearly, this will impose axial stresses in the shaft and nut. However, these will be symmetric about the locking interface, therefore, it is possible to fasten the nut to the moving carriage at this location, resulting in a considerably lower disturbance of the carriage, typically in the range of a few micrometres.

Analogous to a disk brake, the clamp shown in Figure 7.4c is a parallel plate brake that is contacted from one side and clamped using a screw. Such an arrangement results in the clamping forces being effectively separated from both the slideway and feedscrew. Clearly, the action of squeezing the two sides of the plate will produce a small net motion

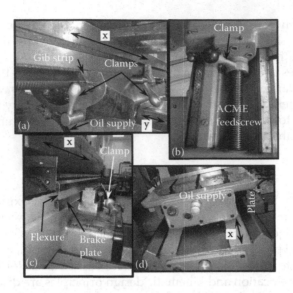

FIGURE 7.4
Methods for clamping linear slideways: (a) clamping by pushing against a gib-strip with piping for boundary lubricant also visible, (b) locking nut on a feedscrew, (c) a brake plate attached to the slideway via a flexure, (d) boundary lubricant oil supply to the ways shown in part c. Note the parallel V-groove slideway in this last image.

in the direction of the screw. However, the purpose of the brake is to prevent motion in a perpendicular direction. Consequently, a thin membrane (called a flexure, see Section 7.2.5) is machined between the brake plate and the strip that is bolted to the carriage. This membrane is relatively compliant in the direction of the clamp while providing a relatively large stiffness in the x-direction. Finally, it is noted that this slideway comprises two V's, and the carriage and ways are directly in contact without an intermediate bearing material. From a kinematic perspective, this is over-constrained and it is necessary to lap and scrape the ways to ensure conformal contact over a large area (see Section 7.2.1). By contacting over a large area, wear of the ways is more uniform and stiffness is maintained, whereas gibs require periodic adjustments to avoid 'play' (Moore 1970).

For positioning or alignments over relatively small displacements, it is often possible to create mechanisms in which a solid body can be adjusted using simple screws. Figure 7.5 shows a flexure-based linear translation stage comprising a double compound rectilinear flexure in which the moving platform is guided by the leaf spring arrangement and positioned by opposing screws. When the platform is close to the desired position, both screws can be tightened. However, in this case, the screws will push against the platform in opposite directions. Upon final positioning, the screws will push against each other resulting in elastic deformation. As the clamp force increases, so too will the stiffness of the contact (see Equation 7.10), leading to increasingly fine motion control. With position feedback, and a little practice, it is often possible to move and clamp such a platform with nanometre control. The use of flexures for motion guidance is discussed in Section 7.2.5. For even finer motion control, it is sometimes beneficial to use set screws to lock the platform and a second set of set screws to both lock the locking screws and achieve ultrafine positioning, as shown in Figure 7.5c.

A variation of the coarse–fine fastening approach is shown in Figure 7.6. In this design, it is desired to create an arc motion at the end of a lever. The coarse screw pushes against the

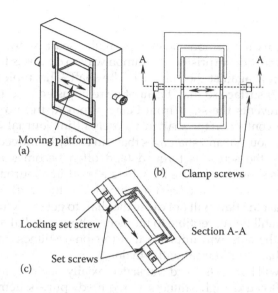

FIGURE 7.5
Linear positioning with fine adjustment and clamping: (a) isometric view of a double compound leaf-type linear flexure mechanism with clamping screws, (b) front view, (c) cross-section of the mechanism showing set screw and locking screws for ultra-fine positioning.

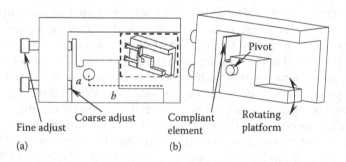

FIGURE 7.6
Angular coarse-fine positioning: (a) front view with isometric cross-section in dashed box, (b) isometric view.

body of the rotating platform below the pivot with an effective lever action b/a for motion at the free end, as indicated in the figure. Fine motion is achieved by pushing against a flexible element with the force being applied above the pivot. Fine adjustment is achieved by the relatively large ratio of deflection of the compliant element to the high stiffness of the surface contact of the coarse screw. This is an example of a soft-spring stiff-spring motion attenuation lever.

All of the aforementioned methods achieve the final position using high clamping forces and rely on friction to prevent motion. Such large residual stresses are sometimes undesirable and can lead to long-term drift or creep. An alternative is to position the two bodies with low clamp forces. Once the final position has been obtained these two bodies are then held in position by filling the surrounding space with an epoxy. This is often referred to as 'potting' for which a variety of stable potting compounds are available. Dimensional stability of these potting compounds is typically achieved by adding metal or ceramic (glass) fillers.

7.1.3.2 Clamping

Commonly, it is necessary to position and clamp circular shafts with the axis being aligned to a datum surface. A bad, but surprisingly common way to do this is to simply drill and tap a hole and use a setscrew against the surface of the shaft. This typically results in plastic deformation of the shaft surface so that it cannot be re-adjusted, and the resultant built-up material sometimes prevents the shaft from being removed from a close-fitting journal. Additionally, the point contact on the shaft in a loosely fitting journal will result in a lack of parallelism between the journal and shaft axes that can easily be moved by an applied force, no matter how tightly the screw is torqued (and often becomes loose thereafter). An improvement on this is shown in Figure 7.7 where a shaft in a journal with two clamping screws bears down on flats that have been machined into the shaft. While built-up edges where the screws contact the flats will not cause the shaft to get stuck, both screw axes being in the same plane will still be susceptible to moments about a radial axis. A second pair of screws further along the axis will improve this torsional stiffness, although the lack of parallelism between shaft and journal will remain.

Typically, the shaft will be rotated and translated axially before clamping, requiring two degrees of freedom. Because high stiffness is required, pure kinematics must often be approximated with semi-kinematic design (see Chapter 6). Consequently, a common solution is to fasten the shaft into a V-groove, see Figure 7.8. Kinematically, the shaft should be contacted at four points. In practice this is approximated as two line contacts. Figure 7.8c and d show a closer kinematic implementation where there are two clamps and the V-groove has been removed in a central region, leaving four, relatively short, line contacts.

To maximise stiffness, it is desirable that the contacting area of a clamp be as large as possible. Clearly, the direct screw is far from this ideal. A common method is to simply squeeze the shaft within a closely fitting journal. Such a squeezing action is achieved by slitting through the journal and clamping from the side. Figure 7.9 shows a more elaborate version of the squeeze clamp in which the journal has been machined with three recesses at 120° pitch to provide more definite contact locations. With computer numerical control (CNC) machining or other complex shape generating methods, these recesses can be achieved by milling four holes, see Figure 7.9c and d. The first hole is close to the diameter of the shaft while the other three holes are smaller and offset from the centre of the journal hole and again at 120° pitch. It can be advantageous to consider the direction of the forces due to the final clamp. For example, a task might require careful positioning and clamping of a shaft without moving it in axial direction. As in Section 7.1.3.1, the final clamping force will result in strain. For the shaft and squeeze clamp, these stresses will be symmetric in the

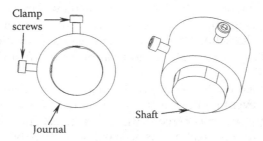

FIGURE 7.7
Shaft held in a journal using two setscrews. The careful reader will observe that the flats are misaligned to the screw axis to illustrate tolerance sensitivity.

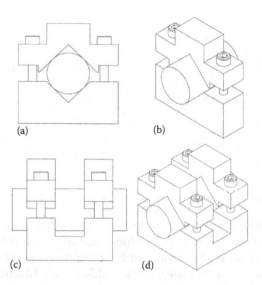

FIGURE 7.8
Shaft clamped into a V-groove: (a,b) single V-groove front and isometric views, (c,d) two clamps and retracted central portion of V-groove.

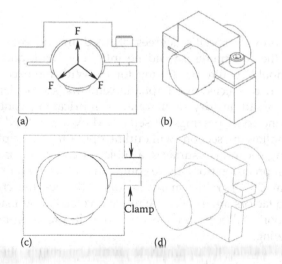

FIGURE 7.9
Fastening of a shaft using a symmetric squeeze clamp: (a,b) front and isometric view of a three area contact, (c,d) three area contact achieved by four overlapping holes.

radial direction. Because of the Poisson's ratio effect, these strains will result in an elongation of the shaft in the axial direction, thereby facilitating a Poisson's ratio actuator with the final clamp forces enabling simultaneous fine-position and hold capability.

The guillotine clamp (see Figure 7.10) is another method for applying the force directly from the clamping screws to a shaft. In this case the 'blade' of the guillotine has a hole closely toleranced to the shaft that passes through it as well as the journal. Clamping of the screws results in a force pulling the shaft against the upper surfaces of the closely fitting

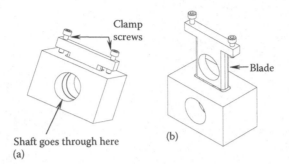

FIGURE 7.10
Guillotine clamp (shaft is omitted for clarity): (a) front view, (b) isometric exploded view.

journal. Consequently, there will be a shear force at either edge of the slot and the forces on the shaft might be considered to occur at half the magnitude and either side of the applied force. These forces would generally result in a bending moment. However, for a blade of thickness smaller than the shaft diameter, this is unlikely to introduce significant bending.

Methods for effective clamping of flat surfaces are discussed in Section 7.2.5.

7.2 Bearings

Bearings are used to provide specific degrees of freedom and constraints between two rigid bodies. Ideally, the constraints should be infinitely rigid, and motions in desired degrees of freedom should require no driving forces. With the exception of flexures, for which the two bodies are connected by compliant elements (see Section 7.2.5), bearings are typically categorised as sliding (dry or marginally lubricated), containing intermediate rolling elements (ball and roller bearings) or separated by a thin fluid film (hydrodynamic and hydrostatic). The following sections will outline operating principles and performance characteristics of sliding and rolling surfaces, a subject area known as tribology, as well as indicating the expected precision (for a more complete introduction to this broad topic, see Hutchins and Shipway 2017). While many factors influence the choice of the type of bearing, the major two factors are usually the relative velocity and load (or pressures) that the bearing must support. All bearing types are found in precision machines and will be discussed in the following.

A first step towards choosing the appropriate journal bearing is often based on required loads and speeds, for which the selector diagram shown in Figure 7.11 can be used. While many other considerations will influence this choice, it is apparent that the rolling elements and sliding bearings both occupy the lower speed region, with the rolling elements extending to higher speeds. Hydrodynamic bearings that operate with a thin fluid separating the two bodies are often the best choice for the highest speeds and loads, but fail to prevent contact at lower speeds, where boundary lubrication leads to marginally lubricated conditions (discussed in Section 7.2.3). Tellingly, externally pressurised journal bearings (hydrostatic bearings) incorporate the best features of all systems. The reason that the other bearing types successfully compete for the vast majority of applications is primarily because of the cost imposed due to requirements of precision manufacturing, external oil pumps and the need to ensure the fluid supply is not interrupted.

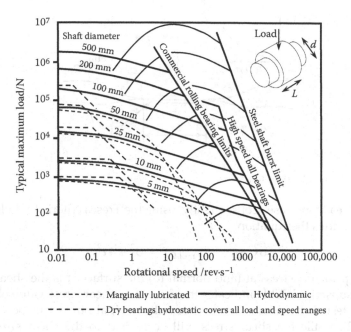

FIGURE 7.11

Bearing selector chart for a rotating shaft. (Modified from Neale M. J. (ed.), 1973, *Tribology handbook*, Butterworths.)

7.2.1 'Dry' or Sliding Bearings

For high stiffness and precise translations at relatively slow speeds, many slideways consist of two solid bodies connected by relatively large area contacts. This section looks into the tribological considerations relevant to the design of sliding bearings, as well as the manufacture of conforming surfaces and some common slideway designs. Although this section is titled 'dry' bearings, simplistic models of the lubricating effect makes it paramount that some form of lubrication is always present and suitable lubrication treatments are also described.

7.2.1.1 Tribology Overview

Dry and marginally lubricated bearings, suitably designed and fabricated, are relatively inexpensive, can handle large loads, have high stiffness and require no active auxiliary equipment other than a rudimentary pump to supply a minimal oil flow. Immediately apparent is the issue of the frictional force required to produce sliding of one surface over the other (often called the sliding pair). For precision instruments and machines, it will be assumed that the sliding will occur under moderate conditions that form the focus of the initial discussion in this section. First, it might be assumed that the contact of two surfaces involves the interaction between surface texture asperities. Considering, for the moment, a single asperity that has formed by, first, contacting at a point and then plastically deforming until the external load is supported. Because of the plastic flow, the stress will be uniform in the contact. It might be considered that the von Mises stress of the asperity interface is at yield. Subsequently, applying a tangential force to the contact will add a shear stress, as shown in Figure 7.12. Assuming that the junction is already at yield, sliding will create an additional stress component that, because the stress cannot increase above its yield value,

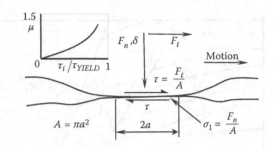

FIGURE 7.12
Stress conditions at an asperity contact.

will require the area to increase. Consequently, using the Tresca criteria for failure (Chapter 3, Section 3.5.2) provides the equation

$$\sigma_{YIELD}^2 = (2\tau_{YIELD})^2 = \sigma_1^2 + 4\tau^2, \tag{7.8}$$

where σ_1 is the principal stress at (and normal to) the surface, τ is the shear stress due to sliding and the subscript YIELD indicates the direct and shear stress value obtained from a uniaxial tensile test. Expressions for the normal principal stress can be easily obtained (see Figure 7.12), while the shear stress will be related to the shear strength τ_i of the interface, which is simple related to the tangential force, F_t by

$$F_t = \tau_i A \tag{7.9}$$

Substituting and rearranging Equations 7.8 and 7.9 provides an estimate of the coefficient of friction μ as

$$\mu = \frac{1}{2\left(\left(\tau_{YIELD}/\tau_i\right)^2 - 1\right)^{1/2}} = \frac{\tau_i}{2\tau_{YIELD}}\left[1 + \frac{1}{2}\left(\frac{\tau_i}{\tau_{YIELD}}\right)^2 - \frac{3}{8}\left(\frac{\tau_i}{\tau_{YIELD}}\right)^4\right]. \tag{7.10}$$

Equation 7.10 is plotted and inset in Figure 7.12 as a function of the ratio of interface shear strength to that of the bulk material at yield. Other junction growth models, based on plasticity using slip line field and asperity ploughing, yield similar results (Bowden and Tabor 1962, Hutchins and Shipway 2017). It is apparent that even for a moderate interface strength reduction to 80% of the bulk material reduces the friction coefficient to less than 1 with a 40% reduction giving a friction coefficient of around 1/3.

For low interface shear strength, the friction coefficient becomes linear and is proportional to the ratio of the shear strength of the interface to the yield stress of the bulk material. There is clearly a major incentive to introduce a treatment that will reduce the strength without compromising the stiffness of the sliding pair. Such a treatment is often provided by coating the surface with long chain molecules (hydrocarbons with reactive end groups) that typically act as boundary lubricants. These tend to be long chain fatty acids (stearic, lauric, palmic, etc.) with reactive end groups that will bond to oxides on the surface of the metal. Typically, this bonding process results in a chemical reaction that results in a strongly adhered film (a metallic soap) with substantially lower shear strength than that of the solid. For bearings in which extreme contact pressures will be encountered, more reactive additives containing sulphur or phosphorus are used (called extreme pressure additives). Boundary lubricants are typically supplied as a suspension in lubricating oils that will coat the surface when they are separated.

So far the stresses in the contact region are assumed to be at the limits of strength of the materials. Typically, such an extreme condition would result in a high wear rate that is unacceptable for a precision slide. At lower stresses, the contacts can be assumed elastic, in which case Hertz contact theory can be used for spheres of combined radius R and elastic modulus E^*. Using Hertz contact theory expressions for the maximum interface pressure σ_1, mutual approach of distant points in the solid δ and contact stiffness k_c are given by (see Chapter 3, Section 3.5.6)

$$\sigma_1 = \frac{3F_n}{2\pi a^2} = \left(\frac{6E^{*2}F_n}{\pi^3 R^2}\right)^{1/3}, \delta = \frac{a^2}{R} = \left(\frac{9F_n^2}{16RE^{*2}}\right)^{1/3}, k_c = \frac{\partial F_n}{\partial \delta} = \left(6RF_n E^{*2}\right)^{1/3}. \tag{7.11}$$

It is immediately apparent from Equation 7.11 that the stiffness is increased and stresses are reduced by increasing the radius of the contact asperities. A remaining issue is the nature of the contact for rough surfaces. This has been studied using statistical approaches by many researchers over the last few decades, the outcome of which is the concept of a plasticity index ψ that, if less than 1, indicates predominantly elastic stress at the contact interface, and if greater than 1, plastic. Based on these models, a number of dimensionless measures have been proposed, all of which have been subject to extensive scrutiny in the literature. Three measures are

$$\psi_{GW} = \left(\frac{E^*}{H}\right)\sqrt{\kappa_s \sigma^*}, \psi_{GM} \approx \left(\frac{E^*}{H}\right)\sigma_m, \psi_{WA} = \left(\frac{E^*}{H}\right)\frac{\sigma}{\beta^*}, \tag{7.12}$$

where H (N $\cdot$ m^{-2}) is the hardness of the material typically measured using Brinell or Vickers indenters, σ^* the rms deviation of peak heights, κ_s is the average curvature of the peaks, σ_m (or $R_{\Delta q}$ in surface roughness standards) the rms slope of the surface, σ (or R_q in surface roughness standards) the rms surface roughness and β^* the exponential exponent of the autocorrelation of the surface profile (represented by τ_o in Chapter 1, Figure 1.7). The subscripts for the different plasticity measures refer to Greenwood and Williamson (1966), Greenwood (2006), Mikic (1974) and Whitehouse and Archard (1970). From the preceding equations, it is again apparent that the stresses can be minimised by having large radii asperities, high hardness and small surface height deviations at the portion of the surface that is significant to interface contact. A series of insights has emerged from these studies. It is found that the average area of contact asperities is constant nearly independent of load. This is explained by the increasing area of contacts existing before the increase in load will be averaged by new, significantly smaller asperities coming into contact. These conclusions from statistical analyses have been evaluated and generally verified using finite element modelling (Saylor 1996, Kadiric et al. 2003). Assuming that the shear strength is uniform, these conclusions are consistent with the Amonton-Coulomb laws stating that friction is a constant independent of load, speed and apparent contact area (some authors also add surface texture to this list). While not generally true, these are reasonable conclusions under the stable conditions discussed later.

To evaluate the stability of frictional behaviour, it is common to measure the wear of a sliding pair, typically using a pin-on-disk apparatus. If friction is constant over time, it is consistent with the Amonton-Coulomb laws (Hutchins and Shipway 2017) to assume that the total volume of material removed per unit sliding distance Q (m^2) as a function of the total normal load F_N can be determined from the linear equation

$$Q = K\frac{F_N}{H}, \tag{7.13}$$

where K is the dimensionless wear coefficient and H is hardness. In practice, the hardness of the interface material is often not known and the dimensional wear coefficient ($=K/H$) is measured. The total volume of material removed is obtained by multiplying Equation 7.13 by the sliding distance. In practice, friction and wear will be a function of the sliding conditions, temperature, geometry and topography of the surface. Wear behaviour of sliding pairs is often visualised using wear maps that are analogous to the bearing selector of Figure 7.11. The wear map shown in Figure 7.13 plots the dimensional wear coefficient (typically measured with a pin-on-disk apparatus) as a function of apparent pressure and velocity, with both axes using logarithmic scales. Typically, the pressure is normalised to the hardness of the material and, sometimes, the velocity is normalised using the Peclet number (radius of the pin multiplied by velocity divided by thermal diffusivity). The Peclet number is used in heat transfer as a measure of the time that a surface is sliding under the pin (and, therefore, receiving heat) divided by the speed that heat can leave the interface. Often, the contours of constant wear are viewed as a two-dimensional plot, with labels identifying different wear phenomena. The significance of the wear map and a considerably simplified wear map for steel (with the wear coefficients removed) are shown in Figure 7.13.

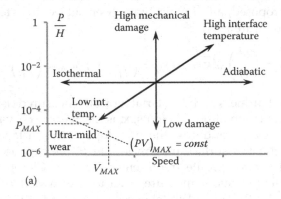

(a)

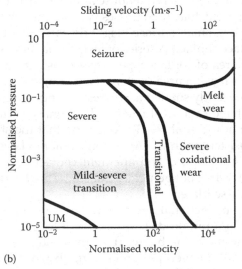

(b)

FIGURE 7.13
Wear mechanism maps: (a) generalised map indicating major influences on wear, (b) simplified wear mechanism map for steel sliding on steel.

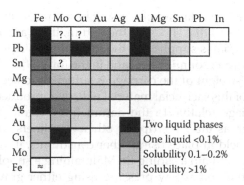

FIGURE 7.14
Mutual solid solubility of pairs of metals. Question marks indicate that solubility is unknown while the approximation symbol indicates the rare tribological properties of iron (see text).

Clearly, the high speeds and velocities will correspond generally to high wear and temperature, with occasional reductions in wear when temperature produces significant changes in surface oxidation rates. For the purpose of precision applications, the lower left region of the map, bounded by maximum pressure and speeds and a loci of constant pressure–speed product, represents a large number of rubbing bearing slideway applications, for which the dimensionless wear coefficient is typically in the region of 10^{-6} to 10^{-4} for many metallic and ceramic pairs (Lim and Ashby 1987, Hsu and Chen 1992). If lubricated, this mild wear region extends to velocities of $1~m \cdot s^{-1}$ and higher (Childs 1988).

A final consideration is the selection of materials for rubbing surfaces. For metals, it is usual to select alloys that do not have a tendency to form solid solution alloys. This is typically gauged from compositional phase diagrams, with the solubility measured as the amount of the second composition that can be absorbed without inducing a phase transformation. This is indicative of the chemical compatibility between the two metals and, conversely, tribological incompatibility. A tribological compatibility map for some selected metals is shown in Figure 7.14; for a more complete map see Rabinowicz (1980). Of particular interest is the tribological compatibility of lead, tin and silver that are frequently used in bearing alloys. Because of its exceptional electrical conductivity and tribological compatibility, silver-coated beryllium copper wires are often used as the flexible element for sliding electrical contacts in slip rings.

7.2.1.2 Slideway Manufacturing

All of the aforementioned considerations of the nature of rubbing surfaces provide guidelines for the design. These considerations may be summarised by five rules:

1. Always use a material for the datum surface of the slideway that is harder than the counterface bearing material.
2. For the softer sliding bearing, use a material that contains tribologically compatible alloying elements.
3. Reduce the contact shear strength using boundary lubricants.
4. Arrange for multiple contacting asperities of large radius and with the peaks being as close to the same height as possible.
5. To optimally preserve the datum, always operate in the mild wear region for both materials.

The first of these rules is typically satisfied using hardened steel for the slideway datum. For the softer bearing material that slides on the datum, there are a number of approaches. Commonly, the slide material is supplied in the form of gib strips that attach to a moving carriage. These are often bronze or mild steel and in the form of wedges that can be adjusted for a sliding fit. If the self-weight of the carriage is sufficient to keep the carriage in contact with the slideway, wear of this sacrificial material will, over protracted time periods, change the position of the carriage relative to the datum (its centroid will move closer to the slideway base). However, as the gib strips or other way-liners wear, a sliding fit will be maintained. With slides held in place using other constraints, it may be necessary to periodically adjust the gib to maintain a sliding fit. Maintaining a supply of boundary lubricant is often achieved by feeding oil at low pressure using either gravity or a simple manual pump into a recessed pocket (called a plenum) in the gib strip.

To maintain a large area contact and correspondingly low pressure, it is necessary that the surfaces closely conform. For large area contact, this again requires plastic or elastic deformation. One method is to utilise low elastic modulus materials that have some intrinsic lubricating properties. These range from polymeric solids with embedded PTFE (polytetrafluoroethylene) and, often, some tribologically compatible metallic particles such as bronze. Typical bearing materials that are self-lubricating are sold under the trade names Rulon™ or Turcite™ (the limiting product of pressure and velocity or PV $\approx$ 0.24 MPa·m·s^{-1} to 0.34 MPa·m·s^{-1}) and these come in thin sheets or bulk solids that, with suitable surface preparations and adhesives, can be glued to the moving carriage. For a bearing material that is likely to provide better stiffness and stability for precision applications, it is possible to use porous bronze that is impregnated with an oil containing boundary lubricant, commonly known by the trade name Oilite™ (PV $\approx$ 1.5 MPa·m·s^{-1}). For thinner films of bearing material and excellent conformability, it is possible to purchase two-part epoxies that contain a variety of proprietary but tribologically compatible materials (a common epoxy is called Moglice™). This epoxy can then be moulded to the datum surface (with a suitable release agent) as a relatively thin film so that even though it has a relatively low value of elastic modulus, the large area and small thickness will result in a high stiffness (simplistically given by the product of elastic modulus and area divided by the thickness). These bearing materials are commonly used for machine tool slideways and other relatively high load applications, as well as some instrumentation bearings.

Polymer bearings are also common for instrumentation bearings of very high precision. These typically comprise a thin layer of acetal or a composite PTFE material physically embedded into a porous bronze matrix that is, in turn, bonded to a steel sheet that is galvanised on the back surface to prevent rust. Acetal requires lubrication, and under light loads and speeds, will tend to operate as a mixed boundary lubricant elastohydrodynamic bearing (see Section 7.2.3) with low friction (0.01 to 0.05) and wear. However, when stationary or at very low speeds, the hydrodynamic lift is inadequate and the conformability of this material can lead to a high static coefficient of friction of around 0.3 to 0.4. Consequently, lubricated acetal bearings are most commonly use in instrumentation systems where the bearing runs continuously at a constant and low speed, as might be encountered, for example, in a roundness testing instrumentation spindle and provide smooth and synchronously repeatable rotations with nanometre repeatability. PTFE bearings are available similarly pressed into a porous bronze matrix on a steel plate and can have a PV value comparable to Oilite™. Typically, the PTFE is also mixed with lead, graphite or other solid lubricants and, when used to slide against a polished glass datum, have been shown to have smooth and repeatable motion with sub-nanometre repeatability (Lindsey et al. 1988). Finally, bearings made from thin ($\approx$ 0.05 mm) layers of ultra high molecular weight polyethylene (UHMWPE), stretched onto a spherical surface, have demonstrated performance comparable to that of the PTFE slideway (Buice et al. 2005).

Finally, for maximum sliding bearing stiffness, it is common to use steel-on-steel slideways. For precision applications, it is desirable that the sliding interfaces are maintained within the ultra-low wear regime of the wear map by ensuring a continuous supply of boundary lubricant and maximising both the number of and the radius of contact asperities. To increase the number of contacts, it is common to 'scrape' the slideway surfaces, effectively using plastic deformation to enhance conformability. Simply, scraping is achieved by painting a thin dye onto a flat surface (called a surface plate). The slideway to be scraped is then slid on this surface so that the dye is transferred to the high spots. A scraping tool is then used to scrape away these high spots, typically removing 1 μm to 5 μm of material from the surface dependent upon how aggressively the scraping tool is applied (see Figure 7.15). Often, each dig of the tool will remove the high spots and then drive further into the surface to create a micrometre scale 'oil well'. Although many slides are hand scraped, there are commercially available powered scrapers. Scraping is repeated until the contact spots uniformly cover the whole way surface. In the process of scraping, a number of small and 'sharp' asperities may be pushed up. It is not unusual to briefly rub a flatstone on the surface after each scraping operation to remove these asperities. These flatstones are typically alumina-based grinding stones that have been ground flat with diamond grinding wheels in a precision grinding machine. The final scraping operation is often performed manually with a simple hand-held scraper, often removing sub-micrometre-thick surface layers. Once complete, a final stoning of the surface ensures multiple large radius contacts to ensure mild wear. An additional advantage of the large number of contacts is that, even though not all asperities will be at the same height, for a carefully manufactured slideway, elastic averaging will result in smooth and repeatable motion with nanometre-level noise.

Effectively, scraping transfers the flatness of a reference surface to the ways of the machine tool. Common references are surface plates, straight edges and angle plates. Straightness and flatness grades will depend on the type of reference, be significantly distorted by temperature gradients and, for granite surfaces, humidity. For surface plates, flatness grade AA is around 4×10^{-6} (or 4 μm·m^{-1}) and the straightness grade is typically twice this value. However, for more demanding applications, laboratory surface plates can

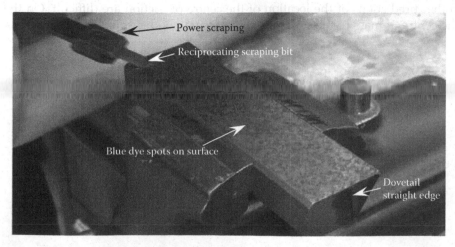

Power scraping

Reciprocating scraping bit

Blue dye spots on surface

Dovetail straight edge

FIGURE 7.15
Photograph of manual scraping of a dovetail straight edge using a power scraping tool.

be produced with flatness deviations of better than a fraction of a micrometre over surfaces measuring more than a metre. For plates less than one metre, this is often achieved by lapping three plates together so that they tend to converge to being flat; basically, a reversal method (see Chapter 5, Section 5.8). In essence, if two plates are lapped to conform, one can be convex while the other would be concave. However, by lapping these two plates against a third, at least one of these will be changed toward a flatter surface. Eventually, by lapping them in sequential pairs, all three will end up with surfaces that will conform. Unfortunately, because 'saddle'-shaped surfaces also conform, this does not ensure a flat surface. To eliminate this saddle effect, it is necessary to rotate the surfaces while lapping (Moore 1970). This same method can be applied for the polishing of optical flats to produce flatness deviations below a small fraction of a wavelength of visible light.

As a postscript to this section, from the criteria for tribological function, polished surfaces might be considered. In practice, highly conforming surfaces do not admit lubrication, and any liquids within a large area with very small surface separation will lead to large viscous forces. Additionally, very smooth surfaces tend to only grow thin oxide layers that can be readily removed during sliding, thereby exposing the clean bulk materials to oxide-free contacts, leading to high friction and, possibly, binding. The exception to this is the use of self-lubricated thin-film polymeric bearings discussed earlier.

7.2.1.3 Slideway Design Considerations

This section provides brief overviews of some representative machine and instrument slideways. A more detailed discussion of precision slideway design and manufacture can be found in Moore (1970). Kinematically, a linear slideway would comprise two rigid bodies joined through five point-contacts as discussed in Chapter 6. However, for large load applications, it becomes necessary to use approximations to this ideal with contacts covering larger areas. Consequently, a single degree of freedom can be achieved with a V-groove and a flat, and such designs are common with many machine tools. Figure 7.16 shows a model of a V and flat slideway. The moving carriage of this slideway shows a hole for the drive screw. In principle, this axis would be located so that the sum of frictional moments will cancel and that the drive axis will also pass through the centroid of the carriage. In practice, the frictional forces are difficult to determine and will often vary over time, so that predictions of the location of the centre of friction are difficult. On the other hand, the centre of mass of the carriage can be accurately determined, although this again

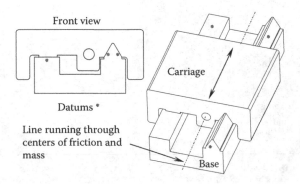

FIGURE 7.16
Front and isometric views a simplified model of a V and flat slideway.

will change if components of unknown shape and mass are attached. To better approximate a kinematic coupling, the contacting surfaces are sometimes shaped so that the area contacts on the V comprise four rectangular surfaces and the flat side opposite the V has a single long contact pad, often in the form of gib strips. For real slideways, these gib strips, and all other surface contacts, would also be supplied with boundary lubrication.

Another, more symmetric design commonly found in machine tools is the dovetail slide, see Figure 7.17. This can be considered as a V and flat with a second V on the other side. The requirement that the second V makes contact under force makes this an over-constrained design requiring fine adjustment of a gib strip to maintain contact and requiring more frequent maintenance to adjust for wear. The additional symmetry of this design does have the benefit of a better defined centre of symmetry and mass centre, and is often used in higher performing machines.

When symmetry is a major objective, the double V slideway shown in Figure 7.18 is sometimes employed. Again, this double V is an over-constrained design and requires some substantial care in manufacturing and assembly, typically involving numerous lapping and alignments procedures (Moore 1970). In what might be considered an extreme case, the underside of the carriage is also contacted with scraped flat surfaces to help reduce distortions of the stage when it has to support moving loads. These designs are usually used for instrumentation bearings and can produce smooth motions with sub-nanometre precision, and are used with measuring machines and for the scribing of gratings using ruling engines. Diffraction gratings comprising flat surfaces with closely spaced, parallel and

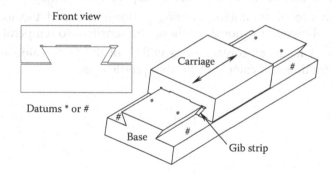

FIGURE 7.17
Front and isometric views of a simple dovetail slideway.

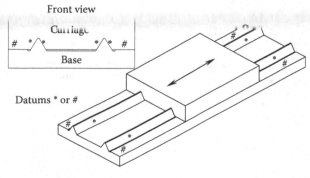

FIGURE 7.18
Double V and/or double V and flats symmetric slideway design for instrument bearings.

straight lines scratched onto the surface were famously produced by Henry Rowland around the turn of the nineteenth and twentieth century (Rowland 1902).

Another common single degree of freedom mechanism is a spindle that is required to rotate a solid body about a fixed axis. Spindles using rubbing bearings, rolling elements and hydrostatic bearings are common. Rubbing-type bearing spindles tend to be used for low load instrumentation applications. A symmetric spindle design comprising two spherical surfaces attached to each end of a shaft is shown in Figure 7.19. The shaft is constrained by three contacts symmetrically disposed on each sphere. Because the surfaces of the spheres are typically polished, this design would utilise polymer or conformal contact bearings. One of these contacts is connected to the housing through a flexure, the deflection of which will provide the force to maintain all six contacts. Kinematically, having three rigid contacts at one end and two at the other is equivalent to a triangular hole and V-groove joint providing the single rotational degree of freedom (Bauza et al. 2009). Runout of this spindle will depend on the deviation from true spherical shape that, in turn, depends on the sphere grade. Sphere grades are usually specified in tolerances of 25.4 nm (i.e. 1 micro inch) increments, hence grade 1 would be toleranced at 25.4 nm, grade 2 at 50.8 nm, and so on.

There are limitations of this symmetric spindle design. Primarily, stick-slip behaviour limits these spindles to constant speed applications. Dependent on the bearing materials, it will also be necessary to incorporate a continuous lubricant supply, and speeds and loads will be limited by the PV values of the bearing and, more likely, limited to a small fraction of the preload of the flexure preload. Advantages of this design include:

- The bearing contacts are all symmetrically disposed about the axis of the spindle.
- As a consequence of symmetry, bearing performance and the location of the rotation axis relative to the housing will not be sensitive to temperature changes.
- Ignoring gravity loads, all contact forces will be of the same value and, therefore, the centre of friction will coincide with the spindle axis.

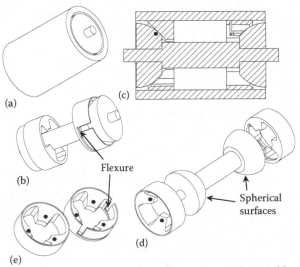

• Indicates contact point where visible

FIGURE 7.19
A symmetric spindle design: (a) assembled spindle, (b) spindle with cover removed, (c) cross section of assembled spindle, (d) exploded view of spindle with cover removed, (e) 'triangular holes' indicating bearing contacts.

- The bearings will traverse the same sphere surface and, therefore, deviations caused by lack of sphericity will be repeatable, typically at nanometre levels, and can, therefore, be compensated.
- The conformability of the contacts and the fact that there are five datum contacts will provide elastic averaging.

Such spindles are used in a variety of measuring machines, typically for measuring of cylindrical features. Details of these designs are often considered intellectual property technology and, therefore, not published in the available literature.

7.2.2 Rolling Element Bearing

The purpose of this section is to explore the precision limitations of rolling element bearings and will not consider the standard design procedures for determination of loads and speeds that are already detailed in machine design textbooks (Juvinall and Marshek 2011, Spotts et al. 2003). An idea of the relevant size of a bearing given the required load and speed can be estimated using the bearing selector in Figure 7.11.

Rolling element bearings typically serve the purpose of providing a joint of defined degrees of freedom between two rigid bodies. They consist of inner and outer races with intermediate, rotationally symmetric rolling elements that provide the guiding mechanism. Rolling friction tends to be substantially lower than sliding. However, it is noted that a ball rolling in a groove must involve some sliding. To understand this, consider the segment of a ball race with a ball of the same radius under a load, as shown in Figure 7.20. It is clear that if the ball is to roll in the groove, it cannot do so without slipping. Under load, the normal stress (indicated as σ_z in the figure) will be a maximum in the central region of the groove, where no slip rolling will occur. Away from this central region, as the contact rises to the edge of the groove, there will be a transition from pure rolling to slipping, a phenomenon called Heathcote slip. Clearly, for rolling element bearings under load, a finite frictional force must always be applied to rotate the spheres. As a result of these frictional forces, the rolling elements will not move until a sufficient force (torque) is applied. However, the torque will often introduce measurable displacements at the joints that will appear to a controller as a significantly different system characteristic. This effect can be

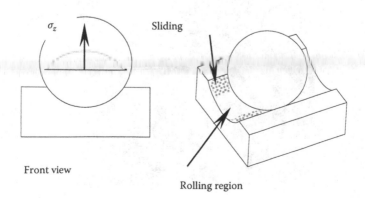

Front view

Rolling region

FIGURE 7.20
A sphere loaded into a circular groove.

problematic or exploited as a fine-adjust mechanism. Some of the controller aspects of this phenomenon are discussed in Chapter 14. Seals are often used to prevent dust from getting into the rolling contacts (and lubricant escaping) and will add friction to a bearing. One compromise is to use a 'shield' that covers the bearing cavity but does not make physical contact on one of the races, typically the inner race. Two common causes of damage observed with failed bearings that will result in additional bearing noise are (1) wear due to Heathcote slip and (2) pitting in the central region of the groove due to the cyclic maximum stresses during rolling. Modern bearings show little wear under normal conditions, with significant damage usually following impact loads (resulting in surface dents called Brinelling) or lubricant starvation (resulting in severe adhesion-based wear or 'galling').

In the absence of loads, other than the preloads for increasing bearing stiffness (and reducing 'play'), bearing noise is typically a consequence of dimensional tolerances and surface texture effects. Figure 7.21 shows a heavy-duty bearing comprising only seven balls. Typically, the bearing will be required to support a radial load between the inner and outer race. As the balls precess around the race, there will be times when a ball is directly in line with the force and other times when the force drives between two spheres. In practice, this produces a small variation in bearing stiffness resulting in a cyclic motion called 'print through' and, because the rotation of the races and spheres are not typically rational number ratios of each other, this is difficult to predict, and will not be synchronous with rotation. Consequently, this deviation will be difficult to predict and compensate. Because of manufacturing tolerances, small changes in the diameter of the rolling elements will cause each element to precess around the race at a different rate. To prevent this, a cage is used. As the bearing rotates, the spheres will push against the cage until the force is sufficient to cause sliding, this being another source of friction in the bearing. It is not unusual for this 'cage-slip' to cause a sudden relaxation of frictional force that can result in unpredictable spikes of noise in a bearing assembly. One method to reduce this precess of rolling elements in a bearing is to make sure that each element in the bearing is as close in dimension as the others. Due to tool wear, environmental changes and many other factors, dimensions of individual rolling elements will vary over time in a manufacturing process. However, successively manufactured rolling elements will be nominally of the same dimension. Hence, each element is loaded into a tube during manufacture so that the balls can then be sequentially assembled into the bearing in the order that they are manufactured (see Figure 7.22).

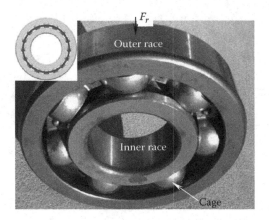

FIGURE 7.21
Photograph of a heavy-duty rolling element bearing comprising seven balls held in a cage. Inset shows an instrumentation bearing comprising sixteen balls.

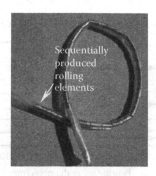

FIGURE 7.22
Rolling element bearing 'tube' for assembling bearing from sequentially manufactured elements.

Another method for reducing the noise is to use as many rolling elements as possible and rely on elastic averaging. One example is the instrumentation bearing shown as an insert in Figure 7.21 that has sixteen balls. Precision instrumentation bearings tend to have lower preloads and are selected for closer tolerances. Specification standards have been developed that stipulate a broad range of, mainly geometric, tolerances for specific classes of bearing, see Table 7.1. Typically, bearings of different precision are categorised by different classes and maximum performance tolerances are specified for each. As an example, Figure 7.23

TABLE 7.1

Bearing Tolerance Classes and Associated National and International Specification Standards

Standard		Tolerance Class					Bearing Types
Japanese Industrial Standard	JIS B 1514	Class 0 (6X)	Class 6	Class 5	Class 4	Class 2	All types
International Organization for Standardization (ISO)	ISO 492	Normal Class 6X	Class 6	Class 5	Class 4	Class 2	Radial bearings (except tapered roller bearings)
	ISO 199	Normal	Class 6	Class 5	Class 4	–	Thrust ball bearings
	ISO 1224	–	–	Class 5A	Class 4A	–	Precision Instrument bearings
Deutsches Institut für Normung	DIN 620	P0	P6	P5	P4	P2	All types
American National Standards Institute (ANSI), Anti-Friction Bearing Manufacturers (AFBMA)	ANSI/ AFBMA Std. 20	ABEC-1, RBEC-1	ABEC-3, RBEC-3	ABEC-5, RBEC-5	ABEC-7,–	ABEC-9,–	Radial bearings (except tapered roller bearings)
	ANSI/ AFBMA Std 12.1	–	Class 3P,–	Class 5P, Class 5T	Class 7P, Class 7T	Class 9P, –	Precision Instrument bearings (metric)
	ANSI/ AFBMA Std 12.2	–	Class 3P, –	Class 5P, Class 5T	Class 7P, Class 7T	Class 9P, –	Precision Instrument bearings (inch)

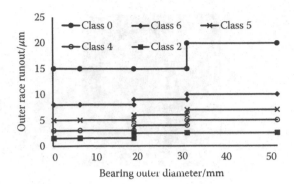

FIGURE 7.23
Outer ring radial runout for different classes of radial bearing as a function of outer diameter.

shows a graph of the maximum radial runout for different classes of bearing. The runout corresponds to the motion in a single direction of the outer race as the inner race is rotated without motion. In practice, the rotation can be achieved using a precision air-bearing spindle that has considerably lower errors than that of the bearing being tested. Alternatively, it is possible to determine the errors using a spindle analyser that utilises the reversal techniques outlined in Chapter 5 (see also Grejda et al. 2005). For a high-quality instrument bearing, a typical runout measurement might be represented by the data plotted in Figure 7.24. This shows a bearing that has a runout that is asynchronous with rotation of the inner race, although there appears to be an almost periodic deviation that could arise from rotation of the balls (or one 'bad' ball) in the bearing. Adding measurements as the bearing is rotated multiple times results in the plot of Figure 7.24b. Over a number of turns, this shows a maximum non-repetitive runout below 1/10 of the requirement for a class 2 bearing of diameter less than 18 mm and hence would be considered an exceptionally good precision instrument bearing. These multiple measurements are also typically averaged to evaluate deviations that are synchronous with the angle of the spindle.

Elastic averaging has already been identified as improving bearing noise performance and this can be exploited more fully with some linear and rotational rolling element bearing implementations. Two popular linear bearing designs are shown in Figure 7.25 and Figure 7.26. Both bearings work well for high-performance applications; the first of these, the crossed roller bearing, appears to be preferred for precision linear slideways. Crossed roller bearings comprise two opposing V's that have precision ground (and often stoned) surfaces on which the rolling elements rotate. Because the bearings will travel one circumference for a

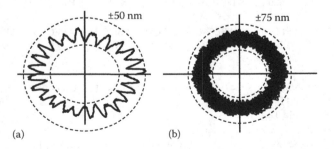

FIGURE 7.24
Polar plot of runout of the outer race of an exceptionally good instrument bearing: (a) single rotation of the inner race, (b) multiple rotations showing non-repetitive runout (NRRO).

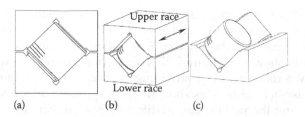

FIGURE 7.25
Small section of a crossed roller bearing: (a) front view that, to the careful observer, shows the rolling contact, (b) isometric view, (c) view of the bearing with the upper race removed. Note that the retainer (or cage) for the rollers is not shown.

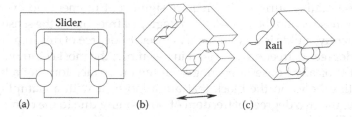

FIGURE 7.26
Linear ball bearing guide rail: (a) front view, (b) isometric view from the underside, (c) view with slider removed.

linear motion of one diameter of the roller, slides can provide a relatively long range without the need to recirculate the bearings as they roll out of the moving race. The addition of a recirculating tube (more common with ball-type linear bearings) enables unlimited translation ranges. A bearing cage, often called a retainer, is used to space the rollers apart and prevent the flat faces of the rollers from contacting the V surfaces. The linear ball bearing rail shown in Figure 7.26 represents a symmetric form that is similar to angular contact bearings clamped together to form a rotary axis. This design has the advantage that a great number of balls can be used, resulting in significant elastic averaging for high precision, stiffness and sharing of loads. Consequently, this type of rail system can provide fast and precise linear motion and is often used for manufacturing and machining slideways. A single row ball variation of this bearing is often used to create a more compact slide for instrumentation applications and is typically less expensive but has lower precision performance.

Both of the designs shown in Figure 7.25 and Figure 7.26 are over-constrained. As a consequence, it is necessary to use precision manufacturing techniques and incorporate substantial alignment in the assembly of these bearing. Under constant load conditions, both the roller and crossed roller bearings can achieve linear motion with deviations less than 1 μm and sometimes lower than 0.25 μm for precision bearings in carefully designed assemblies with controlled operating conditions.

Moore Tools carefully manufactured and assembled a ball bearing spindle. This comprised a honed spindle shaft and housing separated by almost 200 balls with these balls being selected to be closely toleranced to each other. Uniform separation of the balls about the shaft is maintained using a brass retainer (Moore 1970). Such a spindle will provide rotations with runout typically being less than 75 nm with some providing considerably better performance under controlled laboratory conditions (Bosse et al. 1994, Sacconi and Pasin 1994).

In practice, rotating spindles will be assembled from multiple bearings on a shaft. Selection of the appropriate combination of bearings is dependent on the application.

However, some simple rules can guide the designer to a specific choice of bearing. For precision spindles, the choice is narrowed to precision bearings and assemblies that will not result in significant stresses in the spindle. A selection of bearings and their respective degrees of freedom is shown in Figure 7.27. All bearings provide a rotational degree of freedom with the two roller bearings providing an additional freedom along the axis of rotation for small displacements. This short range is typically adequate to accommodate assembly tolerances and thermal expansion differences in operation. The roller bearing of Figure 7.27a is designed to withstand radial loads, while that of Figure 7.27b, being a thrust bearing, is capable of withstanding axial (thrust) loads. Figure 7.27c) shows a self-aligning thrust ball bearing. This is a regular thrust bearing with one race having a convex spherical outer surface that is mated to a further washer having a matched convex surface. These mating spherical surfaces will accommodate angular misalignments during assembly. Under load, these mating surfaces will require significant moments to overcome the frictional forces and these cannot be considered a degree of freedom in the sense of low friction rolling. Consequently, a tilde is used to signify a pseudo-degree of freedom. There are many variants on the degrees of freedom provided in assembly, the most common of which is the pillar block that typically houses a bearing with the outer race forming a ball joint and it provides mounting holes on the block that are a loose fit with a clamping bolt. Hence, during assembly, the two degrees of freedom of positioning due to the clamp holes plus the ball joint in the pillar block provide a total of five degrees of freedom. Once clamped, two of these degrees of freedom are constrained. Figure 7.27d shows a self-aligning ball bearing that provides a full three degrees of freedom that can be considered as being kinematically equivalent to a sphere in a triangular hole.

Clearly, bearings and bearing assemblies can still be considered from a kinematic viewpoint. Figure 7.28a shows the cross-section of a spindle comprising two radial ball bearings. In this particular design the outer race of the left-hand bearing is a sliding fit to the sleeve of the spindle, thereby providing an additional linear degree of freedom. Considering each bearing as a joint and the shaft and sleeve as a link, the Tchebytchev, Gruebler and Kutzbach criterion (see Chapter 6, Sections 6.4.2 and 6.8 for further examples and discussion) indicates

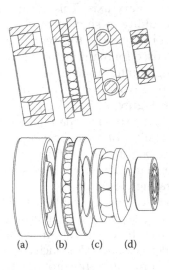

(a)   (b)   (c)   (d)

FIGURE 7.27
Four common rolling element bearings and their respective freedoms: (a) single row roller bearing θ_x, x, (b) thrust roller bearing θ_x, x, (c) pseudo self-aligning thrust ball bearing $\theta_x, \sim\theta_y, \sim\theta_z$, (d) double-row, self-aligning, ball bearing $\theta_x, \theta_y, \theta_z$.

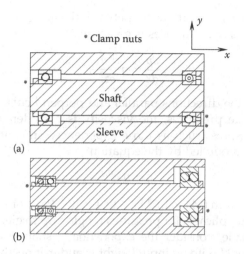

FIGURE 7.28
Cross sections of two rolling element spindle designs: (a) two radial ball bearings having mobility $M = 6(2-2-1) + 3 = -3$, (b) self-aligning bearing and angular contact spindle design $M = 6(2-2-1) + 5 = -1$.

a mobility of -3. Thus, this is over-constrained by a factor of four and, therefore, avoidance of residual stresses (and, therefore, distortions) during assembly will depend on manufacturing tolerances and careful assembly. A second spindle design shown in Figure 7.28b comprises a self-aligning bearing at one end, also with a sliding fit on the outer race and an angular contact bearing at the other end. Angular contact bearings are assembled in pairs and by clamping either the inner or outer races together during assembly. Because one race extends further than the other, when clamped, the bearing will have a defined preload. Additionally, the larger contact angle between the rolling surface and balls provides high stiffness in both the radial and axial directions. Typically, this high stiffness is considered to reduce bearing 'play' and the averaging of the two rows of bearings helps to reduce bearing noise. For this second spindle implementation, the angular contact bearing provides one degree of freedom, while the slide fit and self-aligning bearing provide four, resulting is a spindle mobility of -1. For a desired single degree of freedom, -1 represents a mechanism with an over-constraint of two. For the spindle of Figure 7.28b, the missing degrees of freedom correspond to the additional requirements that the y and z axis locations of the angular contact bearing must line up with the rotational centre of the self-aligning bearing.

7.2.3 Hydrodynamic and Elastohydrodynamic Lubrication (EHL) Bearings

Hydrodynamic bearings are 'self-acting', require little in the way of lubricant supply, have very low friction, support large loads, have high stiffness and provide smooth motion. However, they are not effective at the low speeds typical of many precision processes and a shaft axis will tend to move under varying loads. Consequently, they tend to be less common than the hydrostatic bearings, discussed in the following section.

Section 7.2.1 indicated that sliding contacts substantially benefit from the addition of a boundary lubricant. For two sliding surfaces with an intermediate lubricant film, the fluid being drawn into the contact asperities will be drawn into a thin wedge-shaped region. This wedge action will result in a local increase in pressure while the small distances associated with the contact will result in low Reynolds numbers and, therefore, laminar-type flows.

For two-dimensional flows, the relationship between the gauge pressure p, viscosity η, and relative velocity u_x between two surfaces is given by

$$\frac{\partial p}{\partial x} = \eta \frac{\partial^2 u_x}{\partial z^2}, \tag{7.14}$$

where x and z represent the directions parallel and perpendicular to the plane of the film respectively. For moderate pressures, the viscosity is considered constant but this will increase at the high pressures (such as those experienced in contacts or extremely small separations) and is often modelled by the equation

$$\eta = \eta_o e^{\alpha_\eta p}, \tag{7.15}$$

where α_η typically ranges from 0.44×10^{-8} Pa^{-1} to 4×10^{-8} Pa^{-1} for fluids ranging from water to mineral oils, and being applicable for pressures ranging up to gigapascals and higher. As a somewhat artificial example, consider the exponentially shaped wedge shown in Figure 7.29a, travelling at velocity U with an input height h_i and exit height h_0 above the stationary horizontal surface. Equivalently, it might be considered that the base plane is moving with positive velocity while the wedge is stationary. Integrating Equation 7.14 and including the boundary conditions $u = U$ at $z = 0$ and $u = 0$ at $z = h$, the velocity at any height z is

$$u = \frac{1}{2\eta}\frac{dp}{dx}z(z-h) + \left(1 - \frac{z}{h}\right)U. \tag{7.16}$$

From Equation 7.16, the volumetric flow rate per unit width q is

$$q = \int_0^h u\,dz = -\frac{h^3}{12\eta}\frac{dp}{dx} + \frac{Uh}{2}. \tag{7.17}$$

Selecting a specific location, $\tilde{x}$ along the wedge with separation $\tilde{h}$, where the pressure gradient is zero (corresponding to the peak pressure), Equation 7.17 simplifies to

$$q = \frac{U\tilde{h}}{2} = \bar{U}\tilde{h}, \tag{7.18}$$

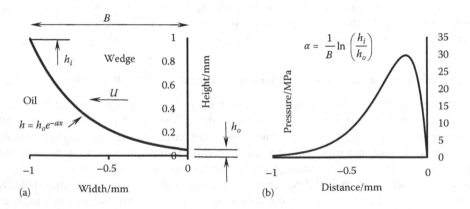

FIGURE 7.29
Model of an exponentially shaped wedge: (a) wedge shape and relevant geometric parameters, (b) pressure distribution $U = 0.5$ m·s^{-1}, $\eta = 1$ Pa·s (SAE 40 oil at room temperature), $h_i = 1$ mm, $h_o = 2$ µm.

where $\bar{U}$ is called the entraining velocity. Substitution of Equation 7.18 into Equation 7.17 yields a simple form of the Reynolds equation:

$$\frac{dp}{dx} = -12\eta\bar{U}\frac{h-\bar{h}}{h^3} . \qquad (7.19)$$

For the exponential wedge profile shown in Figure 7.29, Equation 7.19 can be integrated to give

$$\frac{h_o^2}{6U\eta}p = \frac{e^{2\alpha x}}{2\alpha} - \frac{e^{-\alpha\bar{x}}e^{3\alpha x}}{3\alpha} + C, \qquad (7.20)$$

where C is a constant and α is defined in Figure 7.29. At the entry to the wedge, $x = -B$ and $p = 0$ and at the exit, $x = -B$ and $p = 0$, from which it is possible to solve for the pressure. For a wedge that is considerably wider than the separation, it is possible to change the boundary condition from $x = -B$ to $x = -\infty$, to provide a reasonable approximate solution for a dimensionless pressure $p*$:

$$p* = \frac{ph_o^2}{6U\eta B} = \frac{1}{2\ln H}\left(e^{2\alpha x} - e^{3\alpha x}\right). \qquad (7.21)$$

Because the pressure is considered constant for any given value of x, the vertical load, W, per length, L, of the wedge is the integral of the pressure, from which

$$\frac{W}{L} = \int_{-\infty}^{0} p\,dx = \frac{U\eta}{2h_o^2\alpha^2} . \qquad (7.22)$$

The vertical stiffness k_{hyd} of the bearing can be obtained from the derivative of Equation 7.22 with respect to the separation. This is not trivial but can be approximated by assuming that, for a large ratio of input to output separations, α varies from $1/h_o$ to a constant so that the stiffness will be proportional to $1/h_o^n$, with n typically of the order 2.7.

The horizontal force F required to maintain the velocity of the wedge (or plate) is given by

$$\frac{F}{L} = \int_{-\infty}^{0} \tau_{z=0}\,dx = \int_{-\infty}^{0} \eta\frac{du}{dz_{z=0}}\,dx = \frac{7\eta U}{4h_o\alpha} . \qquad (7.23)$$

Equations 7.22 and 7.23 can be used to obtain the friction μ_{hyd} of the hydrodynamic slider

$$\mu_{hyd} = \frac{F}{W} = \frac{7}{2}h_o u. \qquad (7.24)$$

Clearly, the hydrodynamic friction tends to be linear with the film thickness, while the stiffness increases almost as the inverse cube of this thickness. Also, from Equation 7.22 it is apparent that the ability of a hydrodynamic film of a given geometry to withstand a load will be dependent upon the dimensionless ratio of viscosity, multiplied by sliding speed and divided by the load, that is the parameter $\eta U/W$.

For a journal bearing spindle of a given size and geometry, it is often informative to plot, on logarithmic scales, the coefficient of friction against this dimensionless ratio. The resulting plot was first suggested by Richard Stribeck in 1902 and an example is shown in Figure 7.30. This curve reveals three distinct regions. At the highest loads and lowest speeds, the hydrodynamic pressures will be incapable of separating the contacting asperities resulting

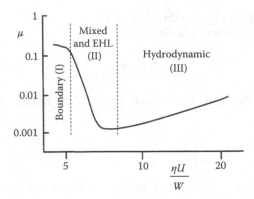

FIGURE 7.30
Stribeck curve for a journal bearing of a given geometry.

in the boundary lubrication region corresponding to the largest friction coefficients. From the previous analysis of the hydrodynamic wedge, it is found that for small wedge angles and sub-micrometre film thickness values, the pressures can approach the elastic limits of the counterface materials. Even for the example given in Figure 7.29, changing the film thickness at the output from 2 µm to 200 nm results in an increase in pressure to 1.5 GPa, which is comparable to the hardness of a medium strength steel. In this region, where the film thickness becomes comparable to the surface texture (of the component of the surface of significance to the contact), these high pressures will result in significant elastic (and some plastic) distortion, as well as a substantial increase in lubricant viscosity. Both effects are known to influence the hydrodynamic behaviour, and the intermediate region of the Stribeck curve represents mixed contact and elastohydrodynamic effects. Beyond this region, full hydrodynamic separation occurs and friction increases almost linearly in agreement with the model given by Equation 7.24.

When frictional losses are a major concern of the designer, it is clear that the optimal operating condition is somewhere at the boundary of elastohydrodynamic and hydrodynamic lubrication. Consequently, it is desirable to determine the region for which the mixed contact finally ceases leaving all asperities separated. This has been explored by Johnson et al. (1972), who conclude, among other things, that the nature of interactions in this region is dominated by two parameters: (1) the plasticity index discussed earlier and (2) the ratio of the theoretically predicted film thickness to surface texture, typically measured by either the R_a or R_q value (see Chapter 5, Section 5.7.5). As a consequence, it is possible to think of the vertical axis of the Stribeck curve as related to the height of the sliding surface and, because this will determine the load experienced by the contacts, this should also correlate with wear. Czichos and Habig (1985) suggested that a plot of the dimensionless wear coefficient against the ratio of film thickness to surface texture parameter, would expose some of the underlying wear mechanisms in a manner analogous to the wear maps discussed in Section 7.2.1 and illustrated in Figure 7.31. While theoretical models and some experimental data appear to corroborate this picture, producing a complete map for these proposed wear regimes has yet to be completed. While such maps have not been fully developed, it is clear that the optimal design again occurs at the final transition from elastohydrodyamic and hydrodynamic separation of the sliding surfaces.

By solving the Reynolds equation for flows between sliding surfaces, full solutions for hydrodynamic bearing designs, including consideration of the end effects, were published by Raimondi and Boyd in 1955 and are still used today. The design procedure outlined in this

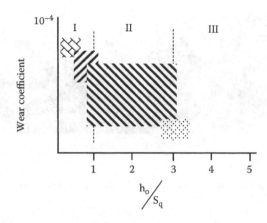

FIGURE 7.31
Map of wear coefficient with proposed wear regimes as a function of the ratio of the theoretical film separation to surface roughness.

work can be found in Juvinall and Marshek (2011). Of importance for precision applications is the determination of the position of the centre of the shaft. One of the charts from the work of Raimondi and Boyd showing the relationship between the ratio of the radius R of a shaft to the clearance c between the shaft and journal and the dimensionless bearing characteristic number (also called the Sommerfeld variable) S, is plotted in Figure 7.32. The different contours on this plot correspond to different ratios of the length L to the diameter D of the journal.

It is clear from the plot of Figure 7.32 that the lift of the shaft is minimised with reduced clearance that results in a reduced wedge angle. As a byproduct, minimal lift will correspond to maximum stiffness. Two parameters are of importance for bearing manufacturing. First, the R_q roughness parameter value of the shaft must be reduced to below the film thickness. Consequently, it is not unusual for hydrodynamic shafts, such as crankshafts in internal combustion engines, to be polished to surface texture R_q values below a few tens of

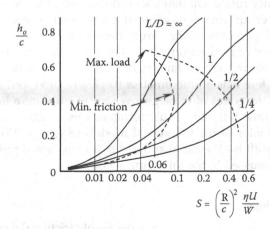

$$S = \left(\frac{R}{c}\right)^2 \frac{\eta U}{W}$$

FIGURE 7.32
Plot of the film thickness variable against the Sommerfeld variable for different journal bearing geometries. Optimal designs typically occur in the region between the two dashed lines.

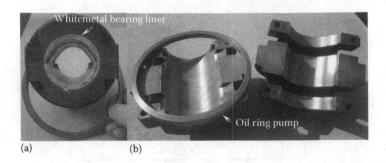

(a)   (b)

FIGURE 7.33
Photograph of a whitemetal journal bearing: (a) assembled bearing, (b) bearing halves and oil ring.

nanometres. Second, to reduce clearance, the material of the journal is often made from a 'white metal'. White metals (also called Babbitt metals) typically comprise soft metal alloys containing tribologically compatible elements such as lead and tin. Today, for modern engine bearings, a large number of alloys are used to withstand extreme ranges of loads, speeds and temperatures (Hutchins and Shipway 2017). All of the white metal alloys also tend to have low hardness and will, therefore, conform to the shape of the shaft, as well as enabling any hard particles in the lubricant to embed into the material. Conformability and embeddability represent examples of plastic design. These types of bearing are often used for dynamically varying loads, such as engine bearings, and are particularly suited for applications where shafts rotate at constant speeds and loads. One example of a hydrostatic bearing that was used to support the rotor in a power generator is shown in Figure 7.33.

Properly designed bearings will require a relatively low volumetric flow rate of lubricant supply. For steady operation with stationary shafts, such as those used in power generators, it is possible to provide adequate lubricant using an oil ring pump. As a solution to the problem of obtaining an uninterrupted lubricant supply, this is an excellent example of reduction in design. Comprising only a bronze ring that rests on the upper face of the bearing, there is only one part to the pump that can operate reliably for almost unlimited durations. In operation, the lower part of the ring is immersed in a sump of lubricant. As the shaft rotates, it drags the ring to similarly rotate. Oil that sticks to the ring in the sump is carried with the rotating ring to the upper surface where some of it is deposited onto the shaft.

It is often beneficial, particularly for thrust bearings, to add patterns of grooves to promote the development of the hydrodynamic layer. In thrust bearings, these grooves are typically some form of spiral pattern (and hence called spiral groove bearings) and only recessed to depths of a few tens of micrometres (typically produced by chemical etching). The step boundaries at the edges of these grooves again produce a local pressure increase creating the hydrodynamic lift, and simple forms of this design are called Rayleigh step bearings after the originator of the theoretical models (Rayleigh 1918). Using grooves to improve hydrodynamic lift is also useful and serves an additional purpose for many of the hydrostatic bearings discussed in the following section.

7.2.4 Hydrostatic Bearings

Hydrodynamic bearings are 'self-acting', have negligible frictional drag and require little in the way of lubricant supply, and cover the complete spectrum of loads and speeds including support of a stationary load. Additionally, suitably designed and manufactured hydrostatic bearings can have the highest stiffness of all bearing types and provide rotations and translations with nanometre-level repeatability, explaining why they are frequently used in demanding precision applications. A major limitation is the cost associated with precision manufacture and supply of a pressurised fluid. Another, potential issue is the work done on the fluid, as its pressurisation is released as it flows through and emerges from the bearing. While this flow rate is relatively small, there will always be heat generated by these flows both for air and oil hydrostatic bearings.

To understand the principles of hydrostatic bearings, consider the simple disk shown in Figure 7.34 that shows a cross-section through its centre. In operation, the disk will support a load. Fluid (air or oil) is fed into that plenum at pressure P_s and passes through the orifice emerging into the second plenum at pressure P_r. For a small gap between the land of the bearing and the flat counterface, the flow resistance in this second plenum is much lower than that between the inner radius r_i and outer radius r_o of the land, and is assumed to be constant. In practice, the thickness t of the step between the land and inner recess will have a significant impact on the speed that the bearing can traverse over the flat counterface. Not shown in Figure 7.34 is the bearing counterface that, for this theoretical development, is considered to be flat and parallel with the lower face of the disk and has a separation h between the lower face of the land and the flat counterface.

Essential to the operation of this bearing is the orifice that acts as a flow restrictor. Under a central load W, the disk will approach the counterface with a reduction in the separation and, therefore, the flow out of the bearing. This reduced flow will result in an increased pressure P_r until the integrated pressure under the complete bearing area is capable of supporting the load. Clearly, a change in this separation will either increase or decrease the flow through the bearing and, therefore, the load. Stiffness of the bearing, being the rate of change in load with separation, will depend on the flow resistance of the orifice and (to a lesser extent for liquids) the volume of the second plenum. Hence, even for this relatively simple design, optimisation is a multi-variable challenge and the interested reader is directed to more comprehensive texts (Rowe 2012). For this circular bearing, Equation 7.17 can be expressed in the form

$$q = -\frac{\pi r h^3}{6\eta}\frac{dP}{dx}. \tag{7.25}$$

Assuming that the restrictor can be modelled as a capillary, equating the total flow across any surface in the land to the total flow through the bearing, that will also be equal to the restrictor flow, results in the equations

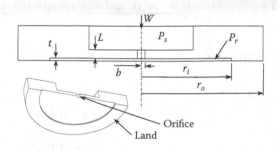

FIGURE 7.34
Cross section through a disk-type hydrostatic bearing.

$$q = \frac{\pi h^3 P_r}{6\eta \ln(r_o/r_i)} = P_r \frac{\bar{B}h^3}{\eta} = \frac{(P_s - P_r)\pi d_c^4}{128\eta L} = \frac{(P_s - P_r)}{K_c \eta}$$

(7.26)

$$P_r = \frac{P_s}{1 + \bar{B}K_c h^3}, d_c = 2b,$$

where the constants K_c and $\bar{B}$ are given by

$$\bar{B} = \frac{\pi}{6 \ln(r_o/r_i)},$$

(7.27)

$$K_c = \frac{128L}{\pi d_c^4}.$$

The bearing constants K_c and $\bar{B}$ are respectively called the capillary factor and flow shape factor and are a major component of hydrostatic bearing analysis. Because this continuity condition must also be true for any integral around the periphery (consider a radius somewhere inside the land), the pressure as a function of radial distance r can be calculated from

$$\frac{P}{P_r} = \frac{\ln(r_o/r)}{\ln(r_o/r_i)}.$$

(7.28)

Given that the total load W on the bearing is given by

$$W = (p_r - p_a)\pi r_i^2 + \int_{r_i}^{r_o} (p - p_a)2\pi r dr,$$

(7.29)

and assuming that the fluid is incompressible (i.e. a liquid), the load can be calculated from

$$W = \frac{\pi P_r (r_o^2 - r_i^2)}{2 \ln(r_o/r_i)} = \frac{\pi P_s (r_o^2 - r_i^2)}{2 \ln(r_o/r_i)[1 + \bar{B}K_c h^3]} = \frac{P_s A_e}{1 + \bar{B}K_c h^3}$$

(7.30)

$$A_e = \frac{\pi (r_o^2 - r_i^2)}{2 \ln(r_o/r_i)}.$$

The constant A_e is called the equivalent area and, like K_c and $\bar{B}$, is also an important parameter for hydrostatic bearing design. Differentiating Equation 7.30 with respect to the bearing separation, the stiffness of this bearing is

$$k_l = P_s A_e \frac{3K_c \bar{B}h^2}{(1 + K_c \bar{B}h^3)^2}.$$

(7.31)

For a gas bearing, the stiffness is given by

$$k_g = \frac{3K_c \bar{B}h^2 (P_s^2 - P_a^2) A_e}{2\left[\frac{P_s^2 + P_a^2 K_c \bar{B}h^3}{1 + K_c \bar{B}h^3}\right]^{1/2} [1 + K_c \bar{B}h^3]^2},$$

(7.32)

where P_a is atmospheric pressure. As an example, Figure 7.35 shows plots of the stiffness and bearing separation for a disk bearing of outer diameter 40 mm with an air supply pressure representative of a typical bearing (oil hydrostatic bearings might typically be two

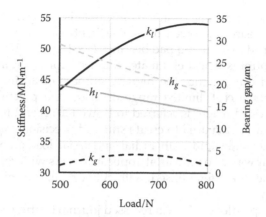

FIGURE 7.35
Plots of bearing stiffness and gap for a simple disk hydrostatic bearing as a function of centrally applied load. The operating parameters for this bearing are $P_s = 550,000$ N·m², $P_a = 100,000$ N·m², $r_o = 0.02$ m, $r_o/r_i = 1.2$, $L = 0.004$ m, $d_c = 0.2$ mm, viscosity of SAE 10-40W oil at 20 °C is 0.2 (kg·m⁻¹·s⁻¹) and viscosity of air at 25 °C is 1.983 × 10⁻⁵ (kg·m⁻¹·s⁻¹).

to five times this pressure and of correspondingly higher stiffness). Even this modest design with supply pressure representative of many bearing applications can support hundreds of newton loads, with stiffness of tens of newtons per micrometre and bearing separations around 10 μm to 20 μm. For comparison, the stiffness values of around 50 N·m⁻¹ to 200 N·m⁻¹ for air bearings, and around ten times this for oil hydrostatic bearings, is not uncommon for precision machine tools. Under no-load conditions, air bearing spindles can have radial error motions of less than 20 nm and are commonly used in diamond turning and roundness measuring machines.

A more feasible air bearing would typically comprise various forms of opposing pads, the simplest of which would be rail or box-type bearing structures. Figure 7.36 shows an opposed pad bearing using a disk-type bearing of the previous example running on a parallel rail. Alignment, preload and setting of the nominal separation h_o are indicated by an adjustable gimbal. A box-type bearing would have pairs of bearings on either side of what is usually a square rail.

The function of the many orifice designs to control flow and influence the stiffness of the bearing suggests that other methods will provide increased gains. Simple consideration of the operation suggests that constant flow control, that represents a form of position feedback, could increase stiffness almost indefinitely. To this end, flow regulators have been designed

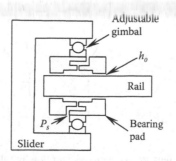

FIGURE 7.36
Opposed pad bearing guided by a prismatic rectangular rail.

to match with air bearings and enhance stiffness. Of note is the spool-based Royle valve and the diaphragm-based Mohsin and Rowe valves. The first two are for single pad bearings, and the Rowe valve is designed for opposed pad bearings (see Figure 7.37). A major advantage of using constant flow control is that it eliminates the need for carefully designed orifices or capillary restrictors and can produce the highest stiffness of all implementations. Disadvantages are the additional cost, limited range of loads and potential for instability. For further details, the interested reader is referred to Rowe (2012) and references therein.

Hydrostatic bearings are often used to create stiff and precision spindles. Typically, such spindles are required to withstand both radial and axial loads, and a broad variety of designs have found their way into industrial applications. As with rolling element spindles, semi-kinematic principles are often utilised, with cylindrical, conical and spherical bearing surfaces being common.

Possibly the simplest spindle would be a recessed journal bearing, shown in cross-section in Figure 7.37. The journal of this bearing comprises four segments that could be supplied through a flow restrictor or, in this case with a constant flow valve, comprising a diaphragm. Flow control is achieved when the shaft in the bearing changes from a concentric position that results in the flow resistance reducing on one side and increasing on the other. This differential resistance change causes a corresponding change in pressure across the diaphragm that responds by reducing the resistance to flow on the high pressure side and increasing it on the opposing pad, to restore the shaft to it concentric position. Suitably designed, such a bearing can maintain concentricity within a limited range of changing loads, thereby giving the appearance of an infinitely stiff bearing. Such constant flow designs typically provide the highest stiffness hydrostatic bearings.

Many implementations of the opposed pad bearings have been designed and used for applications ranging from small high-speed spindles to large telescopes. High stiffness spindles with high radial and axial stiffness are probably one of the most common applications in precision engineering. The high stiffness in both directions is typically achieved by combining thrust and journal bearings on the same spindle, with two simple examples being shown in Figure 7.38. Figure 7.38a is a Yates bearing, where the journal is in the central part of the spindle, while Figure 7.38b shows an inverted design, with two journals at each end of the shaft and a central thrust plate for the axial bearing. Having the two journal bearings at each end of the spindle provides an increased stiffness to moments produced by radial forces offset from the centre of the spindle.

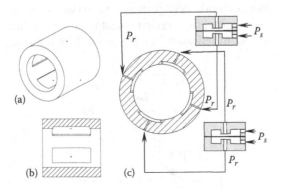

FIGURE 7.37
Four-segment recessed journal bearing with Rowe valve constant flow regulation on opposing pads: (a) isometric view of bearing journal showing orifice feeds and plenums, (b) cross section of journal, (c) schematic diagram indicating pressure supply from Rowe valves to individual plenums of the journal.

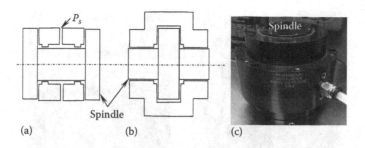

FIGURE 7.38
Hydrostatic spindle bearing implementations for high radial and axial stiffness: (a) Yates bearing, (b) inversion for greater resistance to moments created by radial forces, (c) photograph of a commercial spindle.

In the early 1960s, it was realised that by providing shallow grooves into the bearing surface, it is possible to create flow regulation that would result in larger area pressure gradients, thereby increasing both the load capacity and stiffness of a bearing, as well as reducing flow (Arneson 1967). Depending on the geometry and depths of the grooves, traditional machining and fine finishing processes or chemical etching are often used to create the required geometry. For suitably designed grooved bearings, it is also possible to eliminate the orifice or capillary compensation altogether. One early example drawn from Arneson's patent is shown in Figure 7.39.

For linear translations, box-type aerostatic or hydrostatic bearings are commonly used. An example of a box-type air bearing is shown in Figure 7.40. This box-type grooved air bearing shows *X*-shaped grooves forming opposed pads on either side of the shaft, with similar bearings at the far end of the box to provide stiffness response to moments produced by radial forces applied to the end of the alumina shaft. Since the 1960s, many types of groove bearing designs have been, and continue to be, developed.

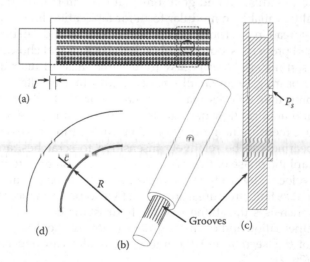

FIGURE 7.39
A grooved air bearing spindle: (a) transparent view of half of the bearing showing the axial grooves that end a distance *l* from the end of the journal, (b) isometric view of the complete bearing with the spindle drawn out of the journal, (c) cross section view of part b, (d) front view of the journal and shaft (the grooves are 0.5 mm deep for this 40 mm diameter shaft and are not visible).

FIGURE 7.40
Box-type linear bearing with X grooves and orifice compensation: (a) disassembled bearing, (b) assembled bearing, (c) close-up view of box bearing showing X grooves and location of orifice/capillary feed (the hole is too small to see).

Instead of feeding the air supply through single or multiple restrictors, it is possible to feed the air through a porous material resulting in what might be considered an infinite number of orifices uniformly distributed over the air bearing surface. For a porous material, the relationship between pressure drop across the material and flow rate per length through it is adequately described by Darcy's law in terms of the ratio of porous permeability to fluid viscosity. Using this approach it is possible to derive models for performance of bearings of relatively simple geometry using compressible laminar flow equations (Plante et al. 2005). Because the porous carbon (and many other porous materials) is not particularly strong, there are design challenges associated with fabricating the material into a bearing housing and supplying pressure to its back face while supporting the material to distribute transmission of forces under load (Yoshimoto and Kohno 2001). Porous air bearings were originally developed in industrial and governmental laboratories starting in the 1960s but are now commercially available in modular form for fabricating linear slides and spindles.

Because it is widespread in electrical and nuclear industries and can be purchased in bulk with precisely defined properties, porous carbon is the material of choice for these bearings. This is a relatively soft material that is easy to machine. Ease of machining also makes it possible to produce bearings with small gaps (<10 μm) that results in high stiffness and squeeze film damping that will dissipate vibrations normal to the bearing surface. The intrinsic damping and uniform flow pattern also has the advantage of an intrinsic stability. Additionally, because there are no grooves or other fine features necessary to obtain stable performance, the bearing can be relatively insensitive to scratches, and the tribological properties of the graphitic phase of this material gives it a resilience to occasional impacts.

At large relative velocity, hydrodynamic pressures will develop that can be comparable or higher than the hydrostatic pressure leading to a hybrid performance that, again for well-designed bearings, increases the stiffness of the hydrostatic bearing. Such bearings often combine orifice compensation and complex groove patterns; the design of such bearings is beyond the scope of this section, but the interested reader can find considerable design guidance in Rowe (2012).

Oil hydrostatic bearings, while stiffer, are slower and require closed systems to contain the fluids but do provide intrinsic shear and squeeze film damping. This relatively high damping (often called viscous friction in theoretical papers) plays a significant part in controller stability (see Chapter 14).

7.2.5 Flexures for Motion Guidance

Flexures represent another method for guiding relative motion between two rigid bodies by exploiting the large differences of mechanical compliance in different directions of simple flexural elements (Smith 2000). Possibly the simplest element is a wide beam attached to separate rigid bodies at each end and subject to specific forces. The goal of flexure design is to create high stiffness in the direction for which constraint is desired and high compliance in the directions of desired freedom. If high stiffness is considered to represent a work-free constraint and high compliance a freedom, the rules of kinematics can be applied to the design of complex flexure mechanisms and generalised design principles have been developed (Hopkins 2015).

Before considering a design, it is informative to review the relative merits and limitations for motion guidance listed in Table 7.2.

The variety of mechanisms and number of degrees of freedom that can be produced with flexures is limited only by the creativity of the designer, and numerous ingenious designs abound in the literature. To keep this section within a reasonable length, only a few examples are presented.

7.2.5.1 Elements

Flexural compliance is usually provided by beams, notches, rods or diaphragms. The simplest of these, the beam, can be analysed using the Euler beam bending equation (see Chapter 3, Section 3.5.4). For the forces applied at the end of a beam (Figure 7.41), by equating moments, the bending equation is given by

$$EI\frac{d^2y(x)}{dx^2} = F_y(L - x) - F_x\left(\delta_y - y\right) + M, \tag{7.33}$$

TABLE 7.2

Merits and Limitations of Flexure Mechanisms

Merits

Motion repeatability can be at the limits of measurement, noise-free throughout the life of the flexure.

Forces and stresses in the flexure can be precisely predicted and used to similarly determine performance.

Because the stresses and stress history can be predicted, S/N and strain-life methods provide reliable estimates of fatigue life.

Monolithic flexure mechanisms can be manufactured, thereby eliminating assembly issues such as interface wear and residual stress from clamping.

Symmetry can be exploited to reduce susceptibility to temperature changes and temperature gradients in specific directions.

The effect of tolerances in manufacture and assembly will not compromise performance. However, the line of motion may deviate from the intended path.

Limitations

For a given range of motion and load capacity, flexures tend to occupy a relatively large 'footprint'.

The actuation force is proportional to displacement.

The ratio of 'constraint' to 'freedom' stiffness are typically lower than that of other bearing types.

Actuation must be aligned to the motion of the flexure, preferably at the centre of stiffness.

Most metallic flexures will exhibit a stress-dependent but finite hysteresis and permanent deformation, particularly with accidental overloads. 'Hard stops' are often used to prevent this.

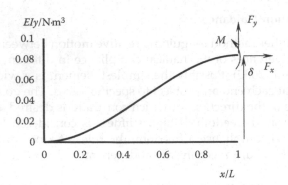

FIGURE 7.41
Deflection of a cantilever beam subject to axial, F_x, and tangential, F_y, forces plus an applied moment, M.

where E is the elastic modulus of the beam, I is the second moment of area about the neutral axis of bending, δ is the vertical deflection at the end of the beam and all other parameters are defined in the figure. Ignoring axial loads for now, the slope and deflection of the beam are given respectively by

$$EI\frac{dy(x)}{dx} = F_y L^3 \left(\frac{x}{L} - \frac{1}{2}\left(\frac{x}{L}\right)^2 \right) + ML^2 \frac{x}{L}$$

$$EIy(x) = F_y \frac{L^3}{2}\left(\left(\frac{x}{L}\right)^2 - \frac{1}{3}\left(\frac{x}{L}\right)^3 \right) + M\frac{L^2}{2}\left(\frac{x}{L}\right)^2.$$

(7.34)

From the first of Equation 7.34, it is apparent that by setting the moment $M = -F_y L/2$, the slope at the free end is zero and the resulting equation has been used to generate the plot in Figure 7.41. Additionally, for this condition of zero slope at the free end, the stiffness $k_{F_y \delta_y}$ of the beam in the y direction is

$$F_y = k_{F_y \delta_y} \delta_y = \frac{12EI}{L^3}\delta_y.$$

(7.35)

Consequently, the maximum bending moment that occurs with equal magnitude and opposite sign at each end of the beam (varying linearly from one end to the other) is

$$M_{max} = -F_y \frac{L}{2} = -\frac{6EI}{L^2}\delta_y.$$

(7.36)

In general, once the deflection curve is known, the important components of displacements and angles between the ends of the beam and bending moments (from which stress can be calculated) can be determined.

When the leaf is subject to an axial load, solutions to Equation 7.33 become more interesting. Defining the dimensionless groups

$$\gamma^2 = \frac{F_x L^2}{EI}, \ m_o = \frac{M_z L}{EI}, \ \varphi = \frac{F_y L^2}{EI}, \ X = \frac{x}{L},$$

(7.37)

Equation 7.33 becomes

$$y = \left(\frac{m_o L + \varphi L}{\gamma^2} - \delta\right)(\cosh(\gamma X) - 1) + \frac{\varphi L}{\gamma^3}(\beta X - \sinh(\gamma X)). \tag{7.38}$$

The slope θ and displacement δ at the end of the beam can be obtained from the matrix relationship

$$\left\{\begin{array}{c} \dfrac{\delta}{L} \\ \tan(\theta) \end{array}\right\} = \left[\begin{array}{cc} \dfrac{(\cosh(\gamma) - 1)}{\gamma^2 \cosh(\gamma)} & \dfrac{\gamma \cosh(\gamma) - \sinh(\gamma)}{\gamma^3 \cosh(\gamma)} \\ \left(\dfrac{1}{\gamma}\right) \tanh(\gamma) & \dfrac{(\cosh(\gamma) - 1)}{\gamma^2 \cosh(\gamma)} \end{array}\right] \left\{\begin{array}{c} m_o \\ \varphi \end{array}\right\} = \mathbf{A}_t \left\{\begin{array}{c} m_o \\ \varphi \end{array}\right\}. \tag{7.39}$$

Cramer's rule can be applied to invert this matrix, after which the solutions are

$$m_o = \frac{\delta}{L}\frac{1}{\Delta_t}\frac{(\cosh(\gamma) - 1)}{\gamma^2 \cosh(\gamma)} + \tan(\theta)\frac{\sinh(\gamma) - \gamma\cosh(\gamma)}{\gamma^3 \cosh(\gamma)}\frac{1}{\Delta_t}$$

$$\varphi = \frac{\tan(\theta)}{\Delta_t}\frac{(\cosh(\gamma) - 1)^2}{\gamma^4\cosh^2(\gamma)} - \frac{\delta}{L}\left(\frac{1}{\gamma}\right)\frac{\tanh(\gamma)}{\Delta_t}, \tag{7.40}$$

where

$$\Delta_t = |\mathbf{A}_t| = \frac{1 + \cosh(2\gamma) - 2\cosh(\gamma) - \gamma\cosh(\gamma)\sinh(\gamma)}{\gamma^4\cosh^2(\gamma)}. \tag{7.41}$$

When the beam is subject to an axial compressive load, the equation for lateral deflection of the beam becomes

$$\frac{d^2y}{dX^2} + \gamma^2 y = m_o L + \varphi L(1 - X) + \gamma^2 \delta, \tag{7.42}$$

from which the deflection is given by

$$y = \left(\frac{m_o L + \varphi L}{\gamma^2} + \delta\right)(1 - \cos(\gamma X)) + \frac{\varphi L}{\gamma^3}(\sin(\gamma X) - \gamma X). \tag{7.43}$$

As for the tensile axial load, the slope and deflection can be obtained from the matrix relationship

$$\left\{\begin{array}{c} \dfrac{\delta}{L} \\ \tan(\theta) \end{array}\right\} = \left[\begin{array}{cc} \dfrac{(1 - \cos(\gamma))}{\gamma^2 \cos(\gamma)} & \dfrac{\sin(\gamma) - \gamma\cos(\gamma)}{\gamma^3 \cos(\gamma)} \\ \dfrac{\tan(\gamma)}{\gamma} & \dfrac{1 - \cos(\gamma)}{\gamma^2 \cos(\gamma)} \end{array}\right] \left\{\begin{array}{c} m_o \\ \varphi \end{array}\right\} = \mathbf{A}_c \left\{\begin{array}{c} m_o \\ \varphi \end{array}\right\}. \tag{7.44}$$

Again, using Cramer's rule, the solutions for m_o and φ are

$$m_o = \frac{\delta}{\Delta_c L} \frac{1 - \cos(\gamma)}{\gamma^2 \cos(\gamma)} + \tan(\theta) \frac{\gamma \cos(\gamma) - \sin(\gamma)}{\Delta_c \gamma^3 \cos(\gamma)} . \tag{7.45}$$

The equation for deflection at the free end is given by

$$\varphi = \tan(\theta) \frac{(1 - \cos(\gamma))}{\Delta_c \gamma^2 \cos(\gamma)} - \frac{\delta}{L} \left(\frac{1}{\gamma}\right) \frac{\tan(\gamma)}{\Delta_c} . \tag{7.46}$$

For compression, the determinant becomes

$$\Delta_c = |\mathbf{A}_c| = \frac{1 - 2\cos(\gamma) + \cos(2\gamma) + \gamma \sin(\gamma) \cos(\gamma)}{\gamma^4 \cos^2(\gamma)} . \tag{7.47}$$

For zero slope at the free end, Equations 7.39 and 7.44 simplify considerably, and it is possible to plot the ratios of the stiffness and bending moments at the free end of the beam to the values for the stiffness and bending moment in the absence of an axial load (Figure 7.42). As might be expected, the stiffness increases with tensile axial loads and reduces when subject to compression. This reduction in stiffness goes to zero at the classical buckling load and even goes negative. Negative stiffness is exploited to cancel the stiffness of springs supporting loads.

One relatively simple way to provide a compliant joint between two rigid bodies is to create a locally thin section connecting the two bodies by drilling two holes closely together, resulting in a localised notch. Often this will enable the manufacture of complex planar flexure mechanisms from a single monolithic piece of material. With advanced manufacturing machines, other notch shapes can be produced, examples of which are shown in Figure 7.43. Because it can be produced by drilling, the circular notch is the oldest and most commonly realised notch hinge, and equations for calculating the compliances were derived by Paros and Weisbord (1965). By drilling holes of radius R at a pitch h, the remaining web thickness t is considered to form the compliant element.

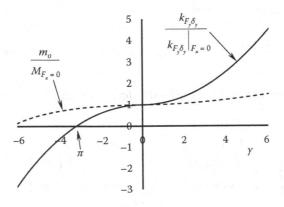

FIGURE 7.42
Ratios of stiffness and bending moments at the free end of a beam to the corresponding values in the absence of an axial load.

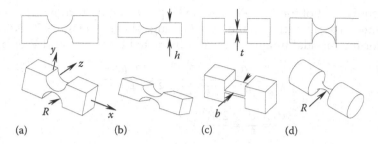

FIGURE 7.43
Notch hinge flexures: (a) a full circular notch, (b) an elliptical notch, (c) leaf-type hinge, (d) toroidal hinge.

The equations to compute the various compliances are relatively lengthy but, because the web thickness is relatively small in comparison to the radius of the notch, reasonable estimates can be calculated from the approximate equations

$$\frac{\theta_z}{M_z} \approx \frac{9\pi}{2EbR^2(2\beta)^{5/2}} = \frac{9\pi R^{1/2}}{2Ebt^{5/2}}$$

$$\frac{\theta_z}{F_y} = \frac{\delta_y}{M_z} \approx \frac{9\pi R^{3/2}}{2Ebt^{5/2}}$$

$$\frac{\delta_y}{F_y} \approx \frac{9\pi}{2Eb}\left(\frac{R}{t}\right)^{5/2}$$

$$\frac{\theta_y}{F_y} = \frac{\delta_z}{M_y} = \frac{\delta_z}{\pi R F_z} \approx \frac{12\pi R}{Eb^3}\left[\left(\frac{R}{t}\right)^{1/2} - \frac{1}{2}\right], \tag{7.48}$$

$$\frac{\delta_z}{F_z} \approx \frac{12\pi R^2}{Eb^3}\left[\pi\left(\frac{R}{t}\right)^{1/2} - \frac{1}{4}\right]$$

$$\frac{\delta_x}{F_x} \approx \frac{\pi}{Eb}\left[\left(\frac{R}{t}\right)^{1/2} - \frac{1}{2}\right],$$

where b is the width of the web and $\beta = t/2R$.

Apart from ease of manufacture, a major difference between the notch- and leaf-type hinge is the distribution of stresses and uncertainty of the pivot points. For the notch hinge, most of the stresses (and, therefore, strains) are concentrated in the central region of the notch. As a consequence, dependent upon the applied forces, the equivalent pivot location will be more or less in this location. However, because the stresses are constrained to a small region, the amount of strain energy that can be stored is limited and this will correspondingly limit either the deflection or stiffness of the hinge. On the other hand, the stresses in a leaf-type hinge are more uniformly distributed along its length, resulting in larger deformation at the expense of a less well-defined pivot location. A compromise between these two extremes is to create a hinge of intermediate shape, such as an ellipsoid, hyperboloid or paraboloid, for which the compliance equations can be found (Smith et al. 1997, Lobontiu et al. 2002).

Modelling the circular notch as a pure hinge subject to a moment M_z, the maximum stress in the web of the notch can be approximated from

$$\sigma_1 = K_t\frac{M_zt}{2I} \approx (1+\beta)^{9/20}\frac{6M_z}{bt^2}. \tag{7.49}$$

A two-axis equivalent of the notch hinge can be produced by turning a circular (or other shape) notch into a rod, resulting in the toroidal hinge shown in Figure 7.43d. The relevant compliances for the toroidal hinge can be approximated by (Paros and Weisbord 1965)

$$\frac{\theta_y}{M_y} = \frac{\theta_z}{M_z} \approx \frac{20}{ER^3(2\beta)^{7/2}} = \frac{20R^{1/2}}{Et^{7/2}}$$

$$\frac{\theta_y}{F_z} = \frac{\theta_z}{F_y} \approx \frac{20R^{3/2}}{Et^{7/2}} \tag{7.50}$$

$$\frac{\delta_x}{F_x} = \frac{2R^{1/2}}{Et^{3/2}} \; .$$

For applied moments, the stress concentration factor can be approximated using the expression in Equation 7.49, so that the stress can be calculated from

$$\sigma_1 = K_t \frac{32M_z}{\pi t^3} \approx (1 + \beta)^{9/20} \frac{32M_z}{\pi t^3} \; . \tag{7.51}$$

7.2.5.2 Mechanisms and Assembly

Flexures are typically used to create linear or rotational translations, typically with a maximum range limited by the strength of the material of construction. To maximise this, although ceramic and even polymer-based materials have been used, it is common to use metal flexures in a maximally hardened condition. Possibly the simplest linear mechanism is the four-bar linkage (sometime called an arcuate flexure), shown as monolithic forms in Figure 7.44 (see also Chapter 6 for in-depth discussions on the four-bar linkage). As illustrated, this flexure is driven through its centre of stiffness at a distance midway between the base and moving platform. For dynamic mechanisms, it is also desirable to have the centre of mass coincident with the centre of stiffness.

For the leaf-type flexure in Figure 7.44a, the deformation shape of the beam is that of the S-shaped flexure and the total stiffness will be the sum of the stiffness of each of the two leafs. For the notch-type flexure, the compliance of the hinges due to the applied moment will dominate and the total stiffness can be calculated from

$$k_{F_y \delta_y} = 4 \frac{k_{M_z \theta_z}}{L^2} , \tag{7.52}$$

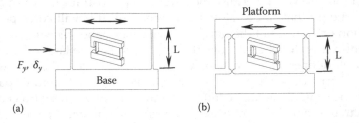

(a)   (b)

FIGURE 7.44
Simple parallelogram type flexures: (a) leaf-type flexure, (b) notch-hinge flexure. Isometric views are shown inset in each flexure.

where the parameters for Equation 7.52 are defined earlier in this section. If the notch hinges are considered to be single degree of freedom joints (four in total), this mechanism comprising four links has a planar mobility M equal to 1 as desired. While the moving platform will remain horizontal for these flexures, as the flexure translates, the platform will follow an arcuate motion. For small translations, this can be computed from

$$\delta_x \approx -\frac{3\delta_y^2}{5L}$$

$$\delta_x \approx -\frac{\delta_y^2}{2L},$$

(7.53)

where the two equations correspond to the leaf- and notch-type flexures respectively shown in Figure 7.44. To eliminate, or at least attenuate, this arcuate motion, consider the compound designs of Figure 7.45. The compound flexure comprises two serial four-bar mechanisms, with one inverted so that if both displace the same amount relative to their respective base, the arcuate motion of each stage will compensate, resulting in a rectilinear motion of the platform. A drawback with this design is the mobility of two reveals that this is a two degree of freedom mechanism. As a consequence, the relative motion of the two four-bar mechanisms will only be the same if either both are individually driven or the platform is driven and both have the same effective stiffness. In practice, if there is any load on the platform directed along the axis of the parallel links (i.e. vertically in the drawing), then one set of links will be in tension and the other in compression, thereby changing the effective stiffness. Additionally, to be collinear with the centre of stiffness, the drive would be placed at a point midway between the platform and upper horizontal member.

The issues with the compound design are somewhat mitigated with the more symmetric double compound flexure shown in Figure 7.45b. This design has three, orthogonal symmetry planes meeting at the centroid of the flexure. Consequently, this design will be stable at this central location in the presence of temperature changes as well as temperature gradients in the planes of symmetry. To visualise this, consider drawing a line on the surface collinear and coincident with a symmetry plane. It should then be apparent that, even in the presence of temperature gradients, the line will not be distorted. While this has a mobility of one, in practice, provided all of the support links are of the same length (a constraint on the geometry) and the platform is in the central position, the outer links can move without displacing the central platform and resonant frequencies associated with this

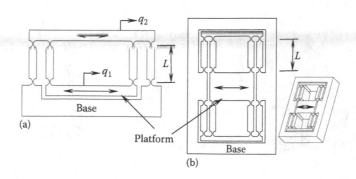

FIGURE 7.45
Compensated linear flexure mechanisms: (a) the compound stage ($M = 3(7 − 1 − 8) + 8 = 2$), (b) double compound rectilinear flexure ($M = 1$).

(forbidden) mode are typically observed; a more detailed discussion of such singularities is provided in Chapter 6, Section 6.4.

While monolithic notch- and leaf-type mechanisms are common, many flexures are assembled from modular parts. Examples of compound and double compound assembled flexures are shown in Figure 7.46. Again, it is apparent that the double compound has three planes of symmetry and the force can readily be arranged to be coincident with the line of action of the platform. It should also be observed that the clamps contain a recess that provides clearance for burrs that might occur during manufacture and also transfers the force from the bolt to two outer surfaces that contact the flexure. Because the force from the bolt is likely to be more evenly split to two regions separated as far as possible, this will provide a defined resistance to moments induced by the flexure. In principle, the clamp plate might have a sharp edge that will create a stress concentration in the flexure. To attenuate this stress concentration, it is necessary that the clamp be made from a material having a lower hardness than the flexure. As a consequence, the clamp will deform to the curvature of the flexure, which, being large, will thereby substantially reduce the local stresses in the flexure to below yield.

To summarise this section, mechanisms containing interesting features are briefly discussed and the reader is left to ponder the relative merits and limitations of each. One common mechanism component is a 'wobble pin' to transmit axial forces, such as the three examples shown in Figure 7.47. A mobility analysis for these three designs indicates only four of the five required degrees of freedom. However, apart from the high axial stiffness, it has been assumed that the hinges will provide torisional resistance along the drive axis. In practice, the torsional stiffness of these elements (particularly the rods) are likely to be relatively low and might be considered as a degree of freedom. In this case, the mobility of the rod wobble pin would be six, of which one of these would be rotation of the central link that would not affect the ability of the mechanism to transmit axial loads. Mobility analysis is always informative, but still requires the designer to determine where the degrees of freedom occur and the impact these degrees of freedom will have on the function of the mechanism.

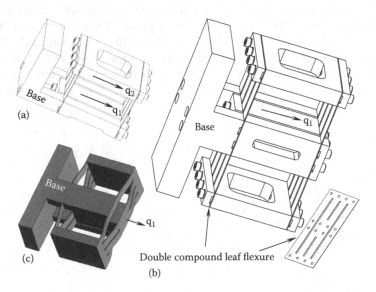

FIGURE 7.46
Compound flexure assemblies: (a) compound flexure, (b) double compound flexure being the mirror of the compound flexure (the leaf for the flexure elements is shown at the lower right corner).

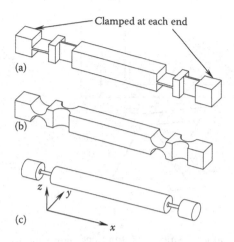

(a)

(b)

(c)

FIGURE 7.47
Three 'wobble pin' designs: (a) leaf-type ($M= 6(5 - 1 - 4) + 4 = 2$), (b) notch-type ($M= 6(5 - 1 - 4) + 4 = 2$), (c) using rod flexures as two axis hinges ($M= 6(3 - 1 - 2) + 4 = 2$).

Combining the rod wobble pins of Figure 7.47, it is possible to create the three-dimensional compound flexure shown in Figure 7.48, for which mobility interpretation is becoming complex and subjective. However, one attribute of this compound design is that the top surfaces of both the base and moving platform are coplanar. Hence, thermal expansion of this mechanism will result in coincident centres and planes for the base and platform. In the absence of loads and for identical flexure supports, this can be thought of as an ideal $xy\theta$ stage.

Another example of a multi-degree of freedom flexure is shown in Figure 7.52 that comprises combinations of both leaf and notch hinges. This flexure is designed to be an xy stage driven by piezoelectric actuators (made from lead zirconate titanate [PZT]) via single bar levers and it is left to the reader to evaluate the purpose of the flexure elements and operation of the mechanism.

It is not possible to leave this section without mentioning two common flexure hinges, these being the cross-strip and cartwheel hinges shown in Figure 7.49. As the names suggest, the cross-strip pivot comprises two or more leaf flexures that are orthogonal to each other creating an axis of rotation where the leafs cross, while the cartwheel is produced by

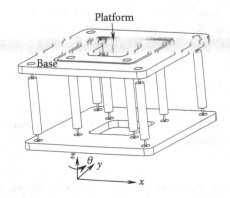

FIGURE 7.48
Three-dimensional compound flexure mechanism.

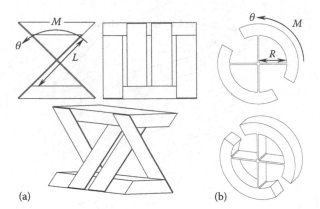

FIGURE 7.49
Leaf spring hinge mechanisms: (a) cross strip pivot, front and side view (top) and isometric view (bottom);
(b) cartwheel hinge, front view (top) and isometric view (bottom).

separating two segments of a wheel so that one of the segments can rotate relative to the other, with the axis of rotation being at the centre (Wittrick 1948, Smith 2000). The angular stiffness values of these two hinges are

$$2k_{cs} = k_{cw} = \frac{M}{\theta} = \frac{4EI}{L_{cs}} = \frac{2EI}{R_{cw}},$$ (7.54)

where the subscripts *cs* and *cw* refer to the cross-strip and cartwheel hinges respectively, R_{cw} is the length of a cartwheel spoke, L_{cs} is the length of the cross strip leaf and *I* is the second moment of area about the neutral axis for the total width of any hinge leg. For example, the cross-strip hinge shown in Figure 7.49 comprises four leafs, with each leg comprising two leafs for which, in this case, the second moment of area is twice the width of an individual leaf multiplied by the leaf thickness cubed divided by twelve.

From the few aforementioned examples, it is clear that this is a broad and actively researched topic and space does not permit a more complete discussion.

Another use of flexible elements is in the design of snap-fit enclosures and assemblies that again represents a field of study extending over many decades. Such applications might be readily appreciated when trying to change the battery of electronic devices for which there are no screws visible from outside of the enclosure. Some resourceful designers have been able to assemble complex machines, such as printers, with only a minimum of screws required.

7.2.5.3 Manufacture of Flexures

Introducing compliance into a material component necessarily requires a localised thinning of the structure. This presents both problems and opportunities for different methods of manufacture, and these are discussed next.

7.2.5.3.1 Machining

Milling and turning represent the most common ways to shape a material. Both methods remove material by physically shearing through it with sharp, hard cutters. This necessarily produces large stresses and forces that will depend upon the size (volume) of the chip being sheared. Reduction of these forces while maintaining productivity can, therefore, be reduced by taking many small cuts as fast as possible. High-speed machining developments over the last few decades as well as research into lightweight monolithic structures has helped develop protocols and models for production of thin-walled sections.

A golden rule for machining of thin or delicate parts is to 'always cut from solid', a simple illustration of which is shown in Figure 7.50. This figure represents a cross-section of the cutting process looking down the axis of the cutter. As illustrated, the web has been machined to produce the lower surface and the cutter is in the process of machining the upper surface of the web using up-cut milling and with the feed being from left to right. It should be apparent from this figure that the chip is being removed from the solid part of the structure and this should in-turn be rigidly clamped into a machining fixture.

Another example of machining from solid material is illustrated by the disk coupling shown in Figure 7.51. This coupling is used to connect two links while providing three degrees of freedom θ_x, θ_y, z. Compliance is provided by the four circular, leaf-type flexures that can be manufactured using a slitting saw (or wire electro-discharge machining, see following section) by cutting four slots perpendicular to the axis of the coupling. The coupling has been machined from a tube with the lower end clamped to the machine table. The first pair of slots should be the upper two shown in Figure 7.51a, so that the material being supported is solid. The second two slots are machined by rotating the tube 90° about the z axis, moving the slotting saw down an appropriate height and slotting the tube as before. Clearly, reversing the sequence of machining would result in the upper pair of slots being machined after the tube has been weakened by the lower slots and is, therefore, 'not as solid'.

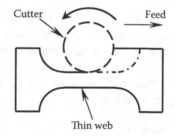

FIGURE 7.50
Milling of a thin web.

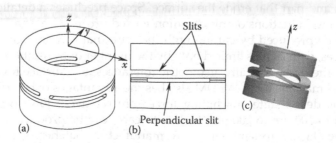

FIGURE 7.51
Monolithic disk coupling, having freedoms z, θ_x, θ_y: (a) isometric view, (b) front view, (c) distortion of the couple when subject to an axial force.

When machining notch flexures, the thin web is often produced by drilling, reaming or otherwise machining two holes closely spaced. However, the drilling or reaming operation produces forces normal to the hole surface that can be sufficient to deform the web. Typically, to prevent this, the first hole can be drilled (and preferably reamed) after which a dowel pin can be inserted to support the web when machining the second hole.

Sometimes it is not possible to machine all of the flexure elements with the remaining material rigidly held by clamps or fixtures. In this case, three approaches can be used. First, it is common that there are surfaces on the rigid links of the mechanism that have no functional use. In these regions, it is often possible to leave a small portion of material bridging across to other rigidly supported links. Once all of the flexures have been machined, these bridges can be gently removed. Alternatively, it might be possible to use proofing alloys, fixturing alloys, optical waxes, adhesives or adhesive tape to temporarily secure the mechanism. Proofing and fixturing alloys are low melting temperature (typically around 70 °C) metals that can be melted and used to fill in pockets and secured after it has solidified. Once the features have been machined into the component, it can be heated to remove it and the alloy can then be reused. Proofing alloys typically have low thermal expansion when solidifying from liquid (at least for a period of many hours), while fixturing alloys have a small positive expansion upon solidification to 'grip' the part.

An alternative to fixturing alloys is to temporarily bond the component into a fixture, typically using optical waxes as an adhesive. Unlike the waxes for candles, numerous optical waxes are available and were originally developed for holding optics during lapping and polishing operations. Typically, these resins have melting temperatures ranging from 60 °C to around 120 °C and have varying strengths, viscosity, adhesive and optical properties. Typically, after the component has been heated and removed, the residual resin can be dissolved with a suitable solvent.

A final method, and probably most common, for holding the mechanism, particularly if batch or large numbers are being manufactured, is to machine custom fixtures. Typically, these fixtures will match the shape of the component at some intermediate stage of manufacture into which the partially machined component is clamped. A well-designed fixture will provide support so that the links on either side and the thin-walled sections of flexure elements will not deflect during machining.

7.2.5.3.2 Wire Electro-Discharge Machining

The two-axis stage shown in Figure 7.52 is designed to be manufactured using wire electro-discharge machining (WEDM). Machining using WEDM uses a thin wire to which high voltage pulses are applied. The wire and component being machined are immersed in a dielectric fluid (typically filtered water or paraffin) and the high fields result in local sparks between the wire and part that erode the surface. Space precludes a detailed discussion of the performance and limitations of this commonly used process; however, the net result is that surfaces can be produced by cutting with the wire that can have diameters down to tens of micrometres, with shapes limited only by the number of translation axes for moving the wire through the part. Typically, geometric tolerances and surface texture are comparable to traditional machine tools. WEDM also has the advantages of being able to machine metals in their hardened state, machining forces are relatively low so that thin-walled structures (down to 100 μm to 200 μm thick with care) can be produced without support and plates can be stacked to simultaneously manufacture batches of components. With suitable finishing cuts, the residual stresses can be negligible, leading to reliable fatigue calculations. Another attribute is the ability to use a U-shaped machining path when cutting material from between links to act as an 'end stop' to prevent over-loading of the flexure

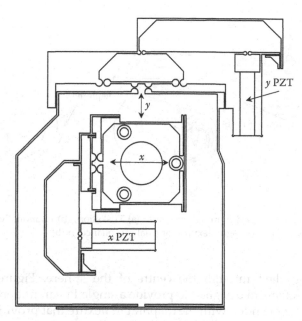

FIGURE 7.52
A levered *xy* stage driven by piezoelectric actuators.

elements. The major drawback of this process is the capital cost of the tool and the relatively slow machining speeds.

7.2.5.3.3 Electrolytic and Wet Chemical Etching

For the manufacture of thin-section flexure components, lithographically based chemical etching must qualify as the highest volume manufacturing tool. Examples of flexures manufactured using variants of chemical etching (including electrolysis, plasma etching and deposition, and microelectromechanical systems [MEMS] processing) include magnetic and compact disk read-heads using electromagnetic drives, MEMS accelerometers and gyroscopes, camera focus and autofocus mechanisms, and virtual reality sensors. Often flexures for these mechanisms are manufactured in production runs in the hundreds of millions. For lower-scale production, there are many companies that can etch patterns using lithographic techniques into thin metal films with almost arbitrary shape, and most will supply comprehensive design guides to determine geometric tolerances. Typically, after paying a relatively small fee for producing the lithographic masks, planar flexures can be produced at costs that reduce with increasing batch size. For these commercial suppliers, geometric tolerances are typically comparable to the thickness of the sheet from which it is etched and limited to around 0.5 mm to 1 mm. There is a large variety of materials available with a correspondingly broad range of heat treatments. Many processes etch the sheet from both sides and this produces a symmetric cusp profile across the etched edge that is called 'feathering', and this is visible by the eye for thicknesses of around 0.4 mm or more.

Figure 7.53 shows two flexures that can be etched from a metal sheet.

Because of its resemblance to a Celtic three-legged design, Figure 7.53a is called a Triskelion flexure and is used to support a coordinate measuring probe comprising a sphere on the end of a rod, the other end of which is attached to the centre of the disk (see Chapter 5, Section 5.6). Suitably designed, this flexure will have the same linear stiffness values in

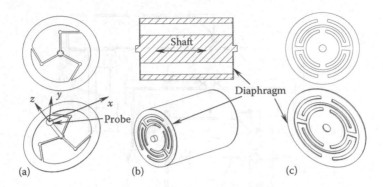

FIGURE 7.53
Two planar flexures that can be etched from a metal sheet: (a) Triskelion, (b) rotationally symmetric flexure for linear motion along the axis, (c) diaphragm flexures for axial compliance in (b).

response to forces applied through the centre of the sphere. Figure 7.53b shows an assembled flexure mechanism designed to provide a single linear motion along an axis that is concentric to an outer cylinder, with the diaphragm flexure that provides the freedom for this mechanism shown in Figure 7.52. Of course, many other designs can be found in the literature on this topic.

Another useful feature of etching processes is the ability to partially etch the sheet. By etching thin channels, it is possible to provide hinge axes in the plane of the sheet and, with large deformation, can also serve as a fold line so that the sheet can be readily folded into three-dimensional shapes.

Having been adapted using semiconductor manufacturing machinery, MEMS processing is now regularly used to manufacture micrometre-scale flexures for applications such as accelerometers, gyroscopes, resonators, atomic force microscope sensors, motor and gear drives, and optical and mechanical switches (Kaajakari 2009). A significant advantage of the MEMS manufacturing method is that electrical circuits and controls can also be integrated with the mechanical device, while a major disadvantage is the high capital cost that makes it only economical for high volume or high value-added devices. Because MEMS processes were developed for the semiconductor industry, most of the components for flexures are made from polycrystalline silicon, silica or (less often) aluminium. While all of the principles and equations for flexure design apply, there are fundamental differences as dimensional scales reduce such as increased strength, reduced self-weight deformation, increased natural frequencies and, therefore, speeds.

Interestingly, many mechanisms use variants of the double compound flexure, although the beams tend to look very slender in comparison to flexures optimised at the more familiar millimetre scales and higher (evident also in nature, think spider legs and elephant legs). Possibly the most significant and ubiquitous application for MEMS accelerometers is to be found in personal communications devices (cell phones) and air bag sensors. Devices are less commonly found in precision machines, although there have been some efforts to use this technology to produce millimetre-scale probe structures similar to the Triskelion probe shown in Figure 7.53 (see Claverley and Leach 2013).

7.2.5.3.4 Electroforming

Electroforming creates thin walled structures by electrochemically depositing a metallic coating onto a shaped form. One particular method is to mould a form in wax followed by

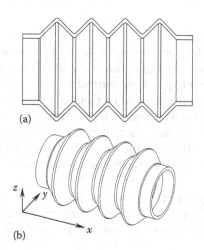

FIGURE 7.54
Bellows coupling providing freedoms $x,y,z,\theta_x,\theta_y,\theta_z$: (a) cross section, (b) isometric view.

electro-less nickel coating. After coating, the wax can be melted and reused. A common form of manufacture by this technique is the bellows coupling that provides five degrees of freedom and has high stiffness to torsional moments about the bellows axis (Figure 7.54). As well as providing a torsion coupling, bellows are commonly used for flexible connections in vacuum applications.

7.2.5.3.5 Additive Manufacturing

Additive manufacturing methods (see Gibson et al. 2014) can be used to make flexures with promising potential for monolithic multi-degree of freedom components. Generally, the polymers (most commonly acrylonitrile butadiene styrene [ABS]) that constitute the most frequently used material for manufacture are not well suited to precision applications unless substantial closed loop controls are included. It is possible that the stability and strength of polymeric mechanisms can be enhanced using more stable fillers and this is being actively researched.

Metal additive manufacturing again looks promising for the generation of interesting flexure mechanism. However, there are substantial technological hurdles with this process for making long, slender structures with differing orientation. Most metal manufacturing machines use a powder bed technology, where the part is made by melting patterns on layers of metal powder. For thin structures, the heat transfer and powder-to-molten liquid interface dynamics cause considerable instability and this is compounded by the residual stresses that remain problematic after more than thirty years of development. In some cases, where the molten layer is supported by powder, a phenomenon of stalactites occurs, where molten liquid penetrates downward into the supporting powder bed. With steadily reducing costs and for manufacture of leaf flexures grown vertically, metal additive processes can produce aspect ratios, minimum thickness and surface finishes that are competitive with WEDM.

Another promising development is the integration of additive manufacturing with other machining processes. These hybrid processes can utilise additive techniques to grow the structure and simultaneously use conventional machining to produce the surface texture and dimensional accuracy necessary for precision flexures.

7.3 Actuators and Drives

Section 7.2 presents methods for constraining solid bodies to motion in specific degrees of freedom. Having created a moving stage, for many instruments and machines it may be necessary to provide an actuator to precisely control the motion. Available sources of energy in most applications are electrical, chemical or mechanical, with electrical energy typically being the most convenient followed by chemical potential utilised by the combustion engine. The first law of thermodynamics teaches that there is a link between mechanical work, electrical work and heat, the latter of which usually dissipates out of the system and frequently shows up elsewhere as undesirable thermal disturbance. The quantity of heat lost is a measure of the process efficiency that contributes to the expense of doing work. However, from a precision engineering perspective, the heat generated in the expenditure of energy is only consequential if it finds its way into the mechanism (see Chapter 11). These and other considerations will be detailed in the discussions of particular actuation techniques in subsequent sections.

An actuator is considered to be the physical means for moving components of a mechanism. In practice, the mechanism for converting an energy source into a force will not be linear and often not repeatable so that a relation between energy provided and displacement cannot be precisely determined. Combining an actuator with a displacement sensor and utilising a suitable control strategy will result in a precise positioning system. Combinations of mechanisms, actuators, sensors and controls are often designed in a modular form for linear or rotational positioning and called translation, or rotary, stages.

Ultimately, the actuator is used to provide work to a mechanism. Doing work fast will increase productivity (always desirable) and therefore not only work but its derivative, power, is of interest to the end user. Another consideration when designing instruments and machines is size, typically with a view to keeping the complete system size to a minimum. In any design, increasing the size of single components, particularly the actuator that is typically near to the core of the machine, often results in larger surrounding components and escalating to a sometimes disproportionate increase in the size of the overall system. Consequently, the work-to-volume and power-to-volume ratios of the actuator are major considerations and are called the work density and power density respectively (Smith and Seugling 2006). Power density is difficult to quantify and often depends on operating conditions and the ability of the process to accommodate the heat generated by inefficiency of the actuator. However, work densities are more readily determined from the product of range of motion and average force divided by the volume of the actuator (see Table 7.3).

7.3.1 Fluid Power (Hydraulic and Pneumatic)

From a precision perspective, the low compressibility of many liquids enables implementation of hydraulic actuators of high efficiency. Lost energy predominantly occurs with the energy source used to generate the high pressures. Hydraulic actuators are most visible as the source that drives mechanisms in many earth moving and other construction machines. The relatively small size of the hydraulic cylinders that can be used to drive large tools makes the high work density apparent. Work done by a pump will typically be stored in an accumulator that comprises a reservoir with an energy storage mechanism, such as a bag of compressible gas (bladder type) or a piston driven by a spring (piston type). Use of an accumulator enables sudden large flow rates over short time periods to be accommodated so that relatively small volume flow rate pumps can be used. Hydraulic pumps commonly

TABLE 7.3

Approximate Work Density Ranges for Different Actuator Types

Actuator Type	$\dfrac{U}{V} \times 10^{-6}$	Equation	Comments
Hydraulic	10	P	Pressure
Shape memory	6	$\sigma\varepsilon$	Cyclic, typically binary
Solid–liquid phase change	5	$\dfrac{\Delta V}{3V}k_{bm}$	Water 8% acetimide
Gas expansion (thermal and pressure)	1	P	Pressure
Thermal expansion	0.5	$\dfrac{E(\alpha\Delta T)^2}{4}$	200 K temperature change
Electromagnetic	0.4 – 0.02	$(BH)_{max}V_m$	Variable reluctance motor $(0.25\ \text{mm})^3$
Electrostatic	0.1 – 0.004	$\dfrac{\varepsilon E_{max}^2}{2}$	Ideal to MEMS comb drive estimates
Piezoelectric	0.05 – 0.01	$\dfrac{(d_{33}E_{max}^2)E}{2}$	PZN – PZT
Muscle	0.02	–	350 kPa at 10%

used include either piston pumps or gear pumps. A piston pump often comprises a number of pistons all moving in the same direction and arranged with their axes forming a cylinder (called an axial piston pump as opposed to the radial piston pump that has similar operational principles and performance but typically occupies a larger physical volume). Motion of all pistons is achieved by a tilted plate that rotates about the axis of the cylinder of the pistons. By changing the tilt angle, different flow rates and pressures can be achieved. This complex pump and control hardware is relatively expensive and the more simple and compact gear pump is a frequently used alternative. The gear pump comprises two meshed gears enclosed in a housing. Rotation of the two gears traps fluid from one side of the housing between the gear teeth and housing surface and transfers the fluid around the outside of each gear to an outlet port. When the fluid arrives in the outlet plenum the meshed teeth of the gears prevent it from getting back to the inlet port resulting in a flow of fluid directly proportional to the rotation of the gears. Gear and piston pumps commonly provide fluids at pressures of around 10 MPa to 30 MPa (1000 psi to 4000 psi) while it is possible to obtain pressures up to 10 times these values. Other pumps such as vane and screw types can provide higher flow rates, albeit at lower pressures, and also share the attribute of flow being directly related to pump speed (piston, gear, screw and vane types all are 'positive displacement pumps').

For standard position control using hydraulic cylinders, the limits on performance depend upon the control of the fluid using flow- and pressure-based control valves. These systems have been extensively developed for control of airplanes and aerospace applications with pressure controls of better than 1 part in 10^4 and flow controls with bandwidths measured in kilohertz being commercially available. Precision positioning typically requires position and pressure feedback. Possibly the most common method for position control of hydraulic cylinders is a flapper-based flow control valve. The flapper is typically controlled using solenoid actuators providing a differential pressure to each end of a spool-type valve that, in turn, provides a channel for pressurised fluid to feed into one side of the hydraulic cylinder while enabling a flow path on the other side of the cylinder to return fluid to a reservoir. Dependent of the position of the flapper restriction of both flows can be controlled thereby enabling both force and velocity control of the piston in the hydraulic cylinder.

There are many designs based on this principle, most of which require elaborate components with high precision tolerances resulting in expensive controls for which applications are predominantly limited to aerospace and advanced manufacturing machines.

Pressure controllers with resolution of a few parts in 100,000 are also available with kilohertz bandwidths at typically hydraulic pressures and can be used to provide large forces for precise control. For example, a force-to-displacement actuator can be made by pushing a stiff flexure with a high force hydraulic piston. One such design was implemented by Tran and Debra (1994) for dynamically controlling a cutting tool during turning and such devices are today referred to as 'fast tool servos'. In another application, pressurized hydraulic fluid has been used to provide pressure to both sides of a uniform cylinder so that the radial stress will induce, via the Poisson's ratio effect, a corresponding strain and, therefore, displacement, along the axis of the cylinder, another example of a Poisson's ratio actuator discussed in Section 7.1.3. The cylinder in this design was also filled by a solid plug and surrounded closely on its outer surface, so that the fluid volume and, therefore, required flow rates were relatively small. This Poisson's ratio actuator is used as the high stiffness control of a novel grinding machine called the Tetraform (Lindsey 1992).

A fine positioning application was developed by Horsfield (1965) for incrementing a scribe in a ruling engine. To create the required fine increments, the cylinder was directly attached to a moving carriage (using dry PTFE bearings on lapped steel datum surfaces) and positioning obtained by injecting small volumes of fluid using a syringe delivery system. With closed loop interferometer control, resolutions of better than 25 nm were demonstrated. Direct use of hydraulic cylinders for positioning with micrometre and sub-micrometre precision does not appear to have been adopted for commercial applications.

For smooth linear actuation, hydraulic cylinders are commonly used to translate the work tables in many conventional and creep-feed grinding machines. For these applications, hydraulic drives have the advantage of providing smooth motions with a broad range of speeds and are capable of delivering high power while housed in a relatively compact space.

7.3.2 Electromagnetic

Electromagnetic drives tend to be at the heart of many precision motion control stages ranging from magnetically levitated lithographic scanning systems to linear and rotational motors driving the slides and spindles in precision machine tools. Diamond turning machines have mostly abandoned conventional feedscrews and now use linear motors with fluid-based temperature control of the windings. Many electromagnetic force actuator designs exploit the force on a charge given by the Lorentz equation:

$$\mathbf{F} = q(\mathbf{E} + \mathbf{B} \times \mathbf{v}), \tag{7.55}$$

where $\mathbf{F}$ is the vector force on a point charge q that is in an electric field of strength $\mathbf{E}$ and magnetic field $\mathbf{B}$ and travelling at a velocity v. Typically, the charge and its velocity are provided by forcing the charge through a wire with an applied voltage, while the fields are often provided by permanent magnets. Actuators directly utilising this relationship include voice coil actuators and dc motors discussed in Sections 7.3.2.1 and 7.3.2.2 respectively.

A related method for producing a force is to use voltages or currents to generate potentials in the form of electrostatic or magnetic fields. Gradients of this potential can then be determined to calculate the force as a function of these voltages or currents. Examples of

this type of actuator include variable reluctance motors, solenoids and electrostatic actuators, and are discussed in Sections 7.3.2 and 7.3.3.

Generally, for electromagnetic actuators it is found that the work density reduces with
size. Consequently, as scale reduces the relative size of the actuator will increase. As an
example, anyone who uses a workshop lathe will have noticed the motor mounted out of
sight somewhere on the floor behind the machine. In contrast, Yuichi Okazaki and co-
workers at the Japanese National Institute of Technology (AIST), in an attempt to make a
suitcase-sized manufacturing facility, built a lathe that could be held in the palm of a hand.
About half of the volume of this lathe was occupied by the motor used to drive the spindle
(Okazaki et al. 2004). In contrast, the slideways for tool positioning utilised piezoelectric
inchworm actuators (see Section 7.3.4) that readily provided the necessary work in a
fraction of the volume.

Before discussing specific actuator designs, a short review of magnetic theory and concepts will provide the necessary background to understand the operating principles and
perform rudimentary calculations. Magnetic fields are typically generated by electrical
currents flowing around a loop, the strength of which can be calculated from the Biot-Savart
equation. The magnetic field **H** (measured in ampere turns per metre, $AT \cdot m^{-1}$) is a vector
quantity and will generate a magnetic flux having a density **B** (webers per metre squared,
$Wb \cdot m^{-2}$ or tesla) at the location of the field where

$$\mathbf{B} = \mu_r \mu_o \mathbf{H} = \mu \mathbf{H} + \mathbf{B}_{rem},$$

where μ_o is the vacuum permeability (= $4\pi \times$ 10-6 $V \cdot s \cdot A^{-1} \cdot m^{-1}$), $\mathbf{B}_{rem} = \mu_o M$ is the residual
magnetisation or remanence, $\mu_r = \mu / \mu_o$ is the dimensionless relative permeability of the space
through which the magnetic flux is passing and M is the intrinsic magnetisation and is a
measure of the magnetic dipole moment per unit volume. For the analysis in this section,
intrinsic magnetisation and magnetic flux densities will generally be coincident and it will
be sufficient to consider only the flux densities. For most calculations it will be assumed that
the remanence is either zero for soft magnetic materials or a fixed value for permanent
magnets. In practice, these rudimentary calculations can provide initial estimates with
subsequent design optimisation usually requiring computational modelling.

In a vacuum, the relationship between magnetic field and magnetic flux density is exact,
while in materials there are numerous complex interactions. To understand the properties
of magnetic materials it is common to measure the flux density as a function of applied field
when the material is initially magnetized and its subsequent response to a cyclic field. A
generic plot, typically called a BH curve, is shown in Figure 7.55. There are a number of
regions of interest that also define key parameters for defining material properties. Measurement starts at the coordinate origin where the slope determines the initial permeability
μ_i of the material. As the applied field increases, the increase in flux density reduces until it
might be considered negligible and is considered to be the saturation flux density. In
practice, the total flux density will continue to increase even though the intrinsic magne-
tization of the material may have reached a limit. Reducing the applied magnetic field from
the saturation value will correspondingly result in a reduction in flux density albeit with a
lag that results in hysteresis. When the applied field is reduced, the remaining flux density is
called the remanence B_{rem} of the material. It is only after the field has been reversed by an
amount H_c, called the coercive force, that flux density is reduced to zero.

Typically, the initial BH curve is used to evaluate the permeability of materials, with
typical curves for the 'soft' magnetic materials (low carbon steel, cast iron and ferrite) are

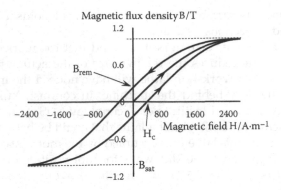

FIGURE 7.55
Generic *BH* curve for a 'soft' magnetic material.

shown in Figure 7.56. Figure 7.56a includes the characteristic curves for two soft magnetic irons—low carbon steel and cast steel—plotted against the magnetic field using a logarithmic scale. It is immediately apparent that the cast steel material has a substantially higher initial permeability and is preferred for larger electromagnets. However, at higher fields, a simple low carbon steel will provide an increased flux density and permeability.

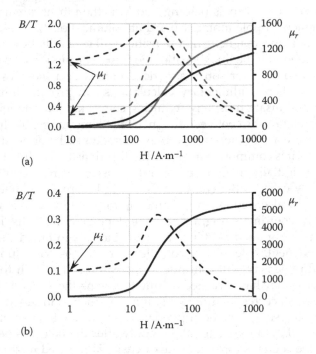

FIGURE 7.56
BH curves and permeability plots: (a) solid line is flux density and dashed line relative permeability for cast steel (black) and low carbon steel (dark grey), (b) *BH* curve and permeability plot for ferrite.

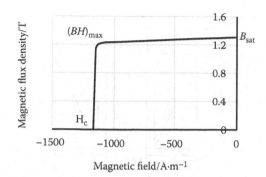

FIGURE 7.57
Left upper quadrant of the *BH* curve for an NdBFe permanent magnet.

Ferrite is a non-electrically conductive material that can be used at high frequencies without power losses due to induced eddy currents. It also has a significantly higher permeability at lower magnetic fields, which ultimately limits the material to lower flux density values compared to most (non-stainless) steels (see Figure 7.56b).

An ideal permanent magnet would have a rectangular *BH* curve bounded by the coercive strength horizontally and saturation flux density at top and bottom. Modern permanent magnet materials such as samarium–cobalt and neodymium–boron–iron show a reasonable approximation to this ideal with saturation flux densities of this latter material typically being 5% to 20 % higher. Many alloy compositions of these and other magnet materials are commercially available and a typical quadrant of NdBFe magnets is in Figure 7.57.

Numerous innovative electrical machines have been invented over the last century and further back. As a consequence only basic principles will be outlined and their applications in precision instruments and machines. More detailed design techniques can be found in Hadfield (1962), Campbell (1994), and Schmidt et al. (2011).

7.3.2.1 Solenoids and Variable Reluctance Actuators

Solenoids are frequently used for binary operations such as fluid control valves or electrical switches (called relays). However, with a feedback sensor and suitable controller, solenoid type actuators can be used for precise motion control. Consider the solenoid actuator shown in cross-section in Figure 7.58 comprising a central, magnetically permeable plunger surrounded by a non-magnetic spool that enables the plunger to slide freely in the vertical direction and provides a pocket for winding a coil of N turns. Both the plunger and spool are assembled in an outer body also made of a magnetically permeable material. Clearance between the outer solenoid body and the plunger where it emerges at the top results in a small air gap x_g. Displacement y is measured from the bottom face of the plunger to the surface of the outer body.

Passing a current I through the coil of length l will result in a magnetic field (analogous to an electric field) along the coil of magnitude:

$$H = \frac{NI}{l}.$$ (7.56)

The flux produced by this field must flow around a closed path and will predominantly flow through the path of least resistance determined by the permeable components of the

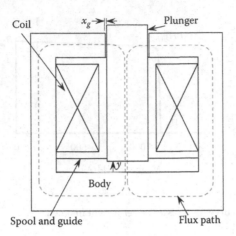

FIGURE 7.58
Cross section of a simple solenoid actuator.

device. Because the magnetic fields are in the form of potentials, their net effects will superpose. As a consequence a coil of N turns will produce a net magnetomotive force (mmf) $\mathfrak{I}$, which is the product of the magnetic field multiplied by the length of the coil:

$$\mathfrak{I} = NI = Hl. \tag{7.57}$$

Because flux is a conserved quantity, the overall path can be modelled as having a total magnetic flux resistance, or reluctance, $\mathfrak{R}$, resulting in a magnetic circuit diagram of the form shown in Figure 7.59. Because the flux path always follows a loop, the total reluctance will be the series sum of reluctances around the loop given by

$$\mathfrak{R} = \frac{l_b}{\mu_b A_b} + \frac{l_p - y}{\mu_p A_p} + \frac{x_g}{\mu_o A_g} + \frac{y}{\mu_o A_p} = \mathfrak{R}_o + y\left[\frac{1}{\mu_o A_p} - \frac{1}{\mu_p A_p}\right]$$

$$\approx \mathfrak{R}_o + \frac{y}{\mu_o A_p}, \tag{7.58}$$

where the subscripts p, b, and g refer to the plunger, body and gap respectively; l is the effective length of the particular flux path; A is the effective cross-sectional area of the flux

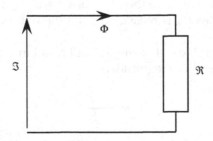

FIGURE 7.59
Circuit representation of the magnetic field, reluctance and flux relation for a solenoid.

path; and it is assumed, precisely for frequencies of mechanical systems, that the permeability of air is the same as that for a vacuum. At any point, the total flux Φ is equal to the integral of the flux density through a surface spanning the flux path. By definition the coil inductance L, which is a measure of the energy stored in the magnetic field as a consequence of the current, is equal to the amount of flux coupled to each turn per unit current, that is

$$L = \frac{N\Phi}{I} = \frac{N\mathcal{J}}{I\mathfrak{R}} \approx \frac{\mu_o N^2 A_P}{\mu_o A_P \mathfrak{R}_o + y} = \frac{\mu_o N^2 A_P}{\frac{l_b A_P}{\mu_{rb} A_b} + \frac{l_p}{\mu_{rp}} + \frac{x_g A_P}{A_g} + y}, \tag{7.59}$$

where μ_{rp} and μ_{rb} are the relative permeability of the plunger and body respectively. Considering the magnetic field and a source of potential energy of magnitude $LI^2/2$, the force along the axis of the plunger is

$$F_y = \frac{\partial L}{\partial y} = -\frac{I^2}{2}\frac{\mu_o N^2 A_P}{(\mu_o A_P \mathfrak{R}_o + y)^2} \approx -\frac{I^2}{2}\frac{\mu_o N^2 A_P}{y^2}. \tag{7.60}$$

The last term of Equation 7.60 is based on the assumption that the plunger gap and the relative permeability of the plunger and body materials is large, and that the gap at the top of the solenoid is small with respect to the plunger-to-body separation. Being based on the change in reluctance with separation, this type of force actuator is often called a variable reluctance actuator. When the plunger contacts the body, the force is equal to $N^2 I^2 / 2\mu_o A_P \mathfrak{R}_o^2$. Substituting Equation 7.56 into the last term of Equation 7.60 provides an alternative form for the force given by

$$F_y \approx \frac{I^2}{y^2}\frac{B^2}{2\mu_o}A_P. \tag{7.61}$$

Figure 7.60 shows a plot of the force on a solenoid plunger as a function of both current and separation for the more complete model in Equation 7.60, with the approximate model being plotted for a current of 0.2 A and shown by the grey dashed line. Parameters for this model are given in Exercise 15.

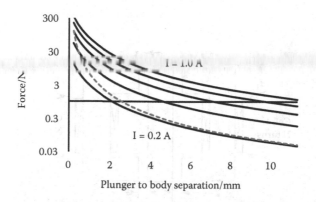

FIGURE 7.60

Plunger force as a function of plunger to body separation for coil currents of 0.2, 0.4, 0.8 and 1.0 A. Grey dashed line shows approximate theory equation (Equation 7.61).

Equation 7.61 can be considered as two terms: one involving the geometry and properties of the solenoid, another by a product of magnetic stress (being attractive) in the air gap multiplied by the area over which is spans. For this particular approximation, it might be considered that the mmf across the gap is equal to that generated by the coil. In this case, the length *l* is equal to the plunger separation, and Equation 7.61 reduces to a simple product of magnetic stress and area. Generally the magnetic forces can be determined from this magnetic pressure that more rigorously should be represented as a vector quantity called the Maxwell stress tensor.

Numerous actuators have been designed based on variable reluctance and a useful implementation is the differential actuator, shown in Figure 7.61, that uses two coils of the same number of turns to create a push–pull effect. Modelling this actuator similarly to the solenoid and ignoring the reluctance of the flux path materials, the force on the moving armature can be calculated from

$$F \approx \frac{N^2 A_g \mu_o}{4} \left[\frac{I_2^2}{(x_o + x)^2} - \frac{I_1^2}{(x_o - x)^2} \right], \tag{7.62}$$

where x_o is the total air gap, A_g is the area of the gap.

It is possible to utilise permanent magnets to provide bias fields in actuator circuits. An important parameter is the field produced by the magnet. In an interesting thought experiment, an ideal permanent magnet is placed in one box, while a coil with a constant current passing through it is placed in another box. There have yet to be any experiments that can be performed to identify which box contains the coil. Hence, an ideal magnet can also be considered as an air (or vacuum) core coil so that its equivalent circuit will be

$$\Phi = \frac{\Im_m}{\mathfrak{R}_m} = B_{rem} A_m, \tag{7.63}$$

from which

$$\Im_m = B_{rem} A_m \mathfrak{R}_m = B_{rem} \frac{l_m}{\mu_o} = M_m l_m. \tag{7.64}$$

Hence for the purpose of modelling magnets, an ideal magnet can always be replaced by an equivalent coil mmf and vacuum reluctance computer from its geometry. Consequently,

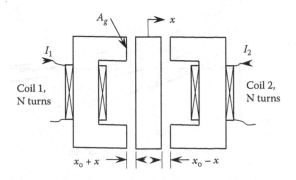

FIGURE 7.61
Differential variable reluctance actuator.

to maximize flux around a circuit, an ideal magnet will have a large area and short length. It will be assumed that applied fields do not approach the coercive strength of the magnets. One such design that exploits the field generated by a permanent magnet is shown in Figure 7.62 and is adapted from Kluk et al. (2012). The design comprises a rectangular loop of magnetically permeable material broken at the centre of the bottom leg and with a moving armature placed in the gap. Coils on either side of the vertical arms can be used to generate a flux, for which the easiest flow path is to circulate around the loop. In the centre of this loop is a permanent magnet connected to the upper leg of the loop, with the other end being close to the moving armature. Flux generated by the permanent magnet will flow across the small, non-working gap and through the armature, where it will split to also flow around the permeable loop. Between the armature and the loop are two gaps, referred to as left and right working gaps, that provide the magnetic pressure to create the actuation force. For the design presented in Kluk et al., the coils are connected and arranged so that a common current passing through both coils will create a circulation of the flux around the loop with the flow direction being dependent on the direction of current. As shown, the flux due to excitation of the coils is of opposite sign to that of the magnet in the right gap and will add to the flux in the left gap. An equivalent circuit for this actuator is shown in Figure 7.63 where Φ, $\mathfrak{I}$ and $\mathfrak{R}$ represent the flux, mmf and reluctance, with the subscripts r,l,n,m,c indicating the right and left components of the loop, the non-working region, and the magnet and coils

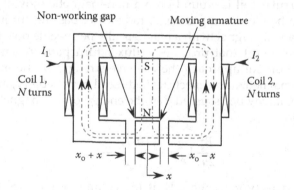

FIGURE 7.62
Differential variable reluctance actuator with permanent magnet bias.

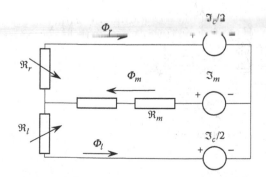

FIGURE 7.63
Equivalent circuit for the actuator of Figure 7.62.

respectively. For this design, Kluk et al. show that the force in the direction of the armature motion for small changes in gap about the central position is given by

$$F = \frac{2ST}{\mu_o A_g},$$ (7.65)

where A_g is the area of all three gaps, and the constant S and variable T are given by

$$S = \frac{\mu_o A_g \mathfrak{I}_m}{2(l_m + l_n)},$$

$$T = \frac{\mu_o A_g \mathfrak{I}_m}{2x_o(l_m + l_n)}\left[\frac{\mathfrak{I}_m}{(l_m + l_n)}x + \mathfrak{I}_c\right].$$ (7.66)

For this design, it is apparent that, for small perturbations about the central position of the armature, the force is linear in both displacement and coil current.

7.3.2.2 *Voice Coils*

Voice coils exploit the Lorentz force produced on a wire that is carrying an electrical current and located in a magnetic field and are thus named because many designs are similar to the actuator for driving loud speaker cones. A simple voice coil design is shown in Figure 7.64. For this design, a circular coil is wound onto a non-magnetic moving element that, for a loud speaker, would be the moving cone. A radial flux passing through the windings is produced by a permanent magnet, with a magnetically permeable pole piece at its top and a similarly permeable support that provides a flux return path. Assuming that the fields produced by the coil do not approach the coercive strength of the magnet, ignoring, for now, flux leakage demagnetization effects (called recoil permeability) and other mmf losses around the flux path, it may be assumed that the energy of the magnet will be available in the gap and, therefore,

$$\frac{B_g^2}{2\mu_o}V_g = k_1^2\frac{(BH)_{max}}{2}V_m,$$ (7.67)

where B_g is the flux density in the gap, V_g is the volume of the gap, V_m is volume of the magnet, $(BH)_{max}$ (J·m^{-3}) is the maximum energy per unit volume of the magnet and $k_1 < 1$ is

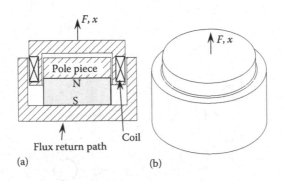

FIGURE 7.64
Voice coil actuator: (a) cross section, (b) isometric view.

a loss factor that, in the absence of a more complete model, would be determined experimentally and would be a function of the reluctance of the gap. Equation 7.67 can be used to determine the flux across the gap. Because the flux will be directed perpendicular to the windings in the coil, removing the electric field component from the Lorentz force equation gives directly

$$F = 2N\pi\bar{r}B_g I = 2N\pi\bar{r}k_1 \left((BH)_{max} \frac{V_m}{V_g} \right)^{1/2} I, \tag{7.68}$$

where $\bar{r}$ is the mean radius of the coil and N is the number of turns that are within the flux field of the gap. From this model, it is apparent that for optimum performance the magnet volume should be as high as possible while the volume of the gap must be reduced to a minimum. Having a large volume of magnet makes this part of the actuator heavy and is why a moving coil actuator is more common than moving magnet implementation. Increasing the number of turns in the coil is also at odds with the desire to minimise the volume of the gap and therefore can only be maximised by reducing the diameter of the wire in the winding (in turn limited by the maximum current density for a particular wire). However, as the diameter of the winding reduces, the resistance of the coil increases, and this will generate heat that is undesirable from a precision perspective. Clearly, design optimization requires consideration of numerous parameters and will depend on the application. An important feature of Equation 7.68 is the linearity of the force with current so that it can provide both push and pull forces, and force resolution is limited by resolution of the control of the coil current.

Absent from Equation 7.68 is any relation between force and displacement and any forces in other coordinate directions. In practice, for very large displacement, the coil will leave the flux region and fewer turns will interact with the field, thereby resulting in a reduced force. However, because the force is only perpendicular to the winding, there will not be forces in other directions. Consequently, provided that the coil is centred axially in the flux field, this actuator has a very low stiffness in all coordinate directions and can often be precisely modelled as a pure force in a specific direction.

X-ray interferometry represents a particularly challenging field of study requiring translation of a single crystal silicon plate with sub-nanometre parasitic deviations from rectilinear motion and rotations limited to less than a few nano-radians in some axes. Implementation of this interferometer is typically achieved by machining the X-ray diffractive elements into a single crystal of silicon followed by cutting away of the support of the moving element to form a flexure (Hart 1968, Basile et al. 2000). This single crystal silicon flexure is then driven by either a piezoelectric actuator or a voice coil design. One interferometer design utilised a moving magnet actuator (Smith and Chetwynd, 1990) and was capable of translating the interferometer distances of 100 nm or more with control at 10 pm precision (Bowen et al. 1990).

For longer-range translations voice coil designs, using numerous coils and periodically poled magnet arrays are common and included as part of the discussion of motors in the following section.

7.3.2.3 Motors

To create a rotational force it is possible to combine variable reluctance and permanent magnet force actuators placed symmetrically about an axis. Figure 7.65 shows a three-pole

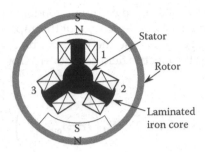

FIGURE 7.65
Three-pole synchronous motor.

motor with a two magnetic poles at 180°. In the current state, energization of coil 1 will create a flux density that will interact with that of the magnet, resulting in an increase or reduction depending upon the direction of current through the coil. If there is an increase in flux density, this will create a pressure to hold the rotor in the position shown. Depending on the desired direction of rotation, to rotate the motor, either coil 2 or coil 3 can be energized to attract the south pole of the magnet to align with the coil. To increase torque, the other two coils can be energized to repel the pole on the other side of the rotor. Consequently, it is apparent that a minimum of three coils are needed as well as a measurement of the position of the rotor magnetic field relative to the coils. Typically, the field of the permanent magnet is measured using three magnetic sensors often based on the Hall effect. In general, driving the coils with and ac current with each phase being at 120° will result in a constant speed rotation of the motor and such an implementation is called a three-phase motor. However, in this case the residual magnetism of the soft iron core will result in some variations of rotational speed called cogging. Cogging can be reduced by adding more magnets as further sets of three phase designs, such as the twelve-coil motor shown in Figure 7.66.

A linear motor can be thought of as a multi-pole motor flattened out. In this case, a simple linear motor would comprise a periodic array of coils with three typically being a minimum to implement three-phase control (Figure 7.67). Because of the iron core there will be an attractive force between the linear magnet array and the forcer (the component housing the coils and Hall effect sensors) that is typically attached to the machine frame. While this force might be undesirable in some circumstances, it is often useful for preloading bearings (particularly air bearings) and for operating in different orientations. An advantage of

FIGURE 7.66
Disassembled twelve-pole synchronous motor with laminated iron cores.

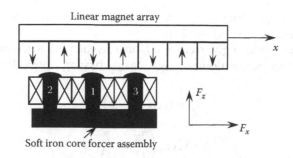

FIGURE 7.67
Iron core, three-pole linear motor.

magnetic stages is that all of the wire feeds go to the coils and these are often attached to the stationary frame that can also provide cooling. Another attribute is that the actuator can be relatively compact, leading to low height linear stages. A cost of moving the magnets is the larger mass that these add to the moving stage. Consequently, when dynamic response is a priority, moving coil designs are often chosen with additional design effort to provide a moving wire harness and, sometimes, cooling. As for the rotational motor, the iron core linear motor will also have some cogging.

It is apparent that the magnet array shown in Figure 7.67 produces equal flux patterns on either side of the array and consequently only 50% of the available flux is utilised with this design. A more efficient use of the magnets is obtained by rotation of the polarisation as a function of position; an idea originally presented by Halbach in 1980. Figure 7.68 shows a three-pole linear motor using a simple, five-magnet Halbach array for which the majority of the magnetic flux is diverted to be from the lower face of the array. While there are many configurations that approximate the ideal, even this simple arrangement can increase forces by around 1.4 and, therefore, the power efficiency by a factor of almost 2 (Trumper et al. 1993). Additionally the flux variation more closely corresponds to a sinusoid as a function of distance along the array and the field on the back face is low and therefore shielding is easier or not necessary, so that the array can be supported by lighter and more stable materials such as ceramics.

Given the increased flux from the Halbach array, ironless (or air-core) coils can often supply sufficient force for dynamic positioning for lower load applications, albeit with forces of about 20% to 30% of an iron core motor of equivalent size. One advantage is that this implementation does not have cogging problems and often provides a smoother drive

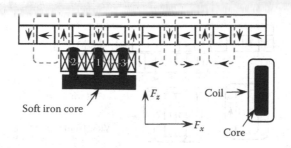

FIGURE 7.68
Three-pole linear motor with Halbach magnet array.

force. Another advantage is that flat 'pancake' coils can be used leading to an even more compact actuator, as shown in Figure 7.69.

For a sinusoidal flux and for a given current I force on each coil in the direction of motion as a function of the displacement x along a magnet array having a periodic length D is given by

$$\frac{F_1}{I} = C \cos\left(\frac{2\pi}{D}x\right),$$

$$\frac{F_2}{I} = C \cos\left(\frac{2\pi}{D}x + \frac{2\pi}{3}\right), \tag{7.69}$$

$$\frac{F_3}{I} = C \cos\left(\frac{2\pi}{D}x + \frac{4\pi}{3}\right),$$

where C is the force constant common to all coils and given by the product of flux density and the length of coil through which the flux passes. The force indices indicate the coil number and the phase of the displacement is measured using three magnetic (Hall effect) sensors. It can be readily shown that the force generated by these coils sums to zero. One method to derive a linear force is to make the current to a nominal current I_o multiplied by the cosine terms in Equation 7.69. In this case, the force from each coil becomes

$$F_1 = I_o C \cos^2\left(\frac{2\pi}{D}x\right) = I_o \frac{C}{2}\left[1 + \cos\left(\frac{4\pi}{D}x\right)\right],$$

$$F_2 = I_o C \cos^2\left(\frac{2\pi}{D}x + \frac{2\pi}{3}\right) = I_o \frac{C}{2}\left[1 + \cos\left(\frac{4\pi}{D}x + \frac{4\pi}{3}\right)\right], \tag{7.70}$$

$$F_3 = I_o C \cos^2\left(\frac{2\pi}{D}x + \frac{4\pi}{3}\right) = I_o \frac{C}{2}\left[1 + \cos\left(\frac{4\pi}{D}x + \frac{8\pi}{3}\right)\right] = I_o \frac{C}{2}\left[1 + \cos\left(\frac{4\pi}{D}x + \frac{2\pi}{3}\right)\right].$$

Summing these forces cancels the cosine terms leaving a constant force of $3I_o C/2$ that can be reversed by changing the sign of the current.

A common linear motor design utilises a U-shaped channel that provides a flux return path and has permanent magnet arrays on the inside of both legs of the U. The three-phase forcer coils are then placed in the channel of the U. The resulting symmetric design efficiently utilises the available magnetic flux, and the motive force is purely along the axis of the channel.

Not mentioned in the preceding descriptions is the fact that the force vectors in many of the linear motors will change with position. Consequently, suitably designed linear motors are also capable of producing combined forces in the linear drive direction and in

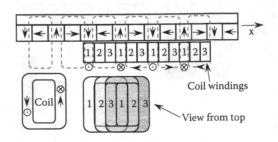

FIGURE 7.69
Air core three-phase linear motor using flat pancake coils. Coil shape is shown at the bottom of this figure.

perpendicular axes to create multi-degree of freedom translation and rotation stages. Such magnetically levitated stages (or 'maglevs') are used in lithography and other tools to create motion control in multi-degrees of freedom with sub-nanometre accuracy and are driving the limits of motion control accuracy (Munnig Schmidt 2012).

7.3.2.4 Stepper Motors

By changing the pole pieces of the three-phase motor shown in Figure 7.65 to include a series of steps, and using a magnetically permeable material for a similarly castellated linear platform, the reluctance is substantially minimised when peaks of the castellations coincide. Energization of any particular coil will produce a variable reluctance-based force that will tend to align the high points of the castellated profiles. By turning off one coil and energizing another, the linear platform can be forced to move in either direction by one-third of the period of the castellation (Figure 7.70). Because coil energization will result in a discrete step, this type of drive is called a stepper motor. In practice, this step action often results in unacceptable transmission of vibration. To attenuate vibration effects, circuits are commercially available that use a sinusoidal drive to the coils, similar to a three-phase amplifier, to provide positioning capability between each step, a technique commonly called microstepping. Consequently, typical castellation distances can be in the range of around 1 mm and micro-stepping controllers are capable of sub-dividing factors of 10 to 100, resulting in substantially smoother motion. Further improvements in driver technology are still needed to reduce the cogging and vibration levels to those achievable by the linear motors of the previous sections. A drawback with stepper motors is the need to maintain a coil current when the platform is stationary and, therefore, generating a continuous heat source.

7.3.3 Electrostatic Drives

Two parallel plate electrodes separated by an insulating medium (considered to be a vacuum for now) and subject to an electric field will result in a total charge density, D (C·m^{-2}), given by

$$D = \varepsilon_0 E, \tag{7.71}$$

where E (V·m^{-1}) is the component of electric field normal to the electrode surface and ε_0 is the permittivity of a vacuum (8.854×10^{-12} F·m^{-1}). Generally, for the dynamic frequencies of most mechanical systems and in this discussion the permittivity of air is considered to be same as that for a vacuum in this discussion. If the electrodes are closely separated, the electric field in the gap will be of uniform density and the fields outside will be considered

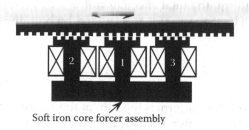

Soft iron core forcer assembly

FIGURE 7.70
Schematic diagram of a linear stepper motor.

negligible. For an electrode area A, separation x and noting that the charge q on an individual plate is half of the total charge density multiplied by the electrode area, the electric field and charge are given by

$$q = \frac{AD}{2} = \frac{\varepsilon_o E}{2},$$

$$E = \frac{V}{x}.$$

(7.72)

Using Equation 7.72, the Lorentz force, considered to act in the negative direction on each electrode, is given by

$$F = qE = -\frac{\varepsilon_o A}{2}\left(\frac{V}{x}\right)^2 = \frac{\partial U}{\partial x}.$$

(7.73)

Integrating Equation 7.73 and ignoring any constant value, the potential energy U is given by

$$U = \frac{\varepsilon_o A}{2x}V^2 = \frac{1}{2}CV^2 = \frac{\varepsilon_o A x}{2}E^2 = \frac{\varepsilon_o V_C}{2}E^2,$$

(7.74)

where C is the capacitance measured in farads (F) and V_C is the volume of the capacitor (m^3). The work density for such an actuator can be determined from the last of Equation 7.74 and is maximum when the electric field is maintained at the field strength E_{max} of the insulating material and is independent of the size of the actuator. Typically air tends to break down at a few volts per micrometre with one volt per micrometre often considered a conservatively 'safe' field level. An important attribute of this actuator is its independence of capacitance in any coordinate direction. Consequently, the electrostatic actuator of suitable geometry will have zero stiffness in off-axis directions and, as with the electromagnetic actuators of the previous section, can be considered as a zero stiffness actuator.

Consider a simple four-bar flexure of stiffness k with a parallel electrode, electrostatic actuator of area A and separation x_o in the absence of a potential difference V (= 0). After application of a voltage to the electrodes, both the platform and the actuator will experience the same force resulting in a displacement x of the moving platform and a corresponding reduction in electrode separation. Hence equating the two forces gives

$$F = -\frac{\varepsilon A}{(x_o - x)^2}V^2 = kx.$$

(7.75)

In terms of the upper platform, the total effective stiffness k_e, given by the derivative of force with displacement, will be the sum of that provided by both the flexure and actuator so that

$$k_e = k - \frac{2\varepsilon A}{(x_o - x)^3}V^2.$$

(7.76)

Consequently, the stiffness of the platform is the sum of the positive stiffness of the flexure plus the negative stiffness of the actuator. Clearly, for small forces the stiffness will remain positive and the platform can be readily controlled. However, above a critical value where

$$\frac{V^2}{(x_o - x)^3} \ge \frac{k}{2\varepsilon A},$$

(7.77)

the system will have zero or negative stiffness, after which the platform will accelerate toward the stationary electrode and often stick to the surface after contact. This sticking is because any finite voltage when the electrodes are in the contacted condition will produce a theoretically infinite electric field and, therefore, force. Such a phenomenon is called 'jump to contact' and can be problematic in MEMS and some fine instrument designs.

In contrast to electromagnetic actuation where work density reduces with scale, electrostatic actuators have found many applications in MEMS, where a voltage source is readily available. Force is provided by an array of interleaved parallel plate electrodes called a 'comb drive'; see Figure 7.71 which shows two combs with interwoven 'teeth'. In practice one of the combs will be attached to a frame while the other is a monolithic part of a linear flexure guideway (not shown in the figure) with linear motion being in the z direction. Typically, the 'teeth' of the comb will produce capacitance, with the majority of the stored energy being in the small gap between the vertical walls. If the two interleaved combs are perfectly centred, the gap h_o between the vertical walls will be the same on each side. For the parameters shown in Figure 7.71 the total capacitance C for a comb having N teeth can be approximated by

$$C = \varepsilon_o b N (z_o - z) \left[\frac{1}{h_o + x} + \frac{1}{h_o - x} \right], \tag{7.78}$$

where b is the depth of the teeth, ε_o is the assumed value of permittivity for an air or vacuum dielectric and z_o is the nominal depth of the interweave of the teeth. For a potential difference V applied across the two combs, the forces in the drive direction z and laterally x are given by

$$F_z = \frac{\partial U}{\partial z} = \frac{V^2}{2} \frac{\partial C}{\partial z} = \frac{-\varepsilon_o b N}{h_o \left(1 - \left(\dfrac{x}{h_o} \right)^2 \right)} V^2,$$

$$\tag{7.79}$$

$$F_x = \frac{\partial U}{\partial x} = \frac{V^2}{2} \frac{\partial C}{\partial x} = \frac{4 \varepsilon_o b N (z_o - z)}{h_o^2 \left(1 - \left(\dfrac{x}{h_o} \right)^2 \right)} \left(\frac{x}{h_o} \right) V^2.$$

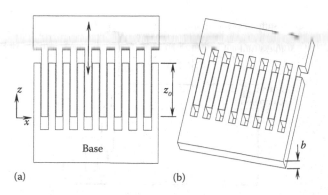

FIGURE 7.71
Comb drive having eight teeth: (a) front view, (b) isometric view.

From the first of Equation 7.79, it is apparent that the drive force is independent of both the displacement and the overlap length $(z_o - z)$ between the comb teeth. In practice, if the overlap length is comparable or smaller than the width of the teeth l_s, fringing fields will significantly impact the capacitance. Forces perpendicular to the drive direction, given by the second of Equation 7.79, indicate an initially linear increase with misalignment in the x direction as well as a linear dependence on the overlap length. Clearly, this undesirable lateral force will be minimised with good alignment and minimal overlap. Both force components increase with the number of comb teeth.

The expression for potential energy in Equation 7.74 is exact when the capacitance value between the two electrodes is used. Consequently, this provides a means to generate forces of known value based only on a knowledge of the capacitance and its variation with displacement in a specific direction. Because of the relatively small value for vacuum permittivity, most electrostatic actuator applications require small force generation, as is the case for electrostatic force balances for small mass and low force calibration (Shaw et al. 2016). To illustrate some interesting design principles that have been utilised, a minimal schematic representation of a force balance is shown in Figure 7.72, with more detailed and complete descriptions of the design assembly and operation being described in Pratt et al. (2002, 2005). The balance comprises a four-bar linkage with cross strip pivots as the hinge elements. An electrostatically generated force is applied to the bottom of the moving platform in response to the force to be calibrated, that can be either upwardly or downwardly directed, being applied to the upper surface. Two concentric electrodes form the actuator on the lower side of the platform. Vertical motion z_m of the platform is measured using a laser interferometer. A counterbalance is used to move the platform to a nominal position with a voltage V_1 being used to apply a bias force. To determine the gradient of capacitance with respect to the vertical motion of the platform, the capacitance between the two electrodes is measured as it is deflected over a short range using a second force actuator (not shown). After the force to be calibrated has been applied, the balance is restored to its initial nominal position and the subsequent voltage V_2 recorded. Hence this is a null type measurement, where the system is nominally in the same state before and after the force is applied to within the resolution of the null control. For more sensitive experiments, this apparatus is operated in a vacuum chamber either with air evacuated or with a controlled gas environment.

Using this technique, masses of up to 10 mg can be measured with deviations between measurements by different researchers around the world being typically less than 1.5 μg. However, to achieve this, two further innovations were necessary. Experimentally it was

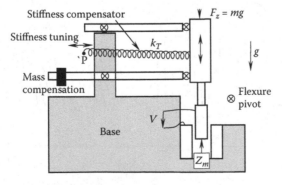

FIGURE 7.72
Schematic diagram of an electrostatic force balance.

determined that due to surface effects and electrical potentials developed in the wiring between the electrodes and measuring circuits, a constant bias voltage V_s would be present with each measurement. To eliminate this as far as possible, a separation method was implemented. This is possible because the force is given by the square of the potential. Hence the same force is generated independent of the polarity of the applied voltage. However, the additional potential will add for one polarity and subtract for the other. Hence by repeating the measurements for both polarities and using the mean value, this bias can be separated from the measurement (Pratt et al. 2005).

A second major limitation is the uncertainty of the force due to the uncertainty of the null location. For a given position uncertainty, the corresponding force uncertainty is directly proportional to the stiffness of the balance. To reduce this stiffness, an interesting compensated spring design is used, and shown schematically in Figure 7.72, as a long coil spring. To understand how this works, consider an equivalent (although artificially simplified) model shown in Figure 7.73. One leg of the four-bar linkage is shown by the line of length l_1 that has a stationary pivot at the left end and connects to the moving platform represented by the vertical line at the other end. Movement of the platform in the vertical direction a distance z results in a rotation ϕ of the four-bar linkage support leg. The coil spring is represented by a straight line and shown to be connected to another stationary pivot at `P that is collinear with the leg of the four-bar linkage in its initial position. The other end of the coil spring connects at the moving platform at a location that is coincident with the linkage pivot. When the platform moves a distance z, the axis of the coil spring rotates an angle θ and shortens from a length $l_o + l_T$ to a length l_3. The length of the coil spring at $z = 0$ is considered to comprise its free length l_o plus its elongation l_T that produces a tension is the spring of magnitude $k_T l_T$. Clearly, by varying any of these lengths, different values for the ratio of the free length to spring extension can be obtained, making this a tunable parameter.

To determine the effective stiffness of this mechanism it is necessary to develop the geometric relationships between these parameters as a function of the platform motion z. Given the model in Figure 7.73, the sum $l_o + l_T$ is a constant and is defined by the parameter l^*. As a first step, consider that the foreshortening effect Δx due to the arcuate motion of the moving platform for small platform displacements is given by

$$\Delta x = l_1(1 - \cos \phi) \approx \frac{z^2}{2l_1}. \tag{7.80}$$

Using this, the length of the coil spring l_3 when the platform is displaced can be approximated from

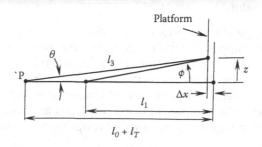

FIGURE 7.73
Mathematical model of the spring compensator.

$$l_3 = \sqrt{l^{*2} - 2l^* \Delta x + (\Delta x)^2 + z^2} \approx l^* \sqrt{1 - \frac{1}{l^{*2}}\left(\frac{l^*}{l_1} - 1\right)z^2},$$

$$\approx l^* - \frac{1}{l^*}\left(\frac{l^*}{l_1} - 1\right)z^2 + O(z^4).$$

(7.81)

Ignoring fourth order and higher terms, the potential energy in the spring as a function of displacement of the platform is

$$U = \frac{K_t}{2}\left(l_T - \frac{1}{l^*}\left(\frac{l^*}{l_1} - 1\right)z^2\right)^2.$$

(7.82)

Differentiating Equation 7.82 gives the force due to the coil spring

$$F_z = \frac{\partial U}{\partial z} = K_t\left(-\frac{1}{l^*}\left(\frac{l^*}{l_1} - 1\right)\right)\left(l_T - \frac{1}{l^*}\left(\frac{l^*}{l_1} - 1\right)z^2\right)z$$

$$\approx -K_t\left(\frac{l_T}{l^*}\left(\frac{l^*}{l_1} - 1\right)\right)z.$$

(7.83)

Hence, to a first-order approximation, the coil spring contributes a negative stiffness that will subtract from the negative stiffness provided by the four cross strip pivots of the four-bar linkage. An additional negative stiffness can be obtained by arranging for the centre of mass of the rotating links to be higher than the pivot constituting an inverted pendulum. The stiffness of an inverted pendulum is the same magnitude as that provided by a regular pendulum only negative (see Chapter 4, Exercise 5).

Typically, electric fields are limited by the material of the electrodes or, more often, the field strength of the dielectric between the electrodes. For air, the breakdown voltage depends on the pressure, humidity and air composition but is typically $1\,V\cdot\mu m^{-1}$ to $3\,V\cdot\mu m^{-1}$, with $2\,V\cdot\mu m^{-1}$ to $10\,V\cdot\mu m^{-1}$ being typical for oils. For vacuum applications it is found that the breakdown voltage reduces with pressure to a minimal value typically around a few hundred pascals and then increases again with increasing vacuum. This breakdown effect as a function of pressure when plotted is called a Paschen curve and can be calculated from Paschen's law. When breakdown occurs the subsequent plasma of the spark often produces carbon and other deposits that contaminate a vacuum chamber and create electrical short circuits. Hence, when using high fields it is important to either turn these off or reduce the field values as the pressure varies.

7.3.4 Piezoelectric Actuators

Piezoelectric actuators typically comprise a thin, solid material with electrodes deposited on opposing faces. A voltage applied to these electrodes will induce a strain that will, in turn, result in displacements in the direction of the electrodes and other directions that depend on the relationship between the direction of the piezoelectric properties and the electrode location. In contrast with electromagnetic and electrostatic actuators, a piezoelectric actuator is rigidly connected to the component being displaced and adds stiffness. In short, this is a stiff actuator and often requires some form of coupling to accommodate misalignment during assembly.

If the dielectric of a parallel plate capacitor is a solid material, it will experience an electric field that, for parallel electrodes with a large ratio of electrode dimension to separation, will be approximately constant. For air or vacuum, the charge density on the surface of the electrodes is simply proportional to the vacuum permittivity. If a dielectric material is placed in the gap, there will be an additional term due to the displacement of charge within the material resulting in a dipole moment (with clear analogy to the moment in a magnetic material). Consequently, the charge density D, considered to be a consequence of dielectric displacement, can be expressed as the sum of the vacuum permittivity and dipole moment per unit volume P given by

$$D = \varepsilon_o E + P. \tag{7.84}$$

Furthermore, with the exception of the vacuum permittivity, the parameters in Equation 7.84 should be vectors and because P ($C \cdot m^{-2}$) can be represented as a line connecting opposite charges it is called the polarisation of the dielectric. Some materials are intrinsically polarised while others can be polarised through application of an applied field when heated above their Curie temperature and are called ferroelectric (Jona and Shirane 1993). For ferroelectric materials, a plot of polarisation on the vertical axis against the electric field will look similar to the *BH* plots for magnetic materials shown in Figure 7.55 and Figure 7.57. By analogy the residual polarisation P after the removal of the electric field can be considered as a residual dielectric displacement in the material, and the reverse electric field required to reduce this polarisation to zero is called the coercive field E_c.

Some crystalline structures do not have a common centre of symmetry. Such non-centrosymmetric crystals that also have a polarisation axis or can be polarised are piezo-electric, meaning that they will develop a charge in response to a stress and vice versa. Such materials include single-crystal quartz (Brice 1985), the ferroelectric barium titanate and, for many actuator applications, the ceramic lead zirconate titanate, often referred to as PZT. The conditions for a material to be piezoelectric are

1. All piezoelectric materials have a non-centrosymmetric structure.
2. Ferroelectric materials are piezoelectric if they are heated above the Curie temperature and cooled in the presence of a high poling field, after which the polarisation will be in the direction of the poling field.
3. All piezoelectric materials are pyroelectric, meaning that they will develop a surface charge with a change in temperature.

All of the piezoelectric materials are anisotropic, requiring tensor mathematics for modelling and are also temperature sensitive with mechanical, electrical and thermal responses being interrelated. Consequently, a complete theoretical framework for analysis of piezoelectric actuators is beyond the scope of this section and only simplified Ideally linear models will be used, with parameters defined in Table 7.4 and typical values for these parameters for a PZT ceramic given in Table 7.5. Consider the rectangular plate piezoelectric actuator shown in Figure 7.74 showing, as is the tradition in this field of study, an orthogonal coordinate system numbered 1 through 3, with the poling direction always along axis 3. The dot shown on the plate in Figure 7.74 indicates the positive electrode. For a piezoelectric element with electrodes in the plane of the poling axis, the free strain ε_x in direction 3 for a field E is given by

$$\varepsilon_x = d_{33}E. \tag{7.85}$$

Using the values for the maximum electric field of 2.5 MV·m^{-1} and $d_{33} = 400$ pm·V^{-1} produces a maximum strain of 0.1%, which is a useful rule of thumb and, consequently,

TABLE 7.4

Definitions of Piezoelectric Modelling Parameters Using Standard Symbols

Constant	Symbol	Unit	Subscripts	Meaning
Piezoelectric constant, strain per field at constant stress	d_{33}	$m \cdot V^{-1}$	Electrodes perpendicular to poling axis. Either the induced strain or stress is in the poling direction.	Ratio of strain to applied field or ratio of charge generated per unit area with short circuited electrodes to applied stress.
Piezoelectric constant, strain per field at constant stress	d_{31}	$m \cdot V^{-1}$	Electrodes perpendicular to poling axis. Either the induced strain or stress is in direction 1.	See above.
Piezoelectric constant, strain per field at constant stress	d_{15}	$m \cdot V^{-1}$	Electrodes perpendicular to direction 1. Shear strain about plane perpendicular to direction 2.	See above.
Piezoelectric constant, field per stress at constant charge	g_{33}	$V \cdot m \cdot N^{-1}$	Electrodes perpendicular to direction 3. Applied stress or induced strain is in direction 3.	Ratio of field to applied stress or ratio of strain divided by charge per electrode area.
Piezoelectric constants	g_{31}, g_{15}	$V \cdot m \cdot N^{-1}$	See above.	
Elastic compliance	s_{33}^E,	$m^2 \cdot N^{-1}$	See above. Superscripts indicate measurement with electrodes connected, E, or open circuit, D.	Ratio of strain to stress. Other components of compliance are needed for different electrode configurations.
Dielectric constant	K_3^T	$= \dfrac{\varepsilon}{\varepsilon_o}$	Superscripts indicate measurement with zero stress or zero strain, S.	Relative dielectric constant.
Electromechanical coupling	k_p		Subscripts, when used, correspond to electric field and strain respectively; indicates ceramic actuators.	Square root of the dimensionless ratio of electrical energy converted to mechanical energy divided by the input electrical energy (<1) or vice versa.

Note: Poling axis is always in direction 3.

TABLE 7.5

Representative Values of the Modelling Constants for a PZT Actuator Ceramic

Density (kg·m³)	Elastic Modulus (N·m²)	Curie Temperature (°C)	K_3^T	k_p	E_{max}, E_c (V·m⁻¹)	d_{31}
7500	66×10^9	350	1725	0.7	2 to 4×10^6, -0.3×10^6	-170×10^{-12}
d_{33}	d_{15}	g_{31}	g_{33}	g_{15}	s_{11}^E	s_{33}^E
380×10^{-12}	584×10^{-12}	-11×10^{-3}	25×10^{-3}	38×10^{-3}	15×10^{-12}	18×10^{-12}

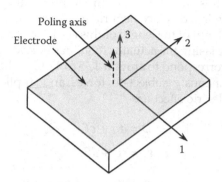

FIGURE 7.74
Rectangular plate piezoelectric actuator with electrodes in the plane of the poling axis.

PZT actuators can be thought of as providing approximately 1 μm translation for every 1 mm of actuator. Displacement x of the actuator will be the product of strain and actuator thickness, while the field is given by the applied voltage V divided by thickness, hence

$$x = d_{33}V. \tag{7.86}$$

Surprisingly, the displacement as a function of applied voltage is independent of the thickness of the actuator, indicating why it is favourable to make the actuators small. In practice the limiting factor on plate thickness is the need to generate large fields. For actuators, plates are rarely more than 0.5 mm to 0.75 mm thick requiring amplifiers capable of providing 1000 V to 1500 V into the capacitive load. To reduce the applied voltage and increase the displacement range, it is common to stack a number of thin plates together into a stack (Figure 7.75). The dark regions in the photographs in Figure 7.75 b represent the active PZT plates with the lighter regions in the middle and at each end being non-active so that the physical connections to the electrical wiring and mechanical contacts do not experience strains during actuation. The observant reader will also note that there are thicker plates spanning the transition between the active and non-active regions designed to

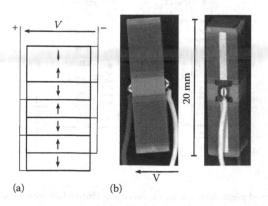

(a)   (b)

FIGURE 7.75
Piezoelectric stack actuator: (a) schematic diagram, (b) two photographs of a 20 mm long co-fired stack actuator.

reduce strain gradients. In this case, the plates are approximately 60 µm thick and can be driven up to 150 V, resulting in a displacement range of around 17 µm, indicating, based on the 0.1 % rule, that around 3 mm of the length of this actuator is not active. Additional range can be obtained by reverse biasing the actuator to values below the coercive field. Typically for PZT actuators this will correspond to around 1/5 to 1/10 of the maximum forward field.

Generally, when simultaneously subject to forces and applied fields, the equation of direct and converse piezoelectric effect are

$$D = d\sigma + \varepsilon E$$
$$\varepsilon_x = s\sigma + dE,$$
(7.87)

where σ is the applied stress. A more complete set of constitutive equations can be found in Jona and Shirane (1993) and Jaffe et al. (1971).

Like most piezoelectric materials, PZT is a brittle ceramic and will fracture under tensile loads or, equivalently, in the presence of high strain gradients at the surface. Consequently, either single element or stack actuators should not be used to produce tensile forces. One way to avoid tensile loads when driving dynamically changing displacements is to provide a preload. If the load is a constant value (often called a blocking force), as would be the case for an applied mass (weight), then the translation range of the actuator is not significantly changed up to its maximum load. Maximum loads for PZT materials are limited to stresses of around 20 MPa, with desirable preloads of about one-tenth of this value. However, if the preload is applied using a spring or the actuator is driving a flexure mechanism, then the stiffness of the mechanism can have a significant impact on the translation range, resulting in a so-called lost motion. Consider the actuator assembly shown in Figure 7.76. For the purpose of this model, the stack actuator is attached to the frame of a flexure mechanism and coupled to the moving platform of the mechanism via a sphere on flat coupling. To evaluate lost motion it is necessary to incorporate all compliance effects around the force loop of the mechanism which, in this case, include stiffness values for the mechanism frame k_b; lower and upper contact interfaces k_{il}, k_{iu}; flexure mechanism k_f; spherical contact (or any other) coupling k_c; and the actuator k_p. Interfacial

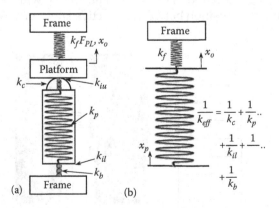

FIGURE 7.76
Schematic diagram of a stacked piezoelectric actuator assembled into a flexure mechanism: (a) schematic diagram, (b) model for assessing the effect of compliances on lost motion.

stiffness depends upon both the texture of the surfaces and their alignment. The compliance due to surface texture can often be reduced by adding an adhesive layer, while the spherical contact stiffness can be increased by using a large radius and high preload, as indicated by Equation 7.11. Because these elements of the mechanism are serially connected, it is necessary to add their compliances as shown in Figure 7.76b. To assess lost motion it is necessary to determine the effective stiffness of the actuator and use this to model the compressions of these components when the actuator is subject to an applied field. In practice, it is not unusual to find that the effective stiffness is a fraction of the stiffness of the actuator with $1/2$ to $1/5$ being typical. In the absence of external stiffness values the actuator will expand by a distance x_p. For modelling purposes this is shown as a displacement of the lower portion of the effective stiffness. From this model it is apparent that the displacement x_o of the platform based on the free displacement of the actuator can be expressed as the ratio

$$\frac{x_o}{x_p} = \frac{k_e}{k_e + k_f} = 1 - \frac{k_f}{k_e} + \left(\frac{k_f}{k_e}\right)^2 - \left(\frac{k_f}{k_e}\right)^3 \dots \tag{7.88}$$

From Equation 7.87 it is apparent that the lost motion depends on the ratio of the external mechanism stiffness to that of the actuator assembly and is directly proportional for a relatively compliant flexure. Clearly, if the flexure stiffness is equal to that of the actuator assembly, then only half of the actuator displacement will be seen at the moving platform.

The preceding linear relationships between applied voltage and strain would imply that a piezoelectric actuator can be used as a positioning mechanism. In practice, PZT and other ferroelectric actuators will have considerable hysteresis and long-term creep. Figure 7.77 shows the output displacement of a flexure mechanism in response to a cyclic voltage ramp on a piezoelectric stack actuator. In this particular mechanism, a lever is used to increase the translation of the actuator from 20 μm to 150 μm, and the staircase effect in the lines of this plot are caused by the steps in voltage from the digital-to-analogue converter. Initially there is an offset of around 10 μm, and the displacement shows a curved path with highest slope at the lower voltage (field); when the voltage ramp is reversed, the displacement shows a different path. In this experiment, after this first cycle, there are a further five cycles during

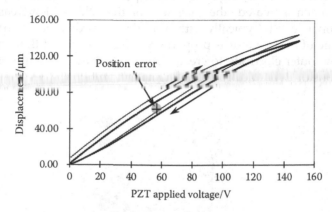

FIGURE 7.77
Output displacement of a piezoelectrically actuated, levered, flexure mechanism.

which the paths are the same to within the discrimination possible in this figure. Although the hysteresis is repeatable and can, therefore, be compensated as a positioning device, any cycles of smaller amplitude and different offsets will result in smaller hysteresis loops within this envelope. Consequently, the flexure mechanism cannot be determined from the voltage, and therefore this mechanism is considered to be an actuator and not capable of positioning. The offset of the initial plot is a consequence of drift and is comparable to the width of the hysteresis loop, both of which result in a variation of position with voltage of around 10% to 20% depending on the actuator composition.

Another consideration is the heat dissipated by the actuator. Typically, losses for PZT actuators tend to result in around 20% of the mechanical power generating heat in the ceramic that will result in a temperature increase with corresponding loss of performance. It is not unusual to include cooling (water or a dielectric oil) when it is desired to extract the highest available power from these actuators. There are other ferroelectric materials that have less hysteresis, most notable ore lead manganese niobate (PMN) and lead zirconate niobate (PZN). PMN is an electrostrictive material for which expansion is more akin to a rotation of electrical dipoles. Consequently, the displacement tends to increase with applied field and will be an expansion independent of field polarity leading to a characteristic parabolic or U-shaped plot of displacement against applied field. PZN is similar to PZT in terms of its crystallographic structure (both are characteristic of 'perovskite' materials) and both derive their large piezoelectric constants by selecting a composition near to a morphotropic phase transition. However, PZN does not have a distinct transition and therefore larger reversible structural changes can occur resulting in achievable strains of up to 0.4% (compared with 0.1% for PZT). This extra strain literally comes at a price (the material is difficult to manufacture and therefore expensive) and, at higher strains, there will be a significant structural change as it crosses the transformation region during which the elastic modulus and strength significantly reduce. However, these materials are sometimes used to drive flexure translation stages (Woody et al. 2004, 2005). Materials that do not have a distinct morphotropic phase transformation and can therefore undergo large structural changes and relax to an original state are called relaxor ferroelectrics and other significant compositions include PMN-PT and PZN-PT.

Because the ferroelectric materials can be poled in arbitrary directions, the arrangement of electrode patterns and poling axes is limited only by the creativity of the designer. A simple actuator can be produced by making a tube and depositing the electrodes on the inside and outside surface. When activated, the thickness of the tube will increase with a corresponding reduction in length (typically with around one-third of the strain). In this case, the displacement range of the actuator is proportional to the length of the tube. It is also possible to pattern the outer electrode to create specific deformation patterns of the tube for creating translations or dynamically oscillating specific mode shapes. Figure 7.78 shows a

FIGURE 7.78
Photograph of a piezoelectric, two-coordinate tube actuator.

two-segmented tube actuator with an inner electrode and two outer electrodes on either side. With the inner electrode connected to ground, applying a potential to both outer electrodes will cause the tube to change length. Alternatively increasing the potential on one outer electrode while reducing it on the other will cause the tube to bend. Consequently, applying suitable voltage to the electrodes enables the tube to move both axially in a straight line and laterally along a curved path (such actuators are commonly applied in scanning probe microscopes).

In contrast to the ferroelectric materials, single-crystal piezoelectric elements such as quartz appear to have no measurable hysteresis. As a consequence, quartz oscillators have been the foundation of timing since the 1930s and are now found in almost all digital electronics and computers. For digital watch and timing circuits, the quartz is machined into a tuning fork shape with electrodes on all four faces of both tines (see Figure 7.79). On a single tine, the electrodes on parallel faces are connected and produce an electric field, as illustrated in Figure 7.80. Because quartz is a permanently polarised crystal, it is tolerant of both forward and reverse bias fields and does not have any measurable intrinsic hysteresis. The electric field vectors produced by the electrodes will be reversed in the region of each common electrode pair. In the polarisation direction, these reversed fields will induce a compression on one side of the tine and an expansion on the other, thereby causing the tine to bend in a shape similar to that of the tines' fundamental vibration mode. The electrodes on the other tine are reversed and, as a consequence, will bend in the opposite direction. Hence, oscillating the tuning fork at its fundamental frequency will result in the tines moving in equal and opposite directions, with a phase shift of 90° between the applied voltage and charge at the electrodes (that translates to a current equal to the product of the magnitude of charge and excitation frequency). Because the tuning fork tines move in equal and opposite directions, the total momentum is zero and the plane of symmetry provides a

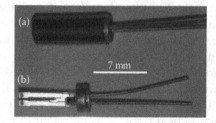

FIGURE 7.79
Photograph of a quartz tuning fork oscillator: (a) inside a vacuum can, (b) with the can removed.

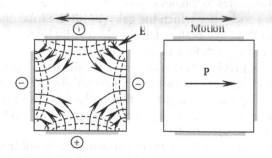

FIGURE 7.80
Cross section of tuning fork tines showing electrodes, polarization direction and electric field lines. Dashed lines are equipotential contours.

nodal point that can be used to hold the tuning fork. In air, the motion of the tines induces circulating airflow patterns that will damp the oscillations. By placing the tuning fork in a vacuum it is possible to achieve Q values of around 50,000 to 100,000, and the natural frequency is typically stable enough to provide the precise timing needed for digital clocking. Because of this sharp resonance, it is possible to detect the effects of materials that might be deposited onto the surface of the oscillator, and this effect is used to measure film growth in vacuum deposition processes, often with sub-nanometre resolution.

Even though the displacement response of quartz is free from hysteresis and creep, because of its low piezoelectric coefficient of around 2.31×10^{-12} m·V^{-1} (about 1/100 that of PZT), while ubiquitous as an oscillator, it is rarely used as an actuator.

Vacuum deposition processes can be used to deposit piezoelectric materials onto surfaces. One popular material is zinc oxide that can be sputter deposited onto surfaces and has been used to excite MEMS oscillators for atomic force probes among other applications.

Polyvinylidinedifloride (PVDF) is a polymer that can be polarised by depositing it as a thin film in a high electric field, by stretching it or by dissolving into a solvent and spinning it onto a surface. It has a piezoelectric coefficient that is around 1/10 that of PZT and an elastic modulus of around 2 GPa to 10 GPa, this latter value for PVDF with PZT filler (1/40 to 1/5 of PZT ceramics) (Jain et al. 2015). The electromechanical coupling of PVDF is also relatively low so that is primarily used as a sensor. PVDF is a cost-effective material that can be readily processed, can be incorporated into additive manufacturing and is likely to find increasing applications.

7.3.5 Thermal Actuators

The high work density of thermal expansion has been utilised to break rocks, clamp tool holders in machine tools and fasten steel railway wheels to axles, and has the potential to change sea levels on a global scale. To utilise this effect, it is necessary to drive heat into, and out of, the volume of material or fluid, usually by increasing the environmental temperature at the surface of the actuator. Methods for heating an object include conduction, convection, radiation and induction (in electrical conductors). Of these methods, induction (when possible) is the fastest method to get heat into a solid, followed by conduction, convection and radiation. Induction cannot be used to remove heat and, therefore, the fastest way is conduction or convection. Under some circumstances, it is possible to use evaporative cooling or accelerate conduction heat transfer using a Peltier cell. Evaporative cooling can be achieved using a self-contained unit called a heat pipe that is commonly used in computers to cool the processing chip. As the name suggests, a heat pipe is a tube that has cooling at one end and attaches to the hot object at the other. Inside this tube is a liquid that will evaporate at the hot end, after which the gas will fill the tube and condense at the cool end. For maximum heat flow in a material, it should ideally have a high thermal diffusivity (see Chapter 12) that is proportional to thermal conductivity and inversely to the heat capacity measured by the product of density and specific heat. Via the Weidermann-Franz law, the thermal conductivity of metals are approximately proportional to their electrical conductivity (Ziman 1972). Consequently, a thermally fast, long-range actuator material will correspond to a maximisation of thermal expansion and conductivity, and a minimisation of density and specific heat.

As the size of an object reduces, the ratio of surface area to volume will linearly increase with scale. Consequently, for smaller devices such as MEMS, thermal actuators can have bandwidths of kilohertz and higher. As scales reduce to molecular dimensions, thermally induced motion is often the most significant mechanism for transport of particles.

Larger scale thermal expansion-based actuators have been produced to provide displacements in the region of 50 μm to 100 μm and capable of positioning loads of up to 1000 N. For example, Figure 7.81 is a plot of the closed loop controlled displacement of a thermally actuated flexure stage driving a load of 1 kN. In this case, the actuator was a 304 stainless steel tube (outer diameter 10.3 mm, inner diameter 6 mm and 110 mm long, expansion coefficient 16.9×10^{-6} K^{-1}) with a nichrome wire wrapped around its outer surface and water coolant being pumped along the inside of the tube. This particular actuator comprised a large thermal mass and used a relatively low power driver for which the response times of 200 s or more are relatively slow. Similar designs using aluminium flexures and higher power amplifiers have produced settling times of around 25 s, with comparable performance, details of which are not included for brevity. It is noted that the response to the first 25 μm set-point demand is considerably faster than the settling time to achieve 50 μm. The reason for this reduced performance at higher temperature is, in part, because even though the actuator can provide sufficient heat to the actuator, at higher temperatures more of this heat is lost by convection, radiation, heating of the wires to the coil, conduction from both the ends of the actuator and into the cooling fluid, and because the actuator temperature is approaching the maximum temperature of the coil that is limited by the amplifier power. Reversing the demand (not shown in the figure), it is possible to run cooling water through the actuator resulting in a considerably faster response. This indicates some of the challenges for implementation of this actuation method. Fundamentally, the control problem tends to be asymmetric and non-linear that requires more elaborate controls involving different dynamics for both absolute displacements and the direction of the changing demand.

Although unexplored at the time of writing, the high work density and large forces that can be generated offer the potential for implementing levered amplification mechanisms to provide long-range motion that might still maintain relatively high output forces.

At the other end of the translation range scale, thermal expansion can be controlled with suitable feedback sensing to provide high-resolution displacements. Thin-film coils attached to flexible Kapton (a thin polyimide film that can withstand temperatures of up to 400 °C) are available that can be adhered onto surfaces. These are used to control the temperature of the glass tubes of gas laser enclosures to maintain frequency stabilities at levels measured in attometres. Actuators comprising a cylinder with a fine winding of

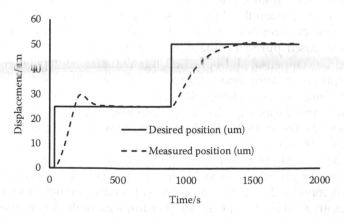

FIGURE 7.81
Output from a thermally actuated flexure driving a load of 1000 N.

copper can produce very precise rectilinear motions with picometre control (Lawall and Kessler 2000).

7.3.6 Screw Drives

It is assumed that the reader is familiar with conventional screw geometries and methods for calculating performance (Spotts et al. 2003). A screw may be thought of as a continuous wedge, with an angle given by the thread pitch divided by the pitch diameter that transforms, via a nut, rotational motion of a drive shaft to linear motion of a moving carriage. For the last few centuries, machine tools have used feedscrews to manufacture more precise screws (Evans 1989). This has been made possible by mapping the errors of an existing screw thread and converting this to a rotation error. Again, this repeatable error in linearity between the rotation of the shaft to translation of the nut, called the pitch error, can be removed using compensators that provide small rotations of the nut as the machine carriage translates. Combining compensation and using nuts that incorporate elastic averaging, manual micrometers and large machine feedscrews provide accuracies of better than 0.5 μm (at 20 °C), and sometimes when controlled by a dexterous user, with resolutions of a few tens of nanometres. With precision measurement for feedback, automated positioning systems can achieve positioning with controller errors within 70 nm. High-precision translation stages with air bearing guideways can, under controlled operating conditions, have accuracies of around 0.25 μm and controller repeatability of better than 50 nm. Controller error refers to the difference between the measured and desired values and should not to be confused with measurement uncertainty or positioning errors that require consideration of the actual translation with respect to a reference value.

Positioning resolution can be improved by mounting a short-range, piezoelectrically actuated translation stage onto the feedscrew-driven, longer-range stage. By stacking these 'coarse' and 'fine' stages, positioning repeatability of a few nanometres or less can be achieved (Buice et al. 2009; see also Chapter 14, Section 14.2.7 for a discussion of controller strategies).

A major limitation of conventional feedscrews is backlash, which occurs when the drive force is reversed. Forces between the nut and screw are transmitted by the compressive contact on one side of the thread. Because the nut must be slightly larger than the screw, there will be a clearance resulting in a small amount of axial motion, often referred to as the 'play' of the nut. Hence, when the direction of the screw is reversed, there will be a finite rotation, as the contact changes from one side of the threads to the other side, when no translation occurs. Based purely on the difference in geometry, this backlash should be constant and can be compensated. In practice, this will also be dependent on loads, error motions of the guiding bearing and with wear of the nut (the feedscrew represents the reference for translation and, therefore, will have a surface that is harder than that of the nut that will be made from a tribologically compatible material). Most automated feedscrew-driven machines will incorporate backlash compensation within the controller strategy (see Chapter 14, Section 14.1.2).

In principle, backlash can be eliminated if the net force on the nut is always of the same sign. A common method to attenuate backlash is to attach two nuts onto a feedscrew. These two nuts are then connected by a linkage to prevent relative motion, and a spring between the two provides an axial load. An even simpler implementation is to use a long nut and introduce a radial slit so that the threads can be clamped and adjusted to reduce backlash to a minimum (see Figure 7.82). A drawback with these preload types of anti-backlash

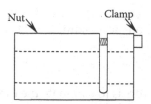

FIGURE 7.82
Nut with clamp for reducing backlash or providing locking capability.

mechanism is the increased friction between the nut and shaft that often renders the feedscrew more difficult to control.

Friction between a nut and feedscrew can be almost entirely eliminated using a hydro-static nut (Slocum 1991) that provides high stiffness and no backlash. However, as stated in Section 7.3.2 for precision motion control, temperature stabilised linear motor drives have replaced screws in many precision machines.

For feedscrews that are not required to provide large drive forces, the use of a low elastic modulus material for the bearing surface of the nut can provide both lower hysteresis and elastic averaging that will reduce pitch errors.

To reduce friction, a 'ballscrew' uses a shaft machine with a spiral ball bearing raceway and with a nut containing ball bearings constrained to roll along a similar spiral path (see Figure 7.83). Because the balls will precess within the nut as the ball rotates, those balls leaving one side of the nut are fed into a tube and recirculated back. Ball screws have lower stiffness than conventional screw and nut combinations, with the advantage of a substan-tially lower friction (although this increases with implementation of anti-backlash mecha-nisms). Dithering using a piezoelectric actuator has been used to reduce friction in a two-nut anti-backlash feedscrew (Chen and Dwang 2000). Another drawback of ballscrews is the relatively large screw pitch of ball screws that must necessarily be larger than the diameter of the recirculating balls in the nut.

Thread-grinding machines can produce threaded shafts with thread surfaces having smooth, almost polished, surface texture (low wear regime sliding pairs). Ground threads will also have sub-micrometre pitch errors, nanometre form deviations (high

FIGURE 7.83
Photograph of a ballscrew showing the nut and short section of feedscrew shaft.

conformability), hardened steel screw surfaces and mild steel nuts for tribological compatibility (almost unique to steel-on-steel sliding pairs), and thermal expansion matching.

7.3.7 Other Drive Mechanisms

One method to eliminate pitch errors is to eliminate the threads. Such an actuator is called a friction drive (or capstan) and comprises a precision ground drive bar or plate that is clamped between rollers, of which one roller is driven (Bryan 1979, Mizumoto et al. 1995). In principle, a friction drive will have little backlash, minimal friction, high stiffness and can provide nanometre resolution, particularly when precision ground air-bearing spindles are used as the rollers. The range of motion for this drive is only limited by the length of the bar making it an economic solution for long-range motion in a compact form. A drawback with this rigid bar is the need for precise alignment or a decoupling mechanism between the friction drive bar and the slideway. All drives that depend on friction will slip at a high enough load, a feature that might be desirable to avoid overloads and catastrophic system crashes.

A ribbon drive may be considered a variant of the friction drive, where the rigid drive bar is replaced by a flexible metal belt. The flexibility of the belt will provide compliance in off-axis directions and, therefore, attenuate any parasitic motions of the drive from being transmitted to the slideway. Two designs are common: the conveyer belt and capstan (see Figure 7.84). As with the friction drives discussed earlier, ribbon and wire drives provide smooth motion with little backlash and can be relatively stiff in the drive axis while providing a relatively high compliance in off-axis directions. Alignment is still necessary with these drives although the capstan and wire drive are more compliant in some coordinate directions than the conveyor ribbon drive.

Somewhat intermediate between the friction drive and feedscrew is the linear actuator, shown in Figure 7.85. This linear actuator comprises a ground shaft that acts as a feedscrew and a split nut that has six ball bearings arranged at the ends. One part of the split nut bearing has four ball bearings at the end, so that contact with their outer race surfaces will form a four-point contact with the shaft (in practice, these crossed cylinders will create an elliptical contact area). The other part of the split nut has two more bearings that will contact the shaft on the opposite side and these two are held together by a force (coil springs in

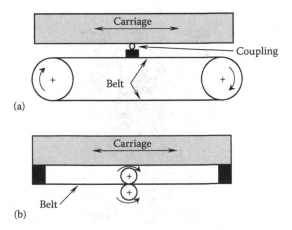

FIGURE 7.84
Ribbon or wire drives: (a) conveyer or pulley, (b) capstan.

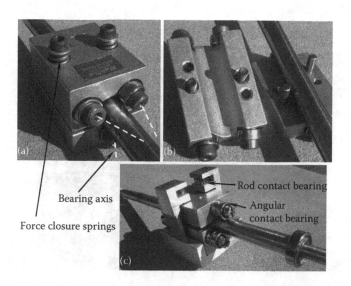

FIGURE 7.85
Feedscrew actuator using a spiral path friction drive: (a) complete assembly, (b) disassembled actuator showing two halves of the nut, (c) assembled linear actuator with angular contact bearing pairs replacing the ball bearings.

Figure 7.85) and maintained in alignment with two dowel pins. All of the bearings are mounted at an angle relative to the shaft axis so that the instantaneous axis of rotation will cause the rollers to follow a spiral path along the shaft. Consequently, this tilt angle will form the pitch angle of an equivalent feedscrew; the greater the angle, the greater the pitch.

Considering the half of the nut that has four bearings to be the driver, four contacts implies two degrees of freedom. The unnecessary freedom in this case corresponds to sliding along the shaft axis. In practice, frictional forces will prevent this unless the load on the nut exceeds the critical value determined by the nut preload. In many applications, the ability to slip at a high load can be used as a safety feature to avoid undesirable loads. Figure 7.85c shows a linear drive designed to be a fine pitch actuator (a pitch of 0.25 mm) by mounting the bearings at an oblique angle to the axis of the shaft. Because the compliance of roller bearings would result in the outer races tilting slightly under load, these are replaced by the stiffer angular contact bearing pairs. However, at these low angles, the pitch of the screw is dependent on the load (Buice et al. 2005).

The friction drives with a split nut and capstan are frequently used to drive hydrostatic bearings and other low load applications. When there is an air supply available, it is common to use pneumatic pistons to supply the force closure. An advantage of this in that the force can be released if necessary for safety or other reasons. It is common to find variations of all of the aforementioned friction drive actuators used for low load precision drive applications, such as probe-type coordinate measuring systems and mechanical measurement instrumentation (see Chapter 5).

A similar pseudo-kinematic feedscrew comprises a precision ground fine pitch screw and a split nut of lubricated acetal bearing pads (Figure 7.86). The acetal pads are used to provide elastic averaging of the threads, and conformance of the thread with the nut is provided by the viscoelastic properties of this bearing material. A drawback of this nut is the frictional forces particularly at low speeds and a time-dependent stick-slip that requires a high starting torque if the screw has been stationary for extended time periods (hours to days).

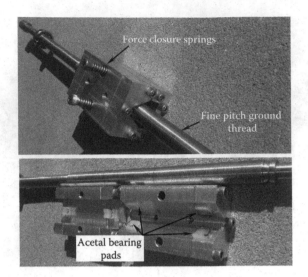

FIGURE 7.86
Split nut feedscrew with fine ground threads and pseudo-kinematic contact of six acetal bearing pads or eleastic averaging.

While piezoelectric actuators can provide large forces with nanometre resolution, their small strains limit achievable translation range. This small range of motion can be overcome by integrating small translations. One mechanism to integrate small motions is the inertial slideway, shown in Figure 7.87, that comprises a V guideway along which an actuator is a sliding fit, so that friction will prevent it moving. Attached to the actuator is a freely suspended rectangular platform that can be translated along the axis by the piezoelectric actuator stage. When the actuator reaches the limit of its extension, it is rapidly retracted. During this rapid retraction, the mass of the platform is sufficiently large that acceleration forces will overcome the friction between the actuator and slideway causing the actuator to move toward the load. Repeated cycles of slow expansion and rapid retraction will result in unlimited motion of the actuator and load along the slideway. Reversing to cycles of a slow retraction and rapid expansion will result in motion of the stage in the opposite direction. A rotation equivalent of this inertial stage is commercially available (called a picomotor™) and can be used for high-resolution rotation of fine thread screws.

Instead of relying on the inertia of the platform to provide the force to overcome the slideway friction, the platform of Figure 7.87 can be replaced by a clamp, shown in Figure 7.88.

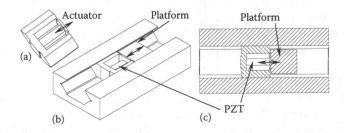

FIGURE 7.87
Piezoelectrically driven inertial translation stage: (a) actuator, (b) isometric view of complete slideway, (c) cross section view.

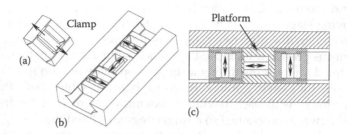

FIGURE 7.88

Inchworm actuated translation stage: (a) isometric view of the clamp, (b) isometric view of the stage assembly, (c) sectioned view of the stage assembly.

In this case, whenever the actuator gets to the end of its range, the clamp can be energised to firmly hold it in place while the actuator is retracted (or extended), after which the actuator can return to a desired extension. Such a mechanism is used to provide the *xy* translations of the miniature lathe presented by Okazaki et al. (2004).

Friction will often vary with time, environment, loads and wear. Consequently, a more reliable method of two clamps and three piezoelectric actuators is illustrated by the 'inchworm' actuator shown in Figure 7.88. As its name suggests, the inchworm actuator is able to move in either direction by applying one clamp and using the central actuator for fine motion control. For this implementation, the central actuator no longer needs to be in contact with the slideway. For long-range motion, the sequence of steps would involve: applying clamp, extending the actuator to its full distance, applying the clamp on the other side of the actuator, releasing the first clamp and retracting the actuator. Repeating this cycle will result in long-range translation, and reversing the cycles reverses the direction of travel.

This grip, contract, grip, and extend methodology is analogous to an inchworm, or monkey-climbing, or even the Michael Jackson moonwalk and hence the name. A number of implementations are marketed commercially. An interesting *xy* stage was produced by Binnig et al. (1981) that had three feet and 'walked' on a flat surface using electrostatic forces to clamp the feet.

Inchworm-type mechanisms are limited to the friction or clamp force, require precision manufacture and fabrication to create the mating surfaces, and can provide nanometre control during actuation, but often create micrometre-scale parasitic motion errors during switching cycles.

7.3.8 Actuator Couplings

Actuators can generally be split into being either stiff or compliant. Compliant actuators, represented by electromagnetic or electrostatic methods, are separated from the guidance mechanism (typically a spindle or slideway) and do not require a physical connection. For these compliant actuators, stiffness is obtained through closed loop control on position and is, therefore, a function of frequency.

Stiffness actuation methods, such as feedscrews, pistons and piezoelectric devices, will require a physical connection to the mechanism. Typically, the motion of the actuating element, such as a nut on a feedscrew, will have larger parasitic error motions than that of the carriage (or spindle). Additionally, there will always be some misalignment between the actuator motion and that of the moving carriage that may also change with externally applied forces and moments. Consequently, it is common to provide a coupling between actuator and moving stage to accommodate the, usually small, deviations.

Couplings also tend to fit within two categories: flexure- or contact-based. Contact-based coupling mechanisms present the simplest approach, with a simple ball on flat providing a single degree of freedom for transmitting a single coordinate motion from one mechanism link to another. Such a system could comprise a base, to which both the actuator and slideway would attach, with each having a moving element (nut for the actuator and carriage for the slideway) constrained to a single degree of freedom. A single point contact would provide five degrees of freedom between the moving elements resulting in a mobility ($M = 6(3 - 1 - 3) + 7$) of one as required. However, a single point contact can only provide a force in a single direction and, in reality, will produce a sliding force in a direction coplanar with the contact due to friction. One method to overcome these frictional forces is to use crossed cylinders, particularly if these cylinders are the surfaces of the outer races of ball bearings (see Figure 7.89). While this crossed roller bearing coupling does provide five relatively low friction degrees of freedom, there will be backlash due to the condition that the central bearing cannot be in contact with both outer bearings simultaneously.

A flexure-based, five degree of freedom axial coupling is shown in Figure 7.90 and comprises four parallel rod-type flexures. All four rods attach to the ends of an intermediate cross-shaped link, with the rod axes being perpendicular to the plane of the link. One pair of flexure rods on opposite arms of the cross is attached to one link, while the other pair of rods attaches to the other link. Because the axial stiffness of the rods will be considerably larger than all bending stiffness values, this can be considered to be a single axial constraint coupling. While this eliminates backlash, there will be off-axis forces and moments proportional to the coaxial deviations between the two links. In principle, a single rod, wire (held in tension) or flexure wobble pins (Figure 7.47) can provide single axial constraints. Traditionally, instrumentation wobble pins can be produced using commercially available conical jewel bearings with a hardened steel needle that are frequently used in mechanical watch movements.

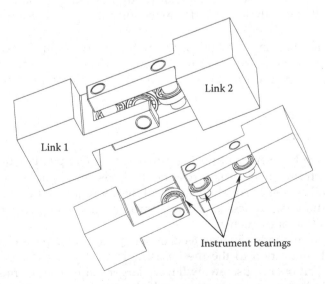

FIGURE 7.89
Single-degree-of-constraint, five degree of freedom axial coupling using crossed cylinders.

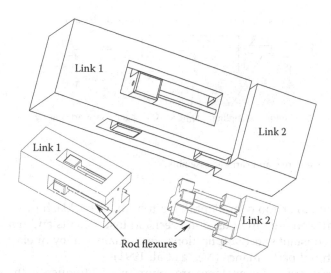

FIGURE 7.90
Single-constraint, five degree of freedom axial coupling based on rod-type flexures.

For transmission of torque from one shaft (motor) to another (feedscrew), it is required to transmit only a rotational motion. The disk coupling shown in Figure 7.51 provides a high torsional stiffness and is compliant in the remaining two rotations and axially. However, this coupling is constrained to the two lateral directions and, therefore, requires precise coaxial alignment of the two shafts (in this case). The collinearity constraint can be removed by having two coaxial disk couplings axially spaced by a short distance, sometimes called multiple membrane couplings (Figure 7.91) (Neale et al. 1991). Bellows couplings, similar to the one shown in Figure 7.54, are often used for low power transmission of rotation, when relatively large angular and axial deviations are necessary, as are double Hooke joints. With the exception of the bellows coupling, most coupling methods will produce periodic variations in angle between the motor and shaft with large angular misalignment (Morrisson and Crossland 1970). More elaborate couplings, called constant velocity joints, are available that will maintain the relationship between rotational angles between two shafts, but manufacturing complexity and necessary precision makes these only economic when produced in large quantities (Morrisson and Crossland 1970).

Besides those mentioned earlier, many other types of couplings have been created to satisfy the demands spanning a broad range of machines and application. Most of these

FIGURE 7.91
Photograph of a five degree of freedom disk coupling connecting a motor to a shaft.

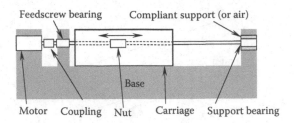

FIGURE 7.92
Schematic diagram of a feedscrew and carriage assembly.

other couplings are either too compliant, have too much backlash or constrain too many degrees of freedom, and are rarely used in precision instruments and machines. However, for high power or constant speed applications, gear, chain, pulley or elastomeric couplings might provide optimal performance (Neale et al. 1991).

Feedscrew and nut-driven linear stages are common, and frequently the use of a coupling between the nut and moving carriage is undesirable, typically because of cost or the, unavoidable, additional compliance of a coupling. The tight tolerance between the thread and nut, as well as the difficulty of precisely controlling straightness of a long feedscrew shaft, results in considerable alignment challenges. While it may be possible to provide adjustments in assembly, so that the feedscrew and nut is aligned at one point along the screw, this may not be possible as the nut traverses along the shaft, resulting in variable frictional forces with travel. One possible, and frequently used, solution is to attach only one end of the shaft into the bearings that will provide a rotation axis and support the axial thrust between the shaft and nut, while the other end of the shaft is relatively free (see Figure 7.92). In this particular implementation, the nut acts as a shaft support, while the bearing or support at the other end functions mainly to stop the shaft from whirling when the nut position leaves a long length of the shaft unsupported. Using a bearing at this unsupported end provides free rotation of the shaft, while the compliant material such as a rubber, or other elastomer, will also introduce damping.

7.3.9 Levers

Precision often requires fine adjustment using convenient and cost-effective actuation methods. Feedscrews are commercially available and fine thread micrometers and custom adjustment screws can provide sub-micrometre control. Sometimes, it is desired to increase this resolution or convert this linear motion to small rotational alignment. Most common, increased resolution is readily achieved using some form of lever mechanism to attenuate the motion of the actuator. Conversely, piezoelectric actuators can have exceptional resolution but are limited in range. When it is desired to increase the range of an actuator, levers are used to amplify its motion. Attenuation and amplification mechanisms are discussed separately in this section. Other examples and more detailed analysis can be found in Smith and Chetwynd (1990) and Smith (2000).

7.3.9.1 Motion Attenuation

Resolution of direct feedscrews is limited by the screw pitch. Fine motion control with moderate loads is often required in instrument mechanisms and test apparatus for which screws having a pitch of up to four threads per millimetre (or 100 threads per inch) are commercially available. By using differential screws, it is possible to substantially reduce the apparent pitch while maintaining the strength of larger pitch threads. Differential screw micrometers are commercially available and typically provide apparent screw pitches of around ten threads per millimetre. Consider the differential screw adjuster mechanism shown in Figure 7.93 that uses two different screw threads to produce small, arcuate motions of a notch flexure hinge. The threaded inserts are free to rotate to accommodate the arcuate motion of the hinge. For two different threads of pitch P_{t1} and P_{t2}, the relationship between the angular rotation θ of the adjuster and motion x at the top of the hinge is

$$x = (P_{t1} - P_{t2})\theta = \left(\frac{1}{N_{t1}} - \frac{1}{N_{t2}}\right)\theta, \tag{7.89}$$

where N is the number of threads per unit length for each thread. For example, if $N_{t1} = 32$ and $N_{t2} = 36$, then the apparent pitch will be 288 threads per length. These two N values are chosen to correspond to unified national fine (UNF) threads numbered 8-36 and 10-32, having 32 and 36 threads per inch that would provide an apparent pitch of approximately 88 µm.

Direct attenuation can be achieved with a lever arm and pivot, as illustrated by the flexure-based coarse–fine adjustment mechanism shown in Figure 7.94. Coarse and fine adjustment is achieved by two levers connected to a linear translation stage via a notch hinge flexure 'wobble pin'. In this particular coarse–fine implementation, the lever pivot of the fine stage is attached to the output of the coarse lever arm. For an ideal lever, the ratio of output to input is the same as the ratio of input to pivot distance divided by the length from the pivot to the output. In practice, forces acting on the hinges will result in deformations that result in a lower effective lever ratio. The reduction in between the ideal output and that observed in a practical mechanism is called 'lost-motion' and is analogous to the similar effect discussed in Section 7.3.4 for piezoelectric actuators. Typically, lost motion represents a significant limit when using levers for amplification and is discussed in the following section.

By pushing a stiff spring through a relatively compliant spring, fine adjustment mechanisms have been implemented to provide sub-nanometre resolution control using a

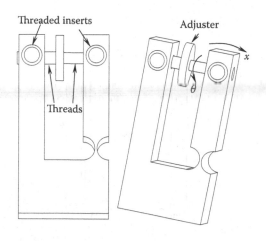

FIGURE 7.93
Differential screw adjustment mechanism.

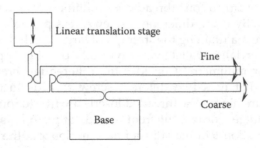

FIGURE 7.94
A flexure-based coarse-fine linear adjuster.

micrometer (Hart 1968). Figure 7.95 shows a monolithic flexure mechanism for angular adjustment of a platform for fine mirror adjustment in optical systems. Soft, or low stiffness, springs are provided by two leaf flexures, while the stiff spring is a notch hinge. For adjustment, the leaf flexures are displaced in direction x to apply a bending moment M to the upper platform that will, in turn, be transmitted to the notch flexure, so that

$$M = \frac{2Ebt^{5/2}}{9\pi R^{1/2}}\theta = \frac{Ebt_1^3}{4L_1^3}x_1, \tag{7.90}$$

where R is the radius of the notch, t is the thickness of the notch, L_1 is the length of the fine adjust flexure, t_1 is the thickness of the fine adjust leaf flexure and b is the depth of the mechanism. In this example, Equation 7.89 can be rearranged to determine the lever ratio defined as the change in angle for a change in displacement of either leaf flexure (replace subscript 1 with 2 for the coarse adjust) given by

$$\frac{\theta}{x_1} = \frac{9\pi}{8}\frac{R^{1/2}t_1^3}{t^{5/2}L_1^2}. \tag{7.91}$$

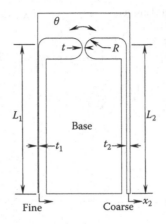

FIGURE 7.95
Flexure-based coarse-fine angular adjustment mechanism.

Considering that the length of the leaf can be made substantially longer than the radius of the notch, and the thickness of the leaf smaller than the notch, this lever ratio can be made almost arbitrarily small. An advantage of the soft-spring stiff-spring approach is that the moving platform has, by definition, a high stiffness.

Self-aligning bearings, ball joints or Heim joints are frequently used to provide rotational alignment about a common axis. Using a long lever arm to control these rotations will enable high-resolution angular alignment.

It was pointed out in Section 7.3.6 that feedscrews produce a continuous wedge action wrapped around the axis of a shaft. For finer resolution adjustments, wedges can be used directly to attenuate linear motion, with the input and output displacements being perpendicular.

7.3.9.2 Motion Amplification

Amplification of motion using levers is, in principle, the reverse of the attenuating lever (note that the soft-spring stiff-spring approach cannot be used for amplification). However, for motion amplification, the design of couplings and pivots can have a more significant impact on performance. Consider the levered single-axis flexure stage model shown in Figure 7.96 that shows the platform of a linear flexure stage being driven via a wobble pin from one end of a lever arm. This model could be used to represent one axis of the levered xy stage shown in Figure 7.52. Also attached to this lever arm are a pivot at a distance b from the wobble pin and an actuator coupling a further distance a from the pivot axis. All rotational degrees of freedom are provided by notch-type flexure hinges.

For modelling purposes, the notch hinges are considered to be ideal. As a consequence, the output motion x_3 of the stage would be related by

$$\frac{x_3}{x_o} = -\frac{b}{a} = -n,$$ (7.92)

where n is the lever ratio. In practice, the notch hinges will have a finite stiffness given by the last of Equation 7.48, and there will be other compliances due to deflection of the lever arm and all other components around the force loop of the mechanism. It is not unusual for the

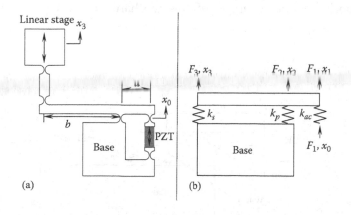

(a)

(b)

FIGURE 7.96

Single-axis linear flexure stage driven by a levered PZT actuator: (a) simple geometric model, (b) equivalent model for static translations.

frame of a flexure mechanism to have significant compliance. A rule of thumb is to make this frame as large as necessary and then double its size to ensure the effects of frame compliance are insignificant. Taking into account all compliances, being serially connected most of which will add, an equivalent model for the relationship between the output stage motion and input piezoelectric free displacement is shown in Figure 7.96b. Using this model, it can be shown that the relationship between actuator free displacement and flexure stage output becomes (Smith 2000)

$$\frac{x_3}{x_o} = \frac{-n}{n^2 \dfrac{k_s}{k_{ac}} + n^2 \dfrac{k_s}{k_p}\left(1 + \dfrac{1}{n}\right)^2 + 1}, \tag{7.92}$$

where k_s is the effective stiffness of the linear translation stage and wobble pin, k_p is the axial stiffness of the pivot and k_{ac} is the effective stiffness of the actuator and couplings (including interface compliances). Because all terms in the denominator of Equation 7.92 are positive, it is clear that the ideal lever ratio is always less in practice, which is a direct measure of the lost motion. It is also clear that lost motion can dramatically increase with the lever ratio and can only be reduced by increasing as far as possible the stiffness of both the pivot and actuator relative to that of the translation stage, which is solely responsible for the lost motion.

In addition to resulting in an increased lost motion, amplifying displacements using levers will also impact mechanism dynamics. In practice, as the lever ratio increases, the forces on the pivot and at the actuator will increase hyperbolically, as can be readily appreciated by attempting to slam a door by pushing against it near the door hinge axis. A simple lever model, comprising an ideally rigid actuator applied to a lever with a mass on the free end, is shown in Figure 7.97. For this lever design, the pivot is placed at the far right end of the lever arm and modelled as an axial stiffness k_p. By placing the pivot at the end of the lever, the input motion x_o at the actuator and the output, x_3 at the mass, are in the same direction with a lever ratio

$$\frac{x_3}{x_o} = 1 + n. \tag{7.94}$$

The kinetic and potential energies of this mechanism, with parameters defined in Figure 7.97, are

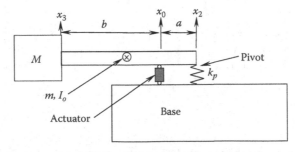

FIGURE 7.97
Model of a levered mass with a pivot of finite stiffness.

$$T = \frac{1}{2}a^2\left[Mn^2 + \frac{m}{3}\left(1 + n^2 - n\right)\right]\dot{\theta}^2$$

$$U = \frac{1}{2}k_p a^2 \theta^2 \tag{7.95}$$

$$\theta = \frac{x_3}{b}\ .$$

Substituting Equation 7.95 into the Lagrange equation (Chapter 4, Equation 4.83) and considering only free motion of the mechanism gives

$$\ddot{\theta} + \omega_n^2 \theta = 0$$

$$\omega_n^2 = \frac{k_p}{\left[Mn^2 + \frac{m}{3}\left(1 + n^2 - n\right)\right]}\ . \tag{7.96}$$

Ignoring the mass of the lever arm, it is apparent from Equation 7.98 that the natural frequency of the mechanism varies inversely and, by implication, the speed of the system will reduce linearly, with the lever ratio.

Large amplifications can be obtained using two-bar levers that comprise two inks connected by a pivot, with both links being constrained in specific degrees of freedom. For two-bar levers, the amplification (or, because these can operate in reverse, attenuation) is best determined from a displacement diagram, as illustrated in Figure 7.98. For this symmetric flexure mechanism, the actuator pushes equally against the two side blocks, so that each block displaces half of the actuator displacement x_o. Motion of the actuator causes the support legs (of length L, mass m_L and second moment of mass about the centre centroid I_o) to rotate an angle α that, because of the constraints of this mechanism, causes the platform link to which it is connected to move in the vertical direction.

From the displacement diagram in Figure 7.98, the relationship between the rotation of the support legs and the link at either end is given by

$$\alpha = \frac{x_3}{2L}\ . \tag{7.97}$$

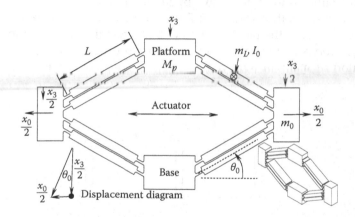

FIGURE 7.98
Symmetric amplification mechanism having twelve links and sixteen single degree of freedom joints with mobility of one.

From the displacement diagram, the lever ratio of this mechanism is

$$\frac{x_3}{x_0} = \frac{1}{\tan \theta_0} \cdot \qquad (7.98)$$

Based on this geometry, the kinetic and potential energy terms for this flexure are given by

$$T = \left[\frac{M_3}{2} + \frac{m_0}{4}\left(1 + \tan^2 \theta_0\right) + \frac{I_0}{L^2} + 10\frac{m_L}{8}\right] \dot{x}_3^2$$

$$\qquad (7.99)$$

$$U = 16\frac{k_{M\alpha}}{2}\left(\frac{x_3}{2L}\right)^2,$$

where M_3 is the mass of the moving platform, m_0 the mass of the side bars and $k_{M\alpha}$ ($N \cdot m \cdot rad^{-1}$) is the bending stiffness of the hinges. From Equation 7.101, the natural frequency of this mechanism is readily obtained. A more detailed discussion of levered mechanisms, including dynamic effects, is presented in Smith (2000).

Exercises

1. Derive the expression for the displacements at the location of the springs shown in Figure 7.2. Use the expressions for the displacements at each end of the bar to derive Equation 7.3 for the displacement and rotation at the point of the applied force. Derive Equations 7.4 through 7.6 in Section 7.1.1.
2. Calculate the stiffness of the two structures shown in Figure 7.99. Assume that these are constructed from prismatic beams of square cross section. Comment on the stiffness, symmetry and structural resemblance to conventional milling machines and coordinate measuring systems.
3. Determine an expression for the stiffness of a sphere held in contact with a flat surface by a normal load. For a steel sphere of radius 10 mm contacting a flat steel surface, determine the stiffness for a normally applied load of 10 N.
4. Derive the friction equation (Equation 7.10) and determine an expression for the coefficient of friction using the ploughing model. Discuss the Archard model from which the wear equation is derived. State how the dimensionless wear coefficient is related to the proportion of material removed as each asperity contact is traversed. A journal

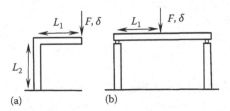

(a)   (b)

FIGURE 7.99
Schematic diagram of two structural frames: (a) dual cantilever, (b) symmetric bridge.

bearing is to support a 25 mm diameter shaft rotating at 600 revolutions per minute with an applied radial load of 400 N. How long would the bearing have to be if it is made of Rulon, Oilite or PTFE composite?

5. Explain the different friction and noise sources from rolling element bearings. It is common to use angular contact radial ball bearings for precision spindle applications. Explain how angular contact bearings are implemented for a precision spindle. State the relative merits of using a roller bearing instead of the angular contact bearings for the spindle shown in Figure 7.28b if this was intended to support a grinding wheel attached to the free end of the shaft. At which end of the shaft would the grinding wheel be mounted?

6. For the exponential wedge discussed in Section 7.2.3, show that the dimensionless pressure is given by

$$p* = \frac{1}{2 \ln H} \left\{ e^{2ax} - \frac{1}{H^2 + H + 1} \left((H+1)e^{3ax} + H^2 \right) \right\}. \qquad (7.133)$$

7. Derive Equations 7.30, 7.31, 7.39 and 7.43. Determine the flow through the bearing, the nominal separation and the bearing stiffness for a circular disk hydrostatic bearing. It may be assumed that the air can be considered as an incompressible fluid. The operating parameters for this bearing are P_s = 550,000 N·m^2, Pa = 100,000 N·m^2, r_o = 0.02 m, r_o/r_i = 1.2, W = 700 N, l_c = 0.004 m, d_c = 0.2 mm. Determine the stiffness and stiffness ratio of liquid-to-gas bearing. Note: The viscosity of air at 25 °C is 1.983 × 10^{-5} kg·m^{-1}·s^{-1}. Remember that the liquid analysis assumes that the fluid is incompressible (i.e. a liquid and not a gas).

8. Determine the vertical force that can be applied to a four-bar leaf-type (with leaf axes being vertical) mechanism at which its natural frequency will reduce to half of its value in the absence of this force.

9. Determine and plot the force characteristic shown in Figure 7.60 for a solenoid actuator given the parameters l_b = 0.2 m, l_p = 0.1 m, A_p = A_b = A_g = 0.0004 m^2, x_g = 0. mm, μ_r = 1000, N = 800, with coil currents 0.5 A to 2.6 A, at five equal increments. Use both the complete and approximate equations for the force and plunger separations ranging from 1 mm to 10 mm. Also produce a plot of force against current at 4 mm separation.

10. Derive expressions for the stiffness components of the assembled rod flexure coupling shown in Figure 7.90. With reference to the two-bar lever in Figure 7.98, state why the flexure of Figure 7.100 might be unstable with rapid temperature changes.

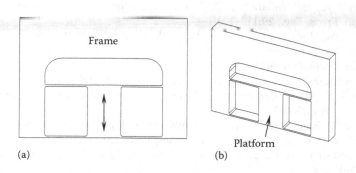

(a)   (b)

FIGURE 7.100
Symmetric over-constrained flexure: (a) front view, (b) isometric view.

References

Arneson H. E. G., 1967. *Hydrostatic bearing structure*, US patent 3,305,282.

Basile G., Becker P., Bergamin A., Cavagnero G., Franks A., Jackson K., Kuetgens U., Mana G., Palmer E. W., Robbie C. J., Stedman M., Stumpel J., Yacoot A., and Zosi G., 2000. Combined x-ray and optical interferometry for high-precision dimensional metrology, *Proc. Roy. Soc. A*, **456**:701–729.

Bauza M. B., Smith S. T., and Woody S. C., 2009. Development of a novel ultra-precision spindle, *Proc. ASPE*, **47**:14–17.

Binnig G., Rohrer H., Gerber Ch., and Weibel E., 1981. Tunnelling through a controllable vacuum gap, *J. Appl. Phys.*, **40**(2):178–180.

Bosse H., Lüdicke F., and Reimann H., 1994. An intercomparison on roundness and form measurement, *Measurement*, **13**(2):107–117.

Bowden F. P., and Tabor D., 1953 and 1964. *Friction and lubrication of solids*, vols. I and II, Oxford University Press.

Bowen D. K., Chetwynd D. G., and Schwarzenberger D. R., 1990. Sub-nanometre displacements calibration using x-ray interferometry, *Meas. Technol.*, **1**:107–109.

Brice J. C., 1985. Crystals for quartz resonators, *Rev. Mod. Phys.*, **57**(1):105–147.

Bryan J. B., 1979. Design and construction of an ultraprecision 84 inch diamond turning machine, *Precis. Eng.*, **1**(1):13–17.

Budynas R. G., and Nisbet K. D., 2014. *Shigley's machine component design*, 10th ed., McGraw-Hill.

Buice E. S., Otten D., Yang H., Smith S. T., Hocken R. J., and Trumper D.L., 2009. Design evaluation of a single axis, precision controlled positioning stage, *Precis. Eng.*, **33**:418–424.

Buice E. S., Yang H., Seugling R. M., Smith S. T., and Hocken R. J., 2005. Evaluation of a novel UHMWPE bearing for applications in precision slideways, *Precis. Eng.*, **30**:185–191.

Campbell P., 1994. *Permanent magnet materials and their application*, Cambridge University Press.

Chen J. S., and Dwang I. C., 2000. A ballscrew drive mechanism with piezo-electric nut for preload and motion control, *Precis. Eng.*, **40**(4):513–526.

Childs T. H. C., 1988. The mapping of metallic sliding wear, *Proc. Inst. Mech. Eng.*, **202**(C6):379–395.

Claverley J. D., and Leach R. K., 2013. Development of a three-dimensional vibrating tactile probe for miniature CMMs, *Precis. Eng.*, **37**:491–499.

Czichos H., and Habig K.-H., 1985. Lubricated wear of metals. In Dowson D. et al. (Eds.), *Mixed lubrication and lubricated wear, 11th Leeds-Lyon symposium on Tribology*, Butterworths, 135–147.

Evans C. R., 1989. *Precision engineering: An evolutionary view*, Cranfield Press.

Gibson I., Rosen D., and Stucker B., 2014. *Additive manufacturing technologies: 3D printing, rapid prototyping, and direct digital manufacturing*, 2nd ed., Springer.

Greenwood J. A., 2006. A simplified elliptic model of rough surface contact, *Wear* **261**:191–200.

Greenwood J. A., and Williamson J. B. P., 1966. The contact of nominally flat rough surfaces, *Proc. R. Soc. A*, **295**:300–319.

Grejda R., Marsh E. R., and Vallance R., 2005. Techniques for calibrating spindles with nanometer error motion, *Precis. Eng.*, **29**:113–123.

Hadfield D. (ed.), 1962. *Permanent magnets and magnetism*, John Wiley & Sons.

Halbach K., 1980. Design of permanent multipole magnets with oriented rare earth cobalt material, *Nucl. Instrum. Methods*, **169**(1):1–10.

Hart M., 1968. An angstrom ruler, *Brit. J. Appl. Phys. (J. Phys. D.)*, **1**(2):1405–1409.

Hopkins J. B., 2015. A visualization approach for analyzing and synthesizing serial flexure elements, *J. Mech. Robot.*, **7**(3):031011.

Horsfield W. R., 1965. Ruling engine with hydraulic drive, *Appl. Opt.*, **4**(2):189–195.

Hsu S. M., and Shen M. C., 1996. Ceramic wear maps, *Wear* **200**(1–2):154–175.

Hutchins I. M., and Shipway P., 2017 *Tribology: Friction and wear of engineering materials*, 2nd ed., Butterworth-Heinemann.

Jaffe B., Cook W. R., and Jaffe H., 1971. *Piezoelectric ceramics*, Academic Press.

Jain A., Prashanth K. J., Asheash Kr., Jain A., and Rashmi P. N., 2015. Dielectric and piezoelectric properites of PVDF/PZT composites, *Polymer Eng. Sci.*, 1589–1916.

Johnson K. L., Greenwood J. A., and Poon S. Y., 1972. A simple theory of asperity contact in elastohydrodynamic lubrication, *Wear*, 19:91–108.

Jona F., and Shirane G., 1993. *Ferroelectric crystals*, Diver Publications.

Juvinall R. C., and Marshek K. M., 2011. *Fundamentals of machine component design*, 5th ed., Wiley.

Kaajakari V., 2009. *Practical MEMS*, Small Gear Publishing.

Kadiric A., Sayles R. S., Zhou X. B., and Ioannides E., 2003. A numerical study of the contact mechanics and sub-surface stress effects experienced over a range of machined coatings in rough surface contact, *Trans. ASME*, 125:720–730.

Kluk D. J., Boulet M. T., and Trumper D. L., 2012. A high-bandwidth, high-precision, two-axis steering mirror with moving iron actuator, *Mechatronics*, 22:257–270.

Lawall J., and Kessler E., 2000. Michelson interferometry with 10 pm accuracy, *Rev. Sci. Instrum.*, 71 (7):2669–2676.

Lim S. C., and Ashby M. F., 1987. Wear-mechanism maps, *Acta Metall.*, 35(1):1–24.

Lindsey K., 1992. Tetraform grinding, *SPIE*, 1573:129–135.

Lindsey K., Smith S. T., and Robbie C. J., 1988. Sub-nanometre surface texture and profile measurement with 'Nanosurf 2', *Ann. CIRP* 37:519–522.

Lobontiu N., Paine J. S. N., O'Malley E., and Samuelson M., 2002. Parabolic and hyperbolic flexure hinges: Flexibility, motion precision and stress characterization based on compliance closed-form equations. *Precis Eng.*, 26(2):183–192.

Mikic B. B., 1974. Thermal contact conductance: Theoretical considerations, *Int. J. Heat Mass Tran.* 17:205–214.

Mizumoto H., Makoto Y., Shimizu T., and Kami Y., 1995. An Angstrom-positioning system using a twist-roller friction drive, *Precis. Eng.*, 17(1):57–62.

Moore W. R., 1970. *Foundations of mechanical accuracy*, The Moore Tool Company, Bridgeport, CT.

Morrison J. L. M., and Crossland B., 1970. *Mechanics of machines*, Longman.

Munnig Schmidt R.-H., 2012. Ultra-precision engineering in lithographic exposure equipment for the semiconductor industry, *Phil. Trans. R. Soc.*, A370:3950–3972.

Neale M. J. (ed.), 1973. *Tribology handbook*, Butterworth.

Neale M. J., Needham P., and Horrell H., 1991. *Couplings and shaft alignment*, Mechanical Engineering Publications, London.

Okazaki Y., Mishima M., and Ashida K., 2004. Microfactory: Concept, history, and developments, *ASME: J. Manuf. Sci.*, 126:837–844.

Paros J. M., and Weisbord L., 1965. How to design flexure hinges, *Mach. Des.*, 11:151–156.

Plante J.-S., Vogan J., El-Aguizy T., and Slocum A. H., 2005. A design model for circular porous air bearings using the 1D generalized flow method, *Precis. Eng.*, 29:336–346.

Pratt J. R., Kramar J. A., and Newell D. B., 2002. A flexure balance with adjustable restoring torque for nanonewton force measurement, *Proc. IMEKO Joint Int. Congress* (Celle, GE, 24–26 September) VDI-Berichte, 1685:77–82.

Pratt J .R., Kramar J.A., Newell D. B., and Smith D. T., 2005. Review of SI traceable force metrology for instrumented indentation and atomic force microscopy, *Meas. Sci. Technol.*, 16:2129–2137.

Rabinowicz E., 1980. In Petersen M. B., and Winer W. O., *Wear control handbook*, ASME, 475–506.

Raimondi A. A., and Boyd J., 1955. A solution for the finite journal bearing and its application to analysis and design, *Trans. ASLE*, 1(1):159–209.

Rayleigh J. W. S., 1918. Notes on the theory of lubrication, *Phil. Mag.*, 35:1–12.

Rowe W. B., 2012. *Hydrostatic, aerostatic and hybrid bearing design*, Butterworth-Heinemann Press.

Rowland H. A., 1902. *The physical papers of Henry Augustus Rowland*, John Hopkins Press.

Sacconi H., and Pasin W., 1994. An intercomparison of roundness measurements between ten national standards laboratories, *Measurement*, 13(2):119–128.

Sayles R. S., 1996. Basic principles of rough surface contact analysis using numerical methods, *Tribol. Int.* 29(8):639–650.

Schmidt R.M., Schitter G., and van Eijk J., 2011. *The design of high performance mechatronics*, Delft University Press.

Shaw G. A., Stirling J., Kramar J. A., Moses A., Abbott P., Steiner R., Koffman A., Pratt J. R., and Kubarych Z. J., 2016. Milligram mass metrology using an electrostatic force balance, *Metrologia*, **53**:A86–A94.

Slocum A. H., 1991. Design and testing of a self coupling hydrostatic leadscrew, *Progress in Precision Engineering: Proceedings of the 6th IPEs conference*, Springer–Verlag, 103–105. Edited by P. Seyfried, H. Kunzmann, P. McKeown, and M. Weck.

Smith S. T., 2000. *Flexures: Elements of elastic mechanism design*, CRC Press.

Smith S. T., Badami V. G., Dale J. S., and Xu Y., 1997. Elliptical flexure hinges, *Rev. Sci. Instrum.*, **68** (3):1474–1483.

Smith S. T., and Chetwynd D. G, 1990. Optimisation of a magnet/coil force actuator and its application to linear spring mechanisms, *Proc. Inst. Mech. Engrs.*, 204(C4):243–253.

Smith S. T., and Seugling R. M., 2006. Review paper: Sensor and actuator considerations for precision, small machines, *Precis. Eng.*, **30**(3):245–264.

Spotts M. F., Shoup T. E., and Hornberger L. E., 2003. *Design of machine elements*, 8th ed., Prentice-Hall.

Tran H. D., and Debra D. B., 1994. Design of a fast short-stroke hydraulic actuator, *Ann. CIRP*, **43**(1): 469–472.

Trumper D. L., Williams M. E., and Nguyen T. H., 1993. Magnet arrays for synchronous machines. Conference Record of the 1993 IEEE Industry Applications Society Annual Meeting, Piscataway, IEEE, 1:9–18

Whitehouse D. J., and Archard J. F., 1970. The properties of random surfaces of significance in their contact, *Proc. R. Soc.*, **A316**:97–121.

Wittrick W. H., 1948. The properties of crossed flexure pivots and the influence of where the strips cross, *Aeronaut. Quart.*, **A1**(2):121–134.

Woody S. C., and Smith S. T., 2004. Performance comparison and modeling of PZN, PMN and PZT stacked actuators in a levered flexure mechanism, *Rev. Sci. Instrum.*, **75**(4):842–848.

Woody S. C., Smith S. T., Rehrig P. W., and Xiaoning J., 2005. Performance of single crystal $Pb(Mg_{1/3}Nb_{2/3})$-32%$PbTiO_3$ stacked actuators with application to adaptive structures, *Rev. Sci. Instrum.*, **76**:075112.

Yoshimoto S., and Kohno K., 2001. Static and dynamic characteristics of aerostatic circular porous thrust bearings, *Trans. ASME: J. Tribol.*, **123**:501–508.

Ziman J. M., 1972. *Principles of the theory of solids*, 2nd ed., Cambridge University Press.

8

System Modelling

Richard M. Seugling

CONTENTS

ABSTRACT   The purpose of this chapter is to illustrate a relatively simple, structured approach for modelling precision systems that can be used to evaluate a broad variety of instrument and machine designs. A number of case studies and examples show how, when used in conjunction with the design principles discussed in other chapters, relatively simple mathematical techniques can be employed to determine critical aspects of a given design. In this chapter it is assumed that mechanisms are comprised from perfect rigid bodies that translate or rotate by means of a guided mechanism (slideways, crossed-roller bearings, air bearings, etc.) that, although not providing ideal rectilinear motions, are repeatable. Specifically, rigid body motions and the inclusion of error sources will be described in detail. With this in mind, the following represents a method for evaluating the combined effects of errors in a mechanism comprised from an assembly of translation and rotation axes. In particular, a method utilising homogeneous transformation matrices for assessing propagation of uncertainties will be developed.

8.1 Basic Rigid-Body Dynamics

Rigid-body dynamics describe the motion of systems subject to external forces. In the case of a rigid body, the force or forces acting on the body do not produce significant deformation of the body itself and represent the reaction motion of the body in the form of both translations and rotations. A rigid body in space has six degrees of freedom (DOF) comprised of three translational and three rotational DOF (McCarthy 1990), and is described by an isomorphic mapping of two or more coordinate systems (McCarthy 1990, Lay 1994). Mathematical models used in this chapter describe the motion of a system combining both the ideal case where every system element and constraint behaves as a perfect system and an estimate of the errors associated with variability of the real system. The mathematical methods used in this chapter are derived from extensive developments in the analysis of robotic systems (see Fu, Gonzalez and Lee 1987; Paul 1981; Tsai 1999).

Figure 8.1 shows a body constrained in-plane with its location being measured relative to an arbitrary reference coordinate system. In this example, the rigid body position is located by the distances to the centre of gravity (CG) of the body relative to the reference coordinate system origin. The relative position of a body relative to a reference coordinate system illustrates a fundamental concept of this chapter, from which the motion and/or performance of a system including errors can be evaluated using a relatively simple mathematical representation.

8.1.1 Cartesian Coordinates

Cartesian coordinates are most commonly used to describe the position of a point in two-dimensional (2D) and three-dimensional (3D) space, where the two or three axes are defined by orthogonal lines as shown in Figure 8.1. The intersection of the axes is defined as the origin of the coordinate system. In the case of 2D, the position of point P can be represented by an ordered pair (x,y). Similarly for 3D, a point P would be represented by an ordered triple (x,y,z), as shown in Figure 8.2.

Cartesian coordinates are the basis for most of the mathematics describing structures and/or moving bodies in space, and will be the basis for all the analysis defined in this chapter. Differences between the Cartesian coordinates of a point of interest in an ideal

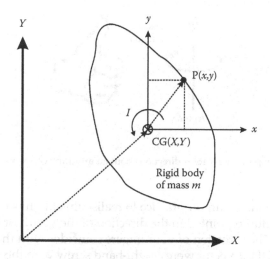

FIGURE 8.1
Arbitrary rigid body in two-dimensional coordinate space. The centre of gravity (CG) of the body relative to the base coordinate system is given by CG(X,Y) and the point P in local coordinates is P(x,y). The point P relative to the global coordinate system is the vector sum of CG(X,Y) and P(x,y).

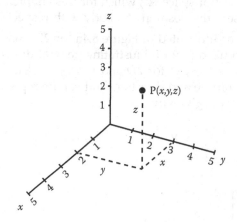

FIGURE 8.2
Three-dimensional Cartesian coordinate system. The location of point P represented by ordered triple (x,y,z).

mechanism and the location of this point in a real machine represent the errors of the system and are the focus of this chapter for calculating quantifiable metrics for evaluating design performance.

Since a rigid body in space is defined by six values, three translations and three rotations representing its six independent DOFs, it is necessary to be consistent with the definition of the rotation direction. To be consistent throughout this chapter, the 'right-hand rule' will be used to define the direction of rotation about a vector or axis. The right-hand rule is often the most common method of defining the rotation direction, but not the only method. The reader should use caution when defining rotation direction as variations or inconsistencies in choosing a rotation direction relative to a vector or coordinate axis can lead to large errors. Figure 8.3 illustrates the proper rotation direction about an arbitrary vector using the

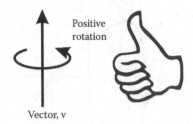

FIGURE 8.3

Right-hand rule showing the positive rotation direction about an arbitrary vector **v**.

right-hand rule definition. Common practice to realise this in physical terms is to use your right hand with your thumb pointed in the direction of the vector or coordinate axis, while the fingers are curled. The direction of your fingers coincides with the direction of rotation, as shown in Figure 8.3. If the vector were a right-hand screw with the arrow being its tip, the direction indicated would correspond to screwing this into an object.

8.1.2 Vectors

Vectors are elements of a vector space $\mathbb{R}^n$, which for this chapter is limited to $\mathbb{R}^2$ (2D) and $\mathbb{R}^3$ (3D) spaces that represent the position of a body with respect to a reference coordinate system. A vector $\mathbf{v} = \overrightarrow{AB}$, as illustrated in Figure 8.4a for 2D and Figure 8.4b for 3D, has a tail point labelled A and a head point B illustrating general direction and magnitude. The vector **v** can be written as $\mathbf{v} = \{x, y\}$ for 2D and $\mathbf{v} = \{x, y, z\}$ for 3D space. By definition, the vector **v** has both magnitude and direction, and can be represented by combination of the two in more general terms given by

$$\mathbf{v} = |v|\hat{\mathbf{v}} = |v|\{\mathbf{i},\mathbf{j},\mathbf{k}\}, \tag{8.1}$$

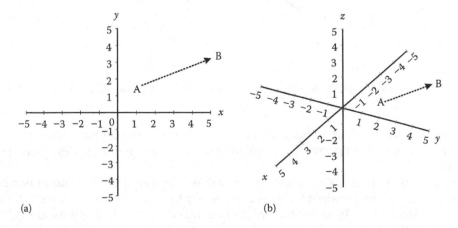

FIGURE 8.4

(a) Two-dimensional vector $\mathbf{v} = \overrightarrow{AB} \rightarrow A(x_A,y_A)$, $B(x_B,y_B)$. (b) Three-dimensional vector $\mathbf{v} = \overrightarrow{AB} \rightarrow A(x_A,y_A,z_A)$, $B(x_B,y_B,z_B)$.

where $\hat{v}$ is a unit vector describing the general direction of the vector in vector space with a magnitude of unity and $|v|$ is the magnitude or length of the vector given by

$$|v| = \sqrt{x^2 + y^2},\tag{8.2}$$

and

$$|v| = \sqrt{x^2 + y^2 + z^2},\tag{8.3}$$

for 2D and 3D space respectively. The notation **i**, **j** and **k** represent the standard basis for a 3D Euclidean space where three independent unit vectors, **i**, **j** and **k** map to the x, y and z Cartesian coordinate axes respectively. A vector can, therefore, be arbitrarily oriented with respect to any coordinate system, but can always be represented by Equation 8.1 for any general case. For a more complete description of vector mathematics (see Shilov 1977, Dettman 1986, Kreyszig 1993, Wylie and Barrett 1995). It is also common in many mathematics texts to represent an arbitrary unit vector using e_n, where n is the number of distinct vectors in the basis. Since the modelling derived in this chapter is based on Cartesian coordinates, the **i**, **j** and **k** notation will be used throughout this chapter.

8.1.3 Direction Cosines

Direction cosines are defined as the cosine of the angle between an arbitrary position vector and each of the axes of the base coordinate system. Figure 8.5 shows an arbitrary position vector **v** and the angles relative to the base coordinate system, where a is the angle between the vector **v** and the x-axis, b is the angle between the vector **v** and the y-axis and c is the angle between the vector **v** and the z-axis. The direction cosines are then given by

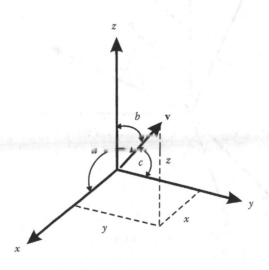

FIGURE 8.5
Direction cosines $\alpha = \cos(a)$, $\beta = \cos(b)$ and $\gamma = \cos(c)$ of an arbitrary vector **v** in a 3D coordinate system.

$$\alpha = \cos a = \frac{\mathbf{v} \cdot \mathbf{i}}{|v|}, \tag{8.4}$$

$$\beta = \cos b = \frac{\mathbf{v} \cdot \mathbf{j}}{|v|}, \tag{8.5}$$

$$\gamma = \cos c = \frac{\mathbf{v} \cdot \mathbf{k}}{|v|}. \tag{8.6}$$

In this case, the unit vector $\hat{\mathbf{v}}$ can be expressed by the following:

$$\hat{\mathbf{v}} = \frac{\mathbf{v}}{|v|} = (\cos a)\,\mathbf{i} + (\cos b)\,\mathbf{j} + (\cos c)\,\mathbf{k} = \alpha\,\mathbf{i} + \beta\,\mathbf{j} + \gamma\,\mathbf{k}. \tag{8.7}$$

Equation 8.7 represents an important concept for mapping the position of an arbitrary vector in space. To represent the position of a point in space relative to a base coordinate system, consider two orthogonal coordinate systems as shown in Figure 8.6. In this case, the point P is considered part of a rigid body represented by the rotated coordinate system C', which is rotated about a common origin relative to a reference coordinate system $\mathbf{C_{ref}}$. The reference coordinate $\mathbf{C_{ref}}$ can be represented by three independent unit vectors $\{\mathbf{i},\mathbf{j},\mathbf{k}\}$ with a common origin and in a similar fashion the rotated coordinate system C' can be represented by three independent unit vectors $\{\mathbf{i}',\mathbf{j}',\mathbf{k}'\}$. Expressing each axis of the rotated coordinate

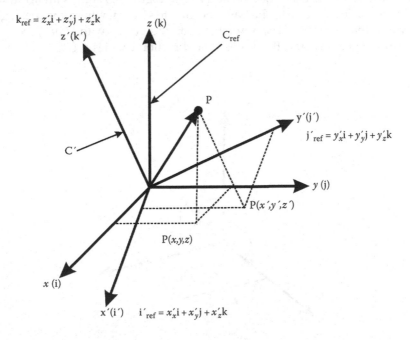

FIGURE 8.6
Point P represented in both the xy and $x'y'$ coordinate systems. The $x'y'$ coordinate system is rotated about the origin of the xy coordinate system.

system as an independent unit vector relative to the reference coordinate system C_{ref} results in the following relationship between the coordinates:

$$\mathbf{i}'_{ref} = x'_x \mathbf{i} + x'_y \mathbf{j} + x'_z \mathbf{k}, \tag{8.8}$$

$$\mathbf{j}'_{ref} = y'_x \mathbf{i} + y'_y \mathbf{j} + y'_z \mathbf{k}, \tag{8.9}$$

$$\mathbf{k}'_{ref} = z'_x \mathbf{i} + z'_y \mathbf{j} + z'_z \mathbf{k}, \tag{8.10}$$

where $x'_x, x'_y, x'_z, y'_x, y'_y, y'_z, z'_x, z'_y$ and z'_z are the scalar coordinates of each of the three unit vectors of the rotated coordinate system $\mathbf{C}'$ relative to the reference coordinate system C_{ref}. The position vector $\mathbf{P}$ relative to both the C_{ref} and $\mathbf{C}'$ can be written as

$$\mathbf{P}_{ref} = x\mathbf{i} + y\mathbf{j} + z\mathbf{k}, \tag{8.11}$$

$$\mathbf{P}' = x'\mathbf{i}' + y'\mathbf{j}' + z'\mathbf{k}'. \tag{8.12}$$

To map the position of point $\mathbf{P}$ that is defined in $\mathbf{C}'$ relative to the reference system C_{ref}, Equations 8.8, 8.9 and 8.10 can be substituted into Equation 8.12, giving

$$\mathbf{P}_{ref} = \left(x'x'_x + y'y'_x + z'z'_x\right)\mathbf{i} + \left(x'x'_y + y'y'_y + z'z'_y\right)\mathbf{j} \\ + \left(x'x'_z + y'y'_z + z'z'_z\right)\mathbf{k}. \tag{8.13}$$

The components of $\mathbf{P}_{ref}$ can be expressed by the following three equations:

$$x = x'x'_x + y'y'_y + z'z'_z, \tag{8.14}$$

$$y = x'x'_x + y'y'_y + z'z'_z, \tag{8.15}$$

$$z = x'x'_x + y'y'_y + z'z'_z. \tag{8.16}$$

Equations 8.14, 8.15 and 8.16 can be compactly represented in matrix form,

$$\mathbf{P}_{ref} = {}^{ref}\mathbf{R}\mathbf{P}', \tag{8.17}$$

where ${}^{ref}\mathbf{R}$ is a rotation matrix defining the location of $\mathbf{C}'$ relative to C_{ref}. Based on Equation 8.17, a general rotation matrix $\mathbf{R}$ can be derived using direction cosines. The direction cosines between each axis of the base coordinate system and each unit vector of the rotated coordinate system can be expressed as

$$\lambda_{i',j} = \cos(\mathbf{i}',\mathbf{j}), \tag{8.18}$$

or expanded for a 3D rotation matrix

$$\mathbf{R} = \begin{bmatrix} \lambda_{i',i} & \lambda_{i',j} & \lambda_{i',k} \\ \lambda_{j',i} & \lambda_{j',j} & \lambda_{j',k} \\ \lambda_{k',i} & \lambda_{k',j} & \lambda_{k',k} \end{bmatrix}. \tag{8.19}$$

Equation 8.17 represents an orthonormal transformation between orthogonal coordinate systems having a common origin. The motivation for deriving the matrix method is to represent the relationship between a reference coordinate system and the coordinate system of interest by the matrix of Equation 8.19. Rotation matrices are a powerful mathematical construct to describe and diagnose motions of systems with almost arbitrary complexity.

8.1.4 Rotation Matrices

In general, a rotation matrix can be developed for an arbitrary vector such as $\mathbf{v} = a\mathbf{i} + b\mathbf{j} + c\mathbf{k}$ using the analysis illustrated in the Section 8.1.3. Figure 8.7 shows a point $\mathbf{P}$ in a 2D Cartesian coordinate system $\mathbf{C}'$ rotated θ about the origin of a reference coordinate system $\mathbf{C}_{ref}$. From basic geometry, the point $\mathbf{P}$ can be expressed relative to the $\mathbf{C}_{ref}$ coordinates as

$$x = x' \cos \theta - y' \sin \theta, \tag{8.20}$$

$$y = x' \sin \theta + y' \cos \theta. \tag{8.21}$$

Equations 8.20 and 8.21 can be written in matrix form expressing the point $P(x,y)$ in $\mathbf{C}_{ref}$ coordinates as

$$\begin{bmatrix} x \\ y \end{bmatrix} = \begin{bmatrix} \cos \theta & -\sin \theta \\ \sin \theta & \cos \theta \end{bmatrix} \begin{bmatrix} x' \\ y' \end{bmatrix}, \tag{8.22}$$

or

$$\mathbf{P} = \mathbf{R}\mathbf{P}'. \tag{8.23}$$

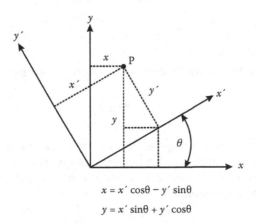

$$x = x' \cos\theta - y' \sin\theta$$
$$y = x' \sin\theta + y' \cos\theta$$

FIGURE 8.7
In-plane rotation of 2D coordinate system relative to reference by angle θ.

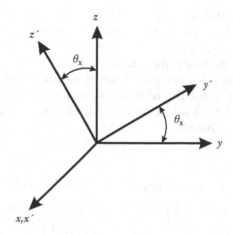

FIGURE 8.8
Three-dimensional coordinate rotation (θ_x) about the x-axis.

Equation 8.23 can be expanded to 3D using the rotation matrix **R** of Equation 8.19. Consider a rotation (θ_x) of **C'** about the x-axis of **C$_{ref}$** as shown in Figure 8.8, by realising that Equation 8.19 can be rewritten in terms of the unit vectors of each of the two coordinate systems being evaluated by

$$\mathbf{R}_{\theta_x} = \begin{bmatrix} \mathbf{i'} \cdot \mathbf{i} & \mathbf{i'} \cdot \mathbf{j} & \mathbf{i'} \cdot \mathbf{k} \\ \mathbf{j'} \cdot \mathbf{i} & \mathbf{j'} \cdot \mathbf{j} & \mathbf{j'} \cdot \mathbf{k} \\ \mathbf{k'} \cdot \mathbf{i} & \mathbf{k'} \cdot \mathbf{j} & \mathbf{k'} \cdot \mathbf{k} \end{bmatrix} = \begin{bmatrix} 1 & 0 & 0 \\ 0 & \cos\theta_x & -\sin\theta_x \\ 0 & \sin\theta_x & \cos\theta_x \end{bmatrix}. \tag{8.24}$$

Similar rotation matrices can be derived independently for rotations about the y-axis and z-axis of a coordinate system given by

$$\mathbf{R}_{\theta_y} = \begin{bmatrix} \cos\theta_y & 0 & \sin\theta_y \\ 0 & 1 & 0 \\ -\sin\theta_y & 0 & \cos\theta_y \end{bmatrix}, \tag{8.25}$$

$$\mathbf{R}_{\theta_z} = \begin{bmatrix} \cos\theta_z & -\sin\theta_z & 0 \\ \sin\theta_z & \cos\theta_z & 0 \\ 0 & 0 & 1 \end{bmatrix}. \tag{8.26}$$

These types of matrices are often referred to as basic rotation matrices and will be employed extensively in the following sections.

8.1.5 Euler Angles

A common method in classical dynamics to determine the orientation of a rigid body is the use of Euler angles (Pars 1965), which describe any orientation in terms of three rotations.

To obtain an arbitrary orientation, this method specifies three sequential rotations starting with a rotation (φ) about the z-axis, then a rotation (ω) about the new y-axis (y') and finally a rotation (ψ) about the new z-axis (z''), as illustrated in Figure 8.9.

Consider Figure 8.9d to represent the *final* orientation of the *uvw* coordinate system relative to the fixed frame of reference *xyz*. A set of rotation matrices can be defined and multiplied together to represent the relative position of *uvw* with respect to *xyz*. From Equations 8.24, 8.25 and 8.26, a set of rotation matrices can be derived representing the rotations described earlier, given by

$$\mathbf{R} = \mathbf{R}_{z,\phi}\mathbf{R}_{y',\omega}\mathbf{R}_{z'',\psi}$$

$$= \begin{bmatrix} \cos(\phi) & -\sin(\phi) & 0 \\ \sin(\phi) & \cos(\phi) & 0 \\ 0 & 0 & 1 \end{bmatrix} \begin{bmatrix} \cos(\omega) & 0 & \sin(\omega) \\ 0 & 1 & 0 \\ -\sin(\omega) & 0 & \cos(\omega) \end{bmatrix} \begin{bmatrix} \cos(\psi) & -\sin(\psi) & 0 \\ \sin(\psi) & \cos(\psi) & 0 \\ 0 & 0 & 1 \end{bmatrix} \quad (8.27)$$

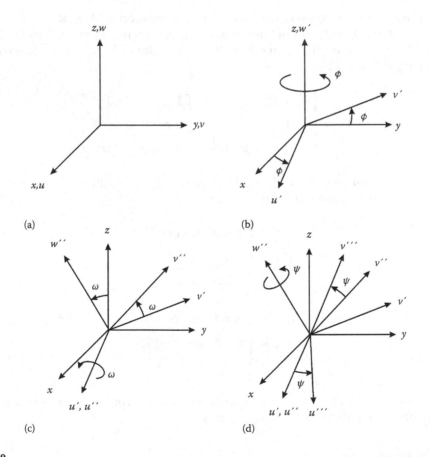

FIGURE 8.9
(a) Initial state of two Cartesian coordinate systems *xyz* and *uvw* where the origins and axes are oriented and aligned. (b) Initial rotation (ϕ) about the z-axis creating the *u'v'w'* coordinate. (c) A second rotation (ω) about the *u'*-axis giving *u''v''w''*. (d) A third rotation (ψ) about the *w''*-axis representing the final location of *u'''v'''w'''*.

$$
= \begin{bmatrix} \cos(\phi)\cos(\omega)\cos(\psi) - \sin(\phi)\sin(\psi) & -\cos(\phi)\cos(\omega)\sin(\psi) - \sin(\phi)\cos(\psi) \\ \sin(\phi)\cos(\omega)\cos(\psi) + \cos(\phi)\sin(\psi) & -\sin(\phi)\cos(\omega)\sin(\psi) + \cos(\phi)\cos(\psi) \\ -\sin(\omega)\cos(\psi) & \sin(\omega)\sin(\psi) \end{bmatrix}
$$

$$
\begin{matrix} \cos(\phi)\sin(\omega) \\ \sin(\phi)\sin(\omega) \\ \cos(\omega) \end{matrix} \Bigg].
$$

(8.28)

Equation 8.28 can be written in the more compact form

$$
\mathbf{R} = \begin{bmatrix} c(\phi)c(\omega)c(\psi) - s(\phi)s(\psi) & -c(\phi)c(\omega)s(\psi) - s(\phi)c(\psi) & c(\phi)s(\omega) \\ s(\phi)c(\omega)c(\psi) + c(\phi)s(\psi) & -s(\phi)c(\omega)s(\psi) + c(\phi)c(\psi) & s(\phi)s(\omega) \\ -s(\omega)c(\psi) & s(\omega)s(\psi) & c(\omega) \end{bmatrix},
$$

(8.29)

where s represents sine and c cosine.

The aforementioned transformation represents one possible sequence of rotations to describe the orientation of a rotated coordinate system relative to a reference frame. In general, any sequential combination of rotation matrices can represent the position of one coordinate system relative to another. However, the derived matrix using a different order of transformations will not be equivalent to Equation 8.29, as matrix multiplication is not always commutative (Shilov 1977, Goldstein 1980, Dettman 1986).

8.1.6 Roll, Pitch and Yaw

The terms roll, pitch and yaw (RPY) are most often associated with aeronautics describing the angular motion of an aircraft as it moves through the atmosphere. In a similar fashion, roll, pitch and yaw in this chapter refer to the angular motion of a rigid body as it moves along its prescribed path, such as the motion of a carriage sliding on ways discussed in the Chapter 7. The RPY angles define the rotations about each axis in succession based on the direction of motion of the system, similar to the Euler angles described in Section 8.1.5. Figure 8.10

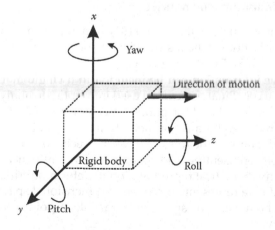

FIGURE 8.10

Rigid body translating along the z-axis as defined by the local coordinate system. In this case roll is about the z-axis (θ_z), pitch is about the y-axis (θ_y), and yaw is about the x-axis (θ_x).

illustrates the RPY of a body relative to its base coordinate system. In Figure 8.10, the roll is defined by the rotation about the z-axis (θ_z), the pitch is defined by the rotation about the y-axis (θ_y) and the yaw is defined by the rotation about the x-axis (θ_x). Although the distinction is minor as to how the roll, pitch and yaw axes are defined, misinterpretation of the rotation operations will have dramatic effects on the resulting values. In this chapter, the combined rotation matrices based on the RPY method will be used unless otherwise stated. However, the definition of the base coordinate system can be arbitrary as will be seen later in this chapter, but the definitions of roll, pitch and yaw are consistent. The roll, pitch and yaw matrices can be written in sequence and are given by

$$\mathbf{R} = \mathbf{R}_{\theta_z}\mathbf{R}_{\theta_y}\mathbf{R}_{\theta_x}$$

$$= \begin{bmatrix} \cos(\theta_z) & -\sin(\theta_z) & 0 \\ \sin(\theta_z) & \cos(\theta_z) & 0 \\ 0 & 0 & 1 \end{bmatrix} \begin{bmatrix} \cos(\theta_y) & 0 & \sin(\theta_y) \\ 0 & 1 & 0 \\ -\sin(\theta_z) & 0 & \cos(\theta_y) \end{bmatrix} \begin{bmatrix} 1 & 0 & 0 \\ 0 & \cos(\theta_x) & -\sin(\theta_x) \\ 0 & \sin(\theta_x) & \cos(\theta_x) \end{bmatrix}$$

$$= \begin{bmatrix} c(\theta_z)c(\theta_y) & c(\theta_z)s(\theta_y)s(\theta_x) - s(\theta_z)c(\theta_x) & s(\theta_z)s(\theta_x) + c(\theta_z)s(\theta_y)c(\theta_x) \\ s(\theta_z)c(\theta_y) & c(\theta_z)c(\theta_x) + s(\theta_z)s(\theta_y)s(\theta_x) & s(\theta_z)s(\theta_y)c(\theta_x) - c(\theta_z)s(\theta_x) \\ -s(\theta_y) & c(\theta_y)s(\theta_x) & c(\theta_y)c(\theta_x) \end{bmatrix}, \qquad (8.30)$$

where c is cosine and s is sine as used in the previous section. This combined rotation matrix will be used extensively throughout this chapter.

8.2 Mathematical Models

8.2.1 Homogeneous Transformation Matrices

The homogeneous transformation matrix (HTM) is a mathematical representation that maps a position vector in homogeneous coordinates between relative coordinate systems (Maxwell 1961, Stadler 1995). By utilising homogeneous coordinates, rotation, translation, scaling and perspective information can be included in a single matrix operation. Homogeneous transformation coordinates are often used in robotics to analyse the kinematics and dynamics of complicated serial and/or parallel mechanisms. HTM modelling techniques are used in mechanism design because of their ability to combine the motions of complex systems in a structured manner. The role of HTMs in precision machine/instrument design is primarily focused on representing and quantifying the influence of error motions of a system and using this mathematical representation to determine critical aspects of a design.

In general, the coordinate representation of an N-dimensional space is transformed to an N + first-dimensional homogeneous space. For example, a point vector $\mathbf{v}$ in 3D Cartesian coordinates is given by

$$\mathbf{v} = a\mathbf{i} + b\mathbf{j} + c\mathbf{k}. \qquad (8.31)$$

Vector **v** can be represented as a column matrix in homogeneous space by

$$\mathbf{v} = \begin{bmatrix} x \\ y \\ z \\ s \end{bmatrix}, \tag{8.32}$$

where

$$\begin{aligned} a &= x/s \\ b &= y/s \\ c &= z/s. \end{aligned} \tag{8.33}$$

In the case of Equation 8.33, a, b and c are the scaled magnitudes of the x-, y- and z-axes respectively. Consequently, a homogeneous representation of a 3D Euclidean point vector is not unique. For example, the vectors $\mathbf{v}_1$ and $\mathbf{v}_2$ represented by $\mathbf{v}_1 = [2\,4\,6\,2]^T$ or $\mathbf{v}_2 = [4\,8\,12\,4]^T$ in homogeneous space are equivalent to the general vector $\mathbf{v} = [1\,2\,3]^T$ in vector space, with scale factors s_1 and s_2 of 2 and 4 respectively.

A point vector in a local coordinate system can be represented in global homogeneous coordinates using a transformation matrix $^A\mathbf{T}_B$ of the following form:

$$^A\mathbf{T}_B = \begin{bmatrix} \mathbf{R}_{3\times3} & \mathbf{D}_{3\times1} \\ \mathbf{P}_{1\times3} & \mathbf{s}_{1\times1} \end{bmatrix}, \tag{8.34}$$

where $\mathbf{R}$ is a rotation matrix, $\mathbf{D}$ is a displacement vector between coordinate system origins, $\mathbf{P}$ is a perspective transformation and $\mathbf{s}$ is the scale factor. For the purposes of modelling kinematic mechanisms, machine tools or robot manipulator motion, the perspective transformation is set to zero and the scale factor $\mathbf{s}$ is set to unity, resulting in the following matrix:

$$^A\mathbf{T}_B = \begin{bmatrix} \mathbf{R}_{3\times3} & \mathbf{D}_{3\times1} \\ \mathbf{0}_{1\times3} & 1 \end{bmatrix} \tag{8.35}$$

The perspective transformation $\mathbf{P}$ and scale factor $\mathbf{s}$ are commonly used for 3D scene analysis as discussed in more detail elsewhere (Duda and Hart 1973). For the purposes of this chapter the general Equation 8.35 will be used to represent homogeneous transformations between coordinate systems, as $\mathbf{P}$ and $\mathbf{s}$ can add confusion and do not add relevant information in most cases.

8.2.2 Fundamental Transformations

Using HTMs, a set of fundamental transformation matrices can be generated to describe the motion of serial and/or parallel systems. Simple translations are given by multiplying the three unit translation matrices where a combined translation matrix can be formed as follows:

$$D_{x,y,z} = T_x T_y T_z = \begin{bmatrix} 1 & 0 & 0 & x \\ 0 & 1 & 0 & y \\ 0 & 0 & 1 & z \\ 0 & 0 & 0 & 1 \end{bmatrix}. \tag{8.36}$$

In this case, the order of multiplication is inconsequential as any combination will yield the same result for this specific case. Along with the three translation matrices, there are three rotation matrices given by

$$R_{\theta_x} = \begin{bmatrix} 1 & 0 & 0 & 0 \\ 0 & \cos(\theta_x) & -\sin(\theta_x) & 0 \\ 0 & \sin(\theta_x) & \cos(\theta_x) & 0 \\ 0 & 0 & 0 & 1 \end{bmatrix}, \tag{8.37}$$

$$R_{\theta_y} = \begin{bmatrix} \cos(\theta_y) & 0 & \sin(\theta_y) & 0 \\ 0 & 1 & 0 & 0 \\ -\sin(\theta_y) & 0 & \cos(\theta_y) & 0 \\ 0 & 0 & 0 & 1 \end{bmatrix}, \tag{8.38}$$

$$R_{\theta_z} = \begin{bmatrix} \cos(\theta_z) & -\sin(\theta_z) & 0 & 0 \\ \sin(\theta_z) & \cos(\theta_z) & 0 & 0 \\ 0 & 0 & 1 & 0 \\ 0 & 0 & 0 & 1 \end{bmatrix}. \tag{8.39}$$

Because matrix multiplication is not always commutative, a combined rotation matrix similar to Equation 8.30 is dependent on the order of each subsequent rotation, as discussed in Section 8.1.6. However, it is common in machine analysis to define angular motions of a system by the RPY, which has a homogeneous transformation matrix given by

$$R_{\theta_z \theta_y \theta_x} = R_{\theta_z} R_{\theta_y} R_{\theta_x}$$

$$= \begin{bmatrix} c(\theta_z)c(\theta_y) & c(\theta_z)s(\theta_y)s(\theta_x) - s(\theta_z)c(\theta_x) & s(\theta_z)s(\theta_x) + c(\theta_z)s(\theta_y)c(\theta_x) & 0 \\ s(\theta_z)c(\theta_y) & c(\theta_z)c(\theta_x) + s(\theta_z)s(\theta_y)s(\theta_x) & s(\theta_z)s(\theta_y)c(\theta_x) - c(\theta_z)s(\theta_x) & 0 \\ -s(\theta_y) & c(\theta_y)s(\theta_x) & c(\theta_y)c(\theta_x) & 0 \\ 0 & 0 & 0 & 1 \end{bmatrix}. \tag{8.40}$$

Combining the translation matrix **D** with the rotation matrix **R** as shown in Equation 8.40, a combined HTM (**H$_{RPY}$**) representing RPY and simple translations (**T**) can be derived and is given by the following:

$$\mathbf{H}_{RPY} = \mathbf{T}_z \mathbf{T}_y \mathbf{T}_x \mathbf{R}_{\theta_z} \mathbf{R}_{\theta_y} \mathbf{R}_{\theta_x}$$

$$= \begin{bmatrix} c(\theta_z)c(\theta_y) & c(\theta_z)s(\theta_y)s(\theta_x) - s(\theta_z)c(\theta_x) & s(\theta_z)s(\theta_x) + c(\theta_z)s(\theta_y)c(\theta_x) & x \\ s(\theta_z)c(\theta_y) & c(\theta_z)c(\theta_x) + s(\theta_z)s(\theta_y)s(\theta_x) & s(\theta_z)s(\theta_y)c(\theta_x) - c(\theta_z)s(\theta_x) & y \\ -s(\theta_y) & c(\theta_y)s(\theta_x) & c(\theta_y)c(\theta_x) & z \\ 0 & 0 & 0 & 1 \end{bmatrix}. \quad (8.41)$$

In general, an arbitrary number of transformations can be applied in succession to model a system. The advantage of this type of analysis is that a single combined transformation matrix for each sub-component of a system can be derived and these can then be combined with relative ease to evaluate the performance of a complex assembly.

This is a good point in the chapter to discuss notation. As the system and subsequent analysis becomes more complicated, the mathematical bookkeeping required to track the sub-system components and their relative position within the complete assembly can be confusing. To help keep the independent parts of the equations unique and distinguishable throughout the process, a standard notation will be used in the rest of the chapter. First, angles of rotation will be defined about a defined axis. As shown in Figure 8.8, the angle θ representing a rotation about the x-axis will be given by θ_x. In a similar fashion, a rotation matrix about the x-axis is given by $\mathbf{R}_{\theta_x}$. Homogeneous transformation matrices will be represented by $^{finish}\mathbf{T}_{start}$, where the subscript on the bottom right is the matrix base system (start) and the superscript on the upper left is where the transformation will be related (finish).

8.2.3 Vector Equivalence

Homogeneous transformation matrices are not the only method that utilises vector and matrix methods to relate the relative positions of systems. Vectors in combination with rotation matrices can also be used to relate a position vector in an arbitrary 2D or 3D coordinate system back to a reference, as illustrated in Figure 8.11. In this case, a point P in C' relative to the reference coordinates C_{ref} can be expressed by vectors given by

$$^{ref}\mathbf{P} = \mathbf{C}'_{ref} + \mathbf{R}_{\theta}\mathbf{P}', \quad (8.42)$$

where $\mathbf{C}'_r$ is the vector between the $\mathbf{C}_{ref}$ and $\mathbf{C}'$ coordinate systems, $\mathbf{R}_{\theta}$ is the rotation matrix given by Equation 8.22 and $\mathbf{P}'$ is the point P relative to the $\mathbf{C}'$ coordinate system. This type of analysis is analogous to the HTM method described in Section 8.2.1. However, as the systems become more complicated, the vector-based approach can become cumbersome. For a more detailed description and use of vector-based approaches to instrument or machine modelling, see Hocken and Pereira (2011).

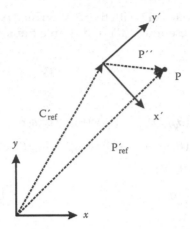

FIGURE 8.11
Schematic representation of a 2D vector of point P in the $x'y'$ relative to the xy coordinate system.

8.3 Error Motions

The analysis described in Section 8.2 illustrates one possible method of quantifying a machine or instrument error budget. Error budgets, in general, are a combination of the system requirements and factors of a particular design. There are many benefits of utilising error budgets throughout the design process. Examples include quantifying required performance specifications of individual parts and diagnosing performance limitations of a complete system.

In this chapter, an error budget is the outcome of a mathematically based model describing the important characteristics of a system design determined by the requirements of the final product combined with the uncertainties associated with the measurement and modelling techniques. To maximise productivity and time to market for design modifications and/or new designs, it is critical that a relatively fast, accurate and directed approach to design evaluation be available to the design engineer. Error budgets offer one of the best methods to predict and/or evaluate critical design requirements without incurring large costs. In this chapter, the definition of 'error' is the difference between the ideal output and what is measured or produced by an instrument or machine, particularly instruments for measuring geometry or machines for manufacturing features in a component. It is recognised that devices incorporated into assemblies are never perfect and hence when systems are fabricated using many sub-components, the individual errors of each will all contribute to errors in the output of the complete system. The analysis of the types of error sources and their influence on the output errors of large systems can seem overwhelming and intractable. This section outlines some of the more common error sources that must be considered when evaluating design of complex precision systems.

8.3.1 Definitions and Assumptions

Models representing any mechanical design can, by applying suitable assumptions, be reduced to a mathematical model for which solutions can be readily computed. As computational power has become ever more cost effective and the ability to analyse a system

TABLE 8.1

Important Definitions and Assumptions for the Evaluation of Errors Associated with Motion Control Devices

Measurement errors	Measured quantity value minus a reference quantity value
Linear displacement error	Difference between actual position or displacement and the commanded position or displacement along the line of motion
Angular error	Difference between actual angle and the commanded angle mainly for the analysis of spindle and rotary encoder errors
Repeatability	Condition of measurement, out of a set of conditions that includes the same measurement procedure, same operators, same measuring system, same operating conditions and same location, and replicate measurements on the same or similar objects over a short period of time
Random error	Component of measurement error that in replicate measurements varies in an unpredictable manner
Systematic error	Component of measurement error that in replicate measurements remains constant or varies in a predictable manner

using finite element methods more available, simple constraining models during the design phase still provide meaningful predictions for evaluating designs. Developing accurate, quantitative and measurable performance models is often one of the most difficult parts of the design process, particularly due to their inherent sensitivity to uncertainties of the boundary conditions, for precision systems. Table 8.1 lists key definitions and assumptions and will used in the development of the analyses outlined moving forward (see ISO/IEC Guide 99, 2007).

8.3.2 Abbe, Cosine and Squareness

Abbe errors are often referred to as the fundamental error in precision engineering. Most instruments and/or machines have some form of Abbe error associated with their inherent design, often as result of practical application, but well-developed precision designs account for and try to minimise the impact of Abbe errors. The Abbe principle is discussed in more detail in Chapter 10. However, it is not prudent to discuss errors and error modelling without introducing the Abbe principle. The Abbe principle has been considered for a long time as a fundamental part of precision machine design (Bryan 1979, Zhang 1989, Slocum 1992, Leach 2014). Evans (1989) expresses the concept simply as 'the Abbe principle requires the line of action of the displacement measuring system be collinear with the displacement to be measured'. Figure 8.12 illustrates the concept where the displacement

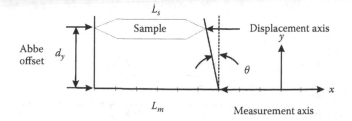

FIGURE 8.12

Illustration of Abbe error where the measurement axis is offset by a distance d_y relative to the displacement axis where θ is the angular deviation between the vertical lines representing the point of contact on the sample and the value of the scale.

axis goes through the sample and the measurement axis is offset by the distance d_y. The subsequent sample length L_s is given by

$$L_s = L_s - d_y \tan(\theta) = L_s - \varepsilon_{Abbe}, \tag{8.43}$$

where L_m is the measured length and θ is the angular deviation between parallel lines defining the measurement planes in two-dimensional space.

The most common example of the Abbe principle in displacement measurement is the difference between a micrometer, where the line of measurement is coincident with the displacement, and a Vernier caliper, where the line of measurement is offset from the displacement measurement (also see Chapter 10). To minimise errors associated with angular influences between the line of measurement and the line of displacement, the distance between the two (d_y) should be zero or as small as possible.

Cosine errors arise when the line of measurement is not collinear with the line of motion as illustrated in Figure 8.13 (also see Chapter 10). In this simplified case the length value along the motion axis is shorter than the length measured by the scale given as

$$L_m = L_s \cos(\alpha), \tag{8.44}$$

where L_m is the length along the axis of motion, L_s is the scale length and α is the angular difference between the line of measurement and the line of translation. The error $\varepsilon_{\cos}$ is then given by

$$\varepsilon_{\cos} = L_s - L_m = L_s(1 - \cos \alpha) \approx L_s \frac{\alpha^2}{2}. \tag{8.45}$$

Cosine errors are often small relative the other error sources. However, for precision machines and instruments, these seemingly small errors are a contributor to the overall error budget of the system and should be included in the evaluation process as needed.

Squareness errors generally correspond to the angular deviation between the axis of motion and the axes defining the reference or base coordinates. The reference or base coordinate system can be defined arbitrarily, but should represent a stable structure. Examples of common features used as reference or base coordinates are structural castings for machine tools or granite tables for measurement systems. As discussed earlier in this chapter, the mathematical approach used to define the motion outlined earlier is based on orthogonal coordinates and deviations from this ideal show up as an error source in the real

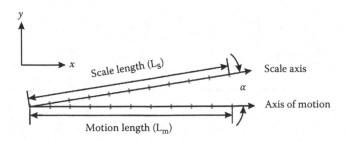

FIGURE 8.13
Illustration of cosine error where the scale axis is rotated by angle α relative to the axis of travel.

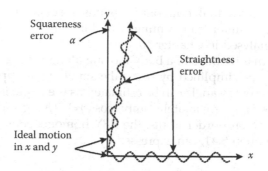

FIGURE 8.14

Illustration of squareness and straightness errors. Squareness is the angular deviation from between the x- and y-axes, and straightness is residual motion perpendicular to the axes of travel.

system. Figure 8.14 illustrates a simple 2D example of an out-of-square coordinate system and associated errors.

Straightness errors are orthogonal deviations from the prescribed motion path. In general, these errors are represented by small linear deviations orthogonal to the prescribed motion path, shown by the wavy dashed lines in Figure 8.14. Straightness errors are often the result of both linear displacement errors and angular errors combined and are described in more detail in Chapter 10. Straightness errors are important to quantify during the modelling or design phase to help guide where design choices, such as specifications of stage quality and alignment procedures, must be specified.

8.3.3 Rectilinear Translations

Rectilinear motion is defined in this chapter as the prescribed path where two independent points on the body remain parallel throughout the motion, as illustrated in Figure 8.15. Physically, rectilinear motion means that there are no rotations of the rigid body as it moves from point A to point B along a straight line. However, in practice all guided motion

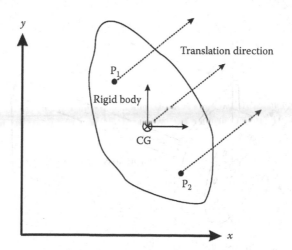

FIGURE 8.15

Illustration of rectilinear motion of a rigid body where two points on the body will remain parallel to the direction of motion.

mechanisms have some residual rotational motions associated with their design and function. Motion systems suffer from a number of different errors as described in Section 8.3.2 and are further analysed in Chapter 10.

Assuming small angular motions of a body moving along the z-axis, as shown in Figure 8.10, Equation 8.41 can be simplified by using the small angle approximation. For small angular errors ε the sine of the angle can be estimated as the angle in radians, $\sin(\varepsilon)\approx\varepsilon$, and the cosine of the angle is approximately unity, $\cos(\varepsilon)\approx1$. Using the small angle approximation and removing second-order terms, the RPY homogeneous transformation matrix $\mathbf{H_{RPY}}$, described in Equation 8.41, can expressed by

$$\mathbf{H_{RPY}} = \begin{bmatrix} 1 & -\varepsilon_z(z) & \varepsilon_y(z) & \delta_x(z) \\ \varepsilon_z(z) & 1 & -\varepsilon_x(z) & y + \delta_y(z) \\ -\varepsilon_y(z) & \varepsilon_x(z) & 1 & \delta_z(z) \\ 0 & 0 & 0 & 1 \end{bmatrix}, \tag{8.46}$$

where $\varepsilon_x(z)$, $\varepsilon_y(z)$ and $\varepsilon_z(z)$ are the angular errors as a function of the displacement in the y-direction about the x-, y- and z-axes respectively, and $\delta_x(z)$, $\delta_y(z)$ and $\delta_z(z)$ are the displacement errors as a function of y in the x-, y- and z-directions respectively.

8.3.4 Axis of Rotation

Rotational systems are common in mechanical systems and are found in almost every mechanical design. They are primarily used to either provide rotational motion or transfer energy, as commonly seen in mechanisms such as drive shafts, robotics joints and machine tool spindles. In the case of a general rotational system, illustrated in Figure 8.16, an axis is defined through the centre of rotation, where the part in motion has a coordinate system and there is a second coordinate system defined at the fixed base or reference. An ideal rotational system has a single rotational DOF defined about an ideal or virtual axis.

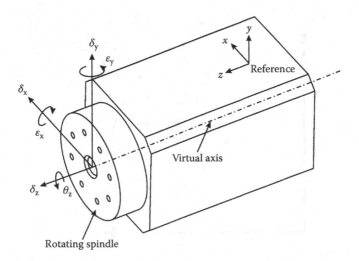

FIGURE 8.16
Sketch of axis of rotation error of a spindle relative to a reference coordinate system.

In the case of a rotational system, where the rotation axis is along the z-axis of the coordinate system, Equation 8.41 can be written to express the error motions by

$$\mathbf{H_{RPY}} = \mathbf{T_z T_y T_x T_{\theta_z} T_{\theta_y} T_{\theta_x}}$$

$$= \begin{bmatrix} \cos(\theta_z) & -\sin(\theta_z) & \sin(\theta_z)\varepsilon_x(\theta_z) + \cos(\theta_z)\varepsilon_y(\theta_z) & \delta_x(\theta_z) \\ \sin(\theta_z) & \cos(\theta_z) & \sin(\theta_z)\varepsilon_y(\theta_z) - \cos(\theta_z)\varepsilon_x(\theta_z) & \delta_y(\theta_z) \\ -\varepsilon_y(\theta_z) & \varepsilon_x(\theta_z) & 1 & \delta_z(\theta_z) \\ 0 & 0 & 0 & 1 \end{bmatrix}, \quad (8.47)$$

where θ_z is the rotation angle about the z-axis, $\varepsilon_x(\theta_z)$ and $\varepsilon_y(\theta_z)$ are the angular errors as a function of the rotation in the θ_z direction about the z-axes and $\delta_x(\theta_z)$, $\delta_y(\theta_z)$ and $\delta_z(\theta_z)$ are the displacement errors as a function of θ_z in the x-, y- and z-directions respectively. For a more detailed evaluation of spindle errors and methods for modelling and measuring rotational systems in practice, see Marsh (2008).

8.3.5 Thermal Contributions

Thermal contributions to the overall performance of any mechanical system have to be considered and can be the single most influential part of the error budget. The performance of precision mechanical systems is often dictated by thermal sensitivity and/or stability. Because of the number of different materials and interfaces found in a mechanical design, the mechanical properties of the independent materials, such as thermal expansion coefficients and mechanical joints can cause stress distributions to arise as the system reacts to the thermal changes. This can cause the mechanism to distort, grow or contract as the temperature changes with time. Mitigating thermal errors can be one of the most difficult aspects of any precision design. There are three general mechanisms (Incropera and DeWitt 1996) of heat transfer that are all relevant to precision machine design. Other material properties and their implications for precision instruments and machines are discussed in Chapter 12.

Conduction is the thermal transfer through a solid or stationary fluid and is fundamentally related to the atomic motion within the material increasing with its energy state (Özisik 1993). For the case of one-dimensional steady-state conductive heat transfer, the heat flux $\ddot{q}_x$ (W·m^{-2}) is given by

$$\ddot{q}_x = k\frac{T_1 - T_2}{L_x} = k\frac{\Delta T}{L_x}, \quad (8.48)$$

where k is the thermal conductivity (W·m^{-1}·K^{-1}), T is the temperature (K) and L_x (m) is the length along the x-direction. Conductive heat transfer is prevalent where heat sources, such as motors and pumps, are attached to frames and structures. In this case, the residual energy of the pump causes a temperature gradient within the structure and can create unwanted expansion and motion of the machine over time.

Convective heat transfer combines the molecular motion within a medium and the macroscopic motion of the medium itself (Bejan 1995). In general, convective heat transfer is described by a fluid moving over and or around a solid, such as air moving over a table. Convective heat flux $\ddot{q}$ (W·m^{-2}) is given by

$$\ddot{q} = h(T_s - T_\infty), \quad (8.49)$$

where h (W·m^{-2}·K^{-1}) is the convection heat transfer coefficient, T_s (K) is the temperature at the surface and T_∞ (K) is the bulk temperature of the fluid. Convective heat transfer is one of the basic methods used for controlling temperature. This is most often seen in hardware, such as electrical cabinets or computer housing, where heat is generated at a nominally constant rate and needs to be removed to keep the system operating at its optimal performance. Convective heat transfer is also used in environmental enclosures to reduce the effects of heat sources in an area where heat sources may move and/or change over time.

Radiation heat transfer is a result of electromagnetic waves interacting within a given medium (Siegel and Howell 1992). For this chapter, radiative heat transfer will be primarily related to solids, but can and does influence fluids. Radiative heat flux can be expressed in general terms by

$$\ddot{q}_{rad} = \varepsilon\sigma\left(T_s^4 -^s T_{sur}^4\right), \tag{8.50}$$

where ε is the emissivity, σ (5.67 × 10^{-8} W·m^{-2}·K^{-4}) is the Stefan-Boltzmann constant, T_s (K) is the absolute temperature at the surface and T_{sur} (K) is the absolute temperature of the surroundings. Radiative heat transfer is prevalent in some form in all environments and is mostly represented by things such as lights and people working within the area.

The three heat transfer mechanisms briefly described earlier are present in all machine and instrument designs, specifically for precision designs where sensitivity to small variations in the environment can have large impact on the overall function of the system. Understanding and mitigating these effects at the design phase is important. Equations 8.48, 8.49 and 8.50 all represent variations in temperature and this will affect the design in any number of ways.

There are two key aspects of thermal effects to keep in mind for the precision designer. The first is that temperature gradients cause expansion of materials and can lead to distortions and strains not accounted for in the design. For a solid volume of constant material, the volumetric expansion due to a change in temperature can be expressed by

$$^{\Delta V}/_{V_o} = \alpha_V \Delta T, \tag{8.51}$$

where ΔV (m^3) is the change in volume, V_o (m^3) is the initial volume, α_V (K^{-1}) is the coefficient of thermal expansion (CTE) and ΔT (K) is the change in temperature. Considering a change in length of a uniform bar of constant cross-section, Equation 8.51 can be simplified to

$$^{(l_f-l_o)}/_{l_o} = \alpha_1\left(T_f - T_o\right), \tag{8.52}$$

where l_o (m) is the initial length, l_f (m) is the final length, α_1 (K^{-1}) is the CTE, T_o (K) is the initial temperature and T_f is the final temperature. This fairly simple linear representation of the change in length given a change in temperature is an important way to represent effects of temperature change given a specific environment without too much complexity.

The second key aspect of thermal effects is conservation of energy, described in Figure 8.17. For a control volume, the energy at an instant in time can be written as

$$E_{in} + E_g - E_{out} = \Delta E_{st}, \tag{8.53}$$

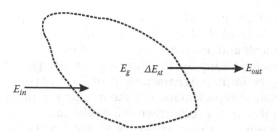

FIGURE 8.17
Energy balance of a body at any instant in time. (Based on Incropera, F. P. and DeWitt, D. P., 1996, *Introduction to heat transfer*, 3rd ed., John Wiley & Sons, New York.)

where E_{in} (J) is the energy going into the system through the control surface, E_g (J) is the energy generated within the body, E_{out} (J) is the energy leaving the system through the control surface and ΔE_{st} (J) is the change in energy stored within the body. In a mechanical system, energy can be generated by a number of different sources, for example motors, pumps, friction due to motion, coolant and many others. Important to the precision engineer is that all these sources can influence the final performance of the system and need to be considered in the error budget as part of the design process. The energy balance described in Equation 8.53 can be used to derive a lumped capacitance model for the body coming into equilibrium with its surroundings, given by

$$\frac{T - T_\infty}{T_i - T_\infty} = \exp\left[-\left(\frac{hA_s}{\rho Vc}\right)t\right],$$ (8.54)

where h (W·m^{-2}·K^{-1}) is the convection coefficient, A_s (m^2) is the surface area, ρ (kg·m^{-3}) is the density, V (m^3) is the volume, c (J·kg^{-1}·K^{-1}) is the specific heat of the material, t (s) is time, T_i (K) is the initial temperature of the body, T_∞ (K) is the temperature of the environment and T (K) is the temperature at a specific time t. Equation 8.54 can be rewritten as

$$\frac{T - T_\infty}{T_i - T_\infty} = e^{-t/\tau},$$ (8.55)

where $\tau = \rho Vc/hA_s$ and is often referred to as the thermal time constant. In practice, modelling thermal effects of heat loads is difficult without the use of finite element analysis (FEA) or computational fluid dynamics (CFD) modelling tools, which require a detailed understanding of the materials and boundary conditions specific to the design. In general, FEA and CFD codes are used to predict thermal effects needed to minimise the error of a specific process or machine design.

8.4 Modelling for Error Budgets

8.4.1 *Guide to the Expression of Uncertainty in Measurement* (GUM)

Measurement uncertainty and the guidelines outlined in the *Guide to the Expression of Uncertainty in Measurement* (GUM) represent a systematic method to predict the performance or performance limits of designs and are discussed in detail in Chapter 9 (see Taylor and Kuyatt 1994, Kirkup and Frenkel 2006). This is an important part of designing and evaluating precision instruments and machines because it defines a standard process based on a statistical approach that can be applied to almost any set of requirements. By utilising these methods for estimating performance characteristics or limits of performance and adding the values into a mathematical representation of the mechanical design using the tools and methods discussed in this text, a comprehensive model can be derived. Consistent treatment of errors and measurement uncertainties is a critical part of any precision design.

8.4.2 System-Wide View

As discussed in Chapter 1, Ishikawa diagrams can be used to represent the important characteristics and requirements of a precision design. Whether evaluating the performance of an existing piece of equipment or developing a conceptual design and requirements for a new apparatus, a holistic view of the system is required. A holistic view of the design is important for precision systems, as they tend to be more sensitive to the various interactions between components and subcomponents, including external influences such as the environment or energy sources.

Considering the diagram shown in Figure 1.3, each box represents a critical requirement or feature that impacts the performance of the overall system. In general, there are two approaches for evaluating a system, bottom-up and top-down. The bottom-up approach looks at the individual parts of the design and evaluates each independently. The individual aspects of the design are quantified and then brought together to complete the system. In the top-down approach, the system requirements are defined and quantified to be broken down into individual pieces based on the overall functional requirements of the system. In practice, both approaches are part of the overall design process and are needed to give the most complete description of the overall system. The important part of the precision design and subsequent analysis process is to derive, quantify and document the requirements as soon as possible. Subsequent modelling and evaluation will be based on these inputs, so unknowns or incomplete definitions will contribute to an inadequate design.

8.4.3 Insight into Limits of Performance

One of the key benefits of developing both a top-down and bottom-up approach is that it allows the engineer to look at the sensitivity of each aspect of the design in a systematic way. By using the modelling methods described in this chapter and Chapter 6, a detailed error budget outlining critical performance criteria can be derived based on estimated errors and specifications from a vendor, as called for by the performance requirements. By deriving a quantifiable model that reasonably represents the requirements of the system, design iterations, part and sub-component specifications can be evaluated without exhausting a lot of resources. Well-documented and detailed error budgets can guide test plans and performance evaluations of the final system enabling rapid updates when there are changes of components or design, without having to redevelop a complete model.

8.4.4 Guide to Where Resources Need to Be Distributed

Every design has multiple parts and assemblies that need to come together as a unified system and perform to a predetermined level. The precision engineer's ability to consider the requirements and distil them into a set of independent specifications or requirements that can be quantified and verified is critical to producing a quality design. In Section 8.7, two case studies are considered as examples of how the methods described in this chapter can be used to understand and evaluate a design. The examples in the following sections show how HTMs can provide a mathematical tool that can be used in conjunction with statistical approaches to derive the sensitivity of the system given a set of performance requirements. As an example, consider three axes stacked as a method to position a point in three-dimensional space. This system can be evaluated using the mathematical models outlined earlier to determine critical performance requirements such as stage straightness or assembly tolerances such as squareness requirements of how well the stages need to be assembled to provide the required performance.

8.5 Software Compensation

Software compensation is common practice for almost, if not all, machine and/or instrument manufacturers to help achieve optimal performance of their systems. Software compensation is a simple concept where the inherent errors of the system are compensated by the computer controller. There are two key requirements that underpin the success of using software compensation. The first key requirement is that the errors are repeatable to a level below the performance criteria and can be measured or quantified using models, look-up tables or other methods. The second key requirement has to do with the calibration routine used to define and quantify the errors. The process by which the system is 'qualified' is essential to achieving the system's optimal performance.

8.5.1 Calibration Routines

Being the foundation for which the baseline performance is evaluated, calibration routines are an important part of software compensation and can be the limiting factor to the achievable performance under a set of specific conditions. In general, calibration routines are a combination of evaluation procedures or processes that attempt to quantify the performance of the system as intended by the design requirements. In this case 'calibration' does not refer to the metrological definition of calibration (see ISO/IEC Guide 99, 2007) used for traceability and the evaluation of measurement uncertainties, although this should be considered as part of the qualification process of any system. The calibration process described here evaluates individual components of the system and the integrated system, including the computer control where applicable.

For precision, an instrument's or machine's calibration routine should evaluate the machine or instrument as close to its final intended use as possible. This includes operating environment, software routines and controller settings, as well as the process itself. In practice, for machine tools, calibration often uses laser interferometric measurement of the translations with a mirror reflector located in the tool holder. Consequently, while calibration occurs at the location where machining will take place, these measurements are taken when the machine is 'cutting air' and the dynamic response of the tool interacting with the workpiece is neglected. Because of the high stiffness of machine tools, under moderate operating conditions, calibrated errors once compensated are shown to significantly improve machining accuracy. One of the fundamental aspects of calibration is documentation of quantifiable measures that represent the process. This can, and should, include environmental conditions; facility inputs such as power, air pressure, humidity and water pressure to name a few; and be documented periodically to maintain the required level of performance. It is also valuable to develop routine checks to be carried out at regular intervals to ensure something has not changed over time and degraded the performance of the machine or instrument. Standard processes and tests have been developed by organisations such as the American Society of Mechanical Engineers (ASME 2016), American Society for Testing and Materials (ASTM 2016) and the International Organization for Standardization (ISO 2016) for a large number of commercial systems and common processes. Specification standards contain a wealth of information that, in some cases, must be used or referenced for developing specific application-based calibration procedures for a wide array of designs, and should be utilised when possible.

8.5.2 Computer Control

Computer-controlled machine tools and instruments have been around since the 1950s and today are part of almost every commercially available tool or instrument. As computational power has increased and become more accessible to consumers, computer-controlled machines now dominate the marketplace. For the purposes of this chapter and its relevance to precision design, computer control can be generally described as either open-loop or closed-loop and some form of electronic communication is used to define input and monitor outputs, including the analysis and storage of information. Figure 8.18a illustrates a simple open-loop control diagram as simple input and output. Figure 8.18b illustrates a closed-loop control process diagram with the dashed lines indicating a feedforward loop. In this configuration shown in Figure 8.18b the feedforward controller is based on a knowledge of the demand that can include both past and future values and therefore can anticipate the optimal control signal to achieve the desired output. Other feedforward models, and there are many, might measure disturbances to the system (sometimes it is possible to measure disturbances before they arrive at the output) and again attempt to modify the controller signal to minimise their impact. Most if not all modern control systems have some sort of closed-loop feedback system and need to be considered when developing the error budget.

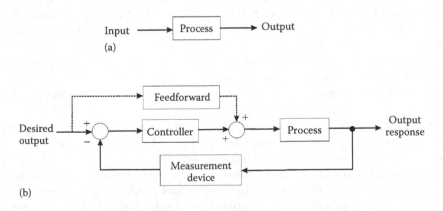

FIGURE 8.18

(a) Open-loop control where the output is only reliant upon the initial directed input. (b) Closed-loop control where the output response is dependent upon the difference between the command and the output monitored by a measurement device as part of the system or feedback. The dashed line is commonly referred to as feedforward, which is common in most high-performance control schemes. See Chapter 14 for more details. (Adapted from Kuo, B. C. and Golnaraghi, F., 2003, *Automatic control systems*, 8th ed., John Wiley & Sons.)

Modern control systems and strategies can be very complex and are discussed in detail in Chapter 14. However, precision engineers often need to understand and consider the numerous options available as part of the design process. It is also important to consider the control system in the design for factors, such as communication, bandwidth, power consumption and data storage, as they are all essential to the final performance and overall functionality. An important benefit to considering the controller at this point in the design process is that the full suite of controller options can be available based on needs and performance requirements. In particular, consideration of control strategies might impact the choice of how many sensors are to be used and where they should be located in the machine. Traditionally, controllers are designed and/or adapted to machines or instruments after the being built or prototyped, which can limit options and overall performance. The performance of the controller must be included as part of the overall performance evaluation of the system. However, a well-performing control system may improve performance, but will not salvage a poorly designed machine or instrument (see Chapter 14).

8.5.3 Algorithms

An algorithm is defined as '(a) method of solving a problem, involving a finite series of steps' (Isaacs 2000). In the case of precision design, algorithms can become a significant percentage of the overall error budget due to the sensitivity of the system to a variety of inputs. The key role of the designer developing new or using available algorithms is to verify the algorithms meet the overall requirements of the system. This seems like an unnecessary statement and should be, but it does not take much for a design to go astray because the level of complexity or number of steps needed was not completely understood to meet the requirements. Algorithms need to be considered as part of the overall design, as they can contribute to large variations in performance that have little to no influence on the mechanical or electrical performance. However, a poorly developed algorithm can have a large impact on the performance.

A further consideration for precision applications is the extent to which values are represented in a computer. Generally, floating point numbers can only represent values to a precision of approximately seven significant figures (some numbers are more difficult to represent than others) as a decimal number, while double precision is reliable to twice this. Additionally, values from analogue sensors have to be converted to binary numbers using analogue-to-digital converters so that it is not unusual that the values are represented by fifteen to seventeen bit numbers, or parts in 10^4 to 10^5 and often, at reasonable sample rates, with one or two bit errors that often represent a significant limit to achievable precision.

8.6 Hardware Compensation

Before the advent of computer control with fast processing speeds enabling error compensation, mechanical performance was dominated by the ability to fabricate parts and assemblies to the required level needed to meet the performance requirements of the system. Manufacturers of precision instruments and machines relied on repeatable and well-characterised manufacturing technologies and the skills of highly trained craftsman to meet the mechanical requirements.

8.6.1 Modifying Geometry to Accommodate Design Parameters

One of the most reliable methods of producing a highly functional and robust system is to design the geometry in such a manner that minimises the overall error within the system. However, this is easier said than done given the stringent requirements for modern precision systems. That being said, techniques used to produce precision instruments and machines in the past are still being used today, and are being combined with modern control systems and strategies to go beyond what was previously achievable.

A classic example of modifying geometry to accommodate design parameters is scraping of cast iron ways for precision machine tools (Moore 1970, Okuma 2013). In this case, hand tools are used in conjunction with master straight edges and flats to produce finished ways that have sub-micrometre straightness errors. Well-designed machines and instruments often have some form of mechanical compensation built into the design. Flats, cylinders and spheres are often ground and hand polished to produce the geometric form required as references for things such as guided motion or as metrology artefacts. Other geometrical forms can be incorporated into the design depending on the geometry. Robust designs frequently combine these types of mechanical artefacts as references or datums. The combination of sound mechanical design with feedback control and well-defined calibration routines are the foundation for high-precision systems.

8.6.2 Mechanical Enhancements to Reduce or Control Error Sources

There are a number of mechanical enhancements that can be utilised to reduce or control error sources. As mentioned in Section 8.3.5, thermal error sources can be one of the most difficult error sources to control in a precision machine or instrument. However, there are mechanical techniques that can be used to help minimise the impact of thermal variations within the environment (Smith and Chetwynd 1992). One such example is to mechanically compensate a system by choosing materials with CTEs that, when matched, compensate each other. Consider a comparator system designed to be used on the shop floor, where it is common to have temperature variations of 15 °C to 20 °C. In this case, there are two vertical columns connected by a cantilever where the stiffness of the system is designed to minimise sag due to gravity. The distance between the reference surface and probe mount needs to be constant to less than 1 μm. A sketch of the system is shown in Figure 8.19. Equation 8.52 can be used to design a matched set of materials considering the CTE of each given by

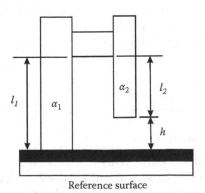

Reference surface

FIGURE 8.19
Comparator where the two vertical posts are thermally compensated to minimize the sensitivity of the distance h due to changes in temperature.

$$\alpha_1 = \frac{\alpha_2 l_2}{l_1},\qquad(8.56)$$

where l_1 and l_2 are the lengths of the vertical pillars and α_1 and α_2 are the CTEs of each column respectively.

Errors caused by temperature variations and/or predictable mechanical performance can be minimised by understanding the material properties and mechanical design. In most cases, where the system performance is critical, designs require some form of low CTE material for structural components. A more detailed description of material properties and how they can be used in precision design is presented in Chapter 12.

8.7 Case Studies

Two case studies are given in the following sections to illustrate the application of the modelling methods developed within this chapter. The first case study looks at a surface profilometer that mechanically traces the surface of a sample quantifying the surface roughness of the sample under evaluation (see Chapter 5). The second case study is a generic 3D moving bridge coordinate measuring system (CMS) commonly used for manufacturing quality control (see Chapter 5). Using a contact sensing probe, a CMS uses three, orthogonal, linear guides to move the probe and measure the coordinates when it contacts the part under evaluation. When sufficient measurements have been taken, the point cloud of data can then be used to compare the shape of the part with the intended manufactured geometry.

8.7.1 Stylus Profilometer (2D Analysis)

Figure 8.20a is a schematic diagram of the major features of a profilometer system. To reduce the complexity of the equations, the instrument is shown as a side view and represented by a 2D model. While extension to a more complete 3D model is relatively straightforward, this first case study will serve to illustrate the information that can be extracted from an HTM approach to performance evaluation. The profilometer comprises a carriage that is driven by a linear motor, the displacement of which is measured with a laser interferometer, as shown. The laser interferometer measures linear displacement of the x-stage (Chapter 5 introduces interferometers). The sample, for which the surface is to be measured, is mounted on a levelling platform, the angle of which is varied using a simple wedge and is static while the sample is under measurement. The stylus assembly is mounted on a vertical column with a coarse-adjust carriage that is translated by the vertical stage (this is also stationary during measurement). Within the stylus assembly there is a linear flexure, the moving platform of which has a diamond stylus on the bottom face and acts as the probe. Also attached to the moving platform of the linear flexure is a capacitance gauge (see Chapter 5) that measures the separation between two electrodes in the direction indicated by the dashed line (being a four-bar flexure this follows an arcuate path; see Chapter 7). An example and detailed description of this type of instrument can be found in Leach (2000).

To model the errors, 2D HTMs will be used to derive equations representing the error motions of the profilometer and an error budget constructed using the data listed next.

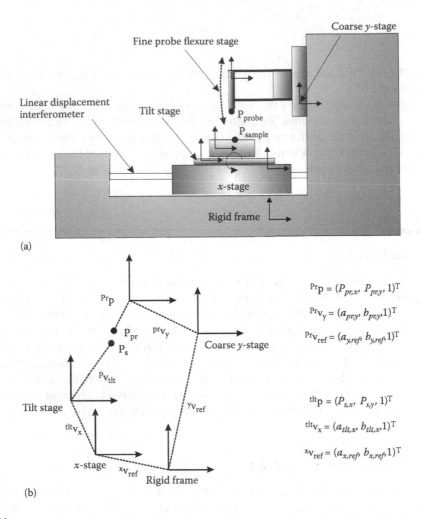

The equations shown in figure (b):

$$^{Pr}p = (P_{pr,x}, P_{pr,y}, 1)^T$$

$$^{Pr}v_y = (a_{pr,y}, b_{pr,y}, 1)^T$$

$$^{Pr}v_{ref} = (a_{y,ref}, b_{y,ref}, 1)^T$$

$$^{tlt}p = (P_{s,x}, P_{s,y}, 1)^T$$

$$^{tlt}v_x = (a_{tlt,x}, b_{tlt,x}, 1)^T$$

$$^{x}v_{ref} = (a_{x,ref}, b_{x,ref}, 1)^T$$

FIGURE 8.20

(a) Sketch of profilometer system where a flexure-based probe scans the surface of a sample. (b) Schematic of profilometer illustrating the key parts of the system. The location of the coordinate systems is arbitrary and can be chosen based on simplicity.

The displacement measuring interferometer (DMI) has a displacement uncertainty of 10 nm, while the horizontal x-stage has a 5 nm straightness error and angular uncertainties of 1 μrad. The flexure is expected to produce rectilinear motion with parasitic errors of better than 0.1 nm and angular rotations of approximately 0.1 μrad. In general, it is considered that the components of the assembly can be aligned to within 10 μrad.

To begin, a schematic representation of the system is constructed as shown in Figure 8.20b, where coordinate systems representing the independent components of each critical subsystem are related to a reference. In this case, the rigid structure of the base represents the reference frame. The reference frame location is somewhat arbitrary and is used as a mathematical construct for the model. However, in reality, the reference frame is a critical component of the overall system and must be stable as compared to the other subsystems. Vectors relating the probe point P_{pr} and the point of interest on the sample P_s are related back to the reference coordinate system through each independent coordinate, as shown in

Figure 8.20b. This simplified representation allows definition of the contributing error sources to be accounted for independently and combined to represent the complete system. For an ideal system, the error between the probe point and the sample point would be the difference between the vectors relating each point through the common reference as represented by

$$^{ref}\mathbf{E}_{ideal} = {}^{ref}\mathbf{P}_{pr} - {}^{ref}\mathbf{P}_s = \left({}^{P}\mathbf{v}_{pr} + {}^{pr}\mathbf{v}_y + {}^{y}\mathbf{v}_{ref}\right)$$
$$- \left({}^{P}\mathbf{v}_{tlt} + {}^{tlt}\mathbf{v}_x + {}^{x}\mathbf{v}_{ref}\right). \tag{8.57}$$

In a similar fashion, HTMs are used to include the rotational components and error terms related to independent coordinate systems as follows:

$$\mathbf{E}_{ref} = {}^{ref}\mathbf{P}_{pr} - {}^{ref}\mathbf{P}_s = {}^{ref}\mathbf{T}_y{}^{y}\mathbf{T}_{pr}{}^{pr}\mathbf{P} - {}^{ref}\mathbf{T}_x{}^{x}\mathbf{T}_{tlt}{}^{tlt}\mathbf{P}. \tag{8.58}$$

The transformation matrices are defined as follows:

$$^{y}\mathbf{T}_{pr} = \begin{bmatrix} 1 & -\varepsilon_{z,pr}(y) & a_{pr,y}(y) + \delta_{x,pr}(y) \\ \varepsilon_{z,pr}(y) & 1 & b_{pr,y}(y) + \delta_{y,pr}(y) \\ 0 & 0 & 1 \end{bmatrix}, \tag{8.59}$$

$$^{ref}\mathbf{T}_y = \begin{bmatrix} 1 & -\varepsilon_{z,y} & a_{y,ref} + \delta_{x,y} + \alpha Y \\ \varepsilon_{z,y} & 1 & b_{y,ref} + \delta_{y,y} \\ 0 & 0 & 1 \end{bmatrix}, \tag{8.60}$$

$$^{x}\mathbf{T}_{tlt} = \begin{bmatrix} 1 & -\varepsilon_{z,tlt} & a_{tlt,x} \\ \varepsilon_{z,tlt} & 1 & b_{tlt,x} \\ 0 & 0 & 1 \end{bmatrix}, \tag{8.61}$$

$$^{ref}\mathbf{T}_x = \begin{bmatrix} 1 & -\varepsilon_{z,x}(x) & a_{x,ref}(x) + \delta_{x,x}(x) \\ \varepsilon_{z,x}(x) & 1 & b_{x,ref}(x) + \delta_{y,x}(x) \\ 0 & 0 & 1 \end{bmatrix}. \tag{8.62}$$

The derived transformation matrices include small in-plane angular errors ε_z (rad) for each component, small displacement errors δ (m) for each component and the vector components a and b connecting each coordinate system. From Equations 8.58 through 8.62, an error matrix can be derived to consider each parameter of the design independently, as shown in Table 8.2. In this case some of the error terms may not need to be included in the compiled error budget based on the overall function of the system.

Applying Equation 8.57, removing second-order terms and considering the function of each component outlined in the error matrix in Table 8.2, the error of the system in the x- and y-directions can be expressed as

$$E_x = \alpha Y + \delta_{x,x}(x) - \delta_{x,pr}(y) + \varepsilon_{z,x}(x)\left(b_{tlt,x} + P_{s,y}\right) + \varepsilon_{z,y}\left(b_{y,pr}(y) + P_{pr,y}\right)$$
$$+ \varepsilon_{z,tlt}\left(P_{s,y}\right) - \varepsilon_{z,pr}(y)\left(P_{pr,y}\right), \tag{8.63}$$

TABLE 8.2

Error Matrix Describing the Error Source for Each Major Component of the Design Including the Quantified Values Where Applicable

y-probe		Measures displacement in y-direction guided by flexure for frictionless motion
$\delta_{x,pr}\,(y)$	0.1 nm	Parasitic error in the x-direction as a function of y-position
$\delta_{y,pr}\,(y)$	0.1 nm	Uncertainty of y-displacement as a function of y-position
$\varepsilon_{z,pr}\,(y)$	0.1 μrad	In plane angular error of the y-probe as a function of y-position
y-stage		Used to move probe into coarse y-position and remains static during measurement
$\delta_{x,y}$	NA	Displacement error in the x-direction
$\delta_{y,y}$	NA	Displacement error in the y-direction
$\varepsilon_{z,y}$	NA	In plane angular error of the y-stage, remains static during measurement
α	10 μrad	Squareness error between the x-axis and y-axis
y-stage		Moves sample along the x-direction monitoring displacement using laser DMI
$\delta_{x,x}(x)$	10 nm	Uncertainty of motion in x-direction as a function of x-position
$\delta_{y,x}(x)$	5 nm	Straightness error in the y-direction as a function of x-position
$\varepsilon_{z,x}(x)$	1 μrad	In plane angular error of the x-stage as a function of x-position
tilt-stage		Used to level the sample under test, provides in-plane angular adjust and is static during the measurement, not applicable unless it is necessary to know the location of the profile trace
$\delta_{x,tilt}$	NA	Displacement error in the x-direction
$\delta_{y,tilt}$	NA	Displacement error in the y-direction
$\varepsilon_{z,tilt}$	NA	In plane angular error of the tilt stage

$$E_y = \delta_{y,pr}(y) - \delta_{y,x}(x) + \varepsilon_{z,y}\left(a_{pr,y}(x) + P_{pr,x}\right) - \varepsilon_{z,x}(x)\left(a_{tilt,x} + P_{s,x}\right)$$

$$- \varepsilon_{z,pr}\left(P_{pr,x}\right) + \varepsilon_{z,tilt}\left(P_{s,x}\right). \tag{8.64}$$

Based on Equations 8.63 and 8.64, an error budget can be derived that allows the designer to consider both geometric constraints, such as offset positions represented by the vector components, and develop performance specifications given those geometric constraints within the system. Negative values represented in Equations 8.63 and 8.64 are based on the mathematics used to derive the error between a point on the sample and the probe. When evaluating the maximum error, the absolute value of the error terms should be used because the actual sign or direction of the error is often unknown. Estimating and evaluating error and uncertainty will be described in more detail in Chapter 9, where the use of Monte Carlo simulations to determine statistical distributions of errors is also presented. The squareness term α is a function of the coarse y-stage, and for the example outlined here is constant as the position of the coarse y-stage does not move during the measurement. However, the squareness α does contribute to x-offset depending on the location of the probe system along the y-axis and does contribute a potential error source if trying to compare the x-position of samples with different heights.

For the profilometer, the x-offsets are given in Table 8.3. The profilometer is primarily a system that relates the y-position of the surface, as represented by the interaction with the probe and the surface of the sample, and the relative scan length in the x-direction, in this case. Because the measures are relative to each other, some of the offset distances are not critical to the output. However, the model includes these potential error sources and the

TABLE 8.3

Vector Matrix Offset Values

$^{Pr}\mathbf{P} = [P_{pr,x}, P_{pr,y}, 1]^{T}$	$[5\ \text{mm}, 5\ \text{mm}, 1]^{T}$
$^{Pr}\mathbf{v_y} = [a_{pr,y}, b_{pr,y}, 1]^{T}$	$[5\ \text{mm}, 25\ \text{mm}, 1]^{T}$
$^{Pr}\mathbf{v_{ref}} = [a_{y,ref}, b_{y,ref}, 1]^{T}$	$[10\ \text{mm}, 50\ \text{mm}, 1]^{T}$
$^{tlt}\mathbf{P} = [P_{s,x}, P_{s,y}, 1]^{T}$	$[0\ \text{mm}, 10\ \text{mm}, 1]^{T}$
$^{tlt}\mathbf{v_x} = [a_{tlt,x}, b_{tlt,x}, 1]^{T}$	$[0\ \text{mm}, 25\ \text{mm}, 1]^{T}$
$^{x}\mathbf{v_{ref}} = [a_{x,ref}, b_{x,ref}, 1]^{T}$	$[0\ \text{mm}, 25\ \text{mm}, 1]^{T}$

TABLE 8.4

Profilometer Error Estimates in x- and y-Directions

E_x	E_y
20.5 nm	5.6 nm

operation of this design in a non-relative or absolute configuration can be evaluated as well with little added effort. Table 8.4 has the estimated errors of the system outlined in Figure 8.20a. The error budget estimates the spatial error to be 20.5 nm with a displacement error in the y-direction to be 5.6 nm.

The ability to understand the design trade-offs with actual functionality and required performance is one of the benefits of doing a detailed error budget at the design phase of any project. A graded approach should be used when detailing an error budget, as the level of detail should be consistent with the requirements of the system, including costs and schedule.

8.7.2 CMS Error Budget (3D Analysis)

Figure 8.21a shows a sketch of a moving bridge CMS capable of locating points within a work volume consisting of 250 mm in x-axis, 150 mm in y-axis and 100 mm in z-axis. This instrument uses Zerodur™ line scales for position measurement in each of the three axes, with a displacement measurement error of 25 nm, and uses a contact probe to locate the surface of the sample being inspected to 100 nm in all three axes. Consider the angular error of each independent axis to be 10 μrad based on manufacturing and assembly tolerances. The straightness and squareness of each axis independently is 200 nm and 1 mrad respectively.

Using HTMs, a model representing the error motions of the system throughout the work volume can be derived. Figure 8.21b shows a schematic representation of the CMS where there is a reference coordinate system and a coordinate system for each of the three motion axes. For the ideal case, the difference in location between point P_{pr} at the probe and the point P_s on the sample representing a part under evaluation can be expressed by summing the vectors given by

$$^{ref}\mathbf{E_{ideal}} = {}^{ref}\mathbf{P_{pr}} - {}^{ref}\mathbf{P_s} = \left({}^{z}\mathbf{P_{pr}} + {}^{z}\mathbf{v_y} + {}^{y}\mathbf{v_x} + {}^{x}\mathbf{v_{ref}} \right) - {}^{ref}\mathbf{P_s}, \tag{8.65}$$

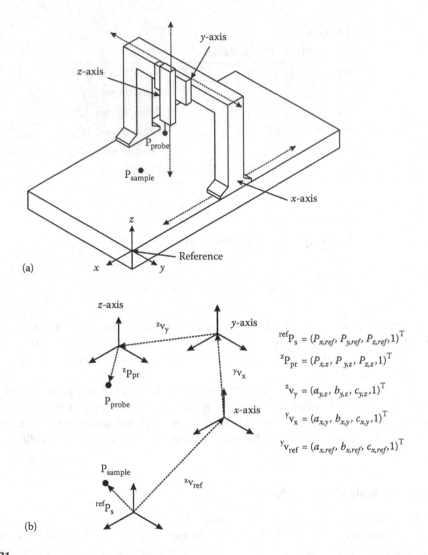

FIGURE 8.21
(a) Sketch of moving bridge CMM for locating points in space. (b) Schematic of CMM illustrating the key parts of the system.

where **v** represent the vectors between coordinates and $^{z}\mathbf{P}_{pr}$ is the vector locating the probe point P_{pr} relative to the z-axis coordinate system. As illustrated in the previous example, HTMs can be derived for each of the independent slides using Equation 8.45 and by including the squareness term α for each of the stacked axes. The error between the point on the sample and the probe can now each be related to a reference coordinate system by

$$^{ref}\mathbf{E} = {}^{ref}\mathbf{P}_{pr} - {}^{ref}\mathbf{P}_{s} = {}^{ref}\mathbf{T}_{x}{}^{x}\mathbf{T}_{y}{}^{y}\mathbf{T}_{z}{}^{z}\mathbf{P}_{pr} - {}^{ref}\mathbf{P}_{s},\qquad(8.66)$$

where the 3D transformation matrices for each coordinate system are given by

$$
{}^{ref}\mathbf{T_x} = \begin{bmatrix} 1 & -\varepsilon_{z,x}(x) & \varepsilon_{y,x}(x) & x + \delta_{x,x}(x) + a_{x,ref} \\ \varepsilon_{z,x}(x) & 1 & -\varepsilon_{x,x}(x) & \delta_{y,x}(x) + b_{x,ref} \\ -\varepsilon_{y,x}(x) & \varepsilon_{x,x}(x) & 1 & \delta_{z,x}(x) + c_{x,ref} \\ 0 & 0 & 0 & 1 \end{bmatrix},
\tag{8.67}
$$

$$
{}^{ref}\mathbf{T_x} = \begin{bmatrix} 1 & -\varepsilon_{z,y}(y) & \varepsilon_{y,y}(y) & y\alpha_{xy} + \delta_{x,y}(y) + a_{y,x} \\ \varepsilon_{z,y}(y) & 1 & -\varepsilon_{x,y}(y) & y + \delta_{y,y}(y) + b_{y,x} \\ -\varepsilon_{y,y}(y) & \varepsilon_{x,y}(y) & 1 & \delta_{z,y}(y) + c_{y,x} \\ 0 & 0 & 0 & 1 \end{bmatrix},
\tag{8.68}
$$

$$
{}^{ref}\mathbf{T_x} = \begin{bmatrix} 1 & -\varepsilon_{z,z}(z) & \varepsilon_{y,z}(z) & z\alpha_{xz} + \delta_{x,z}(z) + a_{y,z} \\ \varepsilon_{z,z}(z) & 1 & -\varepsilon_{x,z}(z) & z\alpha_{yz} + \delta_{y,z}(z) + b_{y,z} \\ -\varepsilon_{y,z}(z) & \varepsilon_{x,z}(z) & 1 & z + \delta_{z,z}(z) + c_{y,z} \\ 0 & 0 & 0 & 1 \end{bmatrix}.
\tag{8.69}
$$

The vectors shown in Figure 8.21b represent the static offsets of the axes relative to each other and are based on the design. At the design stage of a project, schematic representations with estimates of performance can be used to drive the system design to optimise the overall configuration. There are always trade-offs that dictate the final design, but understanding the impact of design choices can result in large savings in both time and resources.

Substituting Equations 8.67 through 8.69 into Equation 8.66 and removing second-order terms, the errors in each of the three directions are given by

$$
E_x = \delta_{x,x}(x) + \delta_{x,y}(y) + \delta_{x,z}(z) + \varepsilon_{y,x}(x)\left(c_{x,y} + c_{y,z} + P_{z,z} + z\right)
$$

$$
- \varepsilon_{z,x}(x)\left(b_{x,y} + b_{y,z} + P_{y,z} + y\right) + \varepsilon_{y,y}(y)\left(c_{y,z} + P_{z,z} + z\right)
\tag{8.70}
$$

$$
\varepsilon_{z,y}(y)\left(b_{y,z} + P_{y,z}\right) + \varepsilon_{y,z}(z)P_{z,z} - \varepsilon_{z,z}(z)P_{y,z} + y\alpha_{xy} + z\alpha_{xz},
$$

$$
E_y = \delta_{y,x}(x) + \delta_{y,y}(y) + \delta_{y,z}(z) - \varepsilon_{x,x}(x)\left(c_{x,y} + c_{y,z} + P_{z,z} + z\right)
$$

$$
+ \varepsilon_{z,x}(x)\left(a_{x,y} + a_{y,z} + P_{y,z}\right) - \varepsilon_{x,y}(y)\left(c_{y,z} + P_{z,z} + z\right)
\tag{8.71}
$$

$$
+ \varepsilon_{z,y}(y)\left(a_{y,z} + P_{x,z}\right) - \varepsilon_{x,z}(z)P_{z,z} + \varepsilon_{z,z}(z)P_{x,z} + z\alpha_{yz},
$$

$$E_z = \delta_{z,x}(x) + \delta_{z,y}(y) + \delta_{z,z}(z) + \varepsilon_{x,x}(x)\left(b_{y,z} + b_{x,y} + P_{y,z} + y\right)$$

$$- \varepsilon_{y,x}(x)\left(a_{y,z} + a_{x,y} + P_{x,z}\right) + \varepsilon_{x,y}(y)\left(b_{y,z} + P_{y,z}\right) \tag{8.72}$$

$$- \varepsilon_{y,y}(y)\left(a_{y,z} + P_{x,z}\right) + \varepsilon_{x,z}(z)P_{y,z} - \varepsilon_{y,z}(z)P_{x,z}.$$

For the preceding CMS, there are a total of twenty-one error terms as defined in Table 8.5, including three angular errors (ε) and three displacement (δ) terms per motion axis and a total of three squareness terms α_{xy}, α_{xz} and α_{yz} for the complete assembly (see Chapter 5). One of the largest contributors to displacement errors in any motion system is the offset distances between the motion axes and the line of measurement, as discussed earlier in Section 8.3. A useful way to help guide a design is to use a matrix approach referred to by Slocum (1992) as an 'error gain matrix', as a way to look at the influence of the stack-up or 'gain' terms that, when combined with angular errors, can have large consequence to the overall performance of the system. Table 8.6 is the error gain matrix for the moving bridge CMS described earlier.

TABLE 8.5

Error Matrix for Moving Bridge CMS Shown in Figure 8.21

x-axis		Defines the x-motion of the system and is the displacement measurement in the x-direction
$\delta_{x,x}(x)$	25 nm	Uncertainty of x-displacement as a function of x-position
$\delta_{y,x}(x)$	200 nm	Straightness error in the y-direction as a function of x-displacement
$\delta_{z,x}(x)$	200 nm	Straightness error in the z-direction as a function of x-displacement
$\varepsilon_{x,x}(x)$	10 μrad	Angular error about the x-axis of the x-stage as a function of displacement in the x-direction
$\varepsilon_{y,x}(x)$	10 μrad	Angular error about the y-axis of the x-stage as a function of displacement in the x-direction.
$\varepsilon_{z,x}(x)$	10 μrad	Angular error about the z-axis of the x-stage as a function of displacement in the x-direction.
y-axis		Defines the y-motion of the system and is the displacement measurement in the y-direction.
$\delta_{x,y}(y)$	200 nm	Straightness error in the x-direction as a function of y-displacement
$\delta_{y,y}(y)$	25 nm	Uncertainty of x-displacement as a function of y-position
$\delta_{z,y}(y)$	200 nm	Straightness error in the z-direction as a function of y-displacement
$\varepsilon_{x,y}(y)$	10 μrad	Angular error about the x-axis of the y-stage as a function of displacement in the y-direction
$\varepsilon_{y,y}(y)$	10 μrad	Angular error about the y-axis of the y-stage as a function of displacement in the y-direction.
$\varepsilon_{z,y}(y)$	10 μrad	Angular error about the z-axis of the y-stage as a function of displacement in the y-direction
α_{xy}	1.0 mrad	Squareness error between the x- and y-axes respectively
z-axis		Defines the z-motion of the system and is the displacement measurement in the z-direction
$\delta_{x,z}(z)$	200 nm	Straightness error in the x-direction as a function of z-displacement
$\delta_{y,z}(z)$	200 nm	Straightness error in the z-direction as a function of z-displacement
$\delta_{z,z}(z)$	25 nm	Uncertainty of z-displacement as a function of z-position
$\varepsilon_{x,z}(z)$	10 μrad	Angular error about the x-axis of the y-stage as a function of displacement in the z-direction.
$\varepsilon_{y,z}(z)$	10 μrad	Angular error about the y-axis of the y-stage as a function of displacement in the z-direction
$\varepsilon_{z,z}(z)$	10 μrad	Angular error about the z-axis of the y-stage as a function of displacement in the z-direction
α_{xz}	1.0 mrad	Squareness error between the x- and z-axes respectively
α_{yz}	1.0 mrad	Squareness error between the y- and z-axes respectively

TABLE 8.6

Error Gain Matrix for Each of the Three Axes That Contribute to the Overall Error

x-axis	$\varepsilon_{x,x}(x)$	$\varepsilon_{y,x}(x)$	$\varepsilon_{z,x}(x)$
E_x	0	$(c_{x,y} + c_{y,x} + P_{z,z}{+}z)$	$(b_{x,y} + b_{y,x} + P_{y,z} + y)$
E_y	$(c_{x,y} + c_{y,z} + P_{z,z}{+}z)$	0	$(a_{x,y} + a_{y,z} + P_{y,z})$
E_z	$(b_{y,z} + b_{x,y} + P_{y,z}{+}y)$	$(a_{y,z} + a_{x,y} + P_{x,z})$	0
y-axis	$\varepsilon_{x,y}(y)$	$\varepsilon_{y,y}(y)$	$\varepsilon_{z,y}(y)$
E_x	0	$(c_{y,z} + P_{z,z} + z)$	$(b_{y,x} + P_{y,z})$
E_y	$(c_{y,z} + P_{z,z} + z)$	0	$(a_{y,x} + P_{x,z})$
E_z	$P_{z,z}$	$P_{x,z}$	0
z-axis	$\varepsilon_{x,z}(z)$	$\varepsilon_{y,z}(z)$	$\varepsilon_{z,z}(z)$
E_x	0	$P_{z,z}$	$P_{y,z}$
E_y	$P_{z,z}$	0	$P_{x,z}$
E_z	$P_{y,z}$	$P_{x,z}$	0

From Table 8.5 and Table 8.6, the predicted error for the machine as defined can be evaluated. In this case, the offset values of Table 8.7 were chosen to be at the limits of the travel indicated earlier.

The values outlined in Table 8.7 are estimates based on the design requirements and practical layout of the hardware. The vector $^{ref}P_s$ represents a sample point anywhere within the work volume of the machine and is represented in Equations 8.70 through 8.73 by the variables x, y and z for each of the three motion axes respectively. The system can now be expressed using the values in Table 8.5 and Table 8.7 to calculate the error as a function of position within the work volume of the instrument. Figure 8.22 shows the error in each of the three travel axes through the work volume of the instrument.

The case studies outlined in this section are practical examples of how mathematical models can be used to derive error budgets for quantifying error at the design stage or when trying to evaluate an instrument or machine as part of a test procedure. The techniques outlined here can be expanded to include dynamics of systems including force and torque components and the more general equations of motion used in modelling control systems (Hale 1999).

TABLE 8.7

Offset Vectors for Moving Bridge CMS

$^{ref}P_s = [P_{x,ref}, P_{y,ref}, P_{z,ref}, 1]^T$	$[0\text{--}250\ mm, 0\text{--}150\ mm, 0\text{--}100\ mm, 1]^T$
$^{ref}P_s = [P_{x,z}, P_{y,z}, P_{z,z}, 1]^T$	$[10\ mm, 10\ mm, 50\ mm, 1]^T$
$^z v_y = [a_{y,z}, b_{y,z}, c_{y,z}, 1]^T$	$[50\ mm, 25\ mm, 0\ mm, 1]^T$
$^y v_x = [a_{x,y}, b_{x,y}, c_{x,y}, 1]^T$	$[50\ mm, 25\ mm, 0\ mm, 1]^T$
$^x v_{ref} = [a_{ref,x}, b_{ref,x}, c_{ref,x}, 1]^T$	$[0\ mm, 0\ mm, 0\ mm, 1]^T$

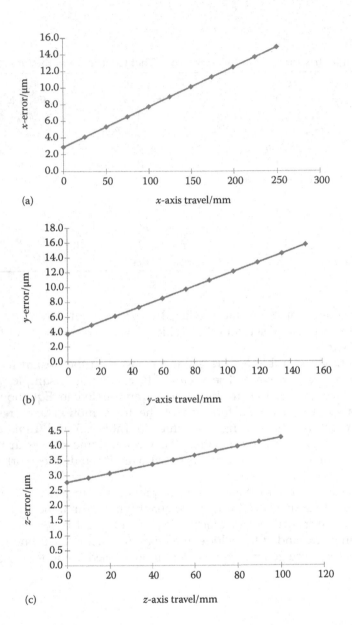

FIGURE 8.22
Plots of error for each of the three axes of the moving bridge CMM. (a) x-axis error, (b) y-axis error, (c) z-axis error.

Exercises

1. Consider a sinusoidal path where the amplitude is 0.10 mm and one period over a range of 100 mm shown in Figure 8.23. Plot the location of point P in the local coordinate system as a function of x-position along the path with and without accounting for the rotation of the local coordinate system. Plot the difference between the x- and y-position for the rotated and non-rotated cases.

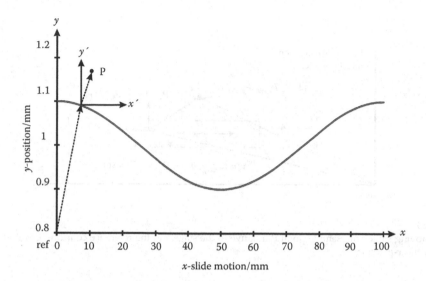

FIGURE 8.23
Plot of y-position error as a function of x-position representing a straightness error over 100 mm of travel.

2. Use the vector method described in Section 8.2.3 to derive the equation for errors of the generic machine tool outlined in Figure 8.24. Include squareness errors and assume small angles for all angular motions. (Note: Remove second-order terms.)
3. Use HTMs and derive the Abbe and cosine errors in the x-direction for a single DOF rectilinear stage, where the point of interest is located at a point P on the stage and the displacement measuring sensor runs through the centre of the stage as shown in Figure 8.25.

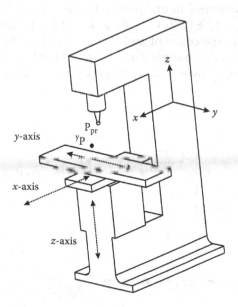

FIGURE 8.24
Illustration of three-axis manual CMM. The system consists of three stacked stages representing each of the three coordinates.

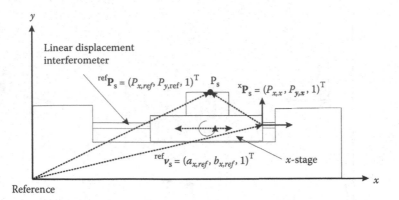

FIGURE 8.25
Sketch of moving stage with sample point P_s relative to the stage and the reference coordinate system represents the error in the x-position.

4. Plot the residual errors in the x- and y-axes of a circular calibration routine with a radius of 40 mm in 2D with a squareness error of 0.1 mrad and straightness error represented by a sinusoidal function with an amplitude of 10 µm and frequency of 0.1 cycle/mm. The x- and y-axes have the same straightness amplitude with a 90° phase difference between them. The 2D work volume of the instrument is 100 mm by 100 mm.
5. For the cantilevered structure shown in Figure 8.26, calculate the displacement error as a function of x-position given a y-slide assembly with a mass of 100 kg running on a cast iron prismatic beam. The cast iron beam has a rectangular cross section 100 mm by 150 mm tall and the length of the beam is 500 mm from the vertical beam. The vertical beam is made from a ceramic that is considered rigid. Plot the shape required to minimise the error of the x-slide as a function of x-position.
6. A precision spindle requires 2 hours to come to thermal equilibrium while running in a laboratory held at 20.0 °C ± 0.02 °C, with an initial temperature of 23.0 °C at start-up. Calculate the thermal time constant and use it to predict the axial error motion of the spindle given a sinusoidal temperature fluctuation of ±1.0 °C and a frequency of 20 hours in the laboratory over a 100-hour period, as shown in Figure 8.27. The CTE of

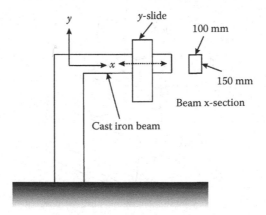

FIGURE 8.26
Sketch of cantilever mechanism with a y-axis that moves along an x-slide.

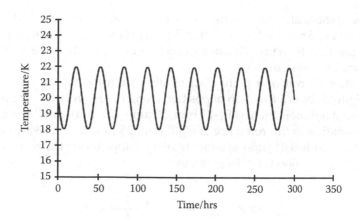

FIGURE 8.27

Axial error motion of a spindle as a function of time given a thermal time constant of 0.4 h and a ±1.0 K° temperature variation.

the spindle was measured to be 2×10^{-6} m/m·K in the axial direction. Assume the spindle remains running after the initial start-up period.

7. Derive a yaw–pitch–roll (YPR) HTM based on Figure 8.10 and show it is not equivalent to the roll-pitch-yaw (RPY) HTM derived in Equation 8.41.

8. Using the small angle approximation discussed in Section 8.3.3, compare the RPY homogeneous transformation matrix and the YPR homogeneous transformation matrix from Exercise 7.

9. Consider a contact-based probe mounted on a spindle as shown in Figure 8.28. The probe point $P(x,y,z)$ is located at $\mathbf{P}(-50$ mm, 0 mm, 50 mm$)$ relative to the spindle coordinate system, and the spindle coordinate system location relative to the reference is given by $\mathbf{O}_{sp}(0$ mm, 0 mm, 25 mm$)$ as illustrated in Figure 8.28. The spindle has a four-lobe sinusoidal radial error motion with a 100 nm (A_e) amplitude described by $\Delta r = A_e \sin(n\theta_z)$, where n is the number of lobes and θ_z (rad) is the angular position of the

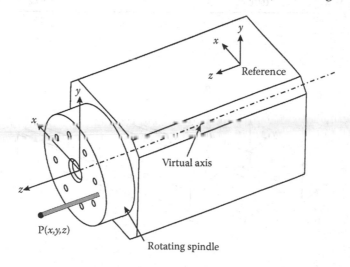

FIGURE 8.28

Spindle with probe mounted on the face. The position of probe point $P(x,y,z)$ relative to the spindle coordinate system is P(−50 mm, 0 mm, 50 mm).

spindle. The spindle also has a constant tilt error $(\varepsilon_x, \varepsilon_y)$ of 10 µrad and an axial error δ_z (θ_z) is constant at 25 nm. Plot the radial and axial position of the probe tip P as a function of angular position θ_z. (Hint: Assume the centre of the circle to be at the origin of the spindle coordinate system.)

10. Plot the positional error of a point P(x,y,z) offset from the top of a rectilinear stage as shown in Figure 8.29 as a function of position along the z-direction. The point P position relative to the stage coordinate is P(−25 mm, 50 mm, −25 mm) and the origin of the stage coordinates relative to the reference is O_{st}(0 mm, 25 mm, z). The stage has an angular error (ε) of amplitude 0.01 µrad as a function of position along the z-axis of motion (z) in each of the three rotational DOFs given by

$$\text{Roll} \quad \varepsilon_z(z) = A_z \sin\left(z\frac{2\pi}{f_z} + \phi_z\right), \tag{8.73}$$

$$\text{Pitch} \quad \varepsilon_x(z) = A_x \sin\left(z\frac{2\pi}{f_x} + \phi_x\right), \tag{8.74}$$

$$\text{Yaw} \quad \varepsilon_y(z) = A_y \sin\left(z\frac{2\pi}{f_y} + \phi_z\right), \tag{8.75}$$

where A (µrad) is the amplitude of the error, f (cycle/mm) is the oscillation frequency over the range of travel and ϕ (rad) is the relative phase of the motion. The displacement errors in all three axes are constant with an amplitude of 0.001 mm.

	Amplitude (µrad)	Frequency (cycles/mm)	Phase (rad)
Roll	0.01	20	0
Pitch	0.01	10	$\pi/3$
Yaw	0.01	5	$\pi/2$

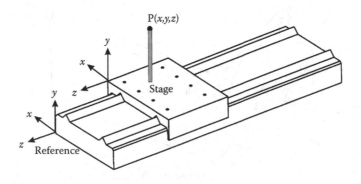

FIGURE 8.29
Rectilinear stage with a probe offset from the stage surface moving along the z-direction. The probe point P(x,y,z) relative to the stage coordinate system is given by P(−25 mm, 50 mm, −25 mm).

References

American Society of Mechanical Engineers (ASME), 2016, About ASME standards and certification, https://www.asme.org/about-asme/standards.

ASTM International, 2016, Standards & publications, https://www.astm.org/Standard/standards-and-publications.html.

Bejan, A., 1995, *Convection heat transfer*, 2nd ed., John Wiley & Sons, New York.

Bryan, J. B., 1979, The Abbé principle revisited: An updated interpretation, *Precision Engineering*, **1**(3), 129–132.

Dettman, J. W., 1986, *Introduction to linear algebra and differential equations*, Dover Publications, New York.

Duda, R. O. and Hart, P. E., 1973, *Pattern classification and scene analysis*, John Wiley & Sons, New York.

Evans, C., 1989, *Precision engineering: An evolutionary view*, Cranfield Press, Bedford, UK.

Fu, K. S., Gonzalez, R. C. and Lee, C. S. G., 1987, *Robotics: Control, sensing, vision and intelligence*, McGraw-Hill, New York.

Goldstein, H., 1980, *Classical mechanics*, 2nd ed., Addison-Wesley, New York.

Hale, L. C., 1999, Principles and techniques for designing precision machines, PhD thesis, Massachusetts Institute of Technology.

Hocken, R. J. and Pereira, P., 2011, *Coordinate measuring machines and systems*, 2nd ed., CRC Press.

Incropera, F. P. and DeWitt, D. P., 1996, *Introduction to heat transfer*, 3rd ed., John Wiley & Sons, New York.

Isaacs, A. (ed.), 2000, *A dictionary of physics*, 4th ed., Oxford University Press.

International Organization for Standardization (ISO), 2016, We're ISO: We develop and publish International Standards, http://www.iso.org/iso/home/standards.htm.

ISO/IEC Guide 99:2007, *International vocabulary of metrology—Basic and general concepts and associated terms*.

Kirkup, L. and Frenkel, B., 2006, *An introduction to uncertainty in measurement using the GUM (Guide to the expression of uncertainty in measurement)*, Cambridge University Press.

Kreyszig, E., 1993, *Advanced engineering mathematics*, 7th ed., John Wiley & Sons, New York.

Kuo, B. C. and Golnaraghi, F., 2003, *Automatic control systems*, 8th ed., John Wiley & Sons.

Lay, D. C., 1994, *Linear algebra and its applications*, Addison-Wesley, New York.

Leach, R. K., 2000, Traceable measurement of surface texture at the National Physical Laboratory using NanoSurf IV. *Measurement Science and Technology*, **11**, 1162–1173.

Leach, R. K., 2014, *Fundamental principles of engineering nanometrology*, 2nd ed., Elsevier, Berlin.

Marsh, E. R., 2008, *Precision spindle metrology*, DEStech Publications, Lancaster, PA.

Maxwell, E. A., 1961, *General homogeneous coordinates in space of three dimensions*, Cambridge University Press, UK.

McCarthy, J. M., 1990, *An introduction to theoretical kinematics*, MIT Press, Cambridge, MA.

Moore, W. R., 1970, *Foundations of mechanical accuracy*, Moore Special Tool Company, Bridgeport, CT

Okuma, 2013, Hand scraping sets the foundation for CNC machining accuracy and long-term stability, http://www.okuma.com/handscraping (accessed 8/10/16).

Özisik, M. N., 1993, *Heat conduction*, 2nd ed., John Wiley & Sons, New York.

Pars, L. A., 1965, *A treatise on analytical dynamics*, Heinemann Educational Books.

Paul, R. P., 1981, *Robot manipulators: Mathematics, programming, and control*, MIT Press, Cambridge, MA.

Shilov, G. E., 1977, *Linear algebra*, Dover Publications, New York.

Siegel, R. and Howell, J. R., 1992, *Thermal radiation heat transfer*, 3rd ed., Hemisphere Publishing.

Slocum, A. H., 1992, *Precision machine design*, Society of Manufacturing Engineers, Dearborn, MI.

Smith, S. T., and Chetwynd, D. G., 1992, *Foundations of ultraprecision mechanism design*, Gordon and Breach Scientific Publishers.

Stadler, W., 1995, *Analytical robotics and mechatronics*, McGraw-Hill, New York.

Taylor, B. N. and Kuyatt, C. E., 1994, *Guidelines for the evaluating and expressing the uncertainty of NIST measurement results*, NIST Technical Note 1297.

Tsai, L.-W. 1999, *Robot analysis: The mechanics of serial and parallel manipulators*, John Wiley & Sons, New York.

Wylie, C. R. and Barrett, L. C., 1995, *Advanced engineering mathematics*, 6th ed., McGraw-Hill, New York.

Zhang, G. X., 1989, A study on the Abbe principle and Abbe error, *CIRP Annals*, **38**(1), 525–528.

9

Measurement Uncertainty

Han Haitjema

CONTENTS

ABSTRACT In precision engineering, the demands for close tolerances and accurate dimensions on workpieces are challenging to manufacture but critical for reliable function when assembled into a mechanical system. These dimensions have to be confirmed by measurement, and such measurements always include disturbances that cause the measurement results to deviate. Typically, the resulting deviations are small, but they can be large enough to prevent a proper decision about a workpiece being taken. Therefore, an appropriately rigorous estimation of measurement uncertainty is essential in the production process. This chapter is about measurement uncertainty and follows the most recent *Guide to the Expression of Uncertainty in Measurement*, commonly called the GUM. From elementary single variable statistics through uncertainty propagation, the chapter shows the reader how to estimate uncertainty in the result of a measurement process. The concept of propagation of uncertainty distributions and Monte Carlo methods of uncertainty estimation are also presented.

9.1 The GUM Uncertainty Framework

In the twentieth century there was quite some confusion in the practical realisation of the concept of uncertainty estimation. Historically, different practices were adopted in different disciplines, such as physics, mechanical engineering or chemistry. A joint effort of harmonisation led to the first edition of the *Guide to the Expression of Uncertainty in Measurement* (GUM) in 1993 and to the most recent version, JCGM 100 (2008). In 2008, Supplement 1 was published titled *Propagation of Distributions Using a Monte Carlo Method*, JCGM 101 (2008). In the meantime, specification standards and guidelines for the practical application of this rather theoretical framework were published, where especially the three guidelines should be mentioned: EA/4-02, 2013 *Evaluation of the Uncertainty of Measurement in Calibration*; ISO 14253-2 (2011) *Guidance for Estimation of Uncertainty in Geometric Product Specification (GPS) Measurement, in Calibration of Measurement Equipment, and in Product Verification*; and *Guidelines for Evaluating and Expressing the Uncertainty of NIST Measurement Results* (Taylor and Kuyatt 1994). This chapter is based on these documents and guidelines.

9.2 Systematic and Random Errors

In the context of measurement uncertainty, the term 'error' means a deviation from a true or average value; this should not be confused with a failure or a mistake. Historically, errors in measurement processes are classified as either systematic or random. In this context, a systematic error is a deviation from the 'true' value in a measurement that remains constant in repeated measurements, and a random error is a component of a measurement error that varies in an unpredictable manner when the measurement is repeated (see Chapter 2 for a fuller description). In uncertainty evaluations, both of these error types may be mixed; when a single measurement is considered, measurement deviations might be assumed to include both systematic and random errors. It is considered more useful to classify uncertainty evaluations as type A and type B. Type A refers to the evaluation of a component of measurement uncertainty by a statistical analysis of measured values obtained under defined measurement conditions. This is related to, but not the same as, random errors. Examples of type A evaluations are

- The standard deviation in a series of measurements
- The standard deviation in the mean of a series of measurements
- The standard deviation from a determined calibration curve

It is common to express the uncertainty resulting from a type A evaluation as a standard deviation.

A type B evaluation encompasses any other method of uncertainty evaluation. Examples are

- The calibration uncertainty of a used physical reference specimen
- The published uncertainty of a physical constant
- The known uncertainty of equations that approximate quantities, for example, for the local gravitational acceleration as a function of altitude and latitude
- A known but uncorrected systematic error

Type B evaluations are commonly related to systematic errors. Such errors can be expressed as a deviation, which may or may not be represented by a standard deviation. Therefore, both type A and type B errors are quantified as standard errors. This commonly means a standard deviation in a type A evaluation and an expected deviation in a type B evaluation.

9.3 Single-Value Uncertainty Evaluation

The most elementary uncertainty evaluation is the single-value uncertainty evaluation, which is the treatment of a direct measurement where the measurement gives a direct estimate of the measurand as it is defined in Chapter 2.

9.3.1 Mean, Standard Deviation and Standard Deviation of the Mean

As an illustration of the basic statistical parameters used in uncertainty analysis, the example of a ball diameter measured using a Vernier caliper is considered. No ball is perfectly spherical, therefore, a different value is measured when the ball is measured at different orientations. For every subsequent measurement, the ball is rotated to another random position. Each diameter measurement is denoted by d_i, the mean value of n independent measurements of the diameter D is given by $\bar{d}$. An example of five measurements ($n = 5$) is given in Table 9.1.

TABLE 9.1

Sphere Diameter Measurement as an Illustration of the Calculation of the Mean and Standard Deviation of a Measurement Series

Measurement Number	Measured Value	Difference from Mean	Squared Difference from Mean
	d_i/mm	$\delta_i = d_i - \bar{d}/\mathrm{mm}$	δ_i/mm^2
1	24.780	+0.045	0.0020
2	24.666	−0.069	0.0048
3	24.726	−0.009	0.0001
4	24.784	+0.049	0.0024
5	24.719	−0.016	0.0003
Sum	123.675	0	0.0096

The mean value $\bar{d}$ for the measurements d_i of the $n = 5$ samples is calculated as

$$\bar{d} = \frac{\sum_{i=1}^{n} d_i}{n} = \frac{123.675}{5} = 24.735 \text{ mm}. \tag{9.1}$$

Because the sample is from a large population of diameters, the standard deviation s is calculated as

$$s = \sqrt{\frac{1}{n-1} \cdot \sum_{i=1}^{n}(d_i - \bar{d})^2} = \sqrt{\frac{1}{4} \cdot (0.0096)} = 0.049 \text{ mm}. \tag{9.2}$$

The mean and standard deviation describe the measurement process well: a diameter of a ball is measured that has a true mean diameter D. However, because the form of the ball is not perfect, a different diameter will be measured, and this dispersion is characterised by the standard deviation s. The measured mean value $\bar{d}$ and the measured standard deviation s can be considered as approximations of the actual mean diameter D and the population standard deviation σ. Also, the single measurements d_i can be considered approximations of D. The population standard deviation σ is a measure of the expected deviation of d_i from the true mean value D. It is logical that the mean diameter $\bar{d}$ is a better approximation to the true mean value D than any single measurement d_i. This is reflected in the standard deviation of the mean σ_m that is approximated by s_m:

$$s_m = \frac{s}{\sqrt{n}} = \sqrt{\frac{1}{n \cdot (n-1)} \cdot \sum_{i=1}^{n}(d_i - \bar{d})^2} = \sqrt{\frac{1}{5 \cdot 4} \cdot (0.0096)} = 0.022 \text{ mm}. \tag{9.3}$$

This value s_m is the estimation of the deviation between the experimentally determined mean diameter $\bar{d}$ and the true mean diameter D. The difference between the experimentally determined standard deviation s and the standard deviation for an infinite number of measurements σ can be determined from the standard deviation s_s in the standard deviation s, and is given by (Squires 2008)

$$s_s = \frac{s}{\sqrt{2n-2}} = 0.016 \text{ mm}. \tag{9.4}$$

The aforementioned statistical parameters are summarised in Table 9.2.

An illustration of how the parameters in Table 9.2 progress, as more and more measurements are taken, is given in Figure 9.1. This shows how the mean, standard deviation and standard deviation in the mean value typically progress when the number of measurements increases.

Note that the standard deviation approaches a constant value and becomes more stable as the number of measurements increases, but it does not decrease. The interval $\bar{d} \pm s$ will more closely approach the interval $D \pm \sigma$ as the number of measurements increases. Also, the measured $\bar{d}$ approaches the theoretical D within the interval $\bar{d} \pm s_m$. Taking more measurements makes the average $\bar{d}$ better defined. It could be considered that the standard deviation in $\bar{d}$ can always be reduced by just taking more measurements, but this method of reducing the standard deviation in the mean is not efficient. To improve s_m by a factor of

TABLE 9.2

Relationship between the Statistical Parameters and Their Approximation by a Finite Number of Measurements

True value	Approximation	Expected Difference: True Value – Approximation
D	d_i	s
D	$\bar{d}$	$s_m = s/\sqrt{n}$
σ	s	$s_s = s/\sqrt{2n-2}$

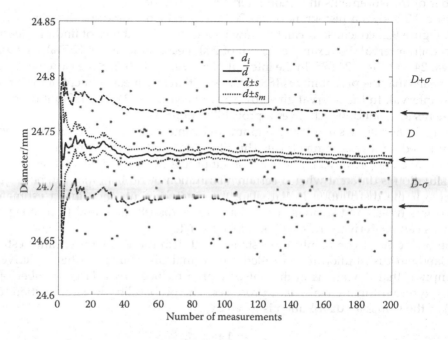

FIGURE 9.1
Measurement series and calculated mean and standard deviations.

two, four times as many measurements have to be taken correspondingly increasing the measurement time, or measurements must be taken at four times shorter intervals. These two alternatives are limited by the time-dependent properties of the measured deviations: the measurements must keep their randomness for shorter measurement times, and slowly varying effects such as temperature drift should not make the measurements slowly drift for longer measurement times.

When reporting a value and an uncertainty based on the statistical variation, it is necessary to consider whether the standard deviation s or the standard deviation in the mean s_m must be used as a basis. This depends on the definition of the measurand. In the preceding example, the measurand can be

1. The diameter measured in any direction—In this case, the ball can be used as a reference diameter measure, where it can be measured once in any orientation. The standard deviation s should be used, as this variation is part of the measurand.
2. The mean diameter measured in all directions—In this case, the ball can only be used as a reference diameter measure if it is measured in many orientations. The standard deviation in the mean s_m should be used, as the mean value is approached by taking more measurements.

9.3.2 Uncertainty Distributions and Confidence Intervals

It may be required to report an interval in which a value can be found with a certain probability. For the example of the ball diameter measurement, establishing such an interval can be illustrated using the 200 values plotted in Figure 9.1. Instead of listing these numbers individually, the spread of values can be more easily visualised as a histogram: the number of measurements in certain intervals is collected.

Figure 9.2 shows a histogram of the 200 measurements used for the plot of Figure 9.1. From Figure 9.2, conclusions can be drawn about the probability of finding measurements in a certain interval. For example, 25 out of 200 measurements (or 12.5%) are found in the interval $24.650 < d_i < 24.675$. In the interval $24.65 < d_i < 24.80$, 188 measurements are found. This means that the probability is $188/200 = 94\%$ that any measurement of the 200 taken are in this interval. In the interval given by $\bar{d} \pm s$, $24.686 < \bar{d} < 24.784$, 137 out of the 200 measurements are found, which gives a probability of 68.5%.

Figure 9.2 also shows that the histogram is asymmetrical, while the statistical distribution of simulated data was symmetrical. This illustrates that a high number of measurements must be taken before anything relevant about probabilities and intervals can be concluded. This situation is different when a certain distribution of probabilities can be assumed. For example if the distribution is Gaussian, the measurements give direct estimates of the parameters (mean and standard deviation) of this distribution, and from this probability intervals can be derived. This will be explained later.

One of the most commonly used statistical distributions for uncertainty estimation is the Gaussian distribution (often called the normal distribution). This is derived on the assumption that deviations in the outcome are produced by a large number of random causes, typical of many real-world processes. The probability density function (described later) of the Gaussian distribution is given by

$$F(q) = \frac{1}{\sigma\sqrt{2\pi}} \cdot e^{-(q-\mu_q)^2/2\sigma^2},$$ (9.5)

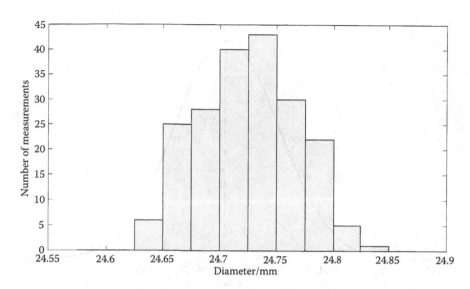

FIGURE 9.2
Histogram of 200 diameter measurements in intervals of 0.025 mm.

where q is the measurement value, μ_q is the expected mean value for an infinite number of measurements and σ is the standard deviation. The function is normalised, so that $\int\limits_{-\infty}^{\infty} F(q)dq = 1$.

Figure 9.2 can correspondingly be normalised by dividing each of the histogram bars by the total number of samples (200 in this example). The probability P to find a measurement in the interval $[q_1; q_2]$ is given by

$$P = \int\limits_{q_1}^{q_2} F(q)dq. \tag{9.6}$$

Given a (theoretical) mean μ_q and standard deviation σ, Equation 9.6 can be used to calculate the probability that a measurement is found in a specified interval. As an example, the probability density function is given for a diameter $\mu_q = 24.725$ mm and $\sigma = 0.043$ mm in Figure 9.3.

For the Gaussian probability density function, the following are the probabilities of finding values in certain areas:

- The probability of finding a value within one standard deviation from the mean is 68.3%.

- The probability of finding a value within two standard deviations from the mean is 95.5%.

- The probability of finding a value within three standard deviations from the mean is 99.7%.

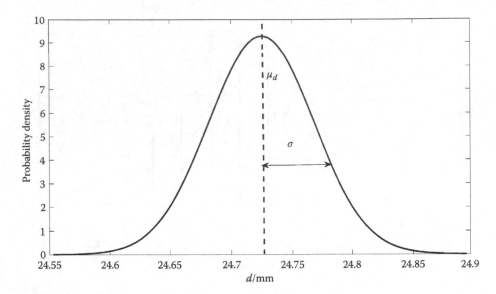

FIGURE 9.3
Gaussian probability density function for μ_d = 24.725 mm and σ = 0.043 mm.

These are basic properties of the Gaussian probability density function. When carrying out an actual measurement, the values of μ_d and σ are unknown and the question is, given a measurement value and its standard deviation, what is the probability of finding the mean in the interval given by the measurement plus or minus its standard deviation, or the mean of the measurements and the standard deviation in the mean? The results for the Gaussian distribution can be used to address this question; however, because of the limited number of measurements, the experimentally determined standard deviation is not so accurately known. To account for the limited amount of measurements, the experimentally determined standard deviation s must be multiplied with a factor t, the so-called Student's t-factor, in order to achieve the same probability as when the standard deviation σ is multiplied with a factor z. This can be written as

$$P(\mu_q \pm z(p) \cdot \sigma) = P(q_i \pm t(n,p) \cdot s). \tag{9.7}$$

This Student's t-factor depends both on the number of measurements n and the probability p, while the factor z depends on the probability only. The t-factor should be applied instead of the factor z when there is a relatively small number of measurements, for example, for $n < 10$. Further considerations on the relationship between s and σ are given in Chapter 3, Section 3.4.11.

An overview of the quantities and probabilities derived so far is given in Table 9.3.

Note that the terms '2 sigma' and '95% confidence interval' are commonly used interchangeably; Table 9.3 indicates the small difference between the two concepts.

TABLE 9.3

Theory and Practice of Confidence Interval Considerations

Theory	Practice	Probability
Measurement q_i in interval $\mu_q \pm \sigma$ Mean of measurements $\bar{q}$ in interval $\quad \mu_q \pm \sigma_m$	μ_q in interval $q_i \pm t \cdot s$ μ_q in interval $\bar{q} \pm t \cdot s$ $t = 1.1$ for $n = 5$ and $P = 0.68$	68%
Measurement q_i in interval $\mu_q \pm 1.96 \cdot \sigma$ Mean of measurements $\bar{q}$ in interval $\quad \mu_q \pm 1.96 \cdot \sigma_m$	μ_q in interval $q_i \pm t \cdot s$ μ_q in interval $\bar{q} \pm t \cdot s_m$ $t = 2.8$ for $n = 5$ ($v = 4$) and $P = 0.95$	95%
Measurement q_i in interval $\mu_d \pm 2 \cdot \sigma$ Mean of measurements $\bar{q}$ in interval $\quad \mu_d \pm 2 \cdot \sigma_m$	μ_q in interval $q_i \pm t \cdot s$ μ_q in interval $\bar{q} \pm t \cdot s_m$ $t = 2.9$ for $n = 5$ and $P = 0.955$	95.5%

For the example of the ball diameter, the following statements can be made:

1. Statement based on the first measurement and 95% probability:

$$\mu_d = d_i \pm t\,(n = 5, P = 95\,\%) \cdot s = (24.780 \pm 2.8 \cdot 0.049)\,\text{mm} = (24.78 \pm 0.14)\,\text{mm}$$

Instead of the first measurement, any other, or any future single measurement can be taken, giving the same uncertainty.

2. Statement based on the mean of five measurements and 95% probability:

$$\mu_d = \bar{d} \pm t(n = 5, P = 95\%) \cdot s/\sqrt{5} = (24.735 \pm 2.8 \cdot 0.049/2.2)\,\text{mm} = (24.74 \pm 0.06)\,\text{mm}$$

More details of the Gaussian and Student's t-distribution are given in Chapter 3.

The mathematical relationship between the mean, standard deviation and confidence intervals is different for different statistical distributions. Other commonly encountered distributions include rectangular, triangular and Poisson (see Chapter 3). For uncertainty estimation, the rectangular (also called 'uniform') distribution is also common because it is often assumed for tolerances, specifications and instrument resolution.

For a rectangularly distributed quantity, the probability of finding this quantity is uniform inside a given interval, and is zero outside this interval. For such a distribution, the relationship between the standard deviation and confidence intervals can be established in a similar way to the Gaussian distribution (see Chapter 3).

In Figure 9.4, both the rectangular and Gaussian distributions are depicted, where both distributions are characterised by the same standard deviation. Similarly, as for the Gaussian distribution, there are relationships between the mean, standard deviation and probability. For the rectangular distribution.

- The probability of finding a value within one standard deviation from the mean is 57.7%.

- The probability of finding a value within 1.73 standard deviations from the mean is 100%.

- The 95% confidence interval is given by the mean plus or minus 1.64 standard deviations.

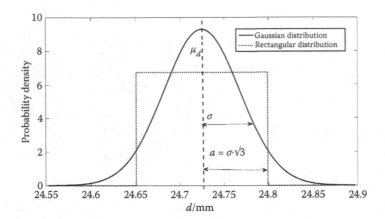

FIGURE 9.4
The Gaussian distribution and the rectangular distribution for μ_d = 24.724 mm and σ = 0.043 mm.

There are two major instances where the rectangular distribution is applied. The first is the resolution of a digital readout where the reading is stable. The second is the knowledge that an instrument or object is specified, that is there is a verified statement that deviations from nominal values are always within some limit.

In general, when the smallest resolution of a digital readout is r, the standard uncertainty due to the resolution of that instrument is

$$u(r) = \frac{r}{2\sqrt{3}} \tag{9.8}$$

As an example, assume the reading d = 24.780 mm is taken. Suppose this reading is stable and the resolution is 0.001 mm, then it can be assumed that the measured value is rectangularly distributed between 24.7795 and 24.7805. The standard uncertainty is $u = 0.001/(2\sqrt{3})$ mm = 0.29 μm.

9.3.3 The Standard Uncertainty Concept

In Section 9.3.2, the meaning of a standard deviation was discussed and how it can be calculated from a series of measurements. In many instances, however, a standard deviation is attributed to a quantity that is not necessarily calculated from multiple measurements. An example is the digital readout as discussed in the example of a rectangular distribution. As many components of an uncertainty calculation use a dispersion measure that is not necessarily statistically established, the more general term 'standard uncertainty' is used for any uncertainty that is expressed as a standard deviation, but without implying it is a result of a statistical process.

Some examples of standard uncertainties are

- Standard deviation s
- Standard deviation of mean value $s/\sqrt{n}$
- Other uncertainty, for example, derived from resolution or specifications

- Uncertainty in calibration
- The published uncertainty of a physical constant
- Error that is not corrected

9.4 Propagation of Uncertainties

Often, a measurement result is used for calculating a different quantity than the one that is measured. That other quantity may be obtained using known relationships, typically in the form of mathematical models or numeric simulations, and its uncertainty is related to the uncertainty in the original measurement, that is the uncertainty of the measurement propagates through the equation of the model and appears as an uncertainty in the other quantity. This is clarified by an example. The diameter measurement that acted as an example in Section 9.3 is now used, but the calculation is for another quantity: the volume of the ball. The uncertainty in the measured diameter will reflect in an uncertainty of the calculated volume. Calculating the uncertainty in the volume can be carried out in a number of ways that are illustrated in the following sections.

9.4.1 Direct Calculation

The method of direct calculation is straightforward and may appear rather unsophisticated. However, in Monte Carlo methods (see Section 9.4.3), direct calculation is used. Also, direct calculation may be the only possibility if the measured quantity is subject to a calculation for which there is no direct expression, for example, as performed by a computer program using an unknown algorithm.

In general, the functional relationship, also called the model function, is given by

$$y = f(x), \tag{9.9}$$

where y is the output estimate of the measurand Y and x is the input estimate for the input quantity X. For the calculation of the volume V from a measured diameter d, the model function is

$$V(d) = \frac{\pi \cdot d^3}{6}. \tag{9.10}$$

The direct calculation of the standard uncertainty in y, $u(y)$ as a result of the standard uncertainty in x, $u(x)$ is given by

$$u(y) = |f(x + u(x)) - f(x)|. \tag{9.11}$$

For the ball volume example with $\bar{d} = 24.755$ mm and $u(\bar{d}) = 0.022$ mm, Equation 9.9 gives $V = 7924$ mm^3 and for the standard uncertainty in the volume

$$u(V) = |V(d + u(d)) - V(d)| = \left| \frac{\pi \cdot (d + u(d))^3}{6} - \frac{\pi \cdot d^3}{6} \right|$$

$$= \left| \frac{\pi \cdot (24.735 + 0.022)^3}{6} - \frac{\pi \cdot 24.735^3}{6} \right| = 21 \text{ mm}^3. \tag{9.12}$$

This method is straightforward and appropriate but gives little insight. For example, it is not immediately clear what happens with the uncertainty in the volume when the uncertainty in the diameter doubles.

9.4.2 Calculation Using Derivatives

More insight into the propagation of uncertainty is obtained when using the derivative of the function $y = f(x)$. This can be derived from the Taylor series expansion of $f(x + u(x))$:

$$f(x + u(x)) = f(x) + \frac{\partial f(x)}{\partial x} u(x) + \frac{1}{2} \frac{\partial^2 f(x)}{\partial x^2} (u(x))^2 + \ldots . \qquad (9.13)$$

Taking the first two terms of this series gives

$$u(y) = |f(x + u(x)) - f(x)| \approx \left| f(x) + \frac{\partial f(x)}{\partial x} u(x) - f(x) \right| \approx \left| \frac{\partial f(x)}{\partial x} \right| \cdot u(x). \qquad (9.14)$$

This is an approximation for small values of $u(x)$. The derivative $\left| \frac{\partial f(x)}{\partial x} \right|$ is called the sensitivity coefficient. It can be useful to substitute the original function $y = f(x)$ in the derivative where possible. This is illustrated for the case of the volume measurement, thus

$$u(V) = \left| \frac{\partial V(d)}{\partial d} \right| \cdot u(d) = \frac{3 \cdot \pi \cdot d^2}{6} \cdot u(d) = \frac{3 \cdot V}{d} \cdot u(d) \qquad \text{or} \qquad \frac{u(V)}{V} = 3 \cdot \frac{u(d)}{d}. \qquad (9.15)$$

Interpreting Equation 9.13, the relative uncertainty in the volume is three times the relative uncertainty in the diameter. This factor of three increase in relative uncertainty is related to the power of three in the equation (see also Table 9.4). The calculation with the values used in Section 9.4.1, $\bar{d} = 24.755$ mm and $u(\bar{d}) = 0.022$ mm, gives $V = 7924$ mm^3 and $u(V) = 21$ mm^3, the same value as in Section 9.4.1.

The factor of three in the relative uncertainty is related to the third power in the volume calculation. For classes of functional relationships, $y = f(x)$, general rules for the uncertainty propagation can be derived. An overview is given in Table 9.4.

TABLE 9.4

Propagation of Uncertainties for Some Functional Relationships

Function $y = f(x)$; c is a Constant	Uncertainty Propagation	In Words
$y = x + c$	$u(y) = u(x)$	Uncertainty is unchanged.
$y = c \cdot x$	$u(y)/y = u(x)/x$	Relative uncertainty is unchanged.
$y = x^c$	$u(y)/y = c \cdot u(x)/x$	Relative uncertainty becomes factor of c larger. This effect often appears and requires attention of the user.
$y = \log(x)$	$u(y) = u(x)/x$	Absolute uncertainty in y is relative uncertainty in x.
$y = e^x$	$u(y)/y = x$	Relative uncertainty in y is absolute uncertainty in x.

9.4.3 Calculation Using a Monte Carlo Method

With a Monte Carlo method, both the propagation of standard uncertainties and the propagation of uncertainty distributions can be simulated. The Monte Carlo method consists of the generation of a large number of simulated measurements according to a defined distribution, applying the calculation using the model function for each individual simulated measurement and considering the distribution of the output values. Simulations are made using the input variable ξ_i that is given by

$$\xi_i = x + u(x) \cdot z_i, \tag{9.16}$$

where x is the best estimate of the input quantity, usually the mean; $u(x)$ is the standard uncertainty of x; and z_i is a random number, taken from a distribution with a mean value of zero and a standard deviation of unity. Most mathematical software packages contain a function to obtain Gaussian distributed random numbers. If a random number generator gives uniformly distributed random numbers r_i between 0 and 1, then a random number z_i with zero mean value and unit standard deviation with a rectangular distribution is obtained from

$$z_i = (r_i - 0.5) \cdot 2\sqrt{3}. \tag{9.17}$$

The probability density function of the output values $f(\xi_i)$ is used to obtain the mean $\bar{y}$, the standard uncertainty in y and the coverage interval that contains the true value Y with a specified probability. This is illustrated for the diameter measurement with $D = 24.725$ mm, $u(d) = 0.022$ mm and 10^6 simulations. This gives as simulated diameters $d_i = D + u(d) \cdot z_i$ and as simulated volume

$$V_i = \frac{\pi \cdot d_i^3}{6} = \frac{\pi \cdot (x + u(x) \cdot z_i)^3}{6} = \frac{\pi \cdot (24.725 + 0.022 \cdot z_i)^3}{6}. \tag{9.18}$$

Figure 9.5 shows histograms for the simulated values of d_i and the resulting values of V_i, where the range of the million simulations is divided into 100 intervals. Note the similarity with Figure 9.4; however, along the vertical axes the number of simulations are given that

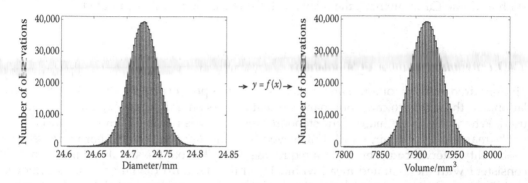

FIGURE 9.5
Illustration of the propagation of an uncertainty distribution in the diameter (left) to a distribution in the volume (right) as a result of a Monte Carlo simulation.

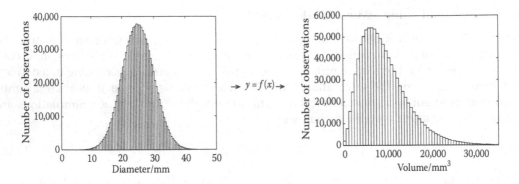

FIGURE 9.6
Illustration of the non-linear propagation of an uncertainty distribution in the diameter (left) to a distribution in the volume (right) as a result of a Monte Carlo simulation.

fall in the interval indicated on the horizontal axis. From the simulated values V_i, the basic parameters V and $u(V)$ are calculated; as expected, the values found are $V = 7914$ mm^3 and $u(V) = 21$ mm^3. These values could be derived with the methods given in Section 9.4.1 and Section 9.4.2 far more conveniently. However, with a Monte Carlo method, fewer straightforward questions related to probabilities can be answered directly, such as: What is the probability that the volume is larger than 7900 mm^3? (Answer: 75%.) The 95% probability interval is taken between the value for which 2.5% of the values are less and the value for which 97.5% of the values are less. This leads to the interval $V = [7873$ mm^3; 7956 mm^3]. Again the result is consistent with the interval $V = 7914$ mm $\pm$ 42 mm (i.e. Mean $\pm$ 2 standard deviations) that follows from Gaussian statistics. For a Monte Carlo method, the assumption of Gaussian statistics is no longer needed and confidence intervals can be found for any distribution.

In Figure 9.5, the Gaussian distribution of the diameter is transformed to a closely similar distribution of the volume, because the deviations are relatively small. For larger deviations, the distribution changes because of non-linear effects. This is illustrated in Figure 9.6, where 10^6 simulations are calculated choosing $d = 24.725$ mm and $u(d) = 5$ mm. In this case, the distribution of the volume is clearly neither Gaussian nor symmetric any more, and the confidence intervals in the volume are far less obviously related to those in the diameter. With a Monte Carlo method, the relevant statistics can be readily evaluated.

9.4.3.1 Required Number of Simulations

The required number of simulations depends on the purpose of the Monte Carlo calculation and the ability to carry out many simulations. As an indication, Equation 9.4 may be used. When only an estimate of the standard uncertainty is required, a relative uncertainty of 5% may already be fair, which is achieved by some 200 simulations. For more detailed probability interval considerations, a requirement may be that the simulated mean value is consistent with the assumed mean within 1% of the standard deviation. This is achieved with 5000 simulations. If calculation time is not an issue, 1 million simulations is usually a safe amount.

9.5 Measurements with Multiple Variables

In Section 9.4, the uncertainty calculation for a single variable and a function of a single variable was treated. In practice, there are often multiple variables that contribute to an uncertainty in a measured quantity. Examples are the influence of measurement force and temperature on a contact length instrument and the influence of air temperature, air pressure and humidity when doing a displacement measurement using a laser interferometer (see Chapter 6). How to perform an uncertainty analysis for such cases is treated in this section, again based on the theory and examples given in Section 9.4.

In general, for N variables, Equation 9.9 can be extended to

$$Y = f(X_1, X_2, ... X_N),\qquad(9.19)$$

where Y is the measurand that is approximated by its measured value y and $X_{1...N}$ are the input quantities that are estimated by their values x_i. As an example, the average density ρ of a ball with volume V is calculated from its mean diameter $\bar{d}$ and its mass m. The model equation becomes

$$\rho = \frac{m}{V} = \frac{6 \cdot m}{\pi \cdot \bar{d}^3}.\qquad(9.20)$$

Functionally, Equation 9.20 can be expressed as $\rho = f(\bar{d}, m)$.

9.5.1 Propagation of Standard Uncertainties

The propagation of standard uncertainties can be carried out numerically as in Section 9.4.1. However, for some more general considerations, the propagation similar to that in Section 9.4.2, using partial derivatives, is considered. The expression for the combined variance $u_c^2(y)$ associated with the result of a measurement is obtained using an extension of Equation 9.12 to multiple variables and is given by

$$u_c^2(y) = \sum_{i=1}^{N} \sum_{j=1}^{N} \frac{\partial f}{\partial x_i} \frac{\partial f}{\partial x_j} u(x_i, x_j)$$

$$= \sum_{i=1}^{N} \left(\frac{\partial f}{\partial x_i}\right)^2 u^2(x_i) + 2 \sum_{i=1}^{N-1} \sum_{j=i+1}^{N} \frac{\partial f}{\partial x_i} \frac{\partial f}{\partial x_j} u(x_i) \cdot u(x_j) \cdot r(x_i, x_j),\qquad(9.21)$$

where y is the estimate of Y, x_i and x_j are the estimates of X_i and X_j, and $r(x_i, x_j)$ is the correlation coefficient between variables x_i and x_j that is given by

$$r(x_i, x_j) = \frac{u(x_i, x_j)}{u(x_i) \cdot u(x_j)}.\qquad(9.22)$$

Variables can be uncorrelated ($r(x_i, y_i) = 0$), fully correlated ($|r(x_i, y_i)| = 1$) or something in between. In general, it can be stated that $-1 \le r(x_i, x_j) \le +1$. Uncorrelated input quantities, for which $r(x_i, x_j) = 0$, mean that a change in one quantity does not imply an expected change in another. In most cases this assumption is justified, and Equation 9.21 reduces to

$$u_c^2(y) = \sum_{i=1}^{N}\sum_{j=1}^{N} \frac{\partial f}{\partial x_i}\frac{\partial f}{\partial x_j} u(x_i, x_j) = \sum_{i=1}^{N}\left(\frac{\partial f}{\partial x_i}\right)^2 u^2(x_i). \tag{9.23}$$

Equation 9.23 is the essential equation for the propagation of standard uncertainties that is the basis of the large majority of uncertainty evaluations. This summation is commonly called a quadratic sum. The derivatives $\left|\frac{\partial f(x_i)}{\partial x_i}\right|$ are called the sensitivity coefficients. The partial derivatives can be omitted by applying the direct calculation similar to Equation 9.11:

$$u_c^2(y) = \sum_{i=1}^{N}(f(x_i + u(x_i)) - f(x_i))^2. \tag{9.24}$$

For the example of the ball density, Equation 9.23 can be written as

$$u^2(\rho) = \left(\frac{\partial \rho}{\partial m}\right)^2 u^2(m) + \left(\frac{\partial \rho}{\partial d}\right)^2 u^2(d) = \left(\frac{\rho}{m}u(m)\right)^2 + \left(3\frac{\rho}{d}u(d)\right)^2. \tag{9.25}$$

As numerical values for the mean diameter d, the values from Section 9.4 are taken: $d = 24.735$ mm and $u(d) = 0.022$ mm.

It is assumed that the mass is determined by weighing the ball on a balance with a digital readout with a limited resolution of 0.1 g. The balance indicates a stable $m = 19.7$ g. With Equation 9.20, this gives a density $\rho = 2.486$ g·cm^{-3}. This value is close to but a bit lower than pure silica glass. For estimating the standard uncertainty, it is assumed that the balance is calibrated using reference masses and has zero errors, and that the limited resolution is the only contributor to the standard uncertainty. The mass can, therefore, be assumed to have a rectangular distribution in the interval [19.65; 19.75] g. The standard uncertainty can be derived from Equation 9.8, giving $u(m) = 0.10/2\sqrt{3} \approx 0.03$ g.

Substituting numerical values in Equation 9.25, the standard uncertainty of the ball density is

$$u(\rho) = \sqrt{\left(\frac{\rho}{m}u(m)\right)^2 + \left(3\frac{\rho}{d}u(d)\right)^2} = \rho \cdot \sqrt{\left(\frac{u(m)}{m}\right)^2 + \left(\frac{3\cdot u(d)}{d}\right)^2}$$

$$= 2.486 \cdot \sqrt{\left(\frac{0.03}{19.7}\right)^2 + \left(\frac{3\cdot 0.022}{24.735}\right)^2} = 0.0031 \text{ g·cm}^{-3}. \tag{9.26}$$

Using Equation 9.24 gives the same result.

9.5.2 Uncertainty Budget

Although Equation 9.23 (or 9.24) is essentially all that is needed for the majority of uncertainty calculations, insight can be obtained more easily if this calculation is arranged in a so-called uncertainty budget. An uncertainty budget is an ordered arrangement of the quantities, estimates, standard uncertainties, sensitivity coefficients and uncertainty contributions used in the uncertainty analysis of a measurement. The uncertainty budget is the representation of Equation 9.23 in the form of a table, as shown in Table 9.5. Sometimes the distribution used is included as an additional column, but this is omitted in Table 9.5.

For the example of the density measurement, such an uncertainty budget is as given in Table 9.6.

From Table 9.6 it can directly be observed that the uncertainty in the diameter measurement is the main contributor to the standard uncertainty in the density.

Note that in the case of direct numerical evaluation, the uncertainty budget can be written as depicted in Table 9.7.

TABLE 9.5

Schema of an Ordered Arrangement of All Aspects of the Uncertainty Analysis of a Measurement, Also Known As an Uncertainty Budget

Quantity X_i	Estimate x_i	Standard Uncertainty $u(x_i)$	Sensitivity Coefficient $c_i = \dfrac{\partial f}{\partial x_i}$	Contribution to Standard Uncertainty $u_i(y) = c_i \cdot u(x_i)$
X_1	x_1	$u(x_1)$	c_1	$u_1(y)$
X_2	x_2	$u(x_2)$	c_2	$u_2(y)$
:	:	:	:	:
X_N	x_N	$u(x_N)$	c_N	$u_N(y)$
Y	y	–	–	$u(y) = \sqrt{\sum_{i=1}^{N} u_i^2}$

TABLE 9.6

Uncertainty Budget for the General Case of a Density Measurement

Quantity X_i	Estimate x_i	Standard Uncertainty $u(x_i)$	Sensitivity Coefficient $c_i = \dfrac{\partial \rho}{\partial x_i}$	Contribution to Standard Uncertainty $u_i(y) = c_i \cdot u(x_i)$
Diameter	$d = 2.4735$ cm	$u(d) = 0.0022$ cm	$c_1 = \dfrac{3\rho}{d} = 3.02$ g·cm^{-4}	$u_1(\rho) = 3\rho\dfrac{u(d)}{d} = 0.0066$ g·cm^{-3}
Mass	$m = 19.7$ g	$u(m) = 0.03$ g	$c_2 = \dfrac{\rho}{m} = 0.126$ cm^{-3}	$u_2(\rho) = \rho\dfrac{u(m)}{m} = 0.0038$ g·cm^{-3}
Density	$\rho = 2.486$ g·cm^{-3}	–	–	$u(\rho) = \sqrt{u_1^2(\rho) + u_2^2(\rho)} = 0.0076$ g·cm^{-3}

TABLE 9.7

Uncertainty Budget Based on Direct Numerical Estimation

Quantity X_i	Estimate x_i	Contribution to Standard Uncertainty $u_i(y)$
X_1	x_1	$u_1(y) = f(x_1 + u(x_1), x_2..x_N) - f(x_1, x_2, ..x_N)$
X_2	x_2	$u_2(y) = f(x_1, x_2 + u(x_2),..x_N) - f(x_1, x_2, ..x_N)$
:	:	:
X_N	x_N	$u_2(y) = f(x_1, x_2, ..x_N + u(x_N)) - f(x_1, x_2, ..x_N)$
Y	y	$u(y) = \sqrt{\sum_{i=1}^{N} u_i^2 u(y)}$

9.5.3 Confidence Interval

In Section 9.3.2 it was illustrated how a standard deviation can be used to define an uncertainty based on a confidence interval. Similar to the Student's t-factor used in Section 9.3.2, it can be stated more generally that a standard uncertainty must be multiplied by a factor k to obtain an uncertainty that can be associated with a defined confidence interval, thus

$$U(y) = k \cdot u(y). \tag{9.27}$$

Here, $U(y)$ is the expanded measurement uncertainty. In general, it is assumed that taking $k = 2$ leads to a confidence interval of at least 95%. This is based on assuming that most quantities will have a Gaussian distribution, and even if this is not the case, the combination of different kinds of distributions will result in a Gaussian or Gaussian-like distribution (this theorem is known as the central limit theorem).

In the example of the ball density, the measurement result could be reported as follows: 'The ball density is given by $\rho = (2.468 \pm 0.016)$ g·cm^{-3}. The reported expanded uncertainty of measurement is stated as the standard uncertainty of measurement multiplied by the coverage factor $k = 2$, which for a Gaussian distribution corresponds to a coverage probability of approximately 95%.'

For by far the most cases, such an estimation is sufficient; however, when the uncertainty distribution and the related confidence intervals need a closer consideration, the Monte Carlo method can be used and is explained in the following section.

9.5.4 Propagation of Probability Distributions

The Monte Carlo method introduced in Section 9.4.3 can be extended to multiple input quantities x_i. This extension has the advantage that each variable can be given its own uncertainty distribution and that the confidence intervals of the output estimate y can be calculated directly from its distribution. The name 'Monte Carlo' means in general that a statistical distribution is generated using random numbers. This means that there is not a single Monte Carlo method, but there are many variations. The most common Monte Carlo method in uncertainty calculations involves the simulation of many measurements while all influencing parameters are jointly varied according to their uncertainty distribution. This means that using the functional relationship as in Equation 9.19, many simulations are carried out using

$$y_k = f(x_1 + u(x_1) \cdot r_1(k), x_2 + u(x_2) \cdot r_2(k), .., x_N + u(x_N) \cdot r_N(k)), \tag{9.28}$$

where y_k is the output value for the kth simulation, k is the simulation number $k = 1...K$, where K is the total number of simulations, $x_{1.N}$ are the input values, r_i is a random number with zero mean and unit standard deviation, that has a distribution according to the input variable x_i. The simulations will give a distribution of the output quantity y. Every simulation gives a direct estimate of the standard uncertainty in y:

$$u(y)_k^2 = (y_k - y(x_1, .x_N))^2. \tag{9.29}$$

After K simulations, the standard uncertainty $u(y)$ can be estimated from

$$u(y) = \sqrt{\frac{\sum_{k=1}^{K}(y_k - y)^2}{K}}.$$ (9.30)

The estimation becomes more reliable with more simulations, similar to the standard deviation which becomes more reliable with more measurements (see Equation 9.4). The distribution of the values of y_k can be used to determine confidence intervals.

The Monte Carlo method is illustrated for the density measurement that was used as an example in the previous sections. The input quantities x_i are diameters d_i are characterised with a mean diameter $\bar{d} = 24.725$ mm and a standard uncertainty $u(d) = 0.0022$ mm with a Gaussian distribution.

Mass m_i is characterised with a mean mass $m = 19.7$ g and a standard uncertainty $u(m) = 0.03$ g with a rectangular distribution.

Multiple density values can be calculated using these input quantities using random generators and the density equation

$$\rho_i = \frac{6 \cdot m_i}{\pi \cdot d_i^3} = \frac{6 \cdot (m + r_{1,i} \cdot u(m))}{\pi \cdot (d + r_{2,i} \cdot u(d))^3},$$ (9.31)

where r_1 is a random number that is taken from a Gaussian distribution with zero mean value and unit standard deviation, and r_2 is another, independent, random number that is taken from a rectangular distribution of random numbers with zero mean value and unit standard deviation. The simulation of these random numbers gives multiple values of ρ_i. The estimation of the standard uncertainty $u(\rho)$ can be illustrated in a form that has a close resemblance to the uncertainty budget. This is illustrated in Table 9.8.

Note the close similarities between the calculation of a standard deviation as it is illustrated in Table 9.1, the uncertainty budget illustrated in Table 9.5, the directly calculated uncertainty budget of Table 9.7 and the Monte Carlo method illustrated in Table 9.8.

The combination of the rectangular distribution of the m_i values and the Gaussian distribution of the d_i values leads to a distribution of the ρ_i values that is neither Gaussian nor rectangular (Figure 9.7). However, when several distributions are combined, they will tend to a Gaussian distribution (the central limit theorem).

The simulated values of ρ_i can be sorted and the interval can be determined where 95% of the values will fit in. Normally this is done by estimating both the lower limit that

TABLE 9.8

Illustration of the Estimation of the Standard Uncertainty in ρ Using a Monte Carlo Method

Simulation Number	d_k/cm	m_k/g	ρ_k/g·cm^{-3}	$\rho_k - \rho$/g·cm^{-3}
1	2.4737	19.696	2.4852	−0.004
2	2.4765	19.726	2.4804	−0.0088
...	...	...	...	...
K	2.4737	19.719	2.5067	0.0175
—	—	—	—	$u(\rho) = \sqrt{\dfrac{\sum_{k=1}^{K}(\rho_i - \rho)^2}{K}}$

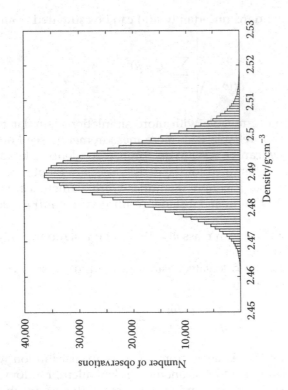

$\rho = f(m, d) \rightarrow$

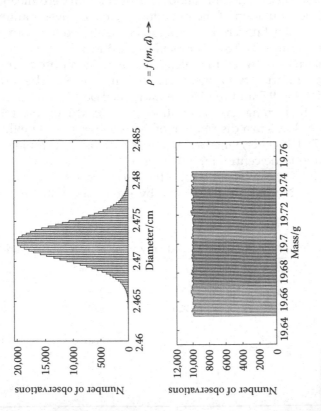

FIGURE 9.7
Illustration of the propagation of probability distributions for the density calculation.

is exceeded by 97.5% of the cases and the upper limit that is exceeded by 2.5% of the cases. In this simulation of 10^6 measurements, the 95% confidence interval is expressed as 2.475 g·cm^{-3} $< \rho < 2.504$ g·cm^{-3}.

This interval can be compared with the interval that is obtained when using the standard uncertainty with a coverage factor $k = 2$. In this case the 95% confidence interval is estimated as $(2.486 - 2·0076)$ g·cm^{-3} $< \rho < (2.486 + 2·0.0076)$ g·cm^{-3} or 2.471 g·cm^{-3} $< \rho < 2.501$ g·cm^{-3}. These values are relatively similar, despite the combination of two different distributions.

Although the results from Monte Carlo simulations may be similar to a more traditional uncertainty budget calculation in many cases, there are apparent advantages to propagating uncertainty distributions by Monte Carlo simulations. These advantages are given next:

- The Monte Carlo method in fact uses the direct calculation method as in Equation 9.24, so the calculation of derivatives is omitted.
- Correlations of input quantities can be incorporated into the simulations by correlating (combining) the random numbers used to characterise the distributions involved.
- The standard uncertainty and confidence intervals are readily obtained; considerations about the appropriate k-factor can be omitted.
- When testing against specifications, the probability of meeting the specifications can readily be calculated.
- The Monte Carlo method is considered as a more rigorous method of uncertainty calculations and provides more insight of how the individual measurements will influence the values and subsequent distributions coming from a particular process (JCGM 101 2008).

9.6 Testing against Specifications

Often a measurement is carried out to verify the conformity or non-conformity with a given tolerance for a characteristic of a workpiece (usually given as an upper specification limit, a lower specification limit or both) or with given maximum permissible errors for a metrological characteristic of a measuring instrument.

Examples are a gauge block with a length that must fall within a certain accuracy class (see Chapter 6), the pitch diameter of a thread plug gauge that must fall within limits that are specified in specification standards, or a coordinate measuring system (see Chapter 6) that is sold to a customer based on a specified maximum permissible error.

The estimated measurement uncertainty is to be taken into account when verifying conformity or non-conformity with specification. A problem arises when a measurement result falls close to the upper or lower specification limit. In this case, verification of conformity or non-conformity with specifications is not possible: the measurement uncertainty includes a finite probability that a true value of the measurand is out of specification even if the measured result falls inside the specification zone or, conversely, is in specification even if the measured value falls outside. The procedure for dealing with this situation is given later and follows the approach detailed in ISO 14253 (2011).

9.6.1 Determination of Conformance with Specification

A product with tolerance limits is taken as an example. On a mechanical drawing a length L of 40 mm is indicated, with tolerances +0.02 mm and –0.01 mm. The lower specification limit (LSL) is thus 39.99 mm and the upper specification limit (USL) is 40.02 mm. Suppose a value $L' = (40.008 \pm 0.002)$ mm is measured; this means that the uncertainty U of this measurement on the basis of $k = 2$ is 2 µm. This situation is shown in Figure 9.8. The measured value including its uncertainty, indicated as L' in the figure, is inside the limits set by LSL and USL, and it can be concluded that the length of the product conforms with specifications.

In the case of an uncertainty of 2 µm, the range of measured values L_m that leads to accepting the workpiece reduces to the interval of an acceptable measured value of L_m to [40 – 0.01 + 0.002 mm; 40 + 0.02 – 0.002 mm] = [39.992 mm; 40.018 mm]. This means that for the measured value L_m the tolerance range is reduced by the measurement uncertainty on both sides. This situation is shown in Figure 9.9.

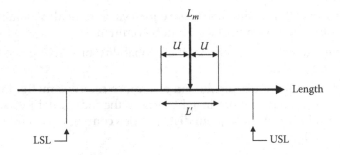

FIGURE 9.8
Situation of measurement that verifies conformance with specifications. LSL: lower specification limit, USL: upper specification limit, U: measurement uncertainty, L_m: measured length, L': measured length with 95% confidence interval.

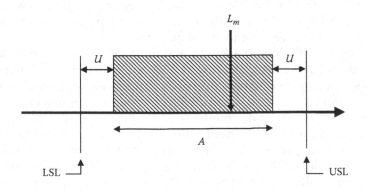

FIGURE 9.9
Display of the range (shaded region) of a measured value L_m from which compliance with specification can be accepted. A is the acceptance zone of measurement values L_m that imply conformance to specifications.

9.6.2 Determination of Non-Conformance with Specifications

A workpiece can only be deemed to be outside of tolerance if the measured value with the uncertainty interval is outside the tolerance limits. For example, a measurement result $L' = (40.028 \pm 0.002)$ mm at the tolerance limits LSL = 39.99 mm and USL = 40.02 mm.

In the case of an uncertainty $U = 2$ μm, the range of measured values L_m that give the conclusion that the non-conformance as a result is: $L_m < (40 - 0.01 - 0.002)$ mm or $L_m > (40 + 0.02 + 0.002)$ mm. This is illustrated in Figure 9.10.

9.6.3 Cases Where Conformance or Non-Conformance to Specifications Cannot Be Concluded

If the measurement uncertainty is relatively large compared to the specification interval, or if the measurement value is close to the LSL or USL, it can happen that the workpiece can be neither accepted nor rejected. For example, $L' = (40.021 \pm 0.002)$ mm, where the tolerance limits are LSL = 39.99 mm and USL = 40.02 mm.

This leads to zones near both LSL and USL where a measurement value L_m that falls inside such a zone; also after taking a measurement it cannot be decided with sufficient confidence whether a product meets the specification or not. This is illustrated in Figure 9.11.

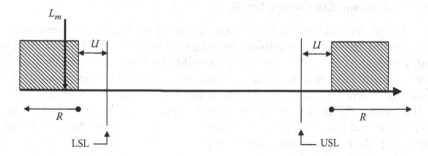

FIGURE 9.10
Display of the range (shaded) of a measured value L_m from which non-compliance with specification can be concluded. R is the rejection zone of measurement values L_m that imply non-conformance to specifications.

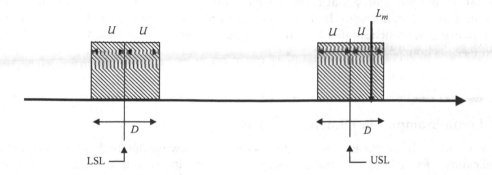

FIGURE 9.11
Display of the range of a measured value L_m from which compliance or non-compliance with specification cannot be concluded. D is the zone of measurement values L_m that imply neither conformance nor non-conformance to specifications.

It is important that manufacturers and their customers for such cases agree who gets the benefit of doubt. Without further agreements, the measurement uncertainty is in principle to the detriment of the one who has an interest in the measurement result. A manufacturer has to measure inside the range as indicated in Figure 9.9 to deliver the product and a buyer should measure in the range of Figure 9.10 to reject the product. This is another illustration of the fact that a measurement is more useful, and less likely to result in disputes, when the uncertainty is smaller. If the uncertainty covers a large part of the tolerance interval, the measurement is not meaningful.

The situation illustrated in Figure 9.11 may be rather unsatisfactory when a customer of a measurement service asks 'Is my product/instrument good or not?' and the only answer is 'We (still) do not know'. In the past, often the rule of thumb was the measurement uncertainty must be smaller than 20% of the tolerance range, and the measurement value L_m must be related to LSL and USL without considering the measurement uncertainty. This has as an advantage that always a yes or no is obtained according to Sections 9.6.1 and 9.6.2 respectively. The disadvantage is that two different measurements, where one leads to accept and the other leads to reject, can be mutually consistent considering their measurement uncertainties. In practice, the customer and supplier may agree on a hard threshold between good/go/accepted and bad/no-go/rejected, but the reader should be aware that a real satisfactory solution for this situation has not yet been found.

9.6.4 Specifications and Confidence Levels

In Section 9.6.1 to Section 9.6.3 it was assumed that the uncertainty represents the 95% confidence interval, that is the probability that the true value is smaller than the lower limit of the confidence interval is 2.5%, and the probability that the true value is larger than the upper limit of the confidence interval is 2.5%. For cases when the uncertainty is much smaller than the tolerance interval, to have at least 95% probability that a measurement classifies the object rightfully inside or outside specifications, it is sufficient that the uncertainty represents the 90% confidence interval, as for this case, the probability that the real value is outside the confidence interval is 5% for both sides. Assuming a Gaussian distribution, this means that instead of $k = 2$, a value of $k = 1.65$ can be taken for the expanded uncertainty U, if the establishment of conformance/non-conformance with specifications is the only aim of the measurement. In Figures 9.9 through 9.11 it is sufficient to take $U = 1.65 \cdot u$, where u is the standard uncertainty. If U is determined by a Monte Carlo simulation, the probability that L_m is inside or outside the acceptance interval determined by LSL and USL can readily be calculated. Also for more complicated cases, where the uncertainty is not much smaller than the acceptance interval, a Monte Carlo method can be used (see Section 9.5.4).

9.7 Some Examples in Dimensional Metrology

To consolidate the concepts developed thus far, the following detailed examples represent applications for which uncertainty estimates are of increasing complexity. The first, involving the measurement of length, considers an idealised situation in which only temperature effects are considered. Subsequent examples incorporate more comprehensive assessments for measurements influenced by multiple sources of errors and more elaborate multi-parameter models.

9.7.1 Length Measurement with Different Materials and Temperatures

A steel bar, with parallel faces and a linear thermal expansion coefficient $\alpha_1 = (11.5 \pm 0.5) \cdot 10^{-6}$ K^{-1} and a nominal length $L_n = 100$ mm, is measured on a length measuring machine that contains a glass line-scale with linear thermal expansion coefficient $\alpha_2 = (5.5 \pm 0.5) \cdot 10^{-6} \ K^{-1}$ as a reference. The temperature of the steel bar is $T_1 = (22.0 \pm 0.5)$ °C and the temperature of the glass scale in the machine is $T_2 = (21.0 \pm 0.5)$ °C. All uncertainties are standard uncertainties. The measurement situation is illustrated in Figure 9.12.

The indication L_2 is $L_{2,20} = 100.005$ mm. The scale has no deviations at 20 °C and the measured length reproduces perfectly. The questions are what is the length of the object at 20 °C and what is its uncertainty.

The model can be derived from the general equation that reduces a measured length to 20 °C that is valid for both L_1 and L_2 (see also Equations 5.1, 8.52 and 11.5):

$$L_{1,T} = L_{1,20}(1 - \alpha_1 \cdot (T_1 - 20\,°C))$$
$$L_{2,T} = L_{2,20}(1 - \alpha_2 \cdot (T_2 - 20\,°C))$$

(9.32)

From this measurement (assuming there are no further error contributions from misalignments and Abbe offsets; see Chapter 10), it follows that $L_{1,T} = L_{2,T}$. Using $1/(1 - x) \approx 1 + x$ for small x, $\alpha_1 \cdot \alpha_2 \approx 0$, and $T_0 = 20$ °C, the model equation can be written as

$$L_{1,20} = L_{2,20}(1 - \alpha_2 \cdot (T_2 - T_0) + \alpha_1 \cdot (T_1 - T_0)).$$

(9.33)

This gives the uncertainty budget shown in Table 9.9.

Note that Equation 9.33 can also be written as

$$L_{1,20} = L_{2,20}\left(1 + \bar{\alpha} \cdot \Delta T + \Delta\alpha \cdot (\bar{T} - T_0)\right)$$

(9.34)

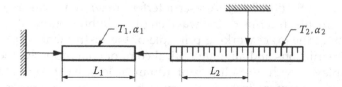

FIGURE 9.12

Length measurement on a steel object with length L_1, temperature T_1 and thermal expansion coefficient α_1 by a length measuring machine with a glass scale that reads the measured length L_2, having a temperature T_2 and thermal expansion coefficient α_2.

TABLE 9.9

Uncertainty Budget for a Dimensional Measurement, Where the Object and the Scale Have a Different Temperature and a Different Thermal Expansion Coefficient

Quantity X_i	Estimation x_i	Uncertainty u_i	Sensitivity $c_i = \dfrac{\partial L_{1,20}}{\partial X_i}$	Contribution to Standard Uncertainty in $L_{1,20}$: $u_i \cdot c_i$
T_1	22 °C	0.5 °C	$L_{2,20} \cdot \alpha_1$	0.6 µm
α_1	$11.5 \cdot 10^{-6} \ K^{-1}$	$0.5 \cdot 10^{-6} \ K^{-1}$	$L_{2,20} \cdot (T_1 - T_0)$	0.1 µm
T_2	21 °C	0.5 °C	$L_{2,20} \cdot \alpha_2$	0.3 µm
α_2	$5.5 \cdot 10^{-6} \ K^{-1}$	$0.5 \cdot 10^{-6} \ K^{-1}$	$L_{2,20} \cdot (T_2 - T_0)$	0.05 µm
$L_{1,20}$	100.00175 mm	$u(L_{1,20}) = \sqrt{\sum (u_i \cdot c_i)^2}$		0.68 µm

where $\bar{\alpha}$ is the mean thermal expansion coefficient of the object and reference, $\Delta\alpha$ is the difference between the expansion coefficients of the scale and object $\Delta\alpha = \alpha_2 - \alpha_1$, $\bar{T}$ is the mean temperature of scale and object, and ΔT is the temperature difference between scale and object $\Delta T = T_2 - T_1$.

From Equation 9.34 the following conclusions can be drawn:

- The difference in expansion coefficient is less important the closer the mean temperature is to 20 °C.
- The mean expansion coefficient is less important the closer the temperatures are to each other.
- The mean temperature is less important the closer the expansion coefficients are to each other.
- The temperature difference is less important the smaller the mean expansion coefficient is.

These are the reasons why dimensional calibration laboratories are kept as close as possible to 20 °C and why it is important to let objects to be measured stabilise in temperature, so that the temperature of object and scale are as close as possible to each other and to 20 °C.

The result of this measurement can be stated as $L = (10.00175 \pm 0.00068)$ mm, where L is the length at 20 °C and the uncertainty is stated as a standard measurement uncertainty multiplied by the coverage factor $k = 2$, corresponding to a confidence level of about 95%.

9.7.2 Displacement Measurement Using a Laser Interferometer

The calibration of the axes of machine tools is commonly carried out using a laser interferometer system (see Section 5.4.5). A laser interferometer is a measurement device that measures displacements in terms of the wavelength of light. Commonly, a user does not need to worry about the exact working principle of the instrument. Typically, by following the manual instructions and aligning optical components properly, the measured displacement is displayed with a resolution of 10 nm or less. Next to the laser head itself, some additional measurement instruments must be used: an air temperature sensor, an air pressure sensor, a material temperature sensor and sometimes a humidity sensor (depending on the target uncertainty). A particular set-up is illustrated in Figure 9.13.

The slide is displaced four times from the zero position to a position of nominally 80 mm. The four read-outs are $L_1 = 80.0005$ mm, $L_2 = 80.0006$ mm, $L_3 = 80.0004$ mm and $L_4 = 80.0008$ mm.

The scale temperature is $T_s = (20.5 \pm 0.2)$ °C, the linear thermal expansion coefficient of the scale is $\alpha_s = (11.5 \pm 0.7) \cdot 10^{-6}$ K^{-1}, the air temperature is $T_a = (21.0 \pm 1.0)$ °C, the air pressure is $p_a = (1020 \pm 2)$ hPa (hectopascal) and the humidity is $H = (50 \pm 20)\%$Rh (percentage relative humidity). All uncertainties are based on $k = 1$.

The read-out of the laser interferometer is already corrected using these values. However, the uncertainties in these values do not appear in the measurement result. This means that in order to estimate an uncertainty knowledge about the measurement process is needed. The laser interferometer detects a number of light–dark transitions (called fringes). These are interpolated and multiplied with the vacuum wavelength of the frequency-stabilised laser and a correction is made for the refractive index of air. Then the indicated displacement is corrected for the material temperature of the measured object, so that the displacement is given for the case that the scale temperature is 20 °C.

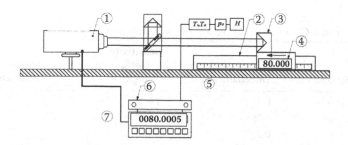

FIGURE 9.13

Set-up of laser interferometer that is used to calibrate a machine tool axis. 1: laser, 2: slide way, 3: corner cube, 4: machine axis read-out, 5: machine bed, 6: compensation unit, 7: laser read-out.

The measured displacement L_i that is indicated by a laser interferometer, without the material temperature correction (Haitjema 2008), is given by

$$L_i = \frac{(N+f) \cdot \lambda_a}{2} = \frac{(N+f) \cdot \lambda_v}{n_a} = \frac{L_v}{n_a},$$ (9.35)

where N is the number of light–dark transitions; f the fraction, that is the interpolation between counts; λ_a is the laser wavelength in air; λ_v is the laser wavelength in vacuum (this is stable and known); n_a is the refractive index of the air in which the measurement takes place; and L_v is the displacement that the system would measure in vacuum.

Because the laser light is kept at a stable frequency (related to the vacuum wavelength λ_v), the wavelength in air must be corrected for the refractive index of air $n_a = \lambda_a / \lambda_v$. For this purpose, lengthy empirical equations are available that have proven to be correct with a relative uncertainty of $2 \cdot 10^{-8}$ (Bönsch and Potulski 1998). Instead of considering these complicated equations for the uncertainty evaluations, it is sufficient to take into account the sensitivities of the indicated length L_i to the atmospheric conditions. These are for general cases indicated in Table 9.10.

Taking these factors into account, the model function can be written as

$$P_{20} = L_{20} - L_s = \frac{L_v}{n_a(T_a, p_a, H)} \cdot (1 - \alpha_s(T_s - 20 \ °C)) - L_s.$$ (9.36)

where P_{20} is the position deviation at 20 °C; L_{20} is the displacement measured by the laser interferometer, reduced to a scale temperature $T_s = 20$ °C; and L_s is the position indication of the scale; other symbols are as noted earlier. The uncertainty budget can be set up as indicated in Table 9.11.

For the standard uncertainty in P_{20} the value $u(P_{20}) = 0.24$ µm. The deviation with an uncertainty U based on $k = 2$ can then be expressed as $P_{20} = (0.57 \pm 0.48)$ µm. Note that this

TABLE 9.10

Influence of Environmental Conditions on a Laser Interferometer System

Environmental Condition	Change or Uncertainty	Change or Uncertainty in the Indicated Length L_i per Metre
Air temperature, T_a	1 °C	0.93 µm
Air pressure, p_a	1 hPa	−0.27 µm
Humidity, H	1%Rh	0.009 µm

TABLE 9.11

Uncertainty Budget of a Calibration Using a Laser Interferometer System

Quantity X_i	Estimation x_i	Uncertainty u_i	Sensitivity $c_i = \dfrac{\partial P_{20}}{\partial X_i}$	Contribution to Standard Uncertainty in P_{20}: $u_i \cdot c_i / \mu m$
L_{lin}	80 mm	0	1	0
L_{20}	80.000575 mm	0.085 µm	1	0.085
T_s	20.5 °C	0.2 °C	$L_{20} \cdot \alpha_s$	0.184
α_s	$11.5 \cdot 10^{-6}$ K^{-1}	$0.7 \cdot 10^{-6}$ K^{-1}	$L_{20} \cdot (T_s - 20\ °C)$	0.028
T_a	21 °C	1 °C	$0.93 \cdot 10^{-6} \cdot L_{20}$	0.074
p_a	1020 hPa	2 hPa	$0.27 \cdot 10^{-6} \cdot L_{20}$	0.043
H	50%Rh	10%Rh	$0.9 \cdot 10^{-8} \cdot L_{20}$	0.007
P_{20}	0.57 µm		$u(L_{1,20}) = \sqrt{\sum (u_i \cdot c_i)^2}$	0.24

uncertainty is considerably larger than the 10 nm read-out resolution suggests. This is caused by the limited uncertainty of the corrections made for temperature, pressure and humidity.

9.7.3 Screw Thread Measurement

In order to measure the pitch diameter of a screw thread calliper, a characteristic value M is measured using three measuring wires. Three measurement wires are used that have a same diameter d_D (Figure 9.14).

The pitch diameter can be calculated using (EURAMET cg-10 2012)

$$d_2 = M + \frac{P}{2 \cdot \tan\left(\frac{\alpha}{2}\right)} - d_D \cdot \left(1 + \frac{1}{\sin\left(\frac{\alpha}{2}\right)}\right) - \left(\frac{d_D}{2}\right) \cdot \left(\frac{P}{\pi \cdot d_2}\right)^2 \cdot \left(\frac{\cos\left(\frac{\alpha}{2}\right)}{\tan\left(\frac{\alpha}{2}\right)}\right) + C, \quad (9.37)$$

where P, α, d_D and d_2 are defined in Figure 9.14. C is a correction for the measurement force.

Note that the fourth term on the right-hand side of Equation 9.37 contains the value d_2, however, as this fourth term gives a small correction, it is convenient to use the nominal value of d_2 instead.

As an example, the calibration of a metric thread limit plug gauge M64 × 6 is taken. The nominal values are $d_2 = 60.1336$ mm, $P = 6$ mm and $\alpha = 60°$ (see ISO 68-1 1998). The gauge is measured using wires with $d_D = 3.4641$ mm and a measuring force $F = 1.5$ N. This gives a correction $C_2 \approx 1.3$ µm.

The measured and assumed quantities and their uncertainties based on $k = 1$ are $M = (65.2993 \pm 0.0004)$ mm, $d_D = (3.4641 \pm 0.0002)$, $P = (6.004 \pm 0.001)$ mm, $\alpha = 60° \pm 1.4'$, and $C_2 = (1.3 \pm 0.13)$ µm.

Note that the flank angle can be difficult to measure. For this example, the tolerance according to the ISO 68-1 is taken (2′) and a rectangular distribution is assumed.

As the partial derivatives in Equation 9.37 are difficult to calculate, for the uncertainty budget the direct calculation as illustrated in Table 9.7 is used. The result is given in Table 9.12

Note that Equation 9.37 can also be used with values for P and α, as these are given in the specification standards that define the nominal thread sizes, and they are assumed to have no uncertainty. In that case, the quantity d_2 is called the 'simple pitch diameter'. Such a measurement is faster, can give a different value and will give a smaller uncertainty.

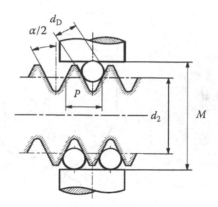

FIGURE 9.14

Sketch of a screw thread measurement using measuring wires. P is the pitch, α is the flank angle, d_D is the diameter of measuring wires and d_2 is the pitch diameter.

TABLE 9.12

Uncertainty Budget for a Thread Measurement Using the Direct Calculation Method

Quantity X_i	Estimation x_i	Uncertainty u_i	$d_2(x_i + u_i)$/mm	Contribution to Standard Uncertainty in d_2: $u_i(d_2) = d_2(x_i + u) - d_2(x_i)$/mm
M	65.2993 mm	0.4 μm	60.1057	0.0004
d_D	3.4641 mm	0.2 μm	60.1047	0.0006
P	6.004 mm	1 μm	60.1062	0.0009
α	60°	1.4'	60.1053	0.0000
C	1.3 mm	0.13 μm	60.1054	0.0001
d_2	60.1053 mm	$u(d_2) = \sqrt{\sum (u_i(d_2))^2}$		0.0012

This illustrates the importance of a thoughtful definition of the measurand before an uncertainty calculation is undertaken.

The result of this measurement can be stated as $d_2 = (60.1053 \pm 0.0024)$ mm, where d_2 is the pitch diameter and the uncertainty is stated as a standard measurement uncertainty multiplied by the coverage factor $k = 2$, corresponding to a confidence level of about 95%.

9.7.4 Polygon Measurement

A polygon with nominal equal angles is one of the basic angular material standards. It can be calibrated with two autocollimators, using the fact that all angles sum up to a multiple of 360° (Yandayan 2002). Figure 9.15 depicts a polygon and gives a schema of the calibration set-up.

Taking autocollimator A as a reference, the n-sided polygon is rotated n times over its polygon angle until it is in the original position again. The reference autocollimator should give a reading of 0, with small differences that are accounted for. The reading of the second autocollimator is a measure for the polygon angle. For a perfect polygon, the difference in reading between the two autocollimators should be constant. A measurement M is the difference in reading between autocollimators B and A. By rotating the polygon, a set of measurements $M_i = B_i - A_i$ is obtained, where $i = 1...n$ and n is the number of polygon angles.

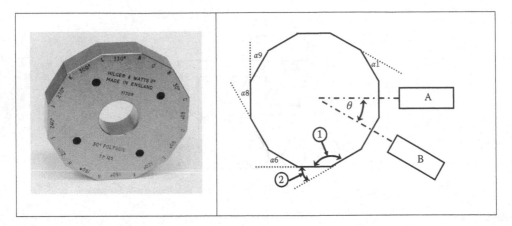

FIGURE 9.15
(Left) Picture of 12-sided polygon. (Right) Schematic of 12-sided polygon with (1) prism angles and (2) α_i, polygon angles. A and B are autocollimators.

For position i,

$$\alpha_i = M_i + \theta, \tag{9.38}$$

where θ is the unknown constant angle between the autocollimators and is calculated from the summation over all measurements:

$$\sum_{j=1}^{n} \alpha_j = \sum_{j} M_j + \sum_{j} \theta \quad \text{or} \quad 360^0 = \sum_{j} M_j + n \cdot \theta \quad \text{that gives} \quad \theta = \frac{360^0 - \sum\limits_{j=1}^{n} M_j}{n}. \tag{9.39}$$

Substituting Equation 9.39 into Equation 9.38 gives the system equation that relates the polygon angles α_i to the measurements M_i:

$$\alpha_i = M_i - \frac{\sum\limits_{j} M_j}{n} + \frac{360^\circ}{n} = M_i \cdot \left(1 - \frac{1}{n}\right) - \frac{1}{n} \cdot \sum_{j \neq i} M_j + \frac{360^0}{n}. \tag{9.40}$$

The power of this calibration is that the sensitivity of the autocollimator, which is typically a fraction of an arc second, is transferred to the polygon angles, which are far beyond its range of a few arc minutes. The uncertainty in α_i, $u(\alpha_i)$, can be calculated from the general Equation 9.23:

$$u^2(\alpha_i) = \sum_{j} \left(\frac{\partial \alpha_i}{\partial M_j}\right)^2 \cdot u^2(M_j) = \left(1 - \frac{1}{n}\right)^2 \cdot u^2(M_i) + \left(\frac{n-1}{n^2}\right) \cdot u^2(M_i)$$

$$= \left(1 - \frac{1}{n}\right) \cdot u^2(M_i). \tag{9.41}$$

Note that the uncertainty is slightly smaller than the uncertainty in M_i. The uncertainty in the angle α, which is the sum of m polygon angles and can be derived from

$$\alpha = \sum_{i=1}^{m} \alpha_i = \left(1 - \frac{m}{n}\right) \cdot \sum_{j=1}^{m} M_j - \frac{m}{n} \sum_{j=n-m}^{n} M_j + \frac{m \cdot 360^\circ}{n}. \tag{9.42}$$

This gives for the uncertainty in α

$$u^2(\alpha) = m \cdot \left(1 - \frac{m}{n}\right)^2 \cdot u^2(M) + \left(\frac{m}{n}\right)^2 \cdot (n-m) \cdot u^2(M) = \left(m - \frac{m^2}{n}\right) \cdot u^2(M), \qquad (9.43)$$

with, as typical results, a zero uncertainty for the trivial case that $m = n$, and a maximum uncertainty for $m = n/2$. This means for $\alpha = 180°$, when n is even, $u(\alpha) = 0.5 \cdot \sqrt{n} \cdot u(M)$.

This is different from the uncertainty in the sum of a number of independent angles, as here the angles are not independent.

Exercises

1. Standard deviation in the mean: Three measurements x_i are used to estimate a mean value y:

$$y = \frac{x_1 + x_2 + x_3}{3}. \qquad (9.44)$$

All three measurements have a standard deviation s.

a. Calculate the standard deviation in the mean value y using the general law of uncertainty propagation (Equation 9.23). Make an uncertainty budget with x_1, x_2 and x_3 as variables.

b. Take the general case for n measurements $y = \sum_{i=1}^{N} x_i/n$ and prove Equation 9.3.

2. Mean and standard deviation: The diameter of a cylinder axis is measured at several positions. The measuring instrument has no errors. The results are as follows.

Measurement Number	1	2	3	4	5	6
Size/mm	2.563	2.542	2.557	2.553	2.560	2.550

a. Calculate the standard deviation in the measurements.

b. Calculate the mean diameter and the standard deviation in the mean diameter.

Assume that these measurements are taken from a Gaussian distribution.

c. Calculate the diameter at any position with a probability of 95%.

d. Calculate the mean diameter with a probability of 95%.

e. Calculate the probability that the diameter at some position is smaller than 2.540 mm. Hint: Use the Matlab function *tcdf*.

Assume that the aforementioned measurements are taken from a rectangular distribution.

f. Estimate the interval of the rectangular distribution.

g. Use a Monte Carlo method to find the distribution of the mean diameter of six measurements. Compare the 95% confidence interval to the 95% confidence interval of a Gaussian distribution.

h. Calculate the confidence interval of 95% for the mean diameter of six measurements. Consider the difference with the answer from part d.

3. Length measurement: A gauge block with a nominal length of 10 mm is measured on a length measuring machine that is read out by a laser interferometer. The uncertainty of the laser interferometer is negligible. The laser interferometer is not corrected for the linear expansion coefficient. The repeated measurements are as follows.

Measurement Number	1	2	3	4	5
Size/mm	10.0001	10.0001	10.0001	10.0002	10.0001

The room temperature is specified as (21 ± 1) °C. Assume a rectangular distribution for the temperature. The linear thermal expansion coefficient of the gauge block is specified as $\alpha = (11.5 \pm 1) \cdot 10^{-6}$ K^{-1}. Assume a rectangular distribution for the linear thermal expansion coefficient.

a. Give the model function for the gauge block length at 20 °C.
b. Give the standard uncertainty caused by the limited resolution of the laser interferometer.
c. Calculate the gauge block length at 20 °C and its uncertainty based on $k = 2$ using an uncertainty budget and the partial derivatives of the model function.
d. Calculate the uncertainty based on $k = 2$ for the gauge block length using an uncertainty budget with the direct calculation method.
e. Calculate the 95% confidence interval for the gauge block length at 20 °C using a Monte Carlo method.

4. Elastic modulus: The elastic modulus according to Hooke's law of elasticity of a certain material is measured by hanging a mass on a wire of this material and measuring its extension. The elastic modulus E is given by

$$E(F, L, \Delta L, A) = \frac{F \cdot L}{\Delta L \cdot A}, \tag{9.45}$$

where F is the force, L is the wire length, A is the area of the wire cross and ΔL is the extension of the wire.

a. Re-write the model function for the diameter d of a cylindrical wire and a mass m that is used to generate the force. Denote the local gravitational acceleration as g.
b. Complete the uncertainty budget for E in the form of a table as give next.

Quantity X_i	Estimate x_i	Standard Uncertainty $u(x_i)$	Sensitivity Coefficient c_i	Contribution to Standard Uncertainty $u_i(E)$/GPa
m	m	$u(m)$	...	
g				
L				
ΔL				
d				
E	E	—	—	$u(E) = \sqrt{u_1^2 + u_2^2 + u_3^2 + u_4^2 + u_5^2}$

    c. Use the following data as input quantities and determine their standard uncertainties:

- m is a 10 kg mass piece class according to OIML class M_1; this means that the maximum permissible error is 50 mg.
- g is the gravitational constant in your area.
- $L = 998.40$ mm, with standard deviation in the mean $s_m = 0.10$ mm.
- $\Delta L = 0.4$ mm, $s_m = 0.05$ mm.
- d the diameter is measured along the wire at five positions. The variation is due to the inhomogeneity of the wire. The result is $\bar{d} = 0.51$ mm, the standard deviation in the measurements is $s = 0.10$ mm.

    d. Make the uncertainty calculation in the form of an uncertainty budget and give the uncertainty based on $k = 2$.

    e. What is the most efficient way to make the uncertainty two times smaller?

5. Elastic modulus: The elastic modulus E of a metal beam is determined by loading it on one side and measuring the deflection using a dial gauge at that place. E is given by

$$E = \frac{4 \cdot m \cdot g \cdot l^3}{u \cdot b \cdot d^3},\qquad(9.46)$$

where

    m is the mass of mass piece, $m = 100$ g, uncertainty negligible.

    g is the gravitational acceleration; make your own estimation of its value and uncertainty.

    l is the length of the beam: 150 mm ± 1 mm ($k = 1$).

    u is displacement, measured with a dial gauge, $u = 13.000$ mm, the maximum deviation of the dial gauge is 17 μm.

    b is the width of the beam, $b = 20.000$ mm, measured with a Vernier caliper with a maximum error of 20 μm.

    d is the beam thickness, measured with a micrometre without errors. At several places on the strip, the measured values are $d = 0.96$ mm, $d = 0.98$ mm, $d = 0.94$ mm and $d = 0.96$ mm.

    a. Calculate E and its uncertainty.

    b. Which measurement should be improved first in order to obtain a lower uncertainty?

6. Radio carbon dating: The most common method to determine the age of organic material is the ^{14}C-method. Because of the solar radiation, a constant part of the carbon in the atmosphere consists of radioactive ^{14}C (^{14}C:^{12}C $\approx$ 1:10^{12}). As soon as an organism stops living, it takes no more CO_2 from the atmosphere, and the ^{14}C decays with a half-life of about 5600 years. This half-life decreases the ratio ^{14}C:^{12}C. The radioactivity of an organic specimen is expressed as pMC; this means percent modern carbon, which is the radioactivity in relation to a specimen from 1950 (since then, this radioactivity has doubled because of atomic bombs and tests in the atmosphere). From this radioactivity A the age t in years before 1950, is calculated from the model function

$$t = -8033 \cdot \ln\frac{A}{100} . \tag{9.47}$$

a. Calculate the age and the standard uncertainty in the age of a specimen with $A = (40 \pm 4)$ pMC (uncertainty based on $k = 1$).

b. A scientific paper gives the age of the leaf of a tree as $t = 13000^{+6500}_{-3000}$ year. The subscript and superscript indicate a confidence level of 68%. Calculate A and the standard uncertainty in A that was the basis of this determination.

c. Assume a Gaussian distribution for A. Use these values in a Monte Carlo method and calculate the 95% confidence interval for this dating. Explain why an asymmetric confidence interval is logic in this case.

d. The limit of this method is an age of about 50000 years. From this fact, estimate the standard uncertainty when an age of 3000 years is established.

7. Torsion wire: A torsion wire with radius r and length l is fixed at one side and is given a torque with moment M at the other side. The angular wrench φ (the total twist angle between the two ends) is given by

$$\phi = \frac{2 \cdot l \cdot M}{G \cdot \pi \cdot r^4}, \tag{9.48}$$

where G is the shear modulus of the wire material.

a. Give the standard uncertainty in φ, $u(\varphi)$, expressed in the standard uncertainties in l, M, G and r.

In order to determine G, a number of measurements of M as a function of φ are carried out. Also the diameter d and the length l of the wire are determined. The results and their standard uncertainties are $\varphi/M = (4.00 \pm 0.12)$ rad·N^{-1}·m^{-1}, $d = (2.00 \pm 0.04)$ mm, and $l = (500 \pm 1$ mm).

b. Give the model function for G, determine G and its uncertainty based on $k = 2$. Proceed as follows:

• Give the model function.

• Make an uncertainty budget in the form of a table.

• Calculate the standard uncertainty in G and the uncertainty based on $k = 2$.

8. Thermocouple: A thermocouple is used to measure the gas temperature in a flame. Because there must be thermal equilibrium between the thermocouple and the environment, the following equation holds for the gas temperature:

$$T_g = T_k + \frac{\sigma\varepsilon_k}{\alpha}(T_k^4 - T_o^4), \tag{9.49}$$

where

T_g = gas temperature in kelvin
T_k = thermocouple temperature in kelvin; this is 600 K, with standard uncertainty 10 K
T_o = environmental temperature in kelvin; this is 300 K, with standard uncertainty 5 K

σ = Stefan Boltzmann constant ($5.67 \cdot 10^{-8}$ W·m^{-2}·K^{-4}), no uncertainty

ε_{κ} = emissivity of the thermocouple; this is 0.20, standard uncertainty 0.03

α = constant, this is 20 W·m^2·K^{-4}, standard uncertainty 4 W·m^{-2}·K^{-4}

Calculate T_g and its standard uncertainty:

   a. Using an uncertainty budget.

   b. Using a Monte Carlo method.

9. Straightness measurement: The straightness of an object can be determined by level measurements as depicted in Figure 9.16. At every position i, an angle measurement α_i is taken using a level with pitch l. As a reference line to determine the straightness deviation, the line through the first point ($h = 0$) and last point at $i = n$ is taken.

   a. Show that the height h of point i relative to the reference line is given by

$$h_i = \sum_{j=1}^{i} \alpha_j \cdot l - \frac{i}{n} \cdot \sum_{j=1}^{n} \alpha_j \cdot l. \qquad (9.50)$$

   b. The measurements α_i are given by $\alpha = (1, 2, 3, 1, 2, 2, -2, -4)$ μrad and $l = 100$ mm. Calculate the heights of point 0, 1, 2, ..., 8.

   c. Calculate the uncertainty in the point with the largest deviation from the reference line. The standard uncertainty in the angular measurements is $u(\alpha_i) = 1$ μrad, and these are not correlated.

Hint: The problem has similarities with the polygon measurement in Section 9.8.4.

10. Gauge block classification: The tolerance of a gauge block of class 0 is $\pm$ (0.12 μm + $2 \cdot 10^{-6} \cdot l$), where l is the gauge block length. In a calibration laboratory, the length of a gauge block having a nominal length of 100 mm is measured l = (100.00025 $\pm$ 0.00010) mm, based on $k = 2$ (95%).

   a. What does the calibration laboratory conclude about the class of the gauge block?

   b. May a reseller sell such a gauge block as a class 0 gauge block?

   c. Can a buyer of such a gauge block send it back to the seller, stating it is outside tolerance?

   d. What should the measurement uncertainty be so that the result would not result in any disputes?

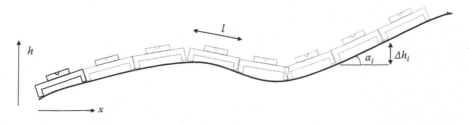

FIGURE 9.16
Straightness measurements using a level.

References

Bönsch G, Potulski E. 1998. Measurement of the refractive index of air and comparison with modified Edlén's formulae. *Metrologia* **35**:133–139.

EA-4/02. 2013. Expression of the uncertainty of measurement in calibration. European Accreditation, http://www.european-accreditation.org/publication/ea-4-02-m-rev01–september-2013 (accessed November 2, 2017).

EURAMET cg-10. 2012. Determination of pitch diameter of parallel thread gauges by mechanical probing. Germany: European Association of National Metrology Institutes.

Haitjema H. 2008. Achieving traceability and sub-nanometer uncertainty using interferometric techniques *Meas. Sci. Technol.* **19**:084002

ISO 68-1. 1998. ISO general purpose screw threads—Basic profile—Part 1: Metric screw threads. Geneva: International Organization for Standardization.

ISO 14253-2. 2011. Geometrical product specifications (GPS)—Inspection by measurement of workpieces and measuring equipment—Part 2: Guidance for the estimation of uncertainty in GPS measurement, in calibration of measuring equipment and in product verification Geneva: International Organization for Standardization.

JCGM 100. 2008. GUM 1995 with minor corrections, 'Evaluation of measurement data—Guide to the expression of uncertainty in measurement. France: International Bureau of Weights and Measures (BIPM).

JCGM 101. 2008. Evaluation of measurement data—Supplement 1 to the 'Guide to the Expression of uncertainty in measurement'—Propagation of distributions using a Monte Carlo method. France: International Bureau of Weights and Measures (BIPM).

Squires G L. 2008. *Practical Physics*, Cambridge University Press, UK.

Taylor B N, Kuyatt C E. 1994. Guidelines for evaluating and expressing the uncertainty of NIST measurement results. NIST Technical Note 1297. Gaithersburg, MD: National Institute of Standards and Technology.

Yandayan T, Akgöz S A, Haitjema H. 2002. A novel technique for calibration of polygon angles with non-integer subdivision of indexing table. *Precis. Eng.* **26**:412–424.

10

Alignment and Assembly Principles

Eric S. Buice

CONTENTS

ABSTRACT In this chapter alignment and assembly principles for the development of precision systems are discussed. The chapter first presents the important alignment principles that can result in Abbe, cosine and sine errors. Assembly and joining methods of both mechanical and optical systems are then presented. A general overview of alignment principles for linear and rotary stages, and optical systems (for example linear encoders, displacement interferometers and lenses) is given. Throughout the chapter, simple illustrations are used to help convey the concepts and where applicable additional references are provided that may be of further interest.

10.1 Design Principles and Considerations for Effective System Alignment and Assembly

Design considerations will be directly affected by project requirements, concept of operations* (ConOps), stakeholders and production volume. Typically, with high-volume production, such as computer hard drives, it is necessary to rely heavily on automated and self-aligning processes during assembly. However, for low-volume production, such as lithography systems used in the semiconductor industry to develop state-of-the-art integrated circuits, it may be acceptable to have an increased number of assembly and alignment steps that are completed by highly trained technicians with very little automation. Although it may be acceptable to utilise skilled technicians in low-volume production, the design of the system should still attempt to minimise the assembly and alignment effort which is in line with the capital investment plan of the project. Therefore, many design for assembly (DFA) methodologies developed for medium- and high-volume production still apply for an optimum and cost-effective product design, even when low-volume production systems are considered.

The following sections will first identify principles that provide tools to determine factors that limit the performance of a system and later basic DFAs will be addressed to improve not only the design but also the costs of the product, with an emphasis on low-volume precision systems.

10.1.1 Abbe Principle and Abbe Error

An important principle when designing a precision system is the Abbe principle, which Bryan (1979) considers to be 'the first principle of machine tool design and dimensional metrology'. The Abbe principle was first formulated by Ernst Abbe in 1890 and states that a measurement system shall be collinear to the line of motion (Abbe 1890, Evans 1989, Leach 2015). To clarify the definition of a line of motion and a measurement system, a two-dimensional (2D) caliper measuring a circular component will be used, as shown in Figure 10.1. Referring to Figure 10.1a, the line of motion is formed by the moving jaw traversing in the horizontal direction. The line of motion will further pass through a point of interest (POI), which is the point where the measurement is desired. In the case of measuring a circular component as depicted in Figure 10.1a, the POI occurs where contact is made between the moving jaw and component. The measurement system also forms a measurement axis, which is generated by two points: the measurement point reference (MPR) and measurement point (MP). The MPR is the point from which the measurement occurs; typically, this would be the zero point and the MP is the location at which the scale of the caliper is read. The difference between MPR and MP is the distance L over which the moving jaw has traversed. Since both the measurement axis and line of motion (passing through the POI) have been defined for a caliper, it can be seen that, while both axes are parallel, the caliper does not obey the Abbe principle. In order for the caliper to obey the Abbe principle, the measurement axis would need to be shifted upwards such that it is collinear to the line of motion. By shifting the measurement axis of a caliper to obey the Abbe principle, it becomes what is known as a micrometer.

---

* The concept of operations is a document which 'describes the way the system works from the operator's perspective. The ConOps includes the User Description which summarizes the needs, goals, and characteristics of the system's user community, including operators, maintainers, and support personnel' (Haskins et al. 2007).

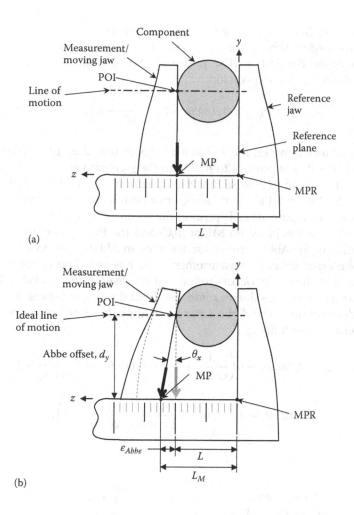

FIGURE 10.1
A 2D depiction of a caliper used to measure the diameter of a circular component to aid in the description of the Abbe principle. (a) A perfect caliper with no errors present. (b) A caliper with the moving jaws having exaggerated motion errors (rotation about the x-axis).

Up to this point, the Abbe principle has been described; however, the consequence of not obeying the Abbe principle has not been discussed. If a measurement tool, such as a caliper, is considered perfect, then there is no consequence of not obeying the Abbe principle. However, this assumption is unrealistic, as there will always be motion errors of any moving object. Thus, by not obeying the Abbe principle, the motion error will result in what is referred to as the Abbe error. Using the 2D example of a caliper depicted in Figure 10.1b, the moving jaw will have a motion error due to a rotation about the x-axis such that the MP deviates from what is considered to be the real location of the MP, causing an Abbe error. The Abbe error can be calculated as follows:

$$\varepsilon_{Abbe} = d_y \tan \theta_x, \tag{10.1}$$

where d_y is the Abbe offset (distance/offset between the line of motion and measurement axis) in the y-direction and θ_x is the angular motion error (or straightness error) of the positioning system (moving jaw in the case of a caliper) about the x-axis. Equation 10.1 only describes the first-order Abbe error; however, there is an additional second-order term present. To determine the first- and second-order terms of the Abbe error, Figure 10.2 is used to derive the following expression:

$$\varepsilon_{Abbe} = d_y \tan \theta_x - L\left(\frac{1}{\cos \theta_x} - 1\right),$$

(10.2)

where L is the distance between the MP at (0,0) (obeys the Abbe principle) and the POI. It should be noted that it is important to maintain the sign convention of θ_x in respect to the defined coordinate system to ensure that the maximum value of the Abbe error is calculated. For example, the largest absolute Abbe error occurs when θ_x is rotated in the negative direction in respect to Figure 10.2. Typically, the second-order Abbe error is neglected for small angles and distances between MP at (0,0) and the POI (Bosmans 2016). The mathematical description of the Abbe error so far assumes an Abbe offset only in one direction. In reality, the Abbe error for each measurement axis has an Abbe offset in two directions (Figure 10.3), so that the errors occur from an Abbe offset in the x- and y-directions along with the angular motion errors due to rotations about the x- and y-axes. Thus, the complete mathematical description of the Abbe error for a single measurement axis (with first- and second-order terms) is as follows:

$$\varepsilon_{Abbe} = d_y \tan \theta_x - L\left(\frac{1}{\cos \theta_x} - 1\right) + d_x \tan \theta_y - L\left(\frac{1}{\cos \theta_y} - 1\right)$$

(10.3)

$$= d_y \tan \theta_x + d_x \tan \theta_y - L\left(\frac{1}{\cos \theta_x} + \frac{1}{\cos \theta_y} - 2\right),$$

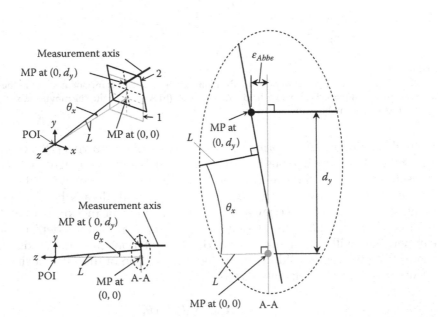

FIGURE 10.2
A one-dimensional depiction of the Abbe principle. (1) Measurement plane (coincident to the MP), (2) the rotated measurement plane (rotation occurs about the POI).

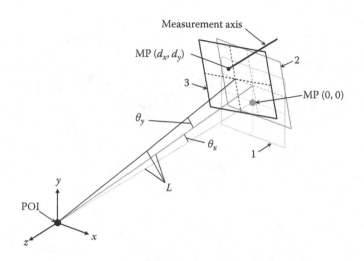

FIGURE 10.3
Depiction of a one-dimensional Abbe error with Abbe offsets in x- and y-directions along with rotations about the x- and y-axes. (1) Measurement plane, (2) the rotated measurement plane about the x-axis, (3) the rotated measurement plane about the y-axis.

where d_x is the Abbe offset in the x-direction and θ_y is the angular motion error (or straightness error) of the positioning system about the y-axis. Again, the importance of maintaining the sign convention when calculating the Abbe error with Equation 10.3 should be noted.

Examples of systems obeying the Abbe principle by design, for example $d_x = d_y = 0$, can be found in Bosmans (2016), Ruijl (2001) and Dai et al. (2004). A good example of a system that does not obey the Abbe principle is a coordinate measuring system (CMS; see Chapters 5 and 11 for detailed explanations and layouts of a CMS).

10.1.1.1 Alternative Method to Reduce the Abbe Error

In practice, it may be difficult to place the measurement system collinear ($d_x = d_y = 0$) to the motion system in order to avoid the Abbe error, and it is unrealistic to expect a motion system to have no angular error motions. To minimise its contribution to the overall error budget, keeping the Abbe offset and angular motion error to a minimum should always be the goal. An alternative method to reduce the Abbe error can be achieved by means of software or physical compensation (requires additional rotational degrees of freedom to the linear positioning system). Both alternatives require an additional two degree of freedom (DOF) measurement capability to measure the angular error motions of each linear displacement axis. By adding two additional sensors to measure the angular error motion, the Abbe error can be calculated at each evaluated position, and a correction in software can be performed to reduce this contribution (Figure 10.4). This is referred to as the software compensation method and is equivalent to performing a coordinate transformation so that the two measurement axes (MP and MPR) coincide. Contrary to the software compensation method, if two additional actuators with rotational DOFs are added, the rotational motion errors can be controlled to a minimum, therefore, proportionately minimising the Abbe error (see Figure 10.5). This is particularly effective for processes in which the MPs might vary depending on the application. In practice, there will always be controller errors and these can be monitored and compensated for a hybrid approach for computer and actuator

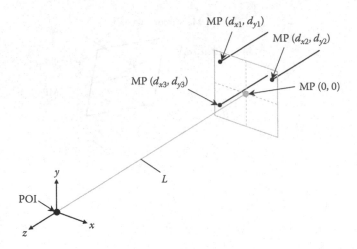

FIGURE 10.4
Depiction of adding two additional sensors measuring the rotation about the x- and y-axes to reduce the Abbe error by means of software correction.

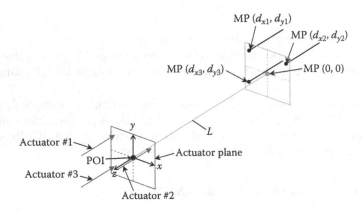

FIGURE 10.5
Depiction of reducing the Abbe error by adding two additional actuators and sensors to correct for any rotation errors about the x- and y-axes.

compensation. Both alternatives achieve the goal of minimising the Abbe error when the Abbe principle cannot be adhered to, albeit at the expense of additional costs, complexity, and potentially compromising the stiffness and dynamic response of the machine.

10.1.2 Cosine Error

In addition to the Abbe error, a cosine error can also affect measurement results. The cosine error occurs when the measurement axis is not parallel to the motion axis or, in other words, the measurement axis is not properly aligned to the motion axis (Figure 10.6). Based on the geometric model shown in Figure 10.6, the cosine error ε_{cos} can be calculated from

$$\varepsilon_{cos} = L_M - L = L_M - L_M \cos \alpha_x = L_M(1 - \cos \alpha_x), \tag{10.4}$$

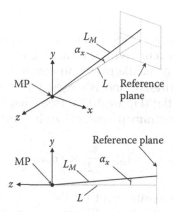

FIGURE 10.6
A one-dimensional depiction of the cosine error.

where L_M is the measured displacement, L is the actual or real displacement between the MP and the reference plane (location where the measurement originates and is used as the reference to determine the measured length), and α_x is the angular misalignment about the x-axis. In practice, it is always necessary to subtract the cosine error from the measured displacement to obtain the real displacement. Again, as with the Abbe error, the cosine error also occurs as a rotational misalignment about the x- and y-axes (Figure 10.7) and is described by

$$\varepsilon_{cos} = (L_M - L_1) + (L_1 - L) = L_M\left(1 - \cos \alpha_y\right) + L_1(1 - \cos \alpha_x)$$
$$= L_M\left(1 - \cos \alpha_y \cos \alpha_x\right), \tag{10.5}$$

where α_y is the angular misalignment about the y-axis. As long as the angular misalignment of the measurement system remains constant, that is there are no mechanical or thermal drifts, the cosine error can be determined to reduce its effect by adjustment (Hale 1999, Lestrade 2010). This, of course, assumes that there is no significant cosine error of the adjustment measurement. Effective alignment methodologies to minimise the cosine error are discussed in the following sections.

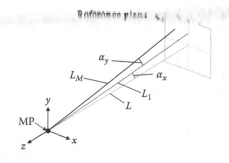

FIGURE 10.7
Showing the cosine error with misalignment occurring due to rotations about the x- and y-axes.

10.1.3 Sine Error

In addition to the Abbe and cosine errors, a sine error can also be present in a measurement system. A sine error typically occurs with a measurement system requiring a mechanical contact, such as a micrometer or tactile probe, to perform a measurement. The sine error occurs when a cosine error is present and is largest when the mechanical contact is over-constrained, that is a flat on a flat (Figure 10.8a) (see Chapter 6 for details on constraints). The sine error with an over-constrained contact can be described by

$$\varepsilon_{sin} = \frac{w_z}{2} \sin \alpha_x,$$ (10.6)

where w_z is the width of the measurement tip (MT) and α_x is the angular misalignment about the x-axis. The sine error also has a 2D error and can be described as

$$\varepsilon_{sin} = \frac{w_z}{2} \sin \alpha_x + \frac{w_x}{2} \sin \alpha_z,$$ (10.7)

where w_x is the width of the MT in the x-direction and α_z is the angular misalignment about the z-axis. Even a single point contact, for example a sphere on a flat, will generate a sine error (see Figure 10.8b) that can be calculated from

$$\varepsilon_{sin} = \frac{w_z}{2} (1 - \cos \alpha_x).$$ (10.8)

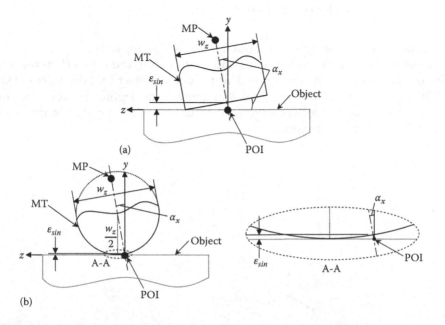

(a)

(b)

FIGURE 10.8
The sine error depicted as a one-dimensional system. (a) Flat on flat contact, (b) sphere on flat contact.

A 2D sine error with a single point contact is given by

$$\varepsilon_{sin} = \frac{w_z}{2}(1 - \cos\alpha_x) + \frac{w_z}{2}(1 - \cos\alpha_z) = \frac{w_z}{2}(2 - \cos\alpha_x - \cos\alpha_z). \tag{10.9}$$

As with the cosine error, the sine error can also be minimised by determining its value, assuming that the angular misalignment error and measurement system remain constant (no mechanical and thermal drifts).

10.1.4 Combining the Abbe, Cosine and Sine Errors

The Abbe, cosine and sine errors are not mutually exclusive in a measurement system. A good example of this is utilising a caliper to measure an object (Figure 10.9). As discussed earlier, the Abbe principle is not obeyed, thus an Abbe error exists. In addition to the Abbe error, a sine and cosine error also exists at the measurement/moving jaw due to angular misalignment. Additionally, the scale also has a misalignment relative to the motion axis causing a second cosine error. The measurement errors in the yz-plane of Figure 10.9 are

$$\varepsilon_{Abbe} = d_y \tan\theta_x - L\left(\frac{1}{\cos\theta_x} - 1\right), \tag{10.10}$$

$$\varepsilon_{sin} = w_y \sin\alpha_{x1} \tag{10.11}$$

and

$$\varepsilon_{cos} = L_M(1 - \cos[\alpha_{x1} + \theta_x]) + L_M(1 - \cos\alpha_{x2}). \tag{10.12}$$

The subscripts 1 and 2 in the cosine error are to indicate the two different alignment errors that are present when making a measurement with a caliper. As can be seen in the caliper example, a detailed analysis of a system is critical in evaluating precision systems.

10.1.5 Assembly against Monolithic Design, Benefits of Reduction

During the design phase of an instrument, machine or device, a decision will be made as to whether a system or subsystem will be comprised of an assembly or made from a mono-lithic component. A monolithic component in this context, is taking an assembly comprising of multiple individual components and turning it into a single fabricated component with all the required functionality of the original assembly. Many factors will inform the decision of which path to follow. In general, it should be a goal to keep the number of individual components, assembly and alignment steps to a minimum. Reducing the number of parts in a system has the immediate benefits of lowering the possibility of components being out of specification; removing assembly errors; lowering the costs of logistics and tracking of components; reducing complexity of the error budget; and minimising the design, align-ment and assembly documentation (Bramble 2012). These benefits will result in lowering the overall cost of the system. The concept of reduction as a design consideration is also briefly discussed in Chapter 1.

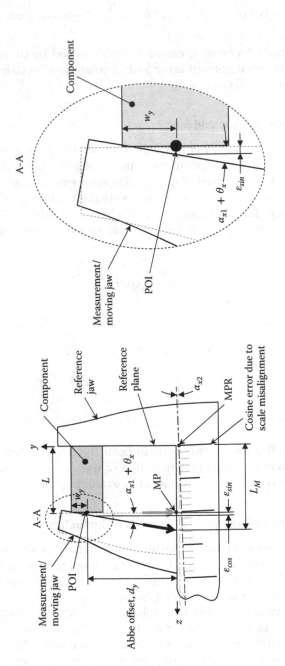

FIGURE 10.9

An example of a caliper measuring a rectangular component, indicating that the Abbe, cosine and sine errors are not mutually exclusive to a measurement system.

It is difficult at the beginning of a project to identify systems or sub-systems which would benefit from being a monolithic component, but introducing intermediate internal design reviews to concentrate on simplifying the overall design is a practical and beneficial process. During these design reviews, the design team should discuss each individual component within the assembly, asking questions, such as the following:

1. Can the adjacent components be combined?
2. Is the assembly process simplified by combining the components?
3. What are the manufacturing consequences of combining components?
4. Can the component be manufactured?
5. Are the required specifications compromised due to reduced adjustment capabilities?

These are just a sample of questions that should be asked during this review stage. Typically, the system requirements and ConOps will provide the necessary guidance and, along with sound engineering judgement, help to determine whether components can be combined.

A major advantage in reducing the number of assembly steps and components is the reduction in the alignment effort. Although self-aligning features can be added, dowel pins or end stops for example, these may result in manufacturing tolerances being tightened to smaller numbers as each component needing to be assembled will require its own sub-error budgeting to meet the overall error budget. The effect of tolerances when multiple components are used in assemblies is called tolerance stack-up and often requires sophisticated analysis tools for this sub-error budgeting. Therefore, a monolithic component which combines several individual components may reduce manufacturing tolerances while maintaining component functionality and specifications. Of course, monolithic components have their disadvantages as well. In particular, the design freedom is limited by manufacturing methods (for instance it may be desirable to have internal cavities to lower the mass of the component) and machining errors (machine geometric errors and alignment errors when the component is handled for subsequent manufacturing steps).

An example that illustrates approaches for the reduction of the alignment and assembly effort is an optical lens module, shown in Figure 10.10. The optical lens module was originally comprised of eighty-four parts, including screws, washers and Belleville washers. The lens module utilised lenses (1A–3A) that were aligned and cemented to individual lens mounts (4A–6A) and each mounted lens was aligned to every other lens. Although the original design met all requirements, it was desired to reduce the overall costs of the optical module. Upon a detail cost evaluation, it was noticed that the largest savings could be achieved by eliminating the lens alignment process. This was achieved by electing to utilise manufacturing tolerances to align the individual lenses to the lens mount and clamping to avoid the application of cement. The second largest cost saving was the alignment of each mounted lens to each other by turning the three individual lens mounts into a single lens mount. By electing for these design changes, the modified optical lens module was developed (see Figure 10.10, right), which included geometrically modified lenses with the same optical characteristics as the original design (1B–3B) and a single lens mount (4B), two preload rings (8B and 9B), a spacer (5B), a spring preload spacer (6B) and springs (7B). In total, the modified design was comprised of only thirty-one parts, while achieving an overall cost saving of approximately 20%, and module throughput was increased by twenty-fold. Although the modified lens module did provide a lower product cost,

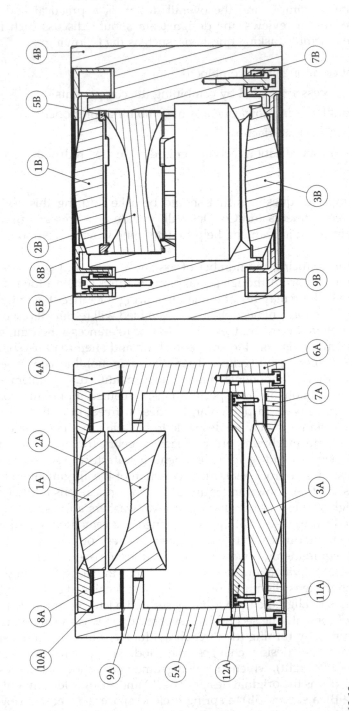

FIGURE 10.10

A cross-sectional view of an optical lens module which was reduced in the number of components in order to reduce alignment and assembly effort. On the left side is the original design (1A–3A: lenses 1–3; 4A–6A: lens mounts 1–3; 7A, 8A and 12A: aperture; 9A–11A: spacer) and the right is the modified design (1B–3B: lens 1–3; 4B: lens mount; 5B: lens 1 and 2 spacer; 6B: spring preload spacer; 7B: preload spring) which had an overall cost savings of ~20%.

with increased productivity, it came at the expense of performance. The performance of the individual modified module was reduced by approximately 5% (not meeting design specification) when compared to the original design specification of this particular module. Since the lens module is part of a larger system comprising multiple different lens modules, it was possible to rearrange the error budget of the entire system, such that the overall system specifications were achieved while reducing the overall costs.

Although it is a goal to have a minimum number of components, each design attribute needs to be evaluated such that the best outcome is achieved. A monolithic design also has its drawbacks, such as increased production costs, lengthy fabrication times, individual component complexity (which may limit the number of companies that can manufacture and assemble the device) and reduced flexibility in terms of future upgrades.

Advancements in additive manufacturing are providing greater design freedoms such that internal cavities and very complex geometries can be included in a monolithic component to help reduce the potential drawbacks of a monolithic design as listed earlier (see Gibson et al. 2015). However, as additive manufacturing is a maturing technology, it may not yet be cost-effective in many situations and often cannot meet the tolerances of precision systems.

10.1.6 Joining of Components

Joining components is an important aspect that is often neglected during the design of systems, particularly for relatively low volume production. In contrast, joining methodologies in high volume production may actually drive the design of a system. There are many different methods of joining two or more components. To limit the discussion, this section will focus on joining methods that are typically used in precision systems. The focus will be limited to threaded fasteners, adhesives and optical contacting. Other joining methods typically employed in assemblies, which are used in lower precision alignment requirements, are not covered in the subsequent sections. These joining methods include welding, spot welding, brazing, press fitting (physical interference of two or more components being assembled), riveting, etc.

10.1.6.1 Threaded Fasteners

The use of threaded fasteners is very common, as they provide an easy means of assembling and disassembling complete systems. Threaded fasteners also allow the clamping force to be controlled during the assembly and alignment process. The clamping force, along with the coefficient of friction between the joined components, enables threaded fasteners to resist shear motions, which is typically the primary function of a bolted joint (Slocum 1992). Other functions of a bolted joint may be to provide a force to generate a seal or to carry direct loads (a bolt joint is in tension). Details of analysis methods for determining the required torque for a given gripping force are not discussed here but can be found elsewhere (for example, VDI 2230-2 2014, VDI 2230-1 2015). When selecting the appropriate threaded fasteners, it is important to consider type, size, applied load and material.

There are a variety of different types of threaded fasteners and the selection of which type to use will depend upon availability at suppliers or what is typically used within an organisation, requirements of standards (for example 'J' threads for high performance applications) and sometimes even cosmetic considerations. From all the different types of threaded fasteners available, socket cap screws are a very popular choice in the development of precision systems. To secure sheet metal covers, it is common to use button socket cap screws. Both the socket cap screw and button socket cap screw can be found with a hex or hexalobular internal (commonly referred to as Torx®) drive. It should be noted that the use of flat socket cap screws should be avoided when it is important to perform an alignment step, as the geometry of the screw will force the component to position itself to its 'natural' position and not necessarily the position that is desired.

The size of the threaded fastener will be dependent on the required applied axial force, available volume and allowable stress. In terms of allowable stresses, it may be beneficial to select a smaller threaded fastener size rather than a larger one. Of course, the consequence of selecting a smaller size will require an increase in the number of fasteners used to have the equivalent clamping force (or 'grip'), as in the case with the larger size.

Choosing the appropriate fastener material is critical to avoid galling or corrosion, and selecting the appropriate coefficient of friction and material strength. Galling, also commonly referred to as cold welding, occurs when a material's (for example stainless steel, aluminium, titanium) oxide layer is removed or damaged during sliding, causing the materials to bond/adhere to each other (Campbell 2011). Galling and corrosion can be mitigated by the use of lubrication or coating, which can also lower the coefficient of friction. Common materials used for threaded fasteners are steel, steel alloy (stainless steel) and aluminium alloy. Stainless steel is often used for structural elements of vacuum systems and, frustratingly, often 'cold welds'. An interesting alternative material to stainless steel and in applications where the use of lubrication is limited or not allowed, is an austenitic stainless steel called Nitronic® (Schumacher and Tanczyn 1975, AK Steel Corporation 1981). Nitronic exhibits good galling resistance, but its implementation as a threaded fastener has been limited due to the lack of commercial fasteners made of Nitronic. However, threaded inserts made of Nitronic commercially exist which can be paired with regular stainless steel fasteners.

The use of washers and preload springs (for example lock washers, Belleville washers) should always be considered. It is advisable to use a preload spring on top of a washer. The washer will provide a larger contact area spreading the contact stress, while the preload spring will ensure that the threaded fastener remains under tension even when thermal cycles and stress relaxation are present. Stress relaxation effects can be reduced by increasing the fastener length, keeping the number of components clamped in series to a minimum and keeping fastener tightening speeds to a minimum. Vibrations are an additional source that may reduce the clamping force over time. Although a preload spring may help, it will not completely alleviate the problem. When loss of clamping force due to vibration is of concern, the use of a serrated face washer, applying an adhesive to the fastener or inserting a locking pin into the fastener may be used to attenuate this problem. When a nut is employed, a second nut (commonly referred to as a jam nut) can be used (the second nut is used to effectively lock the first nut into place) to also alleviate the preload loss due to vibration. It has been recommended that the minimum thread engagement for a threaded fastener shall be at least 0.8 times the fastener nominal diameter (Bhandari 2007, Bickford 2008).

For ultra-precision applications, where sub-nanometre stability is required, Jones and Richards (1973) observed that the stability of a capacitance gauge-based seismometer assembly would vary with clamping forces and even the direction that the screws were last torqued. A substantial increase in stability was also achieved by increasing the compliance by using spring washers and, sometimes, machining slots in the component to locally increase the clamping compliance. A final thermal cycling to 60 °C of the whole instrument resulted in a stability of the seismometer system of better than 10 pm per day; a stability even remarkable by today's standards.

A final comment regarding threaded fasteners is when the clamping force is of concern, it is advisable to perform specific experimentation under the expected conditions in service (for example, fastener type, component materials, lubrication, cleanliness) in the final application, to determine the required nominal clamping force with an acceptable uncertainty to ensure the minimum clamping force is always achieved.

10.1.6.2 Adhesives

The use of adhesives for joining mechanical and optical components is very common, especially in consumer electronics. Adhesives can provide a uniform (often relatively small) stress distribution, ability to join thin and fragile components, allow use of dissimilar materials, possibly prevent or minimise corrosion, possibly reduce overall mass, provide thermal insulation (this can be circumvented by adding fillers to improve the thermal conductivity of adhesives), lead to lower costs, simplify the assembly process and possibly provide damping. Adhesives also have the following disadvantages, as they require surface treatments, can have long curing times, may require fixtures (for holding components during curing), can have limited temperature ranges, lead to outgassing, have limited lifetime (this is especially true for vacuum and higher temperature applications) and may require disassembly effort (may require additional solvents, scraping, tooling etc.) (Kinloch 1987, Yoder Jr. 1993, Banea and da Silva 2009, Ebnesajjad and Landrock 2015). As with threaded fasteners, there is a wealth of information available on how to join two or more components, particularly from adhesive suppliers and textbooks (for example, see Petrie 2009). Adhesive suppliers can advise on the specific adhesives and design attributes that should be implemented to meet the overall design specification. Due to the wealth of information that can be obtained from adhesive suppliers and textbooks, the discussion will be limited to the basics of bond strength, fillers to improve thermal conductivity and coefficient of thermal expansion (CTE), and the volume change of the adhesive. It is strongly advised that an expert on adhesives be consulted to help in selecting the proper adhesive to be used for a given application and adequate testing is performed to ensure specifications are met (Yoder Jr. 1998).

The strength of a bond is directly influenced not only by the type of adhesive but also on the surface area, surface preparation, surface texture, adhesive thickness and joint geometry (Schneberger 1983, Müller et al. 2006, Banea and da Silva 2009, Budhe et al. 2015). In terms of bond strength, the surface preparation and surface texture will have the biggest influence. Surface preparation is performed to remove any surface contaminations (such as particles, lubrications), oxide layers and moisture, all of which will have a negative impact on the bond strength. Quantifying the required surface texture to obtain the maximum bond strength is difficult and will be dependent on the material, operating environment and so forth. Well-controlled surface texture lowers the probability of having voids by trapped air, which improves the bond strength. A well-designed surface texture (often just increasing the average surface amplitudes), however, can provide a larger bonding surface area, which will also improve the bond strength. Clearly, a trade-off is required to obtain the optimal bond strength (Schneberger 1983). As an example of a particular bond, Budhe et al. (2015) showed that an optimal arithmetic average surface roughness Ra (see Chapter 6 for a description of this parameter) for an aluminium to aluminium bond is (1.68 ± 0.14) μm with a commercial epoxy resin Araldite®. Although adhesives provide greater design freedoms, it is important to maintain a proper adhesive thickness and joint geometry. Adhesive thickness is highly dependent on the type of adhesive selected and, therefore, should be experimentally verified or discussed with adhesive experts. The geometry of the joint will be dependent on the forces or stresses applied to the assembly that may be under compression, tensile, shear and any combination of the aforementioned stresses. Typical joint geometries

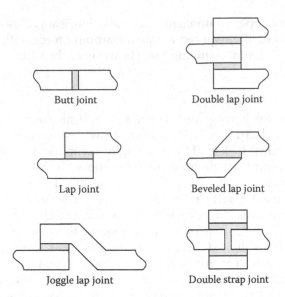

FIGURE 10.11
Typical adhesive joint geometries used to join two components.

are shown in Figure 10.11. The appropriate joint geometry will ensure that minimal stresses are present in the adhesive, thus providing superior bond strength and lifetime (Ebnesajjad and Landrock 2015).

To improve the thermal/electrical conductivity, long-term stability and CTE, fillers may be added to the adhesive. Gold, silver, copper, nickel and single-wall carbon nanotubes have been added as fillers to improve the thermal conductivity of adhesives (Sihn et al. 2008, Ebnesajjad 2010). To improve the CTE of an adhesive, a filler with a low CTE material, such as quartz, silica, silicon carbide or silicon nitride, can be used. Since adhesives typically shrink in volume during curing, the adhesive will force the low CTE filler to support (by means of contact) the component being cemented. This results in a very stable bond with a low CTE. The amount of shrinkage of adhesives that do not contain fillers is dependent on the type, but a good rule of thumb value is a volume shrinkage of 4%.

Section 10.2.1 discusses the use of mechanical alignment features (MAFs) that can be used to cost effectively align and hold the components being bonded while the adhesive is curing. The mechanical alignment features can be in the form of features directly integrated into a component or as a separate alignment fixture.

10.1.6.3 Optical Contacting

Optical contacting is achieved by mating two flat surfaces, such that the intermolecular attraction (van der Waal forces) between the two surfaces provides the bonding force (Holt et al. 1966, Floriot et al. 2006). To produce an optical contact, the surfaces of the two components should be flat, have optical quality surface texture, and be free of scratches and particles. Optical contacting is similar to wringing gauge blocks together as described in Chapter 5. Advantages to using optical contacting over other methods are transparency of bond interface, reduced outgassing, bond stability, and when using identical materials, no stresses due to CTE mismatch. In addition to the surface characteristics requirements, limited alignment time, limited ability to disassemble (disassembly capability is decreased as the quality of the optical contact bond is increased) and cleaning of surfaces are considered to be disadvantages when using optical contacting (Rayleigh 1936, Holt et al. 1966, Kalkowski et al. 2011). To obtain a sufficient bond quality, it has been suggested that the components should have a flatness better than approximately 60 nm (peak-to-valley) and an *Ra* value (see Chapter 5) of better than 5 nm (Elliffe et al. 2005, Floriot et al. 2006). The flatness requirement may be reduced to approximately 600 nm if one or both components being optically contacted is conforming (exhibits low stiffness) (Rayleigh 1936). Bonding of silicon wafers is an excellent example of two conforming components being bonded where flatness of up to 3 µm is considered acceptable (Tong et al. 1994).

Optical contacting is performed by simply bringing the two components into contact by applying a normal force (Holt et al. 1966). Rayleigh added benzene between the two components to aid in achieving the optical contact while applying circular sweeping strokes along with the normal force. As long as the benzene has not evaporated, it is possible to move the two components relative to each other, which can also be used to aid in alignment (Rayleigh 1936). Other evaporating fluids could also be used in place of benzene, such as isopropyl.

Since alignment is critical in many instances and requires time, it is possible to use a hydroxide-catalysed bond (Gwo 2001) as an alternative when silica-based materials are used. The main disadvantage of this method is that a siloxane chain (polymer-like) is present between the two components causing a CTE mismatch. Depending on the hydroxide solution, it can take up to four weeks to evaporate (although significant bond strength is achieved after one day). Evaporation time can be accelerated by applying heat (cooling can also increase the evaporation time) and/or changing the solution pH levels (Elliffe et al. 2005, Douglas et al. 2014). To improve the bond strengths of silica-based materials, the bond may be performed under low vacuum conditions and elevated temperature (approximately 120 °C) (Kalkowski et al. 2013). If direct bonding of silicon wafers is employed, the bond strength can also be enhanced by performing an annealing step with temperatures reaching up to 1400 °C (Masteika et al. 2014).

Examples of optical bonded components can be seen in van Veggel and Killow (2014) with a summary table of achievable bond strengths.

10.1.7 Use of Symmetry

The presence of symmetry in a mechanical assembly can considerably simplify evaluation of performance (Schellekens et al. 1988, Yoder Jr. 1998). Additionally, symmetry in many instances leads to simpler designs, which can also lead to improved manufacturing and metrology methods to be employed (Hale 1999). However, it has also been reported by Schellekens et al. (1988) that the vibrational (modal) energy in the system or component may be enhanced due to symmetry. In this case, breaking the symmetry can reduce susceptibility to vibration. An example of this is the Tetraform grinding machine that has a tetrahedral frame where individual tubular legs have damping elements along the tube centres that could be subject to torsional force and thereby tuned, like a guitar string, to different frequencies (Lindsey 1991). Each of the faces of a tetrahedron is an equilateral triangle, thus each face has three symmetry lines. Symmetry also provides the ability to reduce the calculation effort by reducing the system or component about its axis of symmetry (see Figure 10.12 as an example). Figure 10.12a shows an over-simplified bridge type (also referred to as a gantry type) coordinate measuring system (CMS), where the probe is supported by a bridge (horizontal beam) and is capable of moving in the horizontal direction guided by the bridge. To determine the maximum deflection of the beam due to

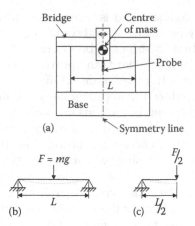

(a)

(b)   (c)

FIGURE 10.12
Example of a bridge-type CMS where symmetry can be used to reduce the analysis effort to determine the maximum deflection of the bridge. (a) Bridge-type CMS with the probe positioned on the symmetry line; (b) simplified schematic of the bridge being supported at both ends with a force applied at the symmetry line; (c) simplification of part b by the use of symmetry.

the mass of the probe supported by the bridge, the CMS can be modelled as a beam supported at both ends with the load applied, in this instance for maximum deflection, at the symmetry line (Figure 10.12b; the vertical supports for the bridge are considered rigid in this discussion). To further reduce the calculation effort, Figure 10.12b can be 'cut' at the symmetry line such that the beam is supported only at one end, with half the load/force applied to the other end (Figure 10.12c). This simplification can also be used to determine the deflection at any point along the bridge, by changing the position of the probe from 0 to $^L/_2$. Since the bridge type CMS in Figure 10.12 is symmetric, the results from the left side (as shown in Figure 10.12c) will be mirrored about the symmetry line to provide the same results on the right side of the symmetry line. By using symmetry, the computation times will also be reduced when performing calculations with finite element analysis (FEA) software (Carpinteri 1997, Qu 2004).

It should be noted that symmetry is considered a design choice and should be evaluated to determine whether it provides the best solution for a particular design. For example, a sub-assembly designed with symmetry in mind may not be the best option if an asymmetrical heat load is applied in the parent assembly. An example of where a symmetric thermal design would have been beneficial was discussed by Stone (1989). Stone reports on a thermal drift of a DSW Wafer Stepper® machine used to produce IC circuits from GCA Corporation. The cause of the thermal drift was traced back to an asymmetric heat load caused by a voice coil actuator used in conjunction with a parallelogram flexure used to focus the objective onto a silicon wafer. In the design, the actuator was mounted off-axis of the optical axis such that the heat generated caused one side of the parallelogram elastic element to thermally expand, causing the objective to rotate around the optical axis of the objective and in turn causing overlay errors. There were several solutions investigated to eliminate the thermal drift and one of the solutions implemented in a later design was to place the voice coil actuator so that the thermal profile spreads symmetrically across the parallelogram flexure and thus eliminates the overlay error caused by the thermal drift reported in the earlier system (Stone 1989).

10.1.8 Assembly Considerations for Optical Components

Mounting of optical components is very common in precision systems, which may contain combinations of optical encoder scales, mirrors, prisms and lenses. For the most demanding applications, low expansion optical glasses (see Chapter 12 on materials) are used, which often demand limited surface deformation and stresses to be present in the sensitive directions. Achieving minimal surface deformations and stresses in optical systems will require analysis of a particular design. Major details to be considered are mounting method, type of material used, operating environment and transport. The following subsections provide a general description of common methods used to assemble optical components.

10.1.8.1 *Mounting of Optical Prisms and Mirrors*

Precision machine tools, CMSs and lithography equipment typically use optical measurement systems that are either optical encoders or displacement measurement interferometers (DMIs; see Chapter 5) as sensors to determine the relative position of a motion stage. Optical encoders employ either a metallic or glass scale with the increments on the scales measured by an optical encoder head. The following discussion will focus on a glass scale of an encoder system (same basic rules will apply to metallic scales) as an example of how to mount optical prisms and mirrors. The mounting methods discussed can be employed on any optical prism or mirror.

In general, the desire is to kinematically constrain all six DOFs (see Chapter 6 on kinematics) of a glass scale as shown in Figure 10.13. In Figure 10.13, the glass scale is resting on three points (A, B and C) and restricts movement by friction only. The spacing of the three vertical constraints should be chosen to minimise deformation of the upper surface to avoid any additional measurement errors. To accommodate position stage accelerations, transport shocks and so forth it is desirable to apply an additional holding force to the glass scale. Again, to minimise surface deformation of the glass scale, the force should be applied such that it is normal to the glass scale horizontal surface and is positioned directly in line to the horizontal constraint points A, B and C. The holding force may be achieved by compliant elements (discussed in Chapter 7), adhesives or, less commonly, magnets. Up to this point, thermal expansion of the glass scale relative to the component on which it is mounted has been ignored. Taking the configuration shown in Figure 10.13, the glass scale has a fixed point in the horizontal plane at point D. This assumes that the relative thermal expansion between the glass scale and the component on which it is mounted can overcome the frictional forces present at points A, B and C. If the frictional forces cannot be overcome, then additional deformation of the glass scale will occur, resulting in a measurement error.

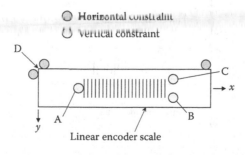

FIGURE 10.13

Exact constraint of a linear glass scale. Each constraint provides a single point contact (for example sphere on flat contact interface) and it is assumed that the scale is fixed to the constraints at D.

468

Basics of Precision Engineering

To overcome or minimise this additional deformation to the glass scale as shown in Figure 10.13, it is necessary to provide two additional linear DOFs at each vertical constraint point, including the holding forces, to accommodate expansion in the horizontal plane. For this reason, it is preferable to use an adhesive (with a filler to minimise the CTE of the adhesive) to provide the holding force. Alternatively, it is possible to use symmetry about the x-axis by removing the horizontal constraints and utilising point A to be the fixed point, while still providing two linear DOFs at each point B and C. The advantage of using symmetry in this instance is that the mounting of the glass scale is simplified at the expense of requiring an additional alignment step, which is discussed in the following section.

Although it is desirable to have a kinematic mount design with six point contacts, in many cases the stiffness or dynamic specifications are not achievable, such that a semi-kinematic approach is required (see Chapter 6 for more on semi-kinematic design). When forced into such a design decision, it is advisable to keep the number of contacts (and if possible using point contacts) as close to six as possible. Additionally, it should also be the goal to keep the newly implemented surface contacts as small as possible. A common solution is to convert the vertical constraints in Figure 10.13 into three small surface contacts, while maintaining the three horizontal point constraints. Alternatively, to further venture away from a kinematic design would be to reduce the three small surface contacts into one or two surface contacts, as depicted in Figure 10.14. When the highest performance demands are required, it is advisable to utilise optical contacting to mount the linear glass scale, as this method will eliminate the CTE mismatch (if identical materials are used), and improve the stiffness or dynamic requirements. Of course, this method will also incur the highest cost, as minimal surface specifications (as described in Section 10.1.5.3) are required, and the potential increase in part rejection due to not meeting the alignment criteria. The same methodology used for mounting a linear encoder scale can be applied to prisms, mirrors and rotary encoder scales.

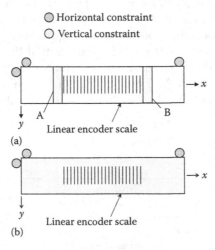

FIGURE 10.14
Over-constraint mounting configuration of a linear glass scale. (a) Linear encoder scale is mounted with two vertically constrained surfaces (A and B). (b) The linear encoder scale is mounted with only one vertically constrained surface.

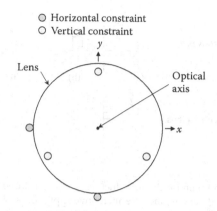

FIGURE 10.15
Kinematically constrained lens mount configuration.

10.1.8.2 Mounting of Optical Lenses

As with the assembly of a prism or encoder scale, it is desirable to design a lens mount using six point contacts to form a kinematically supported lens. Unlike a prism, a rotationally symmetrical lens about the optical axis only requires five DOFs to be constrained as shown in Figure 10.15. However, if lenses with aspherical* or freeform[†] surfaces are to be mounted, then an additional constraint is necessary to ensure that the lens is orientated properly. As with linear encoders, it is desirable to include clamping forces to accommodate operating and transport conditions, which should be applied directly in line with the constraint points. Depending on whether the optical system is polarisation sensitive, it is necessary to avoid stress-induced birefringence due to the introduction of mechanical stress in a lens (Yoder Jr. 1993). Therefore, it is advisable, if an optical system is sensitive to stress-induced birefringence, that the applied forces be kept to a minimum and the force vector is either parallel to the optical axis or pointing radially away from the optical axis (see examples in Figure 10.16). Radial forces should be avoided if stress-induced birefringence is of concern. A force can be applied through a compliant or non-compliant component and/or adhesives may also be used to restrict movement. The advantage of using a compliant mechanism is that the force (or stiffness) can be tailored to minimise lens surface distortions. As with mounting prisms and encoder scales, typical mounting methods for a lens can also employ a semi-kinematic approach. Mounting approaches for lenses are dependent on performance and cost requirements. Examples of common methods to mount a single lens into a mount are shown in Figure 10.17 (Yoder Jr. 2008) Hard clamping, as shown in Figure 10.17a, is one of the most cost-effective methods of mounting a lens, as it utilises the smallest number of components and the alignment is achieved by means of machining tolerances. To avoid the stresses introduced by hard clamping, an elastic element, such as a spring or elastomer, can be introduced and is shown in Figure 10.17b. The advantage of using an elastic element is that the applied force can be tailored to the allowable lens surface distortions and stress induced birefringence requirement. The limitation of knowing the applied force is

---

* An aspherical lens has surface with a varying radius of curvature and is rotationally symmetric about the optical axis.

† A freeform lens with a surface does not exhibit any symmetry.

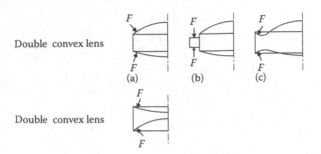

Double convex lens

Double convex lens

FIGURE 10.16
Application of clamping forces to minimise the introduction of stress-induced birefringence in a lens. Column (a) shows typical clamping forces; (b) and (c) indicate alternative applications of a clamping force to minimise the introduction of stress-induced birefringence with an annulus and a non-optical functioning concave surface added respectively. Note: A concave surface is ideal in this case as the clamping force vector points away from the optical axis. A radial force should be avoided if possible.

dependent on the geometric tolerances that can be achieved during manufacturing and the knowledge of the material characteristics of the elastic element. If the available volume (size) for a lens mount is limited, adhesives may be used to restrict the movement of a lens, as shown in Figure 10.17c. The advantage of this solution is that it also provides the opportunity to perform active alignment to improve the optical performance of a system. When volume is very limited and manufacturing tolerances do not provide adequate lens alignment, the lens can also be constrained in the radial direction, as shown in Figure 10.17d and e. The difference between the two constraint methods is that one uses adhesives (Figure 10.17d), while the other uses an elastic element (Figure 10.17e) to hold the lens in position. Both of these designs require additional alignment tooling and, as previously stated, care is

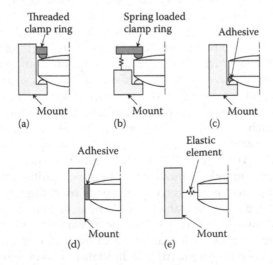

FIGURE 10.17
Examples of common methods to mount a single lens element. (a) Lens clamped (axial direction) with a threaded clamp ring (hard clamping), (b) lens clamped by an elastic element (spring, axial direction), (c) adhesive used to hold the lens in place, (d) adhesive (radial direction) used to hold the lens in place, (e) elastic element used to clamp the lens in the radial direction.

required for optical elements that are susceptible to stress-induced birefringence variations. These examples can be combined to obtain further design options. As an example, Figure 10.17d and e can be combined so that higher levels of axial shock can be accommodated, since the shock resistance of Figure 10.17e is no longer limited by a combination of the coefficient of friction and the radial applied force.

10.2 Alignment Principles

Either mechanical alignment features (MAFs) or metrology equipment can be used to align mechanical and optical components to meet system requirements. MAFs can be either machined directly into a component, an alignment tool or sub-assemblies. An example of a sub-assembly would be to insert dowel pins into a component so that the dowel pins act as the MAF. An alignment tooling is utilised to transfer the MAF into a dedicated tooling that is referenced to an additional separate MAF or placed into a defined location with the aid of metrology equipment.

Utilising an alignment tooling typically requires the geometric and alignment tolerances of the tooling itself to be better than the components being aligned, since the location features of the tooling are used to position the components into an assembly. While this tooling might be expensive, it is removed and reused to align multiple assemblies, thereby not significantly contributing to the cost of an individual assembly. The advantage of using MAFs to align a component is that the component can be simply contacted at defined reference points that will determine the position of the component and, therefore, meet the alignment requirements (by means of manufacturing tolerances) at reduced alignment costs. However, MAFs can (depending on design and alignment requirements) add complexity to the manufacturing process and increase manufacturing costs.

Metrology equipment, on the other hand, provides the ability to actively perform the alignment process with the potential of reducing the manufacturing costs. However, metrology equipment does require capital investment (can be very expensive if, for example, a CMS is required) and typically higher hourly labour costs during the assembly. The decision as to which method is to be used should be made during the early part of the design stage to ensure that proper interfaces and geometric features can be implemented into individual components. A trend that can be easily envisaged is that the use of MAFs for alignment and assembly increases as the production volume increases. It should also be noted that both methods can be combined to achieve different goals during the alignment process. Sections 10.2.1 through 10.2.3 provide basic 2D case study scenarios to illustrate principles of mechanical and optomechanical alignment. Similar principles will often apply for extension of these concepts to three dimensions, although the difficulty of implementation often increases substantially.

10.2.1 Alignment of Mechanical Components

One of the easiest and cost-effective methods to align mechanical components is to utilise MAFs. MAFs can be in the form of datum planes and cylindrical/curved or flat surfaces (for example holes, dowel pins, planes). When designing MAFs into a component or sub-assemblies, the goal should be to have a kinematic (or at least a semi-kinematic) configuration to meet the required alignment specification (see Chapter 6 on kinematics). The quality of the alignment will be determined by the machining tolerances of the MAFs. When using MAFs, friction is present between the interfaces and, therefore, it is necessary to ensure that the component achieves and maintains contact with the intended reference surfaces. To determine the appropriate force vector when pushing a component into contact with three MAFs (in this example dowel pins are used, but the same methodology applies for other reference geometric features), the following steps are necessary (Figure 10.18) (Koster 2005):

1. Add a centre line for each dowel pin (1, 2 and 3) normal to the component surface at the contact interface between the dowel pin and component (Figure 10.18b).
2. Determine the required rotation at each centre line intersection to keep the component in contact with the three dowel pins. The intersection of the centre lines provides a pivot point P_{12}, P_{13} and P_{23} to where the rotation is applied. For example, a counter-clockwise rotation is required at P_{12}, such that dowel pin 3 remains in contact with the component (Figure 10.18b).
   a. If the required rotation directions at each pivot point are in the same direction, an additional external moment is required to keep the component in contact with all three dowel pins. Otherwise, a single force is sufficient to maintain contact between the component and the dowel pins.
3. Add exclusion zones which violate the required rotations determined in step 2 (Figure 10.18c). If the applied force to a pivot point generates a rotation in the opposite direction as determined in step 2, then this becomes an exclusion zone, where a force may not be applied, as this would result in the component not being in contact with a dowel pin (Figure 10.18c).
4. Add a line from the point of contact between each dowel pin and the component at an angle of $\beta_f(=tan^{-1}\mu)$ to include the coefficient of friction μ (Figure 10.18d).
   a. Friction between the component and the dowel pins further restricts the applicable force that may be applied to the component to ensure contact is achieved/maintained at all three dowel pins.
5. Shade the areas from the lines added in step 4 with the sides where the friction would resist rotation at each rotation point (see Figure 10.18e). The friction force is denoted as F_f in the figure.
6. Combining the restricted areas from steps 3 and 5 results in a limited area where the force may be applied to ensure that all three dowel pins remain in contact with the component. The force vector that may be applied is limited in location and angle, denoted by the non-shaded areas in Figure 10.18f. As an example, a force F is shown in Figure 10.18e, which ensures contact is achieved and maintained between the component and dowel pins.

From this example it can be seen that it becomes difficult to obtain the proper force vector if the contact areas are increased (for example a plane on plane contact). Therefore, it is advisable to use point contacts where possible to avoid limiting the potential solutions to where a force may be applied.

An alternative use of dowel pins to perform an alignment would be to press two dowel pins into a reference component, and a second component, to be aligned to the reference component, would have a hole and slot to accommodate the two dowel pins (Figure 10.19). The diameter of the hole in the second component should be slightly larger than the dowel

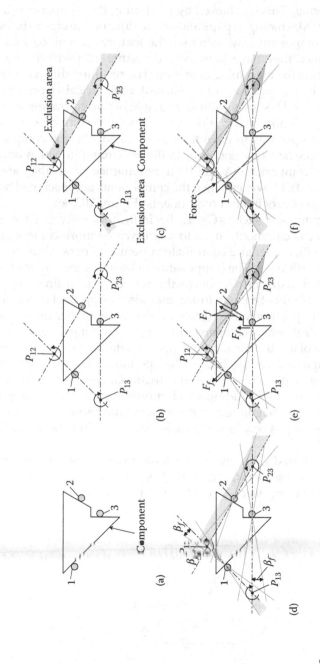

FIGURE 10.18

Method of determining the appropriate force vector to be applied to a component to ensure that contact is made and maintained with three horizontal dowel pins. (a) Ideal state, (b) addition of centre lines and determination of the required rotations such that contact is made with the third dowel pin, (c) determination of exclusion zones, (d) addition of the friction lines, (e) determination of friction exclusion zones, (f) final configuration indicating appropriate area where a force is to be applied to ensure all three dowel pins make and maintain contact with the component.

pin (such as a sliding fit) and will determine the position of the component, while the slot will restrict the rotation about the hole; the uncertainty of the position and orientation will be determined by the tolerances of the dowel and manufactured features. By using the method depicted in Figure 10.19, it is possible to not only perform an alignment step but also aid in the manufacturing. This is achieved by machining the components as pairs in a single manufacturing step. Machining in pairs allows a feature or features to be machined at the same time for both components. By doing so, the features machined into both components are 'identical' since they have nominally the same geometrical and alignment errors that are present during the machining operation. The resulting difference between the two machined pairs will be dependent on the inherent geometrical errors present in the machine tool. Figure 10.20a and b shows a simple example of two components machined in two step operations, while Figure 10.20c shows the same two components machined in pairs. The goal in this example was to have the holes drilled into the components to be concentric. When the spindle has a motion error in the horizontal direction only, it can be seen that the two holes in components A and B are not aligned when they are machined independently (Figure 10.20d). However, when the components are machined as a pair, the motion error of the spindle occurs in both components (Figure 10.20e).

The use of metrology equipment, such as CMSs, dial indicators and capacitance gauges, is an additional method that is commonly used to align two or more components to each other. When utilising a CMS, a reference coordinate system is generated to indicate where the component should be positioned. The component can be positioned by means of tooling or by applying a force manually in relation to the reference coordinate system. If the component/sub-assembly is robust enough to accommodate an applied shock, the operator can tap the component into position (often using a polymer or rubber mallet), which provides better position control when compared to simply pushing a component. When tapping a component into place, the friction force plays a vital role to improve the position sensitivity for a given impulse force applied by the operator. Since an operator is limited in the ability of applying a particular force, the frictional force may be increased by increasing the clamping force, simply adding an additional mass to the component or using a smaller mallet to reduce the magnitude of the incremental motion for a given shock/impulse. Generally, positioning is considerably easier with real-time visual feedback of the displacement.

The following steps are based on a one-dimensional example, shown in Figure 10.21, where a CMS and displacement sensors (such as a dial test indicator or capacitance gauge) are used to align and position a component. The goal, in this example, is to align and position

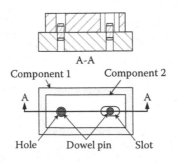

FIGURE 10.19
An example of using two dowel pins to align two components to each other.

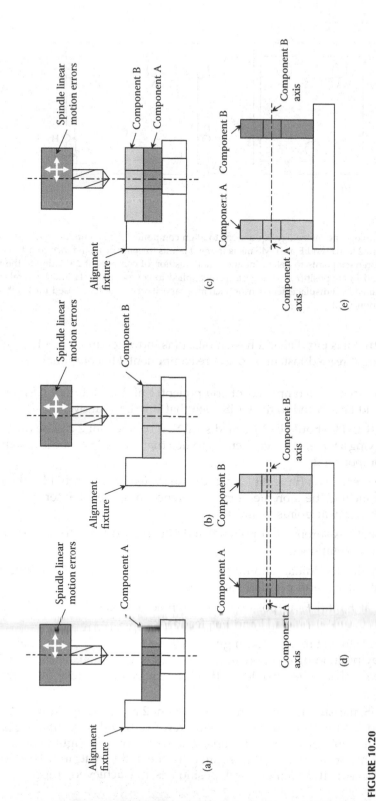

FIGURE 10.20

An simplified example showing how machining two components as a pair can reduce the misalignment error between the two components. (a and b) Drilling of a hole in component A and B respectively, (c) drilling a hole into components A and B as a pair, (d) indicating the misalignment between the two components, (e) indicating that the two components share the same alignment errors when they are machined as a pair.

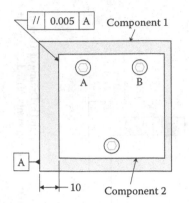 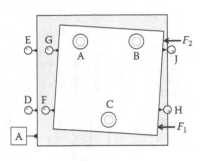

FIGURE 10.21
A one-dimensional example using metrology equipment to align component 2 to component 1. A, B and C are screws to hold component 2 to 1; D and E are CMS measurement points to generate a reference coordinate system; F and G are CMS measurement points to determine the actual position of component 2 relative to the reference coordinate system; F_1 and F_2 are positions where a force is applied to component 2; H and J are relative measurement points determined by a displacement sensor such as a capacitance gage (only used for feedback when applying a force to component 2).

component 2 such that it is parallel and has an offset (as indicated in Figure 10.21) to component 1, while using threaded fasteners to secure component 2 to component 1.

1. Place the sub-assembly (comprised of component 1 and 2) onto the CMS. Component 1 should be secured to the CMS such that it will not move.
   a. Screws A, B and C should be loosened such that component 2 can be moved when applying a force, but secure enough such that the CMS probe is not able to push component 2.
2. Generate a reference axis (in reality a coordinate system is generated by the CMS which would include the aforementioned reference axis; see Chapter 5) by measuring with the CMS at points D and E.
3. Perform the first measurement at points F and G using the CMS to determine the initial position of component 2.
   a. This will define the initial amount that component 2 needs to be shifted to meet the alignment and position requirement.
4. Apply forces at F_1 and F_2 to position and align component 2 to 1, utilising the displacement sensors at points H and J as feedback.
   a. It is advisable to first remove the angular error between components 1 and 2, followed by positioning component 2 to the desired position. This can help reduce the number of iterative loops that are required to achieve the correct alignment.
5. Using the CMS, measure the position of component 2 as described in step 3. If the alignment and position is achieved, tighten screws A, B and C (tightening is best performed in multiple incremental rounds, where one round is tightening A, then B, then C at 20% for example of the required torque and repeat until 100% of the torque is achieved). If alignment and position is not achieved, repeat steps 3 through 5.

This methodology can be adapted in many ways, including alignment of position stages, which will be described in the following section. It should also be noted that the use of a displacement sensor is optional, but without the use of the sensor, the force (strength of the impulse) an operator applies is performed without active feedback. Thus, the required time to align and position the component is increased, as the CMS can only provide feedback after an alignment cycle is performed (steps 3 to 5).

10.2.2 Alignment of Linear and Rotation Stages

Linear and rotary stages may also be aligned with the use of MAFs and metrology equipment. The same process steps can be adapted from the previous section to align linear and rotary stages. If angular alignment is the only concern, then displacement sensors (for example, dial indicators, capacitance probes, DMIs) can be used along with an alignment artefact. An example of using a displacement sensor, with a mechanical square as an alignment artefact, to align an x-axis stage to a y-axis stage (to achieve a perpendicularity between the motion axes) is described in the following steps (Figure 10.22).

1. Mount a calibrated mechanical square to the x-axis stage (Figure 10.22a). The mechanical square will act as the alignment reference and must have a perpendicularity better than the required perpendicularity between the x- and y-axis position stages.

2. Align the mechanical square with the use of a displacement sensor, such that the surface of the mechanical square is parallel to the motion axis of the x-axis stage (Figure 10.22b). This is achieved by traversing the x-axis stage back and forth while rotating the mechanical square until no change is measured by the displacement sensor.

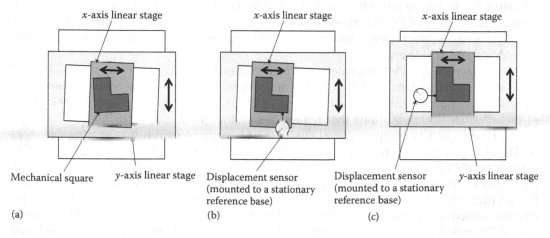

FIGURE 10.22
Alignment of an x- and y-axis stage such that the two stages are perpendicular to each other using a displacement sensor with a calibrated mechanical square. (a) Initial alignment of the mechanical square and x-axis stage in relation to the y-axis stage, (b) mechanical square is aligned to the x-axis stage, (c) x-axis stage is aligned to the y-axis stage.

3. Reposition the displacement sensor to measure the face of the mechanical square that should be parallel to the *y*-axis stage (Figure 10.22c). The *y*-axis stage is traversed back and forth while monitoring the indication on the displacement sensor. Again, when adjusting the alignment of the *x*-axis stage, until the displacement sensor measures no change, the two stages are considered perpendicular to within the uncertainty of these measurements.

Such an approach is limited by the quality of the mechanical square and the measurement capability of the displacement sensor. To remove the influence of the measurement square, a reversal technique can be employed to remove any systematic errors (Whitehouse 1976, Chapter 5). An alternative method to measure the perpendicularity is to use an interferometer in an optical square configuration as described by Bewoor and Kulkarni (2009). Aligning a rotary stage axis is also possible with the same measurement methodology using a variety of artefacts, such as mechanical squares, cylinders and master spheres.

10.2.3 Alignment of Optomechanical Systems

Although the alignment techniques described in the previous sections can be employed in aligning optomechanical systems, there are additional techniques that can be useful for the alignment of DMIs, encoders and optical systems with lenses. These are described in the following sections.

10.2.3.1 Displacement Measurement Interferometer

DMIs (see Chapter 5) are commonly used in precision systems. Aligning a DMI system can be challenging, as the DMI laser source may be mounted outside of the machine/instrument, there may be long optical paths, beams may be steered through different levels or environments, and so forth. It should be noted that if the laser source is mounted outside of the machine/instrument, care is required to minimise the disturbances that may be introduced into the measurement loop. Disturbances may be in the form of machine distortions, movement of the vibration isolation mounts and environmental effects, to name a few. Machine distortions and movement of the vibration isolation mounts can also introduce Abbe and cosine errors.

Basic alignment of an interferometer can be performed with a master target, usually with a cross or circular pattern, such that the beam is steered to a point at the centre of the target. An iris or aperture may also be used as a target where the laser beam is pointed to the centre of the aperture. In applications that require the best alignment of a DMI, an active measuring target, such as charge-coupled device (CCD) cameras or position sensing detectors (PSDs) are used to measure the position of the beam. The main advantage in using a CCD camera or PSD is that the position of the beam can be determined to the measurement capability of the device (such as pixel resolution, signal-to-noise ratio) and software capabilities (determination of centre of laser beam). When using a passive target with a cross, circular pattern or an iris, the resolution is dependent on the user's visual ability to interpolate the position of the laser beam. Regardless of the type of target being used, the alignment procedures are the same and are described in the following steps to align a DMI beam steering optic with a right-angle bend (see Figure 10.23 as reference; beam steering optics encompass all optics that are used to direct and/or split the laser beam from the DMI laser source to the interferometer head) (Ellis 2014).

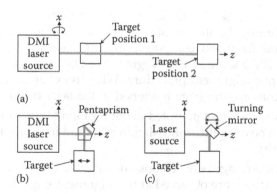

FIGURE 10.23
Alignment of a DMI laser source and beam-steering optic. (a) Aligning a DMI laser source to be parallel to the reference surface (*xz*-plane) using a target, (b) bending of laser beam with use of a pentaprism to set the initial position of the target, (c) aligning a turning mirror to the target (position of target was set in part b).

1. The DMI laser source is aligned to be parallel and to the correct beam height above the reference surface (this is the surface in the *xz*-plane on which all components are mounted).
   a. This is achieved by setting the target to the desired beam height and simply placing the target near the DMI laser source (target position 1 in Figure 10.23a). The DMI laser source is aligned so that the laser beam is centred to the target at position 1. The target is then moved to the furthest possible point away from the DMI laser source (position 2). Adjustments are made until the DMI laser source at both positions 1 and 2 is in the centre of the target.

2. Placing a pentaprism (an optic having five sides that is designed so that light entering into a specific face will be reflected and emerge from another face at 90° to the incoming beam) into the beam path to achieve a right angle bend, it is possible to position the target such that the beam is in the centre of the target, as shown in Figure 10.23b. (Note that the DMI laser source shall not be adjusted at this point, as the alignment performed in step 1 would be lost.) The target should be fixed as this becomes the new reference point.
   a. It should be noted that the pentaprism should be pre-aligned prior to performing this step in the process to ensure that the reflected beam is parallel to the reference surface (this is the surface in the *xz*-plane on which all components are mounted) using the DMI laser source from step 1. This pre-alignment is performed as described in step 1, with the exception that the DMI laser source is fixed and the pentaprism is adjusted.

3. Replace the pentaprism with a turning mirror. The turning mirror is then adjusted so that the laser beam is centred to the target (see Figure 10.23c).

4. Repeat steps 2 and 3 for each additional right-angle bend in the system.
   a. Note: Steps 2 and 3 can also be used for a beam splitter or any other optical element/system with a right-angle bend being introduced into the beam path.

After aligning the beam-steering optics of a DMI system, it is required to align the measurement mirror mounted to the position stage. This is achieved with the following steps (Ellis 2014).

1. Align the position stage so that the laser beam and motion axis are parallel. This is achieved by mounting the alignment target to the positioning stage and traversing the stage back and forth between position 1 and 2 (see Figure 10.24a). The alignment procedure is the same as with aligning the DMI laser source in step 1 of the beam-steering optic alignment procedure. When no changes are observed at the target at both positions, the stage is aligned to the laser beam.

2. Replace the alignment target mounted to the position stage with the measurement mirror and add a beam splitter and alignment target in front of the position stage (see Figure 10.24b).

3. Traverse the position stage between positions 1 and 2 and adjust the measurement mirror until no changes are observed at the alignment target (similar to step 1, see Figure 10.24b).

With the DMI laser source, beam-steering optics, position stage and measurement mirror aligned, it is sufficient to nominally align the interferometer laser head with the use of MAFs or special targets typically supplied with the interferometer.

From the aforementioned alignment steps, it can be seen that, if the system has multiple measurement axes, the alignment procedure can be time consuming. To avoid the lengthy alignment time, it is possible to introduce a shear plate and Risley prism (explained next) into the optical path to perform the alignment, while using MAFs (such as dowel pins) to nominally position the DMI source, steering optics and DMI laser head. To laterally displace a laser beam, a shear plate is used (Figure 10.25), which works on the principle of Snell's law of refraction. The lateral displacement—with the subscripts indicating the direction in which the displacement occurs, δ_x and δ_y—is achieved by simply rotating the shear plate such that the beam is no longer normal to the optical entrance surface. The resulting lateral displacements of a shear plate are determined by

$$\delta_x = t\left(\tan \theta_y - \tan\left[\sin^{-1}\left(\frac{n_e}{n_g} \sin \theta_y \right) \right] \right) \cos \theta_y, \tag{10.13}$$

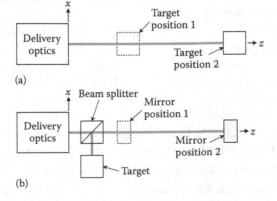

(a)

(b)

FIGURE 10.24
Alignment of the position stage and measurement target to the laser beam of a DMI. (a) Alignment target mounted to the position stage to align (parallel) the motion axis of the position stage to the laser beam axis, (b) alignment of the measurement mirror such that the optical surface of the mirror is perpendicular to the laser beam axis.

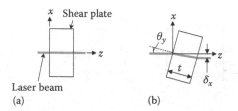

FIGURE 10.25
A schema of a shear plate. (a) Shear plate with no effect on a laser beam lateral displacement, (b) laser beam sheared or displaced by rotating the shear plate about the y-axis.

and

$$\delta_y = t\left(\tan\theta_x - \tan\left[\sin^{-1}\left(\frac{n_e}{n_g}\sin\theta_x\right)\right]\right)\cos\theta_x,$$ (10.14)

where t is the thickness of the shear plate, n_e is the refractive index of the environment, n_g is the refractive index of the glass, θ_x is the rotation about the x-axis and θ_y is the rotation about the y-axis. A Risley prism is used to perform angular adjustments to the laser beam pointing, and is comprised of two identical wedge prisms, as shown in Figure 10.26 (Yoder Jr. 2008). When the wedge angle of the two wedge prisms are parallel, the laser beam will pass through the optics with a linear offset (see Figure 10.26a), where the offset δ_x in the x-direction can be determined by the following:

$$\delta_x = \left(z_w + \frac{D}{2}\tan\alpha\right)\tan\left[\alpha - \sin^{-1}\left(\frac{n_e}{n_g}\sin\alpha\right)\right] + z_s\tan\left[\sin^{-1}\left(\frac{n_g}{n_e}\sin\left[\alpha - \right.\right.\right.$$

$$\sin^{-1}\left(\frac{n_e}{n_g}\sin\alpha\right)\right]\right) + \left[z_s + \left(\frac{D}{2} - \left(z_w + \frac{D}{2}\tan\alpha\right)\tan\left[\alpha - \sin^{-1}\left(\frac{n_e}{n_g}\sin\alpha\right)\right] - \right.$$

$$z_s\tan\left[\sin^{-1}\left(\frac{n_g}{n_e}\sin\left[\alpha - \sin^{-1}\left(\frac{n_e}{n_g}\sin\alpha\right)\right]\right)\right]\right)\right]\tan\left[\sin^{-1}\left(\frac{n_e}{n_g}\sin\left[\sin^{-1}\left(\frac{n_g}{n_e}\sin\left[\alpha - \right.\right.\right.\right.$$

$$\sin^{-1}\left(\frac{n_e}{n_g}\sin\alpha\right)\right]\right)\right],$$ (10.15)

where z_w is the width of the wedge prism, z_s is the spacing between the two wedge prisms, D is the diameter of the wedge prism and α is the wedge angle. If one of the wedge prisms are rotated about the z-axis by π radians (see Figure 10.26b), then the maximum beam deviation angle is achieved. The maximum beam angle θ_x can be determined by

$$\theta_x = \sin^{-1}\left(\frac{n_g}{n_e}\sin\left[\alpha + \sin^{-1}\left(\frac{n_g}{n_e}\sin\left[\sin^{-1}\left(\frac{n_g}{n_e}\sin\left[\alpha - \sin^{-1}\left(\frac{n_e}{n_g}\sin\alpha\right)\right]\right)\right]\right)\right]\right) - \alpha.$$

(10.16)

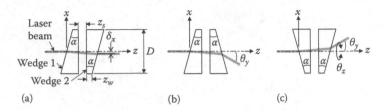

FIGURE 10.26
A schema of a Risley prism. (a) Risley prism with the wedge angles of the two wedges aligned such that they are parallel resulting in a linear shift of the beam in the x-direction. (b) Second wedge is rotated about the z-axis resulting in the maximum beam deflection θ_y. (c) Both wedges are rotated together.

Finally, the beam can be rotated about the z -axis by rotating both wedge prisms together by θ_z and is shown schematically in Figure 10.26c. The adjustment ranges required for the shear plate and Risley prism are dependent on the mechanical alignment tolerance stack-up. Introducing the shear plate and Risley prism to a DMI system is beneficial, as it reduces overall alignment times and complexity by providing the opportunity to use MAFs to place beam-steering optics into their nominal positions. However, care also needs to be taken on the impact that the coatings will have on the measurement error. Coatings are typically designed for angles of incidence of 0° and 45°. By deviating from these angles, the transmission efficiency and polarisation states will be negatively affected (Ellis 2014). To avoid this negative impact on performance, custom-designed coatings may be required. For additional alignment techniques for other optical components, such as waveplates, see Ellis (2014).

10.2.3.2 Encoders

Alignment of an encoder system is critical to not only obtain a measurement signal but also to keep the measurement errors to a minimum. The simplest method to align an encoder scale and head is by using MAFs, as discussed in Section 10.2.1. If the required alignment tolerances are not achievable with the use of MAFs, the difficulty of the alignment process is significantly increased. When this occurs, it is advisable to use MAFs for either the encoder scale or head, and manually align the other encoder component, with feedback on the alignment quality given by the electronics of the encoder system. Aligning both the scale and encoder head without a reference is not advisable.

Misalignment of a linear encoder will result in an Abbe and cosine error as discussed Sections 10.1.1 and 10.1.2 respectively. Misalignment of a rotary encoder will also result in cosine error due to tilt errors (motion in the axial direction during rotation) and be affected by the decentring of the rotary encoder. Decentring of the rotary encoder will lead to multiple harmonics during rotation leading to undesired angular measurement errors (Wilson 2016). Under closed loop control, these errors can result in additional noise.

10.2.3.3 Lens System

As discussed in Section 10.1.8.2, there are several methods that can be employed to mount a lens. As with any component that is produced, manufacturing errors will be present, as shown in Figure 10.27 for a double convex lens. When choosing to actively align a lens to a lens mount, the geometrical manufacturing errors depicted in Figure 10.27 can be compensated during the alignment process. To perform an alignment of a lens to a lens mount, a rotation stage, sensors to measure the optical surfaces of the lens and displacement sensors

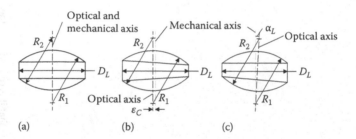

FIGURE 10.27
Geometrical errors of a double convex lens due to manufacturing errors. (a) Double convex lens without any manufacturing errors, (b) schema of the linear offset error between the optical and mechanical axis, (c) schema of the optical axis tilted in respect to the mechanical axis.

to measure the position of the lens mount are required. The following steps describe an alignment procedure for a double convex lens fastened into a lens mount (the lens is constrained in the radial direction using an elastic element and adhesive, combining the mounting concepts shown in Figure 10.17d and e).

1. Align the radius of curvature of R_1 such that it is coincident with the rotation axis (Figure 10.28a). The centre of R_1 can be determined with the use of sensor S_3, while S_2 provides the capability to monitor the alignment in subsequent steps to ensure that the tooling did not move.
2. Replace the lens with the lens mount, so that the lens mount is mounted to the tooling and aligned using sensor S_4 to the rotation axis (Figure 10.28b).
3. With the lens mount aligned and secured, the lens is placed again onto the tooling to align the optical axis of the lens to the rotation axis (Figure 10.28c). This is achieved by applying a force F to the radius of curvature R_2, while using sensor S_5 as feedback. The lens will rotate about the centre of R_1 ensuring that the centre of R_1 remains aligned with the rotation axis. When no measurement change is observed at S_5 and S_3, then the optical axis of the lens is aligned to the rotation axis.
4. Apply adhesive between the lens mount and lens (see Figure 10.28d).

The resulting alignment error can be calculated for a double-convex lens by

$$\varepsilon_{cx} = (R_1 + R_2 - t)\tan\left(\tan^{-1}\frac{S_5}{D_{S5}}\right), \tag{10.17}$$

where t is the thickness of the lens, R is the radius of the lens, subscripts 1 and 2 differentiate between the two radiuses of the lens, S_5 is the amplitude of R_2 moving in the axial direction during rotation, and D_{S5} is the radial distance sensor S_5 from the rotation axis. A similar equation can be formulated for a double concave lens

$$\varepsilon_{ce} = (R_1 + R_2 + t)\tan\left(\tan^{-1}\frac{S_5}{D_{S5}}\right). \tag{10.18}$$

The preceding alignment steps can be used for any combination of lenses and lens mounts. The advantages of implementing such an alignment technique are compensation of manufacturing errors present in the lens and lens mount, symmetrical lens mounting and the quality of the alignment is limited to the measurement capabilities of the sensors. With

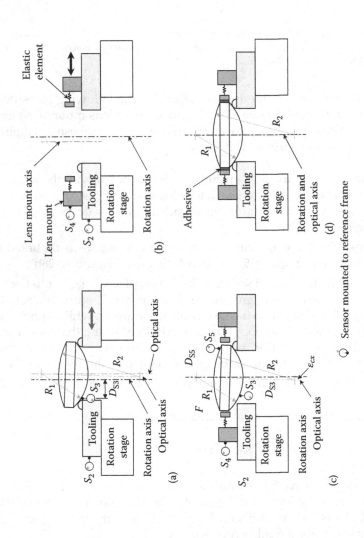

FIGURE 10.28

Schemas used to describe a methodology of aligning a double convex lens to a lens mount. (a) Alignment of the centre of curvature of R_1 such that it is coincident to the rotation axis, (b) alignment of the lens mount to the rotation axis, (c) alignment of the lens mount to the rotation axis, (d) application of adhesives to secure the aligned lens to the lens mount.

these advantages also come disadvantages, such as increased cost, decreased productivity and limited lens sizes (limited by the axial stiffness of the lens mount). The preceding alignment procedure can also be applied to lens mounts that support the lens in the axial direction, as shown in Figure 10.17c. When such a lens mount is chosen, the tooling is integrated into the lens mount, therefore no additional tooling is required.

Another method to align a lens to a lens mount is called centre lathing. Centre lathing involves machining the lens mount reference surfaces such that the optical axis is aligned to the lens mount mechanical axis. In this process, the lens is arbitrarily aligned to a lens mount. The preferred method in this alignment process is to use one of the following lens mounting methods shown in Figure 10.17a, b and c, or a combination thereof. Regardless of the type of mounting method chosen, it is critical that the lens does not change its position/orientation during the machining operation (Beier et al. 2012). The advantages to centre lathing are geometrical tolerances of the lens are compensated and the lens can be coarsely aligned to the lens mount. The disadvantages of centre lathing are the potential of damaging the lens coating and potential of contaminating the lens and lens mount. As with the lens centring method, the alignment is limited to the ability to measure the lens orientation, but in this case, it will also include the accuracies of the machining process.

Aligning multiple lenses to each other with the use of individually mounted lenses can be achieved as follows:

1. Align the optical axis of the first mounted lens to the rotation axis (Figure 10.29a). This is achieved by using sensor S_2 to determine the misalignment, while sensor S_1 can be used as feedback, and to determine the offset between the mechanical and optical axis of the first mounted lens.

2. After aligning the first lens to the rotation axis (Figure 10.29b), the second mounted lens is added and aligned following the same procedure as in step 1 (Figure 10.29c). It is typical to add a third sensor during the alignment process to monitor the position of the first mounted lens. This extra sensor is optional, but increases the confidence in the alignment process that nothing has changed during the subsequent alignment steps.

3. Repeat step 2 for each additional lens that is required to be aligned.

The goal of these steps is to align the optical axis of each individual lens to a reference axis (seen as either the rotation stage axis or the first mounted lens optical axis) of an optical system. Alternatively, it is also possible to align the mechanical axis of each mounted lens instead of the optical axis. The advantage of aligning to the mechanical axis is simplified measurement requirements (no need to measure the optical surface of the lens) and reduced alignment time. However, the misalignment between the mechanical and optical axes of a mounted lens will result in an alignment error.

An even easier method to align multiple individual mounted lenses is to drop the individual lenses into a barrel and use the outer diameter of the lens mount and the inner diameter of the barrel to passively align the optical system (Figure 10.30 shows only a single mounted lens for simplicity). The achievable alignment is dependent purely on mechanical tolerances. This method can be improved if centre lathing is used to align a lens to a lens mount. During centre lathing it is possible to machine the outer diameter of the lens mount to a specific diameter, with the misalignment between the optical and mechanical axes being limited by the measurement and machining capability of the lathe. Once all individual mounted lenses have been aligned by centre lathing, the inner diameter of the barrel

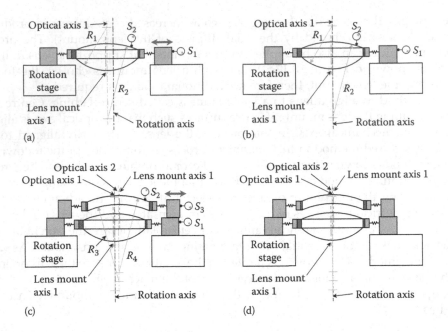

FIGURE 10.29

Schemas indicating the process of aligning lenses that are mounted in individual lens mounts. (a) Initial setup of placing the first mounted lens onto the rotation stage and aligning the optical axis to the rotation axis, (b) the first mounted lens' optical axis is aligned to the rotation axis, (c) mounting and aligning a second mounted lens' optical axis to either the rotation axis or the optical axis of the first mounted lens, (d) second mounted lens aligned to the first mounted lens.

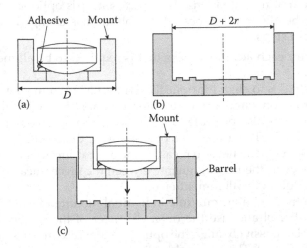

FIGURE 10.30

A single lens mount that is dropped into a barrel for a passive alignment. This example shows only a single mounted lens; in practice multiple mounted lenses would be mounted to the barrel with each lens having a different diameter (first lens mount would have the smallest diameter with each successive lens mount diameter being larger). (a) Lens mount with known diameter, D. (b) A barrel which will house the mounted lens from part a. The inner diameter of the barrel is machined to the outer diameter of the lens mount plus a radial clearance, r. (c) A mounted lens is slid into the barrel.

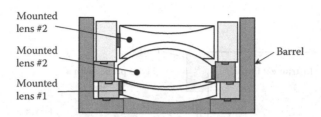

FIGURE 10.31
Multiple mounted lenses inserted into a barrel that has the inner diameter of the barrel machined to match the outer diameter of the lens mounts. Note: A radial clearance is required to ensure that the lens mounts can be inserted into the barrel. Radial clearance between the lens mount and barrel is exaggerated to indicate that the alignment is purely based on the mechanical tolerances.

can be machined to the specific required radial clearance between the outer diameter of the lens mount and the inner diameter of the barrel (Yoder Jr. 2008). Figure 10.31 shows multiple mounted lenses inserted into a barrel that has the inner diameter machined to match the outer diameter of the lens mounts.

Exercises

1. Derive the 1D Abbe error equation, $\varepsilon_{Abbe} = d_y \tan \theta_x - L(\frac{1}{\cos \theta_x} - 1)$.
2. Derive the small angle approximations for the 2D Abbe, cosine and sine errors.
3. Identify the Abbe, cosine and sine errors that are present in the profilometer instrument shown in Figure 10.32. Background information: A profilometer is capable of measuring the surface texture and form of a specimen (see Chapter 5). The measurement is performed by moving the specimen in the x-direction while a stylus probe measures the vertical deviations in the y-direction, thus providing the surface texture and form topography of the specimen as a function of x-displacement. The measurement system in this example uses a displacement measurement interferometer (DMI) to determine the relative position change of the positioning stage and stylus.
4. How can the Abbe, cosine and sine errors of the profilometer instrument in Exercise 3 be reduced?
5. Show the area where the appropriate force vector can be applied for the system shown in Figure 10.33.
6. Provide a methodology of how the rotation stage, shown in Figure 10.34, can be aligned such that it is parallel to datum A and the xy-plane, using a cylindrical artefact.
7. List the possible error sources that can occur in Exercise 6.
8. Generate a basic procedure to align the two mirrors in Figure 10.35, such that the reflective surfaces are perpendicular using a linear stage, rotation stage, beam splitter, PSD sensor and single DMI measurement axis. Assume that the linear and rotation stage are perfect.
9. Prove that the input and output beam angle of incidence are equal in a shear plate. Hint: Snell's law states $n_1 \sin \theta_1 = n_2 \sin \theta_2$, where n is the index of refraction and θ is the angle of incidence, and the subscript 1 indicates the input beam and 2 the output beam.
10. Derive the equation to determine the linear offset generated by a shear plate $\delta_x = t(\tan \theta_y - \tan[\sin^{-1}(\frac{n_z}{n_g} \sin \theta_y)]) \cos \theta_y$. Hint: Use Snell's law as in Exercise 9.

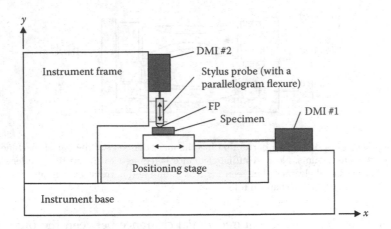

FIGURE 10.32
Profilometer instrument capable of measuring the surface texture and form of a specimen.

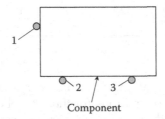

FIGURE 10.33
A component that is to be aligned with the use of three horizontal dowel pins (1, 2 and 3).

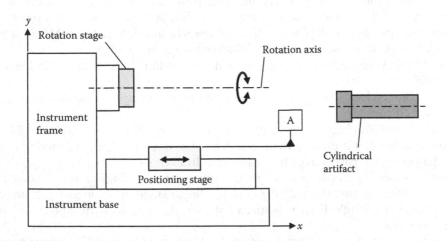

FIGURE 10.34
Schema of a system comprising of a linear and rotation stage. The rotation axis of the rotation stage is to be aligned such that it is parallel to datum A and *xy*-plane.

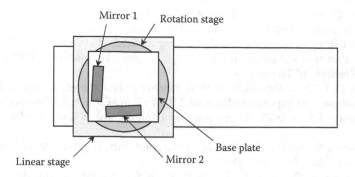

Mirror 1   Rotation stage

Base plate

Linear stage   Mirror 2

FIGURE 10.35
Image of the two mirrors that are mounted to a base plate and require alignment.

References

Abbe, E. 1890. Messapparate für Physiker. *Zeitschrift für Instrumentenkunde* 446–447.

AK Steel Corporation. 1981. Nitronic (R). Trademark 1177800.

Banea, M. D., da Silva, L. F. M. 2009. Adhesively bonded joints in composite materials: an overview. *J. Mater. Des. Appl.* **223**:1–18.

Beier, M., Gebhardt, A., Eberhardt, R., Tuennermann, A. 2012. Lens centering of aspheres for high-quality optics. *Adv. Opt. Tech.* **1**:441–446.

Bewoor, A. K., Kulkarni, V. A. 2009. *Metrology and measurement*. New Delhi: Tata McGraw-Hill.

Bhandari, V. B. 2007. *Design of machine elements*. New Delhi: Tata McGraw-Hill.

Bickford, J. H. 2008. *Introduction to the design and behaviour of bolted joints: Non-gasketed joints*. Boca Raton, FL: CRC Press.

Bosmans, N. 2016. Position measurement system for ultra-precision machine tools and CMMs: Prototype development and uncertainty evaluation. PhD dissertation, University of Leuven.

Bramble, K. L. 2012. *Engineering design for manufacturability*. Engineers Edge.

Bryan, J. B. 1979. The Abbe principle revisited: An updated interpretation. *Precis. Eng.* **1**:129–132.

Budhe, S., Ghumatkar, A., Birajdar, N., Banea, M. D. 2015. Effect of surface roughness using different adherend materials on the adhesive bond strength. *Appl. Adhes. Sci.* **3**:20.

Campbell, F. C. 2011. Joining: Understanding the basics. Materials Park, OH: ASM International.

Carpinteri, A. 1997. *Structural mechanics: A unified approach*. London: Taylor & Francis.

Dai, G., Pohlenz, F., Danzebrink, H.-U., Xu, M., Hasche K., Wilkening, G. 2004. Metrological large range scanning probe microscope. *Rev. Sci. Instrum.* **74**(4):962–969.

Douglas, R., van Veggel, A. A., Cunningham, L., Haughian, K., Hough, J., Rowan, S. 2014. Cryogenic and room temperature strength of sapphire jointed by hydroxide-catalysis bonding. *Class. Quantum Grav.* **31**:1–10.

Ebnesajjad, S. 2010. *Handbook of adhesives and surface preparation: Technology, applications and manufacturing*. William Andrew.

Ebnesajjad, S., Landrock, A. H. 2015. *Adhesives technology handbook*. Amsterdam: Elsevier.

Elliffe, E. J., Bogenstahl, J., Deshpande, A. *et al.* 2005. Hydroxide-catalysis bonding for stable optical systems for space. *Class. Quantum. Grav.* **22**:S257–S267.

Ellis, J. D. 2014. *Field guide to displacement measuring interferometry*. Bellingham, WA: SPIE Press.

Evans, C. 1989. *Precision engineering: An evolutionary view*. Bedford, U.K.: Cranfield Press.

Floriot, J., Lemarchand, F., Abel-Tiberini, L., Lequime, M. 2006. High accuracy measurement of the residual air gap thickness of thin-film and solid-spaced filters assembled by optical contacting. *Opt. Commun.* **260**:324–328.

Gibson I., Rosen D. W., Stucker B. 2015. *Additive manufacturing technologies: 3D printing, rapid prototyping, and direct digital manufacturing*. New York: Springer.

Gwo, D.-H. 2001. Ultra precision and reliable bonding method. US Patent 6,284,085 B1. 4 September.

Hale, L. C. 1999. Principle and techniques for designing precision machines. PhD dissertation, Massachusetts Institute of Technology.

Holt, R. B., Smith, H. I., Gussenhoven, M. S. 1966. *Research on optical contact bonding*. Technical Report, Device Development Corporation, Bedford, UK: Air Force Cambridge Research Laboratories, 43.

Jones, R.V., Richards, J. C. S. 1973. The measurement and control of small displacements. *J. Phys. E.* 6:589–600.

Kalkowski, G., Risse, S., Rothhardt, C., Rohde, M., Eberhardt, R. 2011. Optical contacting of low-expansion materials. *Proc. SPIE* **8126**:81261F.

Kalkowski G., Fabian, S., Rothardt, C., Zeller, P., Risse, S. 2013. Silicate and direct bonding of low thermal expansion materials. *Proc. SPIE* **8837**:88370U.

Kinloch, A. J. 1987. *Adhesion and adhesives*. Springer Netherlands.

Koster, M. P. 2005. *Design principles for precision mechanisms*. Eindhoven, Netherlands: Philips Research, Centre for Technical Training (CTT).

Leach, R. K. 2015. Abbe error/offset. In *CIRP Encyclopedia of Production Engineering*, edited by Laperrière, L., Reinhart, G. Berlin: Springer.

Lestrade, A. 2010. *Dimensional metrology and positioning operations: Basics for a spatial layout analysis of measurement systems*. CAS–CERN Accelerator School Magnets, 273–333.

Lindsey, K. 1991. Tetraform Griding. *Proc. SPIE* **1573**:129–135.

Müller, M., Hrabě, P., Chotěborský, R., Herák, D. 2006. Evaluation of factors influencing adhesive bond strength. *Res. Agr. Eng.* **52**(1):30–37.

Masteika, V., Kowal, J., Braithwaite, N.St.J., Rogers, T. 2014. A review of hydrophilic silicon wafer bonding. *ECS J. Solid State Sci. Technol.* **3**(4):Q42–Q54.

Petrie, E. M. 2009. *Handbook of adhesives and sealants*. New York: McGraw-Hill.

Qu, Z.-Q. 2004. *Model order reduction techniques: With applications in finite element analysis*. London: Springer-Verlag.

Rayleigh, F. R. S. 1936. A study of glass surfaces in optical contact. *Proc. R. Soc. Lond. A Math. Phys. Sci.* **156**(888):326–349.

Ruijl, T. A. M. 2001. *Ultra precision coordinate measuring machine: Design, calibration and error compensation*. PhD dissertation, Delft University of Technology, Netherlands.

Schellekens, P., Rosielle, N., Vermeulen, H., Vermeulen, M., Wetzels, S., Pril, W. 1988. Design for precision: Current status and trends. *Ann. CIRP* **47**(2):557–586.

Schneberger, G. L. 1983. *Adhesives in manufacturing*. New York: Marcel Dekker.

Schumacher, W. J., Tanczyn, H. 1975. *Galling resistance austenitic stainless steel*. US Patent 3,912,503. 14 October.

Sihn, S., Ganguli, S., Roy, A. K., Dai, L. 2008. Enhancement of through-thickness thermal conductivity in adhesively bonded joints using aligned carbon nanotubes. *Compos. Sci. Technol.* **68**(3–4):658–665.

Slocum, A. H. 1992. *Precision machine design*. Dearborn, MI: Society of Manufacturing Engineers.

Stone, W. S. 1989. Instrument design case study flexure thermal sensitivity and wafer stepper baseline drift. *Proc. SPIE* **1036**:20–24.

Tong, Q. Y., Schmidt, E., Gösele, U., Reiche, M. 1994, Hydrophobic silicon wafer bonding. *Appl. Phys. Lett.* **64**(5):625–627.

VDI-Fachbereich Produkentwicklung und Mechatronik. 2014. *Systematic calculation of highly stressed bolted joints: Multi bolted joints. VDI-Standard: VDI 2230 Part 2, VDI-Gesellschaft Produkt- und Prozessgestaltung*, The Association of German Engineers (VDI).

VDI-Fachbereich Produktentwicklung und Mechatronik. 2015. *Systematic calculation of highly stressed bolted joints: Joints with one cylindrical bolt. VDI-Standard: VDI 2230 Part 1, VDI-Gesellschaft Produkt- und Prozessgestaltung*, The Association of German Engineers (VDI).

Whitehouse, D. J. 1976. Some theoretical aspects of error separation techniques in surface metrology. *J. Phys. E.* **9**:531–536.

Wilson, C. S. 2016. Automatic error detection and correction in sinusoidal encoders. Proceedings of ASPE 31st Annual Meeting, 155–160.

Yoder Jr., P. R. 1993. *Opto-mechanical systems design*. New York: Marcel Dekker.

Yoder, P. R. 1998. *Design and mounting of prisms and small mirrors in optical instruments*. Bellingham, WA: SPIE Optical Engineering Press.

Yoder Jr., P. R. 2008. *Mounting optics in optical instruments*. Bellingham, WA: SPIE Press.

11

Force Loops

Niels Bosmans and Dominiek Reynaerts

CONTENTS

ABSTRACT This chapter introduces the basics of force loops by considering conceptual structural components for the construction of machine tools and dimensional measuring machines. In particular, different design configurations are presented to illustrate the issues for each configuration and explain the concept of functional loop separation that provides a means of assessing different designs in terms of achievable precision and significant limiting components within the assembled mechanism. The following sections focus on positioning and machining tasks, describing structural loops in more detail and providing design guidelines for optimal behaviour. Having established tools for identification and characterisation, the definition and composition of a typical metrology loop is given, together with general design guidelines for optimising the metrological performance of the machines. Another important design consideration is the identification and analysis of thermal loops in the system. Following the definition of thermal loops, several design concepts are introduced for the reduction of thermal disturbances in instruments and machines. To illustrate application of the principles developed in this chapter, examples of state-of-the-art machines and instruments that apply these concepts are discussed.

11.1 Basics of Force Loops

Ultimately force loops can be obtained using Newton's law stating that for all systems in static equilibrium the sum of all forces is zero. This turns out to be a valuable concept to evaluate the performance of instruments and machines. The basic ideas behind the force loop concept can be best illustrated by answering the following two questions: What is the force loop of mechanical machines? and What are the paths of typical force loops in machines?

11.1.1 What Is the Force Loop of Mechanical Machines?

A force loop can be defined as a closed path in a machine, which runs from the tip of the end effector through multiple machine components to the effective point on the workpiece, and which determines the relative position of the end effector with respect to the workpiece. The end effector can be a tool or a measuring probe. The effective point is then the point that is machined by the tool or measured with the probe. Forces entering the force loop can be

- Process forces
- Forces resulting from machine dynamics
- Forces due to thermal stresses

These forces will deform the components in the force loop, which displaces the end effector with respect to the workpiece. In a Cartesian coordinate system, separate force loops can be considered for every coordinate x, y and z. Moreover, every direction could have multiple force loops determining the position of the tool and workpiece. The following section will discuss typical force loops in machine tools and measuring machines, and will identify the effect of forces on their precision (see Chapter 2 for a definition of precision).

11.1.2 What Are the Paths of Typical Force Loops in Machines?

11.1.2.1 Subtractive Machine Tool

In a subtractive machine tool, a tool cuts away material from a workpiece. In these machines, the movement of the tool with respect to the workpiece is, therefore, copied into the workpiece. A positioning error results directly in a dimensional error of the workpiece.

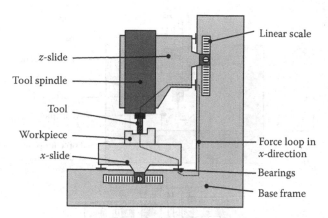

FIGURE 11.1
One of the force loops that determines the x-position in a subtractive machine tool (milling machine).

A simplified schematic drawing of a machine tool with two motion axes (x and z) is shown in Figure 11.1. The y-slide has been omitted to simplify the representation. Nevertheless, in real systems it should be included in the analysis.

One of the force loops that determines the x-position in this machine is also indicated in Figure 11.1. This force loop runs from the tip of the tool (the end effector) contacting the workpiece, through the tool spindle and the bearings of the z-slide, the base frame and the x-slide bearings, to the worktable and the workpiece. Deformation in any of these components will cause the relative position of the tool and workpiece to be changed in the x-direction. If the linear encoders that measure the position of the slides relative to the base frame, the output from which is fed back to the computer numeric controller (CNC), do not measure this deformation, a machining error occurs. Although Figure 11.1 only shows one force loop, in practice, multiple force loops spread out over all joints that transfer stresses between the components.

11.1.2.2 Coordinate Measuring Systems

A contact coordinate measuring system (CMS) has a touch probe as an end effector, which signals contact with the workpiece (see Chapters 5 and 8). Typically, three orthogonal slideways with scales to read position are used to move the probe around a three-dimensional rectangular volume. To measure a part, the probe is moved into contact with the part at multiple points and, at the moment of contact, the positions of the axes are recorded. As such, probing multiple points on a workpiece provides surface locations that can be used to determine the shape of a part. A typical configuration of a CMS is shown in Figure 11.2a, again with only two of the three axes being shown.

One of the force loops of the x-direction in this CMS runs from the tip of the probe contacting the workpiece, through the z-slide and the x-slide via the base frame to the workpiece. In a CMS, the process force, that is the contact force between the probe and the workpiece, is of the order of 0.1 N. However, there will also be changes in dimensions of the beams of the base frame due to the moving mass of the x- and z-slides, although these deformations can be measured. The machine components are also subjected to heat loads from the surrounding environment and motors used to move the slideways. A gradient in the room temperature, for instance, could cause a temperature difference between the left and right vertical beam of the base frame. Consider a lower temperature in the left beam.

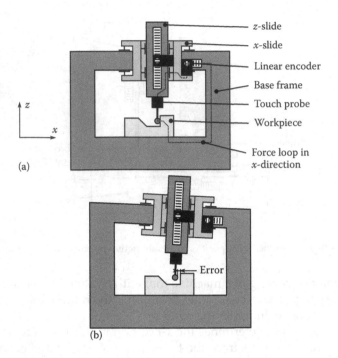

FIGURE 11.2
Example of one of the force loops that determines the x-position in a coordinate measuring system (CMS). (a) Force loop. (b) Thermal deformation of the base frame in the force loop.

This creates thermal stresses in the beam, which makes the beam contract, as shown in Figure 11.2b. Since the beam is in the force loop that determines the x-position, the contact point between the touch probe and the workpiece is shifted in the x-direction, resulting in a measuring error.

11.1.3 Functional Independence Design Principle

Ideally, every function of a machine should be designated to only one controllable design parameter (Kroll 2013). This parameter can be optimised for its task without degrading the performance of other functions. Often, this means that every function is assigned to a different component, although this is not a general rule. For instance, in a milling machine, the function of movement in the x-, y- and z-directions is dedicated to separate axes. This is known as the functional independence design principle.

The functional independence design principle is explained by considering that the main function of the motion system in a precision machine is to establish an accurate relative position between the tool and the workpiece. An important boundary condition is that this position should be established within certain time limits, to ensure productivity. Additional constraints such as, in the case of machine tools, no significant overshoots, may also be necessary. This main function can be further broken down into two important sub-functions:

- Minimising deformation due to process forces
- Minimising deformation due to thermal loads

The process forces mostly result from the cutting forces between the tool and workpiece and the dynamics of the moving slides. Stiff frames with solid cross-sections generally keep

TABLE 11.1

Functions of a Precision Machine, Associated Design Parameters, and Example Implementation of the Design Parameters

Function		Design Parameter	
		Frame Cross Section	Frame Material
Minimising deformation due to process forces	Cutting forces	Solid	Steel
	Inertial forces	Hollow	Aluminium
Minimising deformation due to thermal loads		Solid	Invar, Zerodur

positioning errors due to cutting forces within limits. Hollow cross-sections reduce mass, reducing inertial forces and dynamic errors. Thermal stability is optimised by application of low-expansion materials, such as Invar or Zerodur (see Chapter 12); by solid cross-sections in the frames that enhance thermal conductivity (at the expense of a longer time constant); and by using thermal symmetry between the heat sources and structures. The design should also take into account the effect of the weight of the moving elements, changes in the weight of the workpiece and the fact that the stiffness of the slides is dependent on their position.

Table 11.1 gives an overview of the sub-functions and the design parameters that fulfil them. A design conflict occurs if the two sub-functions are to be carried out by the same machine components. Therefore, it is better to apply the functional independence principle by assigning these sub-functions to different components. In practice, this can be achieved by having a structural loop that carries the process forces and a separate metrology loop that ensures thermal and temporal stability. Both loops can be individually optimised for their function, which avoids design conflicts. Nevertheless, it is difficult to entirely separate the structural and metrology loop. Mostly, the number of components in which both loops pass through is minimised, and these components are made as stiff and as thermally stable as possible. The following sections will elaborate on the design of these loops.

11.2 Structural Loops

11.2.1 Definition and Function of a Typical Structural Loop

ASME B5.54 (2005) defines a structural loop in a machine tool as 'an assembly of mechanical components, which maintain relative position between specified objects. A typical pair of specified objects is a cutting tool and a workpiece: the structural loop includes the spindle shaft, the bearings and housing, the slideways and frame, the drives, and the tool and work-holding fixtures'. The same definition can hold for a CMS, in which the tool is a touch probe. Consistent with the ASME definition, the tool and probe may be considered more generally as the end effector of the process. A machine contains one or more structural loops.

The function of a structural loop is to keep the following errors to a minimum:

- Kinematic errors
- Thermomechanical errors
- Errors due to loads

- Dynamic errors
- Errors in motion control and control software

The following sections explain the origin of these errors and how to mitigate them.

11.2.2 Kinematic Errors

Kinematic errors are due to imperfect geometry and assembly of the machine's components. Examples are straightness deviations of the guideways or misalignment of the axes. This is not to be confused with kinematic design which considers the constraints and freedom of mechanisms, and is discussed in Chapter 6.

A common way to reduce the kinematic errors in machine tools and CMSs is by calibration. First, dedicated equipment that has considerably lower measurement errors than the performance of the machine is used to determine the error motions. The most popular method is the use of laser interferometers (see Chapter 5). Next, these measured errors, if repeatable, are compensated using a look-up table in the numerical control of the machine. More ways to reduce kinematic errors by calibration can be found in Schwenke et al. (2008).

Another way to reduce kinematic errors is using aerostatic or hydraulic bearings, because their error motion (particularly non-repeatable 'noise') is smaller in magnitude than that of ball or roller bearings (see Chapter 7). Aerostatic or hydraulic bearings are also less influenced by wear, which makes their motion more repeatable over time, allowing for more effective calibration and numerical compensation. The influence of kinematic errors can also be reduced by applying the Abbe principle (see Chapter 10).

11.2.3 Thermomechanical Errors

Thermomechanical errors arise due to the thermal expansion of the structural loop components caused by heat loads. Section 11.4 will elaborate on this subject.

11.2.4 Errors Due to Loads

Process forces, accelerations, gravitation and friction in the guides all act on the machine frames and cause them to deform. If the structural loop is part of the metrology loop (see Section 11.3), errors are reduced by constructing the structural loop to be as stiff as possible (see Chapter 7). This includes the use of stiff bearings for the guides, the avoidance of large overhanging structures or the application of materials with a high ratio of Young's modulus to density (see Chapter 13 for further discussion of materials selection).

11.2.5 Dynamic Errors

Dynamic errors are caused by, for example, dynamic process forces, jerk, stick-slip effects in the guides, floor vibration, and backlash in the drive spindles. These errors can be reduced in the same way as the errors due to loads. Moreover, it is advised to reduce the error sources as much as possible.

Dynamic process forces are reduced by making use of the filter effect principle (Nakazawa 1994), which states that dynamic errors or noise should be reduced as close as possible to the source. Dynamic process forces could be reduced by making use of a hydraulic tool spindle.

The fluid film in the bearings creates a damping effect, which dissipates vibration energy that would otherwise be transmitted to other components of the structural loop, which could create unwanted resonances.

If the numeric controller (NC) generates trajectories without limiting the jerk (derivative of acceleration), the slides are subjected to a high impulse load, which triggers the eigenmodes of the structures and causes vibration. Jerk limitation in the controller is one way to reduce these errors. However, a lower jerk also leads to a reduction of velocity, resulting in lower productivity. In the end, there is always a trade-off between higher velocities and lower dynamic errors.

Stick-slip results from the Stribeck effect (Olsson et al. 1998; see Chapter 7). This effect is not present in aerostatic or hydrostatic bearings, in comparison with slideways, ball bearings and roller bearings. If mechanical bearings are preferred, the effect can be, at least partially, compensated by intelligent NC schemes.

Floor vibrations can be reduced by placing the machine on passive or active vibration isolators (see Chapter 13). Passive isolators support the instrument or machine on spring elements which, together with the high mass of the machine's base frame, attenuate floor vibrations from entering the structural loop components. The lower the spring stiffness of these elements and the higher the mass of the base frame, the lower the influence of vibration. However, lower spring stiffness also makes the machine more susceptible to the acceleration of the slides. Therefore, precision machines often employ active vibration isolators. These isolators are, in principle, actuators that reduce the effect of floor vibration by active damping of the base frame accelerations. On the other hand, these actuators also keep the machine level, such that accelerations of the slides do not disturb the base frame's alignment. These isolators, however, come with a higher cost.

Backlash is the mechanical effect that, during reversal of motion, the output motion momentarily does not follow the input motion. This is caused by play or reduced stiffness of the drive components and is typical for gear or lead screw transmissions. Backlash in the drive spindles can be avoided with the use of direct drive motors. However, should this not be an option, the use of hydraulic drive spindles will minimise backlash because of increased stiffness and damping. Preloading of ball bearings could also reduce the backlash errors. Multiple ways of preloading exist and differ from one manufacturer to another.

11.2.6 Errors in Motion Control and Control Software

Motion control errors can easily be studied by modelling a translation stage for which position is controlled by applying a force to a mass on a spring. The mass m is the mass of the slide and the spring equals the virtual control stiffness k_c (Rankers 1997). More detailed discussions of controller design and methods for error calculation are presented in Chapter 14. If the control stiffness is known, the control frequency bandwidth f_c is then calculated by

$$f_c = \frac{1}{2\pi} \sqrt{\frac{k_c}{m}}. \tag{11.1}$$

Forces F acting on this slide give rise to a displacement error e equal to

$$e = \frac{F}{k_c} = \frac{F}{m \cdot (2\pi f_c)^2}. \tag{11.2}$$

500

Basics of Precision Engineering

Motion control errors are then reduced by increasing the control bandwidth f_c. The control bandwidth is typically limited by the first eigenfrequency of the open-loop transfer function of the system, which mostly corresponds to the first eigenfrequency f_b of the mechanical structure that can be readily determined from the system model. If the position is measured close to the actuator, a so-called collocated configuration of the sensor and actuator is present and, by rule of thumb, the control bandwidth is limited to (Rankers 1997)

$$f_c \leq f_b \tag{11.3}$$

When the actuator and position sensor are located apart and have a mechanical flexibility between them, a non-collocated configuration occurs, which drastically limits the control bandwidth to (Rankers 1997)

$$A \cdot f_c \leq f_b, \tag{11.4}$$

with A typically between five and ten, corresponding to a dramatic reduction in the control bandwidth. To keep the control errors to a minimum, the following guidelines can be followed (Rankers 1997).

- Keep measurement and actuation close to each other. Be aware that this conflicts with measuring as close as possible to the end effectors, such as the tool and the workpiece. A trade-off is always needed.
- Always drive the slide as close as possible to its centre of mass (COM; see Chapter 7). An offset invokes a rocking motion and is similar to non-collocated control behaviour. Alternatively, the position sensor can also be placed as close as possible to the COM.
- If the above is not possible, always keep the sensor and actuator on the same side of the COM.
- If these measures are already considered, design the motion system for highest stiffness and eigenfrequency. Unwanted resonance peaks should include sufficient damping.

The need for damping is also apparent by considering a simple spring–mass–damper model. If a simple integrator is used, then the 180° phase shift is at the undamped resonance with an integral control (90° for integrator, 90° at resonance), hence the integral gain is directly limited by the damping ratio of the first eigenfrequency.

It is important to reiterate the points made in Chapter 1 that the field of controls typically focuses on the optimal design of control strategies based on the assumption that the 'plant' is a fixed system that cannot be altered. When designing instruments and machines, the parameters of the plant can often be varied thereby providing considerable flexibility to alter all models of the system.

11.3 Metrology Loops

The metrology loop is a structural loop of all elements from end effector to workpiece for which dimensional changes would not be detected by the measurement process.

For example, if a bearing element is part of the metrology loop and the bearing deflects, this would result in a measurement error. On the other hand, if deflection of this bearing is measured by a displacement sensor, deflection of the bearing is detected by the measurement process and hence the bearing is not part of the metrology loop anymore. By definition, the displacement sensor is then part of the metrology loop. A typical metrology loop for a CMS consists of the touch probe, the slides, the bearings and the workpiece.

The dimensional changes of the elements in a metrology loop should be limited. If not, an apparent relative motion between end effector and workpiece occurs. As the metrology loop is a structural loop, the types of errors in a metrology loop are similar to those in a structural loop (see Section 11.2) and will not be discussed again in this section.

An example of a metrology loop is shown in Figure 11.3. Here, a laser interferometer is used to measure the error motion of a milling machine in the x-direction during milling. Mirror unit 1 is connected to the tool spindle, while mirror unit 2 is connected to the workpiece table. The laser interferometer measures the x-displacement between both mirror units. Figure 11.3 indicates the structural loop, the metrology loop of the milling machine based on the integrated encoder and the metrology loop using the laser interferometer. The structural loop runs through the tool, the spindle, the base frame, the x-slide bearings, the x-slide and the workpiece. Examples of forces through this loop in the x-direction are process forces, x-slide acceleration and friction forces, which produce deflections in all the components of the structural loop. The metrology loop of the encoder has the same components as the structural loop, including the encoder itself. Note that the bearings are also part of the metrology loop and, therefore, a deflection of the bearings results in a measurement error. For simplicity, this has not been indicated in Figure 11.3. The metrology loop of the laser interferometer consists of the tool, tool spindle, mirror unit 1, the laser beam path, mirror unit 2, the x-slide and the workpiece. Measurement errors in the laser interferometer measurement are attributed to tool deflection, tool spindle bearing deflection, thermal deformation of mirror unit 1, changes of refractive index in the beam path (see Ellis 2014), thermal and dynamic deformation of the mirror unit 2, and deformation of the x-slide and workpiece. Nevertheless, the sum of deviations in the metrology loop is smaller than that of

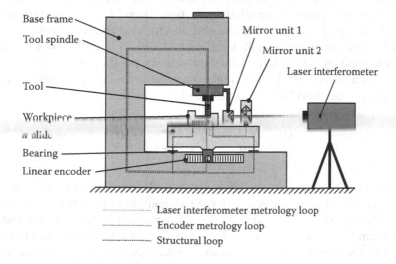

FIGURE 11.3
Schematic representation of a machine tool with a horizontal x-slide, in which position is measured by a linear encoder and a laser interferometer. The structural loops and metrology loops are indicated.

the encoder metrology loop, as the deflections of the base frame and the slide bearings are eliminated. Therefore, separation of the structural loop and metrology loop results in an increase of the position measurement accuracy. Utilising the interferometry measurement in the CNC control of the milling machine could also increase machining accuracy if care is taken to ensure the stability of the environment.

Although the previous example is only applicable in one direction, functional separation can also be applied for three degree of freedom (DOF) systems by using metrology frames. Section 11.5 gives examples of the design of such metrology frames.

11.4 Thermal Loops

11.4.1 Definition

Every component in a mechanical assembly is subjected to heat loads, whether originating from internal heat disturbances such as motors, friction, expansion of gas in air-bearings, manufacturing processes, sensors and electronics, or from external disturbances such as ambient temperature changes or operator heat. As a result of these loads, the temperatures of the components will change by an amount ΔT, which results in a thermal expansion according to

$$\Delta L = L\alpha\Delta T, \tag{11.5}$$

where ΔL is the length change over a distance L and α is the linear coefficient of thermal expansion (CTE). In practice, there will be effects due to differential expansions of constrained mechanisms as well as temperature gradients. Both can result in angular rotations and distortions of shape. The components that are important for the thermal stability of a machine are those that establish the relative position between the end effector, for example the tool, a touch probe or an optical sensor, and the workpiece. A thermal expansion of these components will change their relative position. Therefore, these components are said to belong to the thermal loop between the end effector and the workpiece.

A thermal loop is defined in Schellekens et al. (1998) as 'a path across an assembly of mechanical components, which determines the relative position between specified objects under changing temperatures'. The thermal expansion of the components in a thermal loop changes the relative position between two objects in one direction by

$$\Delta L = \sum_i L_i\alpha_i\Delta T_i - \sum_j L_j\alpha_j\Delta T_j,$$

in which components i cause the two components to drift apart in the considered direction and components j result in an approach of the two components in the considered direction for the temperature change ΔT. Figure 11.4 clarifies Equation 11.6 by an example instrument used for measuring the length of gauge blocks (see Chapter 5). The instrument consists of a workpiece table that holds the gauge block, a touch probe, a z-slide that moves in the vertical direction, a scale indicating the position of that slide and a base frame, again in the z-direction. Assume that the z-slide remains stationary at the position indicated by the

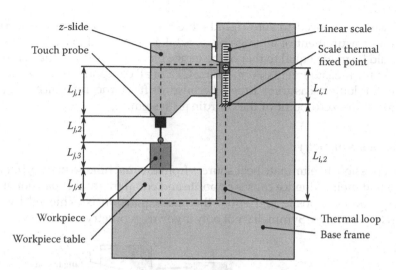

FIGURE 11.4
Example of a thermal loop in an instrument for the measurement of gauge blocks.

linear scale. The components i that cause the workpiece and the touch probe (the end effector) to drift apart are the base frame and the linear scale. The thermal fixed point of the scale is at the bottom of the scale and the scale is allowed to expand freely in the other direction. Expansion of the z-slide, the touch probe, the workpiece table and the workpiece itself will cause the workpiece and touch probe to approach one another (j components).

The heat loads in this example are the heat from the environment, the reading head of the linear scale, the z-slide actuator and the operator. Their heat input should be minimised and the instrument should be designed such that these heat loads have minimum effect. Several design guidelines and principles exist to achieve this, which are elaborated in the next sections.

In the aforementioned example, the force and metrology loops are coincident with the thermal loop. For other designs, each of these three loops may follow different paths. In such cases, the resulting errors due to thermal effects will be mainly confined to distortions of the elements that are also in the measurement loop. Consequently, for many machines, the thermal loop follows the components of the measurement loop.

As a final comment, thermal effects can cause errors in electronic signal conditioning electronics by changing the reference values that are compared to the value of the sensing element. Examples of this would be thermal sensitivity of reference capacitors, Schottky diode voltage references and quartz oscillators for timing references. These reference also used in capacitance, inductance and resistance sensing as well as analogue-to-digital converters.

11.4.2 Design Guidelines and Examples

11.4.2.1 Matching Coefficients of Thermal Expansion

Equation 11.2 shows that for a uniform and equal temperature change ΔT in all components of the metrology loop, choosing the right $\alpha_{i/j}$ and $L_{i/j}$, results in $\Delta L = 0$. For instance, for the instrument shown in Figure 11.4, this could mean that for the measurement of steel gauge blocks, all components in the metrology loop should have a CTE equal to that of steel. If the system is always referenced to zero right before the measurement at the workpiece table surface, and a measurement of a gauge block is taken shortly afterwards, only the CTE of the scale should be made equal to that of the gauge block. The drift of the zero-point is then cancelled from the measurements. A matching CTE of all components in the metrology loop is important for long measuring times, because drift of the reference point causes an apparent shift of the zero-point of the coordinate system.

11.4.2.2 Thermal Symmetry

Often it is not possible to eliminate heat sources from the machine structure, but it is possible to make sure that their influence causes opposite and equal thermal expansion in all paths of the metrology loop. Opposite and equal thermal expansion is achieved by introducing symmetry in to the design. Symmetry not only involves geometry but also symmetry in heat

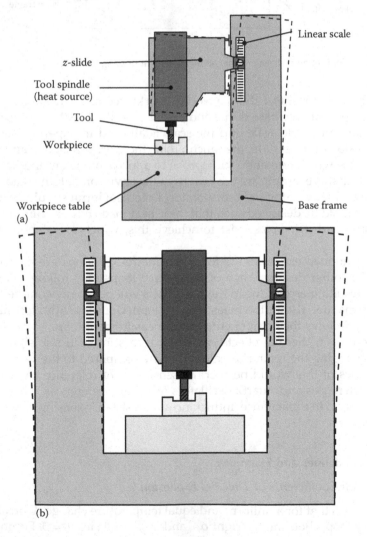

FIGURE 11.5
Schematic representation of a milling machine with and without thermal symmetry. (a) Milling machine without thermal symmetry. (b) Milling machine with thermal symmetry.

transfer, thermal capacity and CTEs. Figure 11.5a shows a milling machine in which the spindle acts as a heat source that thermally deforms the base frame. The tool drifts away from the workpiece, causing a machining error. Machining errors by thermal deformation of the base frame are reduced if the base frame and z-slide are constructed in a symmetric way, as shown in Figure 11.5b.

11.4.2.3 Thermal Centre Principle

By mounting components in a kinematic way (see Chapter 6), and by correctly choosing the directions of the constraints, two components can be made thermally invariant at one point, which is called the thermal centre. In the thermal centre, for a uniform thermal expansion of the two considered components, they will experience no movement with respect to each other. Figure 11.6a depicts an example of a kinematic mount that illustrates the thermal centre principle. Again, the example of the gauge block measurement system is taken, but the workpiece table is now mounted in a kinematic way to the base frame using three ball-in-V connections (see Chapter 6). The centre lines of the ball-in-V connections all coincide at the base of the workpiece table, such that a uniform thermal expansion of the workpiece

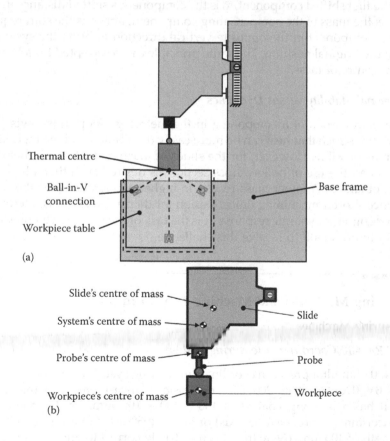

FIGURE 11.6
(a) Measuring instrument in which the workpiece table is kinematically constrained to the base frame using ball-in-V connections. In this way, a thermal centre is created at the intersection of the planes normal to the constraint directions. The workpiece table and the base frame will not displace in the thermal centre for uniform temperature changes. (b) Demonstrates the centre of mass principle for a metrology system that consists of a slide, a probe and a workpiece.

table, or a uniform thermal expansion of the base frame, does not result in a shift of the base of the workpiece table with respect to the base frame.

The general design rule for kinematic connections to create a thermal centre is that all planes normal to the direction of the kinematic constraints should intersect at the thermal centre. By using a thermal centre, zero expansion at one point between two components can be obtained without the use of low-CTE materials. In cases where all of the normal planes of the kinematic constraints do not have a common intersection point, the components would experience a rotation.

Please note that the thermal centre is different than the centre of mass. The latter represents the mean position of matter in a system. Figure 11.6b shows an example of the centre of mass of a metrology system that consists of a slide, a probe and a workpiece, and how it differs from the centre of mass of the individual system components.

A shift in the vertical direction of any of the system's components due to thermal expansion will result in a vertical shift of the system's centre of mass in the same direction. One way to bring the system's centre of mass back to its original point will be to shift another system component a total distance of $x_j = \dfrac{m_i}{m_j} x_i$ in the opposite direction, where m_i is the mass of the first shifted component, x_i is the component's shifted distance in one vertical direction, m_j is the mass of the compensating component, and x_j is the shift required by the compensating component in the opposite vertical direction to bring the systems centre of mass back to the original position. The same principle can be adopted for a horizontal shift in the system centre of mass.

11.4.2.4 Thermal Stability versus Dynamics

To minimise deformation of a component in the metrology loop, it is advisable to apply solid cross-sections such that heat can be more easily distributed and the thermal expansion is as uniform as possible. However, for the slides of machine tools and other dynamically moving machines, the use of hollow cross-sections is desired such that a high stiffness-to-mass ratio is obtained. A lower mass results in higher accelerations for the same applied force and hence shorter machining times. Design for thermal stability is, therefore, often in conflict with design for dynamic response. It is the task of the precision engineer to come up with an optimum trade-off to resolve this challenge.

11.5 Measuring Machines and Machine Tool Examples

11.5.1 Measuring Machines

11.5.1.1 ISARA 400 Coordinate Measuring System

The ISARA 400 is an ultra-precision coordinate measuring system developed by IBS Precision Engineering BV (Donker et al. 2009) for dimensional measurement of micro-components and optics. It has a working volume of 400 × 400 × 100 mm. IBS PE quotes a 1D measurement uncertainty for one axis (x-axis) of 52 nm (95% confidence interval) over a measurement length of 400 mm. (Note that it is not strictly correct to quote an uncertainty with a measuring instrument; normally a maximum permissible error would be quoted. See Chapters 5 and 9).

Figure 11.7 shows the concept of the ISARA 400. The touch probe is mounted on a metrology frame that also holds three interferometers for measurement of the x-, y- and

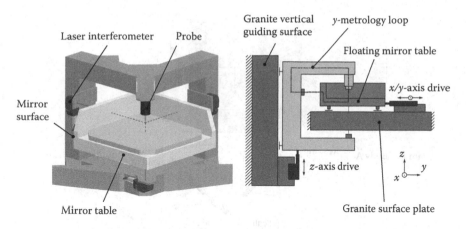

FIGURE 11.7
Concept of the ISARA CMS. (Courtesy of IBS Precision Engineering BV.)

z-positions respectively. The three laser beams of the interferometers intersect at the centre of the touch probe, in this way minimising the effect of Abbe offset (see Chapter 10). The metrology frame is actuated in the vertical direction by the z-axis drive and is supported by a vertical granite surface. The workpiece is positioned on a mirror table that holds the three mirrors that act as a reference for the laser interferometers. The mirror table is actuated in the x- and y-directions and is supported by a horizontal granite surface plate.

The metrology loop of this CMS consists of the touch probe, metrology frame, laser interferometer, mirror table and the workpiece. The position of the floating mirror table is directly measured with respect to the metrology frame by the laser interferometers. Therefore, the bearings are not part of the metrology loop and introduce no measurement errors due to error motions.

11.5.1.2 Vermeulen CMS

This example, which was developed by Vermeulen (Vermeulen et al. 1998) and later commercialised by Zeiss under the name F25, makes use of three linear encoders of which the x- and y-encoder are configured such that Abbe offsets correspond only to the vertical distance of the probe tip from a horizontal mid-plane of the measurement volume. Figure 11.8 depicts the layout of the machine and the configuration of the x- and y-encoders. The probe is attached to a beam Z which moves in the vertical direction through the body PL. PL holds the x- and y-scale, while the reading heads Mx and My are attached to intermediate bodies A and B. The latter can move along their respective guiding beams I and II, simultaneously guide the body PL in the x- and y-direction and make sure the reading heads move along with their respective scales. This configuration ensures that the measurement probe is always in line with the scales when the probe is in the mid-plane of the measurement volume but is not compliant with the Abbe principle over the whole work volume. When measuring out of this mid-plane, the maximum Abbe offset is 50 mm for the $100 \times 100 \times 100$ mm measurement volume. For the commercialised version of this machine by Zeiss, the maximum permissible error (MPE) is quoted as $0.25 + (L/666)$ µm (L is the measuring range in millimetres). The machine is designed such that thermal expansion of all components in the metrology loop are matched, making Equation 11.6 equal to zero. However, the metrology loop and the structural loop are not completely separated, as motion errors of bodies A and B at the location of the reading heads still results in a measurement error.

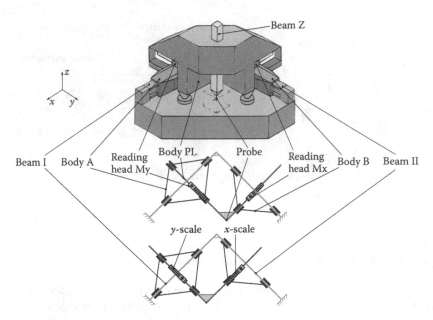

FIGURE 11.8
Layout of the CMS developed by Vermeulen et al. (1999). (Courtesy of TU Eindhoven).

11.5.1.3 Leuven mAFM

At KU Leuven, a metrological atomic force microscope (mAFM) was developed which has a configuration of laser interferometers similar to the ISARA 400 concept. It has a working volume of 100 μm × 100 μm × 100 μm and can achieve a single axis displacement measurement uncertainty of around 1 nm for a displacement of 100 μm (95% confidence interval) (Piot et al. 2013).

The concept and the layout of the mAFM are shown in Figure 11.9. The laser interferometers are configured orthogonally and intersect at the tip of the AFM probe, in this way being compliant with the Abbe principle. The sample is mounted in a sample holder that contains the target mirrors for the laser interferometers. The sample holder is actuated by a fine positioning unit. The metrology loop consists of the AFM probe, the metrology frame, the laser interferometer, the target mirrors, the sample holder and the sample. The metrology frame and the sample holder are made of Invar. The positioning units are excluded from the metrology loop by the functional separation principle, which explains how a 1 nm measurement uncertainty over a length of 100 μm can be attained in an environment controlled to ±0.1 °C. Such a low measurement uncertainty would be very difficult to attain when relying on the straightness of the guides.

11.5.2 Machine Tools

11.5.2.1 Ultra-Precision Five-Axis Grinding Machine

Figure 11.10 shows a machine for the grinding of freeform optics with a maximum peak-to-valley form deviation of 0.3 μm (developed by Hemschoote 2008). Figure 11.11 shows the same machine with the master metrology frame (MMF) removed and indicates the direction of the axes, while Figure 11.12 only shows the metrology loop components of the machine. The tool spindle, holding a spherical grinding wheel, resides on a yoke swivelling about

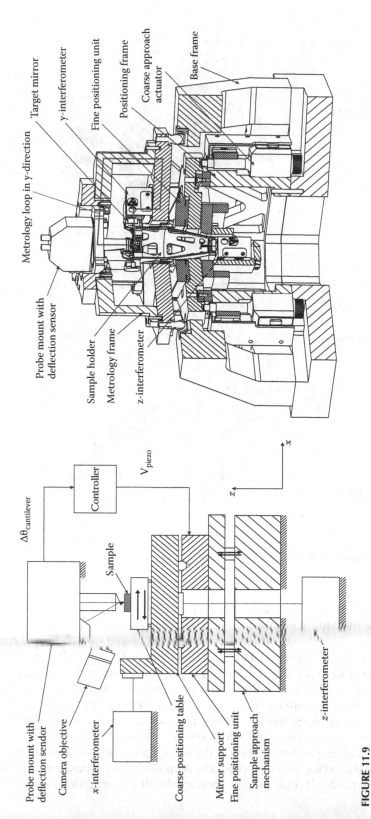

FIGURE 11.9

Conceptual design and mechanical layout of the mAFM developed at KU Leuven. (From Piot, J. et al., 2013, Design of a sample approach mechanism for a metrological atomic force microscope, *Measurement* 46 (1): 739–46).

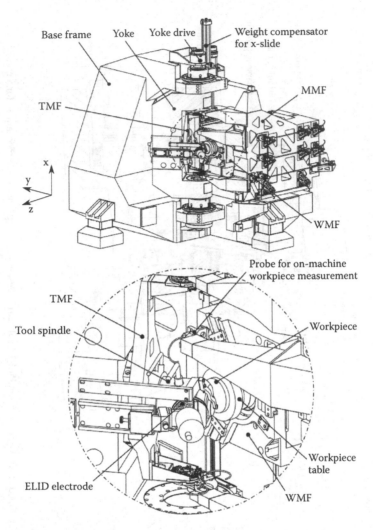

FIGURE 11.10
Layout of the five-axis grinding machine with metrology frames.

the x-axis. The workpiece is located on a workpiece spindle which has an axis of rotation parallel to the z-axis. This workpiece spindle is connected to a three-DOF slide system that translates in x-, y- and z-directions over distances of 395 mm, 225 mm and 107 mm respectively, as shown in Figure 11.11. A tool metrology frame (TMF) made from Invar is connected to the yoke axis and monitored by a rotary encoder and five capacitive sensors that are connected to the Invar MMF. This provides simultaneously measurement of the rotation and the error motions of the yoke. The workpiece spindle is surrounded by a Zerodur workpiece metrology frame (WMF), the position of which is measured by seven laser interferometers connected to the MMF. Five of the interferometers are used simultaneously during machining and measuring of the workpiece. The two laser interferometers in the x-direction are shared for both purposes. The on-machine workpiece measurement probe is also connected to the MMF, but could be connected to the TMF. Using this configuration of

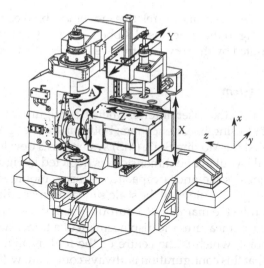

FIGURE 11.11
Layout of the five-axis grinding machine without metrology frames.

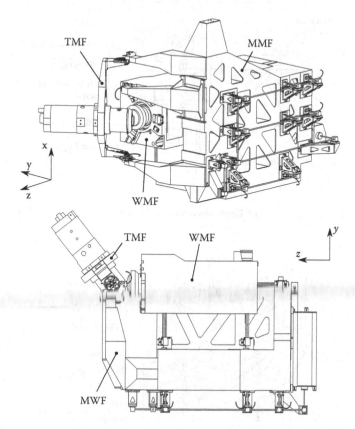

FIGURE 11.12
The metrology frames of the five-axis grinding machine.

metrology frames and position sensors, the relative movement between the tool and workpiece can be measured according to the generalised Abbe principle (see Chapter 10) and all error motions can be compensated by the servo control system (see Chapter 14).

11.5.2.2 Moving Scale System

The same configuration as the laser interferometer configuration explained in Section 11.5.1.1 can be thought of using linear encoders; both can be thought of as periodic scales. This type of system, called the Moving Scale (MS) system, was developed by Bosmans et al. (2016) and its conceptual layout in two dimensions is depicted in Figure 11.13. Each encoder module consists of a linear scale and a capacitive sensor, mounted in line with the end effector on an interface which is guided in the scale's measurement direction and driven by a linear motor that is controlled to maintain the output signal of the capacitive sensor at a null value. The capacitive sensor measures the displacement of a target surface on the workpiece table. The functional point, which is the centre of the tool, is aligned with the scale and capacitive sensor such that this configuration is always compliant with the Abbe principle. A three-DOF layout in a machine tool is depicted in Figure 11.14. The structural loop in this

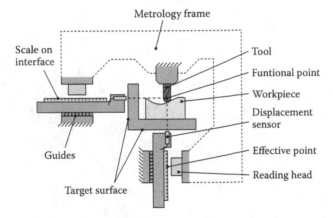

FIGURE 11.13
Conceptual layout of a configuration of linear encoders with metrology frames and Abbe-compliant linear encoders.

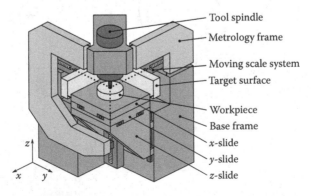

FIGURE 11.14
Conceptual layout of the Moving Scale system and metrology frames on a three-axis machine tool.

machine tool consists of the tool; the tool spindle; the base frame; the x-, y- and z-slides, including the bearings; and the workpiece. The metrology loop consists of the tool, the tool spindle, the metrology frame, the MS system, the target surface, the x-slide and the workpiece. The base frame and the stack of x-, y- and z-slides are not included in the metrology loop. This configuration increases the accuracy of the machine tool, as process forces only go through a small part of the metrology loop: the tool, the tool spindle, the workpiece and part of the x-slide.

Exercises

1. Figure 11.15 gives an example of an instrument for measuring the stress–strain relationship of a material sample. The sample is clamped at both ends, with one end attached to the base frame and one to a vertically movable slide. The slide is actuated by a linear motor and guided by a linear bearing. The displacement of the slide is measured with a linear encoder on the base frame, which is used to calculate the strain of the sample. A force sensor below the sample measures the applied force. What is an important source of displacement measurement error in this configuration of the instrument? Draw a conceptual layout of an improved version of this instrument with separation of the structural loop and the metrology loop. Assume that the force sensor is infinitely stiff.

2. Figure 11.17 illustrates a milling machine. The slide moves in the horizontal direction and its position is measured 120 mm below the tool centre point. The slide is actuated by a force $F = 50$ N which is applied 100 mm below the position measurement and counteracts the process forces in the x-direction at the tool. The slide is guided by two bearing surfaces, 320 mm apart, with a vertical stiffness of 200 N·μm^{-1}. Calculate the measurement error by the position measurement of the tool centre point that is caused by the process force.

3. The measuring instrument in Figure 11.4 is dedicated to the measurement of AISI steel 430F (CTE = 10.5×10^{-6} K^{-1}) and aluminium AISI EN-AW-5083 (CTE = 24.5×10^{-6} K^{-1}). Which of the following materials is most appropriate for the scale of the linear encoder, assuming the scale is mounted kinematically and is not influenced by thermal expansion of the base frame. The scale is also calibrated at 20 °C.

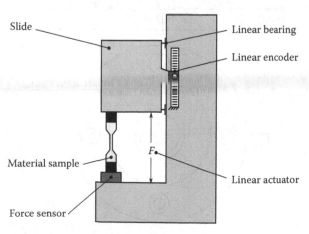

FIGURE 11.15
Instrument for measuring the stress–strain relationship of a sample.

Material	CTE [× 10⁻⁶ K⁻¹]
Zerodur	0
Glass	8
Stainless steel	10.6

4. Figure 11.18 shows a vacuum workpiece chuck. Design it in such a way that the bottom face, which resides on the workpiece table, does not expand with respect to the top face, which contacts the workpiece. It can be constructed from separate components and it is only allowed to make use of materials 1 and 2. The CTE of material 1 is three times that of material 2.

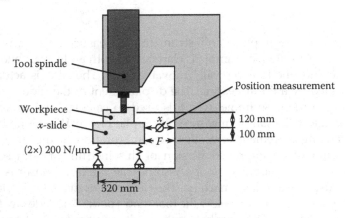

FIGURE 11.17
A milling machine with a slide moving in horizontal direction.

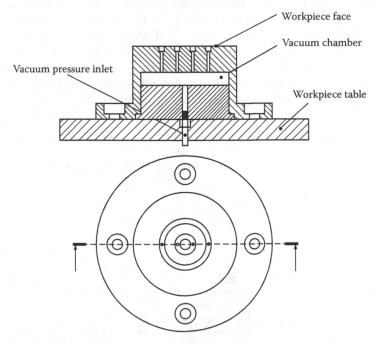

FIGURE 11.18
Vacuum workpiece chuck.

5. Figure 11.20 shows a wafer inspection system. An aluminium stage that holds the wafer moves in two linear directions to scan the surface of the wafer with an inspection camera. The wafer is connected to the stage in a kinematic way with a thermal centre in the centre of the wafer. The position of the stage is measured using two laser interferometers, directed towards the centre of the inspection camera lens. Two Zerodur mirrors serve as targets for the laser interferometer beams. These mirrors should be kinematically mounted such that a thermal expansion of the aluminium stage does not displace the mirrors with respect to the centre of the wafer. Draw the kinematic connections that achieve this.

6. Figure 11.22 shows a platform that is kinematically constrained by three flexures in the direction of the length L, (the z-direction is not considered for this example). The platform is made of aluminium (CTE = α_{Al}) and the flexures are made of Invar (CTE = α_I). Due to temperature change ΔT, the platform and the flexures will expand. The expansion of the flexures causes a shift of the theoretical thermal centre that is determined by the configuration of the flexures. Moreover, the expansion of the horizontal flexure results in a small rotation of the stage. Determine the theoretical location of the thermal centre, the shift of the thermal centre and the rotation of the platform due to the expansion of the flexures. Assume that the flexures are infinitely stiff in the constraint direction and have zero stiffness in the perpendicular direction.

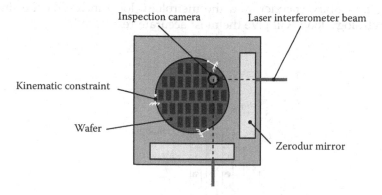

FIGURE 11.20
Wafer inspection system.

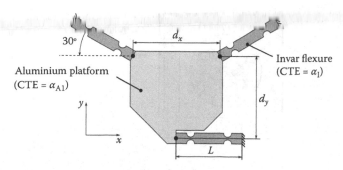

FIGURE 11.22
A platform, supported by flexures.

7. In clocks, the time scale depends on the oscillation frequency of the pendulum. This oscillation frequency is determined by the distance between the pivot point and the centre of mass of the pendulum. However, thermal expansion causes this distance to change, which results in a change of the time scale and an error in the time indication. This error can be eliminated by a special configuration of the pendulum and using multiple materials, which is shown in Figure 11.24. The pendulum consists of a rod (length L_R, mass m_R and CTE α_R) and a support (length L_S, mass m_S and CTE α_S) attached to the rod. The support holds two containers with mercury. The mass of the containers can be ignored, while the mercury volumes both have a height L_m, mass $\frac{m_m}{2}$ and CTE α_m. Determine the unexpanded height $L_{m,0}$ such that a uniform thermal expansion of the pendulum will not result in a change of the distance between the centre of mass of the pendulum and the pivot point.

8. The wafer inspection system of Exercise 5 is now equipped with two integrated aluminium mirrors instead of Zerodur mirrors. Determine the orientation of the three constraints that hold the wafer such that there is no thermal drift of the target surfaces with respect to the wafer, assuming that the wafer itself is not expanding. The position of the constraints is indicated with three dots on Figure 11.25.

9. A simple representation of a machine tool is shown in Figure 11.27. It consists of one slide that holds a tool spindle and is driven by a rotary motor through a lead screw. The position of the slide is measured by a rotary encoder on the motor and a linear encoder. Draw the structural loop that carries the process forces. Draw the metrology loop including the rotary encoder, and the metrology loop including the linear encoder. Which metrology loop will give the most accurate result? Why?

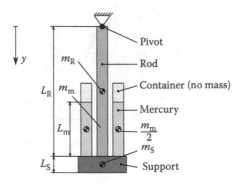

FIGURE 11.24
Conceptual example of a pendulum with thermal expansion compensation.

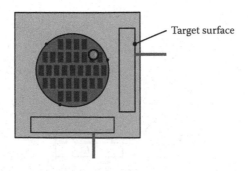

FIGURE 11.25
Wafer inspection system of Exercise 5 with aluminium target surfaces.

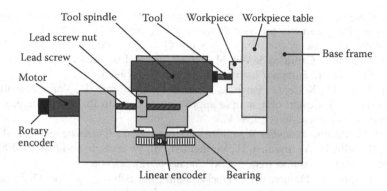

FIGURE 11.27
Conceptual layout of a 1-DOF machine tool with lead screw drive.

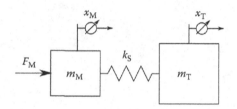

FIGURE 11.28
Simplified lumped-mass model of the 1-DOF machine tool of Exercise 9.

10. A simplified lumped mass model of the machine tool of Exercise 9 is shown in Figure 11.28. The model consists of the equivalent mass of the motor and lead screw m_M, a spring representing the stiffness of the lead screw drive train k_S and a mass that represents the slide and the tool spindle m_T. The motor force F_M is applied on the motor mass. The displacements of motor and tool masses are x_M and x_T respectively. Determine the transfer functions $\dfrac{x_M}{F_M}$ and $\dfrac{x_T}{F_M}$ and estimate the maximum theoretical control loop bandwidth for both systems.

References

ASME. 2005. Methods for performance evaluation of computer numerically controlled machining centers (B5.54 - 2005).

Bosmans, N., Qian, J. and Reynaerts, D. 2016. Design and experimental validation of an ultra-precision Abbe-compliant linear encoder-based position measurement system. *Precision Engineering* 47:197–211.

Donker, R., Widdershoven, I. and Spaan, H. 2009. Realization of Isara 400: A large measurement volume ultra-precision CMS. Asian Symposium for Precision Engineering and Nanotechnology.

Ellis, J. 2014. *Field guide to displacement measuring interferometry*. SPIE.

Hemschoote, D. 2008. *Ultra-precision five-axis ELID grinding machine: Design and prototype development*. KU Leuven.

Kroll, E. 2013. Design theory and conceptual design: Contrasting functional decomposition and morphology with parameter analysis. *Research in Engineering Design* **24** (2): 165–83.

Nakazawa, H. 1994. *Principles of Precision Engineering*. Oxford University Press, New York

Olsson, H., K. J. Åström, C. Canudas de Wit, M. Gäfvert, and P. Lischinsky. 1998. Friction models and friction compensation. *European Journal of Control* **4** (3): 176–95.

Piot, J., Qian, J., Pirée, H., Kotte, G., Pétry, J., Kruth, J.-P., Vanherck, P., Van Haesendonck, C. and Reynaerts, D. 2013. Design of a sample approach mechanism for a metrological atomic force microscope. *Measurement* **46** (1): 739–46.

Rankers, AM. 1997. *Machine dynamics in mechatronic systems: An engineering approach*. TU Delft.

Schellekens, P., Rosielle, N., Vermeulen, H., Vermeulen, M., Wetzels, S. and Pril, W. 1998. Design for precision: Current status and trends. *CIRP Annals* **47** (2): 557–86.

Schwenke, H., Knapp, W., Haitjema, H., Weckenmann, A., Schmitt, R. and Delbressine, F. 2008. Geometric error measurement and compensation of machines-an update. *CIRP Annals* **57** (2): 660–75.

Vermeulen, M., Rosielle, N. and Schellekens, J. 1998. Design of a high-precision 3D-coordinate measuring machine. *CIRP Annals* **47** (1): 447–50.

12

Materials Selection in Precision Mechanics

Derek G. Chetwynd

CONTENTS

ABSTRACT Materials science and technology is a major field of study, still rapidly developing, and clearly a book such as this one cannot and should not try to address it in great breadth or depth. Yet, it is not possible to design and build any physical structure without influence from its materials' properties. Choice of materials can be a major factor in the success or otherwise of design ideas for precision engineering systems. The approach will be to provide a brief and general exploration of how options for materials choices affect high-precision mechanical systems. Properties that particularly influence ultra-precision devices, miniature mechanisms and so on are considered, noting that they sometimes differ from those governing other, less demanding applications. Graphically oriented methods for data visualisation that help to systematise materials selection processes are used to explore classes of materials that might satisfy the constraints presented by typical precision engineering applications. The higher value associated with precision enables speculation about the relative merits of a design if access to more esoteric materials became economic. Readers of this chapter should gain an insight into the options available to designers and why or when they might be effective choices, sufficient to enable effective discussions of the issues with materials specialists. In other words, the chapter offers not a 'scholarly review' of ultimate performance or limiting applications, but rather seeks to provide guidance of where to start the selection process.

12.1 Drivers for Materials Choice in Precision Engineering

12.1.1 Introductory Comments

A brief glance at a materials selection database or a supplier's catalogue will confirm that there is a vast number of choices on offer to mechanical designers. A closer look shows that much of this commercially oriented choice still comprises subtle variations of just a few key material families such as steels and aluminium alloys. Most elements, even most metals, do not appear in their own right. The large number of alloys and processing variables (hardness states, hot or cold work, thermal cycling, etc.) reflects a highly successful evolution and diversification from the few relatively common ores from which technologically useful metals could be extracted at reasonable cost. First to emerge were copper, tin and lead in the Bronze Age, then the Iron Age, and then the efficient smelting of aluminium in the nineteenth century. Tellingly, these metals can form alloys that offer enhanced behaviour for specific applications, and modern materials scientists and technologists have learned how small additions can selectively and favourably alter one property of a material, while leaving others much as previously used. There are cost associated with such changes, but there are also commercial incentives to research and supply them. To take just one example, a slight change in the composition of a steel alloy might leave strength, stiffness, thermal expansion and so on essentially unchanged, but increase long-term dimensional stability, while reducing corrosion resistance; this could be an attractive advantage for some precision engineering applications, but a distinct disadvantage in other fields.

This chapter will not consider either the design of, or choice between, alloys (or the equivalent subtleties in, say, ceramics) at this level. It would take much more space than is available here. Moreover, there are positive reasons for taking a generic, high-level view of the topic. First, the study of the best alloy for a new design only becomes relevant once a decision is taken that 'steel', say, is generically the right choice for the particular application. Second, precision engineering brings its own needs and constraints. Physically small devices or precision machines that are made in small numbers will tend to be less sensitive to the higher costs of less common materials, which might, singly or in combination, offer unusually high performance in one application-critical aspect. Hence, this chapter explores the background to questions about when, and why, different types of materials might be used. A typical example is posed in the caption of Figure 12.1, which shows two instruments of quite different designs, both of which move a lightly contacting small stylus across an almost flat surface in order to measure the small height differences of features on it (Chapter 5 presents such stylus instruments in some detail); this specific question will be further examined in Section 12.4.5. Having set out some general requirements for the materials of structural mechanical components, this chapter will briefly visit a few issues in materials technology and design to provide insight into how new materials might be devised and why existing ones do not always behave as expected. Mostly, the chapter concentrates on properties that are more specifically relevant in precision engineering applications, while also introducing tools to aid comparisons between candidate materials when exploring the strengths and weaknesses from a wide range of materials choices.

FIGURE 12.1
Views of two late-twentieth century nanometre-sensitivity instruments for stylus-contact measurements of step heights on 'flat' surfaces. One is apparently made mostly of aluminium, the other (shown with outer polymeric covers removed) largely of a glass-like material. Why might designers have made such different choices of materials, and which, if either, is the better choice?

12.1.2 Top-Level Requirements

The major goal for this chapter is to address the selection of materials to facilitate the integrity and necessary performance of the mechanical structures within high-precision electro-mechanical systems. The force loops and metrology loops (see Chapter 11) within high-precision electro-mechanical systems provide the framework for this task. In essence, the loops must support the loads imposed upon them without any notable changes to their definitive geometric parameters, both in the short term and over the design lifetime. The greater challenges in precision systems will commonly lie more with maintaining the shape and size of the loops (stiffness) than with their load-bearing capacity (stress), while vibration sensitivity might often prove to be the main dynamic constraint. Selection criteria will then tend to focus, although not exclusively, on some readily measured elastic and thermal properties that will be explored later. Care should be taken not to place too much emphasis on such parameters just because their well-defined numerical values make comparisons much easier. In practice, selections will be constrained by other top-level requirements that are not so easily quantified.

Cost is always a factor in design, but the materials aspect is rarely simply one of euros or dollars per kilogram. The real question is how the choice of material will affect the overall cost of manufacturing a complete device or, even better, the whole-life cost of manufacture, ownership and disposal. A qualitative view of manufacturability can be readily obtained; for example, any skilled lathe operator will agree that it is 'easier' to cut aluminium, brass or mild steel than it is to cut stainless steel or titanium. While a numerical value might be associated with this concept for comparison purposes, closer consideration reveals that its definition involves further subjective (qualitative) judgments. Not least, the cost to a particular organisation could well depend on what machine tools and other production equipment they happen to own. Ease of recycling means that some materials have higher inherent scrap value than others, which can affect the overall manufacturing costs for designs that require a lot of material removal, as well as potentially affecting the cost of ownership. Another closely related viewpoint is that of the significant cost of change. When the task is one of reviewing an existing design for upgrade to a new model, it might well prove more effective to retain the original material even though an alternative, 'better', one is available, if the latter offers only a small improvement in performance. All selection criteria should, therefore, be judged in terms of whether a big enough difference can be achieved. Note, however, that there could also be a considerable market disadvantage associated with not changing the material when other competing designs have done so. Commercial success might even depend on customer desire, possibly driven by social and other market forces, for a material that is technically inferior. Thus, while technical performance must always be considered rigorously, it must also be recognised that it might not, in practice, always dominate the final design decisions taken.

Many designs will involve consideration of materials compatibility. Most directly, this could include constraints imposed by external regulators, where one obvious example is the need for bio-compatibility in medical devices. The possible performance benefits from a material not on a list of certified options will often be totally negated by the high costs of proving that the new choice is safe to use in the application. Many industry sectors have either formal regulations or widely held conventions about acceptable materials that must be given high weighting in the final decisions; even unwarranted prejudice against a novel use of material might mean that an otherwise good design becomes a commercial failure. Electro-chemical compatibility will be a factor at any interface between materials. Although high-precision systems are likely to be located in relatively benign environments, there could still be cases where, say, two metals create an electrical cell (via galvanic action) leading to slight but steady corrosion and degradation at the interface, with consequent effects on the overall loop stability. Corrosion resistance in a more general sense might sometimes be a significant factor. Other compatibility factors might arise in particular applications. There are no simple, consistent rules here, with there being no substitute for experience in the application field and testing under service conditions.

A wide range of materials will be specified based on exceptional properties that they offer to the final application, even though other selection criteria are relatively unfavourable. Examples range from gold for certain types of mirrors to super-hard 'diamond-like' tribological surfaces. Very often materials will be used in thin coatings, with the underlying loop integrity given to more conventional choices. The additional constraint on those choices is then of compatibility (mechanical, chemical, etc.) for achieving a high-quality, stable coating. On a few occasions, the overall function of a machine might be enhanced by restricting choices for its structural loops to electrically conducting or insulating materials (perhaps when electrostatics are being exploited or for general safety) or to non-magnetic materials. Commercial aesthetic considerations might sometimes become a significant secondary factor; for example, a natural granite base might be used on an instrument, even if it offers no specific advantage and incurs modest cost penalties, because potential users will associate it with high-quality machines. Occasionally, the usual guidelines for selection might be countermanded by specific features of an application; for example, it might generally be preferred that metrology loops have low thermal expansion, but a more efficient design for a measuring instrument targeted solely at high expansivity specimens might involve matching expansion coefficients and thermal time constants. The assumption for this chapter will be that such broad and varied constraints associated with special cases can be easily incorporated alongside other selection procedures when the specific application requires them; consequently, they are rarely treated explicitly in the following discussions.

It is, naturally, only properly valid to compare property values that have been measured under comparable conditions. A considerable number of national and international standard tests have evolved to meet this requirement, and published databases will typically draw on a few of them. There is, though, a potential trap here for precision engineers. Testing protocols will specify specimen sizes that are convenient to handle but large enough that manufacturing variations in the shape and texture of their surfaces have limited influence and that there are statistically consistent averages accounting for the internal variations of microstructure. Objects that are considerably smaller than these specimens will not necessarily exhibit the same behaviour. A long-known example is the high tensile strength of glass fibres. The drawing process ensures that extant fibre will have exceptionally low numbers of significant internal and surface defects that would locally raise stresses and initiate crack propagation and so the fibre more nearly approaches the theoretical limits toward the ideal strength of the material. Figure 12.2 illustrates how glass fibre strength increases as diameter is reduced, using, for historic interest, data from Griffiths' (1920) original paper on his theory of fracture. This behaviour is widespread and is also seen in ductile materials, such as thin metallic wires. Other materials might have quite large minimum sizes for features of their microstructures resulting in defects being an increasingly important factor within small cross sections and small components performing less well than expected from test data on larger ones. The apparently simple expedient of running materials tests on suitably small specimens falls foul of the high costs of doing so at the requisite scale and also of the lower statistical reliability implied as the manufacturing tolerances become larger compared to the overall dimensions. In the case of micro-electromechanical systems (MEMS), attempts to estimate, say, the strength or stiffness of a material from the actual behaviour of real devices, often give numbers very different to normally accepted bulk values; it is sometimes difficult to say whether this is a real materials scaling effect or merely a reflection of low accuracy in measuring dimensions and other parameters inherent to the estimation. It is intuitively obvious that materials are likely to behave differently at nanometre scales and it ultimately makes little sense to attribute most conventional materials properties to a single atom or molecule. Clusters with dimensions of tens of nanometres exhibit technologically important differences from both smaller and larger objects, but their implications are not so relevant to the dominantly mechanical theme of this chapter. It is, therefore, especially important when considering test data measured at a scale different to the intended use or measured from very small specimens that designers

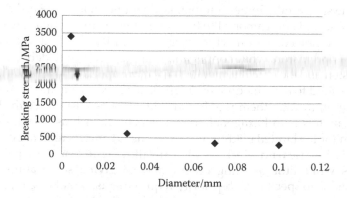

FIGURE 12.2
The variation of tensile breaking strength of glass fibres as their diameter changes, taking illustrative data from the original work by Griffiths (1920).

adopt a critical attitude towards all quoted property values based on their phenomeno-
logical understanding of why, and when, such values might be unrepresentative.

Despite all the qualitative factors, such as those outlined earlier, it is still common to find
that optimising materials choices (perhaps from an already restricted range of candidates)
will most often rely on numerically expressed mechanical and thermal properties. Hence,
this chapter will concentrate quite heavily on them, but it is, therefore, important to
emphasise that its examples may well not tell the complete story; on the other hand, the case
that none of the 'good' choices indicated by the quantitative properties is acceptable against
the qualitative factors suggests a more fundamental problem with the original design
concept. First, a brief review will discuss some background about the structure of solid
materials, intended as a guide toward how and where a desirable combination of properties
might be found and even to how specialists might start to think about designing a new
material for a new application.

12.2 Design of Materials: A Brief Overview

12.2.1 Major Classes of Materials

Solids relevant to structural loops can be broadly grouped into four families, recognising
that composites are also used that draw from more than one family: metals, ceramics,
glasses and polymers. The fundamental behaviour in each family relates to the position of
their elements in the periodic table (their chemistry). Atoms form into crystals, and in some
cases they can be a physically stable mix of more than one element (as opposed to a
chemical compound). Small crystals bind together to produce the larger masses that
provide much of our common experience of solids.

Metallic elements are in the majority and appear in the lower-numbered groups of the
periodic table. The outer-shell electrons are relatively loosely bound to their nucleus, and
within crystals they are readily shared across the material (sometimes called an 'electron
gas', 'electron sea' or even 'jellium'). This facilitates the transport of charge and kinetic
energy, making metals generally good conductors of electricity and heat. Metallic elements
are chemically reactive, some violently so, and only a few of the less reactive ones are
mechanically (structurally) important materials; others appear, often in small proportions,
within alloys. The structurally important metals are found mainly toward the central region
of the periodic table and often among the transition metals. Iron remains generally the most
important, often combined with its close neighbours: nickel, cobalt and chromium. Aluminium ranks second to iron, while copper and titanium are also widely used. Some others,
from tin to tungsten and the precious metals, have more specialised application. Structurally
important metals tend to have moderately high elastic moduli, be fairly strong and show at
least some ductility, making them tough and quite shock-resistant. With the exception of
aluminium and metals with hexagonal crystal structure, such as beryllium and titanium
(also magnesium for the limited fields in which it is used), metals tend to be relatively dense.
A major feature is the ease with which metals form alloys with many other elements, altering
some properties while little affecting others, and so giving some ability to optimise a
material's behaviour to specific, high-priority requirements. Metals are mostly amenable to
primary manufacturing processes such as melting and then casting or forging, followed by
conventional machining. Additive manufacturing techniques are becoming increasingly
used for many applications (see, for example, Gibson et al. 2015 and Gardner et al. 2001,

Chapter 7), while, especially at smaller scales, 'micro-machining', etching and energy beam methods can be applied to metals.

'Ceramics' actually covers a very wide field, but this chapter follows the common practice of using it as equivalent to 'engineering ceramics'. These can be broadly thought of as materials created using heat and pressure to bond together powders of small crystallites of stable compounds, although they are also produced by casting from liquid suspensions and by deposition processes, typically using microsystems technology. The technologically most common forms of engineering ceramics tend to be compounds of elements from the centre, spanning across to the right columns and higher rows of the periodic table, typically of a metal and one of carbon, nitrogen or oxygen. Ceramics tend to be hard, strong and stiff but quite brittle; to have moderate to low density; to be electrical insulators at normal temperatures; and to have good resistance to chemical attack in many environments. The sintering processes typically used to make ceramics offer good prospects for near-net shape manufacture, and ceramics are increasingly being used with additive manufacturing.

Glasses offer interesting optical properties, but are otherwise still often underestimated as a material for precision mechanics, despite plenty of evidence in modern architecture of its glass being used to make floors, staircases and so forth. Actually, a distinction needs to be made between the common meaning of glass, essentially a material based on silicon dioxide, and the class of glassy structures. The latter indicates amorphous solids in which the molecules exhibit no crystalline ordering; they are often, if not always helpfully, described as or compared to extremely viscous liquids. This group covers a lot more than just silica-like materials; even some metals can appear in a glass phase under some circumstances and many common plastics exhibit a so-called glass-transition temperature, at which their properties change significantly. Conventional glasses tend to have modest strength and stiffness, and quite low density. They cast and polish well and also machine more readily than many suppose.

Polymers, in which a small molecule (the monomer) links repeatedly in a chain-like manner to produce large and often entangled super-molecules, is another large class in which only a restricted range should be thought of as engineering materials. Polymers are also widely used as the matrix for artificial composites, carrying a stronger fibre or particle reinforcement. While there are a few instances based on silicon (the silicones and siloxanes), carbon chemistry totally dominates this group. Polymers have low densities, but even the more robust examples exhibit low stiffness and strength. Most have quite large thermal expansion and often a susceptibility to humidity. Thus, they tend to be restricted to specialised roles within precision engineering. On the other hand, polymers can yield high-precision mouldings at very low cost in mass production. New small-scale applications may arise with the continued development of additive manufacture (for example, micro stereolithography).

A few other materials that do not quite fit into the aforementioned categories are relevant to precision mechanical applications. Carbon exists in several different crystalline forms (allotropes) with remarkably different properties: among them, graphite can be used in different circumstances either as a structural material or as a solid lubricant; diamond is extremely strong, stiff and hard; macroscopic fibres and even nanotubes can be used in high-performance composites; nanotubes and graphene offer many interesting properties to nanotechnology. The relevance here of semiconductor materials lies heavily in their use in MEMS, exploiting processes derived from their dominant use in microelectronics; silicon is currently the most important. There are also specialist applications such as for infrared lenses that incorporate a variety of metallic and semiconducting elements as part of their crystalline composition.

12.2.2 Microstructure and Properties

The balance of the various short-range inter-atomic forces causes most materials to solidify with preferential, ordered arrangements of their atoms, that is, as crystals. The magnitudes of these forces together with the specific atomic layout and spacing within the crystal lattice govern much of the mechanical behaviour theoretically achievable with a particular material. Of course, in practice it is extremely rare for a solid component of any significant size to be composed simply of one near-ideal crystal. A crystal will have internal imperfections in its lattice arrangement and materials generally start to solidify from many points, so creating a multi-crystalline material. Impurities will also distort lattices, with effects on bulk properties. Many materials can exist in more than one (ideal) arrangement (polymorphism) or without even any short-range order (amorphism). The study, understanding and exploitation of such microstructures forms a major branch of materials science and technology and is, therefore, well beyond the scope of serious exploration in this book (many textbooks cover this topic, for example, Askeland 1996). This section will simply summarise a few insights about microstructure relevant to precision mechanical systems.

Crystal lattices comprise a repeated unit cell of just a few entities organised into a simple three-dimensional (3D) geometric arrangement. The word 'entity' emphasises that crystals can form from things other than just atoms, for example molecules, but it will be simpler here to think in terms of atoms. Only a few such unit cells are stable, with the most common ones built around sets of square, rectangular or hexagonal grids. The simplest is the cubic arrangement, where eight atoms have equilibrium positions centred on the corners of a cube; this often occurs with simple compounds such as sodium chloride (common salt). If the atoms are visualised as solid spheres, only one-eighth of each lies within any given cube. The unit cell, therefore, 'contains' just one atom, which, given the size of the unit cell, defines a density. This low packing density is generally inefficient; if a layer of touching spheres is set out in a square arrangement, a second identical layer would naturally be placed offset laterally by half the lattice spacing. One way to improve the packing factor is to place another atom at the centre of the cubic cell, probably slightly increasing its lattice dimension: a body-centred cubic (BCC) crystal. There are now two atoms per cell. It can be even more favourable instead to expand the basic cubic cell and place another atom at the centre of each of its surfaces to form a face-centred cubic (FCC) crystal. Half of each atom at a face centre lies within the unit cell, so there are now four atoms per cell. It can also be more efficient for some types of atoms to stack planes of a hexagonal grid, notably in a hexagonal close-packed (HCP) crystal in which the lattice spacing is notably greater between these planes than within them. Some other classical crystal types remain very convenient for general discussion, but tend now to be treated theoretically as topological variants of these types. The best known is the diamond lattice, where it is easily visualised that the four covalent bonds of the carbon atom should align at maximal angular spacing in a tetragonal grid having four nearest neighbours. The tetragonal cell has a lower packing factor than the FCC structure to which it relates, but the latter has a larger overall unit cell, so both imply the same density.

Different materials have different inter-atomic bonding forces (or energies) which directly influence the equilibrium lattice spacing and the amount of work that must be done to alter that spacing. Features such as the strength and stiffness of a piece of bulk material subjected to a force relate directly to these bond energies, summing over vast numbers of individual bonds acting in parallel; this is why stress, which expresses a spatial distribution of force, is the proper parameter for expressing materials behaviour (note that elastic moduli are expressed in units of stress). Elastic properties measured by applying a force to a single crystal will vary with the orientation of application relative to the unit cell, because the number of bonds per unit area will vary with direction, as will the overall vector component of those bond forces. Many properties of single crystals will be anisotropic, even though their lattice is symmetrical. Changing the internal thermal energy (temperature) slightly shifts the equilibrium positions of bonds and so most materials tend to expand and become less stiff and less strong as temperature rises. There are many other such dependencies, but these examples will suffice.

Real crystals can have many types of imperfections. Even if the material is pure, there will be various types of dislocation within a single crystal. At its simplest, there might simply be a void where an atom should be, giving a point defect (point dislocation) around which the lattice remains in proper order, perhaps with a little localised distortion. More likely, an initial defect during crystal growth leads to a spatially extended error in the lattice. Notable are line dislocations, where in effect the original lattice appears to be slightly peeled open from some internal point to allow an 'extra' line of atoms to be introduced, and fully 3D screw dislocations. Dislocations provide a means for material to move around relatively easily within the crystal; an atom can jump into an adjacent point defect, leaving a new defect behind and so on. Generally, the effect on average bond energy is too small to affect significantly a bulk property such as elastic modulus, but dislocation mobility will affect other properties by, for example, enabling material movement that can introduce ductility and affects useful strength. Defects can also involve additional material. An impurity atom might replace the one that should be at a specific lattice site to form a substitutional defect. If the effective atomic size is different, this will cause some distortion to the lattice. It is also possible for atoms to become stuck between the regular lattice points, forming an interstitial defect. This is more likely to happen if the effective size of the interstitial atom is smaller than those in the main lattice, but even then, it can cause significant lattice distortion. One practical effect of lattice distortion is that it introduces strain energy to the lattice, shifting equilibrium points and tending to increase brittleness and commonly strength. Again, it tends to have little effect on elastic modulus, but the distortion can, for example, impede electron flow in metals and so reduce electrical and thermal conductivity.

Polymorphism is quite common in engineering materials and, as indicated by the above discussion, is likely to affect the bulk properties in important ways. Allotropes of carbon, briefly mentioned earlier, where diamond and graphite (a hexagonal layered lattice) have remarkably different mechanical properties and can readily co-exist over a wide range of conditions. Iron is stable as a BCC crystal (often called ferrite) at room temperatures but as an FCC one (austenite) above about 900 °C. The term 'allotrope' is usually reserved only for variations within a single element. Molecules containing more than a few atoms often have more than one stable way of arranging their internal bonds to create isomers, materials with the same chemical composition but topologically distinct molecules that lead to (sometimes very) different physical and chemical behaviours. Isomers are less immediately relevant in a precision engineering context, except perhaps for a few instances of polymeric materials. Materials formed from the mixing of two or more elements may be able to crystallise in different ways, or phases, depending on the relative amounts of those elements present in the lattice. Bulk solids, even at small scales, will not usually crystallise from a single point and so real materials are poly-crystalline: they comprise many individual small crystals, or grains, that have grown into contact from random points with random lattice orientations. The grains might be of one phase or several. Clearly, the atomic structure must be at least partially disrupted at such grain boundaries. The effect on bulk properties is not always so easy to predict. For example, the imperfect alignment at grain boundaries means they tend to be weaker than a pure crystal but the boundaries can also act as a barrier to dislocation movements and so help to prevent ductile failures. The random orientations of grains lead to most bulk materials having almost isotropic properties.

Grain boundaries represent non-ideal microstructure in the sense that they imply more internal energy than needed for a perfect lattice, suggesting that there is a non-equilibrium state in which diffusion of atoms should occur to grow one grain at the expense of an adjacent one. Diffusion of specific atomic species can also lead to changes in the phases and/or unit cells present, again with resultant changes to the bulk properties. Thermally driven diffusion in solids is always very slow, especially at room temperatures or lower. It is to be expected, therefore, that many, probably a large majority of, materials will contain phases that are theoretically unstable at the prevailing temperature. The materials are not in thermodynamic equilibrium and, generally, their microstructure will be changing very slowly. The increase of one allotrope or phase at the expense of another can affect density as a result in changes to the lattices and so, for a constant amount of material, it will also affect the external dimensions of a bulk object. The materials used even to make precision mechanical loops are not, in theory, dimensionally stable. The design question is then really of whether the rate of change is completely negligible in a particular context; this question cannot be ignored, but there seems rarely to be a problem in practice.

The design of alloys is largely explainable in terms of controlling or exploiting the sorts of microstructural behaviour discussed earlier. As a first example, consider a simplified description of some features of low-carbon steel. Small amounts of carbon dissolve (as a 'solid solution') into interstitial sites in both FCC and BCC iron lattices. The FCC lattice has larger interstices and so can dissolve a rather larger amount without becoming unstable. Any carbon in excess of the amounts that can be so dissolved is accommodated in the solid as an extra phase of an iron carbide compound (Fe_3C). This description is really for an equilibrium condition, that is, at some temperature that has been held for an adequately long while. If a steel object is held at high temperature for a significant time, the material will tend to become a mix of austenite saturated with interstitial carbon and of Fe_3C. Rapid cooling (or quenching) will drive the lattice to its BCC form, which can take less interstitial carbon, but without allowing time for excess carbon to fully diffuse into the carbide phase. Carbon thus gets trapped at inappropriate sites in the ferrite phase, resulting in high lattice strains and a brittle, strong material. There is, of course, much more subtlety to the design and behaviour of steels, including the use of other elements to help control phase change processes, but even this description makes it clear that this alloy is designed to allow control of ductility, strength, hardness and so on by means of heat treatment. In other materials, different alloying strategies could be used to give better control over properties. For example, another approach to controlling ductility and strength is to add alloying elements that, on suitable heat treatment, lead to a reasonably uniform distribution of small grains of the main material interspersed with very small grains of a much harder phase. The idea is largely one of heavily restricting dislocation movement. A few more examples will appear in context elsewhere in this chapter. For now, it suffices to mention that many important alloys have one dominant element, with just a few percent of others to control specific behaviours. In such cases, bulk properties that depend more on a notional average bond energy than on detailed microstructure tend to vary relatively little across whole families of materials that vary widely in other respects; elastic modulus and thermal expansivity tend to fit this category. There are a few important cases, mostly of iron, with other transition metals closely adjacent to it in the periodic table, where there are large proportions of alloying elements, but these bulk properties still vary only a little; here it is because the underlying properties of the elements themselves are similar. Other cases of high alloying tend to be concerned with features such as suppressing melting temperatures for ease of processing by casting, and so on, while small amounts of other alloying elements provide controlled variation of some properties; brasses and bronzes are typical examples.

Amorphous materials tend to be of two classes. In terms of structure, glasses have some similarities to liquids, having randomly placed atoms or molecules but with much closer mean spacing and much higher inter-atomic forces. They tend to be brittle, while remaining workable, at lower temperatures and to soften and become extremely ductile at higher ones. They also tend towards highly isotropic behaviour and, despite not being in thermodynamic equilibrium, often exhibit good practical levels of dimensional stability. Polymers are created by the linking up by chemical bonding of many copies of one, sometimes two, relatively small molecules (the monomers). With some materials, there is extensive, but fairly randomly organised, cross-linking in all three dimensions, but with fewer bonds between the molecules than within them. Since it is the former that ultimately govern overall bulk properties, these polymers tend to be moderately stable but much less stiff and strong than atomic crystals. Other polymers form as long chains, almost linear, sometimes with shorter side branches, that tend to curl up and entangle with others in the bulk material under the action of inter-molecular forces, mostly electrostatic, or van der Waals, forces. These forces are weaker than inter-atomic ones and so the materials tend to be very ductile (as the chains straighten out) but of low stiffness and strength, high thermal expansivity and low stability. Chemical cross-linking of long-chain molecules is possible in some such materials to improve the mechanical properties.

While molecules are electrically neutral overall in their base forms, that is, unless specifically ionized, it is quite common to find that the mean positions of the negative and positive charges from their constituent electrons and protons do not exactly coincide. This is known as polarisation. In some crystalline materials, the internal ordering causes the individual polarised constituents to line up to create a polarised effect in the bulk material. This is the cause of piezoelectric behaviour. Applying a suitable potential difference across a piezoelectric material encourages the centres of negative and positive charge to move very slightly closer or further apart, so creating a displacement actuator that is also a stiff and strong piece of solid material (see Chapter 7). There is a tendency for electrons to drift towards the more positively charged end of a piece of piezoelectric material, so that a net charge can be detected across it: straining the crystal slightly alters the spacing of the internal charge centres, modifying the externally detectable field and creating a sensor. Polarised molecules are the source of the electrostatic bonding forces important in many ceramic and polymeric materials, where they are randomly oriented and show no bulk electrostatic behaviour. The sites of hydrogen atoms are commonly associated with polarisation effects, both in causing it and being the points on a molecule that are attracted to the more negatively charged regions of other molecules. This is given the specific name of hydrogen bonding, which is, for example, responsible for many of the unusual properties of water. The stable structure of a water molecule has its two hydrogen atoms offset from a diagonal across the oxygen atom. A hydrogen bond, much weaker than the intra-molecular chemical bonds, can then form as one of the hydrogens of a molecule that is electrostatically attracted towards the more 'open' side of the oxygen atom of another molecule. This extra level of bonding explains, for example, why such a small molecule as water is in the liquid phase at room temperatures and pressures. While obviously of little relevance to mechanical structural loops, hydrogen bonding in water is of potential interest because it is a strong source of driving forces for self-assembly in nanotechnological manufacture, as well as in many natural biological systems.

12.2.3 Non-Mechanical Properties

While this chapter primarily explores material properties related to the stability of mechanical loops in precision systems, it would be misleading to imply thereby that such properties are always the only, or even necessarily the dominant, ones for precision design. This section considers just a few non-mechanical properties, concerned mostly with electrical and optical behaviour.

Many of the mechanical properties of materials derive quite directly from the nature of inter-atomic bonds, which are strongly influenced by the distribution of electrons in the constituent atoms. In terms of a basic visualisation, this bonding relates to electrons considered as local to a 'parent' atom. However, electrons are relatively mobile even in solids and many other properties arise from this. The electrons in a solitary atom occupy a set of orbits, corresponding to physically permitted energy levels. In the ground state, these levels would be filled sequentially from the lowest upwards, but at higher energies (temperatures) some electrons tend to jump into higher energy levels, where they may or may not be able to remain stably. In a crystal, the Pauli exclusion principle ensures that no two electrons throughout it can have exactly the same state and so a set of permitted and prohibited energy bands develop. It is the outermost (highest energy) of the stable bands, called the valence band, that dominates bonding and much chemical behaviour. As the internal energy rises, increasing numbers of electrons can be (temporarily) excited into yet higher bands, where their degree of connection to a specific atom becomes weaker. There can then be a significant number of electrons that are free to wander with little impediment anywhere within the confines of the crystal, as an 'electron gas'. The electrons carry a negative charge and so the drift of this electron gas transports charge through the material, so providing the major source of electrical conduction. At the same time, the electrons can transport energy (essentially as kinetic energy) and so can be a major contributor to thermal conductivity.

In some materials, there is a relatively large gap of forbidden energies between the valence band and the next available conduction band in which an electron gas could exist. If this gap is significantly greater than the thermal energy associated with room temperatures (actually any specified temperature, but day-to-day descriptions tend to focus on 'normal' conditions), there is little chance of an electron gaining enough energy to make the jump, so few conduction electrons are present at any instant and the material is an electrical insulator. In some other materials, the permitted levels of the two bands overlap, so large numbers of electrons will always be in the conduction band and the material is a good electrical conductor. This is the usual behaviour of metals, whereas good insulators are typically ceramics, glasses and polymers. If the band gap is of comparable size to the thermal energy available, a smaller but significant number of electrons will be excited into the conduction band and so the material will have a reasonably good conductivity that is likely to be quite sensitive to external environmental stimuli. Such materials are called semiconductors, where the boundary between semiconductor and insulator is ill defined and context dependent.

Alongside the thermal energy of electrons, internal (kinetic) energy is also present in the form of vibration of atoms within the lattice. Such kinetic energy couples, or transmits, quite readily across a lattice should one area become especially excited and is the major mechanism for thermal conductivity in non-metals. Conversely, larger lattice vibrations, or any deviation of the lattice from its ideal structure, will tend to impede the flow of electrons. So, in metals, for which temperature makes little difference to the effective availability of conduction electrons, electrical conductivity generally reduces as temperature rises and alloys tend to have poorer conductivity than their major pure metal constituent. Also, for example, strain distorts the lattice and so it tends to reduce conductivity a little. In semiconductors, even small increases in temperature can significantly increase the availability of conduction electrons, which overwhelms any increased vibrational impediments, and conductivity tends to rise quite rapidly with increasing temperature. Trace amounts of different materials in an almost pure semiconductor have almost no effect on basic mechanical properties but can have profound effects on other behaviour; the amounts are so small that they are called 'doping' rather than being thought of as a form of alloy. To take just one example, silicon has four electrons in its outer, valence shell and might be doped with boron or phosphorus atoms, which have respectively three and five. The effect of the boron is that the ideal bonding structure has gaps ('holes') where an electron can more easily hop from association with one atom to an adjacent one, so modifying properties that depend strongly on electron activity. Doping with phosphorus leads to some electrons being somewhat less bound because there are too few stable bonding sites to accommodate them. These holes or excess electrons lead to p-type and n-type semiconductors respectively. They are absolutely key to the operation of all semiconductor electronic devices, but are also exploited in the mechanical design of MEMS because their varying chemical properties enable manufacturing routes by exploiting various etching processes.

The electron gas of a metal makes it very difficult to maintain an electric field in the material and so electromagnetic waves propagate very poorly through metals. Bulk metals are almost total reflectors of light, although photons can penetrate through thin metallic films; in passing, note also that nanoparticles have much more complicated interactions with light. More generally, bound electrons interact with light and can absorb energy from a photon, so being promoted to a higher energy shell. This is a less stable state and so at some later point the electron will fall back to its lower energy level, emitting a photon of energy (which is also effectively wavelength and colour) matching the difference between the two levels. Thus, the colour of a material is largely concerned with what photon energies can be absorbed eventually to become internal thermal energy and what will be re-emitted from electron transitions. Insulators will be transparent over ranges of wavelengths for which the relative energy levels lead to only infrequent interactions. Even under these conditions, various interactions between photons and atoms will occur (broadly classified as 'scattering' phenomena and not further discussed here) so that energy is gradually absorbed and real materials are not totally transparent. Different refractive indices arise for similar reasons, making it of little surprise that they show an approximate correlation with density. Small particles and other material microstructures, especially those with nominal diameters below the relevant wavelength will also scatter light leading to greater energy absorption, some colouring of transmitted light and opacity. Indeed, the majority of insulators are opaque to visible light. The optical effects of materials and their surface structures on the aesthetics of a design should not be neglected, even in apparently function-driven precision applications; users will tend to trust more and use with more care a device that looks and feels 'right'.

12.3 Classifications for Materials Selection Criteria

12.3.1 Setting Up Comparison Schemes

One characteristic feature of precision mechanical design is a special emphasis on the integrity of metrology loops (see Chapter 11). High-precision devices tend also to carry relatively small forces (and stresses) and to operate in fairly benign environments. Across the whole broad range of mechanical engineering, from transportation to consumer goods, there are likely to be more regular concerns about factors such as strength, weight and robustness. Sensitivity to cost is likely to be considerably higher for a mass-produced consumer product than it is for a high-precision device or machine made in smaller numbers. The majority of available data and advice on materials choice is focused on these general (and historically dominant) markets; while certainly relevant, it has often been poorly suited to precision engineering applications. Furthermore, the nature of materials selection seems to have been seen historically as mostly a craft, with choices evolving based on the experience of designers and prototype testing.

This situation was shaken up in the mid-1980s by a new interest in applying analytical ideas to the selection process. One reason was that demands for ever-higher performance and efficiency were leading increasingly to mechanical systems that pushed materials to their limits. In turn, there was increasing recognition that, while the key selection indicators had to become more application specific, a common core still underpinned them. Less emphasis was placed on individual easily measured and documented properties, and more on the role of functionally defined combinations of those properties. This concept was not completely new, with the most obvious example perhaps being the need for high strength-to-weight (more strictly in terms of material properties, density) ratios in aerospace applications. The major impetus, though, was that easily accessible computing and data-bases provided, for the first time, prospects of wide-ranging searches for improved options. The seminal work came from Ashby's group at the University of Cambridge (see for example Ashby 1989, 1991), which has since developed into an evolving series of very well-regarded commercial tools, Waterman and Ashby (1991) being the first to include a really large database. Their first step was to propose property charts (often now called Ashby charts) in which two materials properties were set along orthogonal axes and any specific material would be represented by one point in the space so formed; this idea is familiar in thermodynamics and also in the phase-spaces used in systems engineering. Plotting these positions for a large range of materials gives a very clear visual image of which are similar in particular aspects and which are very different. Naturally, this makes practical sense only if the properties represented along the axes are ones that tend to interact functionally, suggesting a significant mathematical relationship between them (for example, their ratio might matter more than the absolute value of either). There needs, also, to be an easy way to judge what area of the chart represents the 'best' combination of those properties for a particular application. Both of these needs were met by using logarithmic axes. Log scales can accommodate a much wider range of values than linear scales; such large ranges are often needed to cover all plausible engineering materials, although this may be more important for the study of materials science than to design, where a limited range could be predefined. Crucially, though, logarithmic axes mean that constant values of all multipli-cative and power relationships between the properties plotted appear as straight lines. It is then conceptually equivalent to sliding a ruler over the chart, keeping it parallel to a line of constant function, to successively filter out materials showing higher or lower values in that chosen function.

Figure 12.3 illustrates the major features of an Ashby chart, showing the Young's modulus and density clusters for some types of material. The ratio of these two properties is relevant to the degree of distortion from self-weight loading that will occur in a structural loop of given dimensions; it is intuitively obvious that a structure made from an inherently high stiffness material that is also low density (i.e. having a high value for this ratio) will suffer relatively little self-weight sag. Comparing the slope of the line of constant E/ρ to the

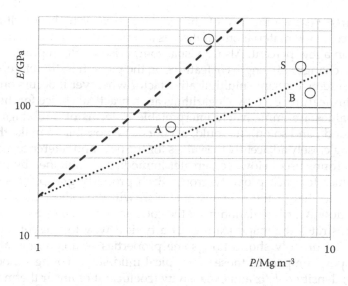

FIGURE 12.3
Typical regions for aluminium alloys (A), brasses (B), alumina ceramics (C) and mild steels (S) on a logarithmic plot of Young's modulus E against density ρ, illustrating the major features of an Ashby plot. A line of constant E/ρ (dotted) and a line of constant $E^{0.5}/\rho$ (dashed) are shown. Higher values of these ratios lie to the upper-left region of the plot. Further discussion of these types of property groupings is provided in Sections 12.3.2 and 12.3.3.

positions of the clusters indicates that steel and aluminium are very similar in this respect, while engineering ceramics might offer metrology loops that are less susceptible to gravitational effects (assuming, of course, that the extra costs, brittleness and so forth are acceptable for the application). Further details of what constitutes sensible pairs of properties and how this is established will be deferred until the following sections. For completeness, note that the fully developed ideas of the Ashby materials selection schemes attempt also to capture time-varying or less tangible factors, such as cost and even the effects of external constraints on the geometrical aspect ratios of structures.

By the mid-1980s questions were being raised more specifically about which combinations of materials properties were most often influential in high-precision mechanical systems; this was independent from the Ashby group, who raised some similar questions later (Cebon and Ashby 1994). The first attempts to do this developed graphical approaches not unlike the Ashby charts (Chetwynd 1987). However, a new approach emerged rapidly (Chetwynd 1989) through an inversion of the original question; rather than comparing many materials for two properties, this approach offers a direct graphical comparison of several property groups between two materials. A property group is here taken as any multiplicative relationship (for example, products, ratios or reciprocals) of one or more basic properties. The rationale for this approach is partly that a conveniently small number of property groups tend to occur very frequently when investigating the requirements for precision applications and partly that many new designs start from a concept based upon a previous device (more generally, experience of devices offering similar challenges), including the materials used. The methods for defining appropriate property groups will be discussed in Sections 12.3.2 and 12.3.3. In order to give an easily assimilated 'material profile' of the ways in which a material of interest is superior or inferior to a reference material, the values for each group are plotted side by side in a thermometer-style plot or a bar chart. The comparison scales are logarithmic, but for reasons rather different to those with Ashby charts. Because the cost of change is often very high, it will normally make sense to introduce a new material only if there is a very marked improvement, say a factor of two rather than a few percent. Also, a visual comparison will, in practice, be dominated by judgments of (changes of) lengths of features on the plots. It is, therefore, helpful that the same length represents the same multiplicative factor wherever it occurs on the plot: a log-scale provides this feature. Base-two logarithms are a practical choice for the scales because the numerical values are immediately interpretable in terms of doublings ('four times as good', 'half as good', and so on). It would, of course, be possible to make the comparisons by simply plotting closely adjacently the values for the new and reference materials for each of the property groups used. However, greater immediacy is obtained by plotting the ratio of these two values for each property group. Each property group for the new material (now normalised to that of the reference material) will then show as an upward (better functionally) or downward deviation from the constant baseline of the reference.

Figure 12.4 illustrates the major features of a typical way to present a property group profile. Figure 12.4 broadly shows how some properties of a common aluminium alloy (around 4% copper) compare to those of a typical mild steel: Young's modulus E; tensile yield strength Y; density ρ; E/ρ; and expansivity (coefficient of linear thermal expansion) α. Each bar represents the ratio of the value for the alloy to that for steel (which is here considered the reference material). An upward bar indicates that the value for the alloy is numerically larger, a downward bar that it is numerically smaller. The vertical scale is the base-two logarithm of the ratio, so zero represents equal values (i.e. the ratio is unity), while ±1 indicates that the value for the alloy is twice or half as large. Thus, a quick glance at the plot confirms that aluminium expands about twice as much as steel ($\sim 2^1$, or one doubling), strengths are closely comparable, while aluminium has roughly one-third ($\sim 2^{-1.5}$, or halved 1.5 times) the stiffness and density of a steel, leading to their ratio being quite closely the same. This presentation varies in some details from what has commonly been used previously, in which the scales tended to show the numerical ratio with values plotted by point markers connected by straight lines. There are advantages and disadvantages with each approach. Reading true numbers might be slightly more immediate, especially for occasional users, but reading the log-value as the number of doublings (halvings) helps to stress

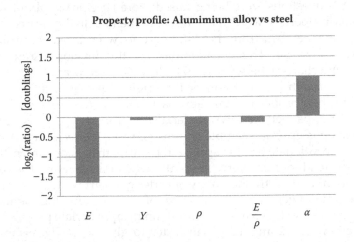

FIGURE 12.4
An illustration of the main ideas for constructing a profile of materials properties, comparing the ratio of values for those of a typical aluminium alloy with respect to a typical mild steel.

the amount of difference needed to drive change. Also, given plots for several materials against the same reference material, comparisons of any two of them are available by simple subtraction of the log values. Connecting points by lines is artificial in the sense that no physical function is attributed to them, but was previously done to help guide the eye and make it easier to spot similarly performing materials through their similarly shaped profiles. On the other hand, plotting as bars gives a stronger visual image without any mathematical artificiality and potentially allows denser presentations involving several materials. It is important to note that no special relevance is claimed for the set of parameters used in Figure 12.4, although they are all ones that will sometimes be important in precision structures. The whole point is that the method is intended as an aid to discussing choices and on many occasions designers will be able to specify a set of parameters known to be a high priority for their application. A discussion of how suitable property groups might be generated and justified, together with a modest-sized recommended set for general use, will follow shortly.

The stability of structural loops in high-precision devices will generally be affected by the forces carried within them and by their immediate thermal environment (see Chapter 11). Sections 12.3.2 and 12.3.3 explore each of these in turn, based on the underlying use of materials properties that are usually easily accessible: Young's modulus E; tensile yield strength Y; density ρ; coefficient of linear thermal expansion (expansivity) α; thermal conductivity k; and specific heat c. There then remains the question of what constitutes a good choice of reference material for a general-purpose study. While various choices (not least, steel) could be justified, this chapter follows an earlier convention by using an artificial set of property values that might each be thought quite good, but not outstanding, for an application involving precision mechanics. Arbitrary values are taken that broadly represent a material having the strength and stiffness of mild steel, the density and expansivity of an engineering ceramic, the thermal conductivity of brass and a specific heat in the range typical of many metals. This leads to the Model Material, summarised in Table 12.1, that will be used as the reference in the discussions here, except where specifically stated otherwise. For consistency, the model parameter values are the same as those used in earlier sources (Chetwynd 1989, Smith and Chetwynd 1994). The effect of using this good-but-not-excellent reference is that most materials to come under serious consideration are likely to have property groups that are at most a few times better or worse. This, in turn, tends to make property profile plots reasonably compact and easy to place on a common scale that aids comparison between them. As with much of this approach, details and layouts are mostly about the ergonomics of an efficient comparison and selection process, and users are encouraged to modify them when application-specific data provides a particular focus.

TABLE 12.1

Base Properties to be Used for Generating Property Groups and the Values Attributed to the Model Material Used to Normalise Data in This Chapter

Property	Model Value	Typical of
Modulus of elasticity	200 GPa	Mild steel
Maximum (allowable) strength	300 MPa	Mild steel
Density	4000 kg m^{-3}	Oxide ceramic
Thermal expansion	7×10^{-6} K^{-1}	Oxide ceramic
Thermal conductivity	150 W m^{-1} K^{-1}	Aluminium alloy
Specific heat	750 J kg^{-1} K^{-1}	Many metals

12.3.2 Properties and Behaviour under Stress

Materials choices will nearly always interact with other aspects of design in complicated ways. Considerable caution is needed with any attempt to compare the effectiveness of materials using only numerical values derived from readily quantified properties. Nevertheless, it can still be illuminating to compare materials under closely controlled, somewhat artificial conditions. So, one question to ask is how the mechanical performance of a structure would vary if the material were changed while everything else is kept exactly the same. Under these circumstances, Figure 12.4 indicates that a change from steel to aluminium alloy would make the structure considerably less stiff, while its natural frequency would be reduced only a little. Designers are, however, rarely so tightly constrained. If there were no effective restrictions on the cross-sections of the structure, the lower density suggests that overall stiffness might be recovered (possibly even increased) by enlarging the sections and keeping the overall mass similar. It would be quite rare to find this level of freedom to adjust dimensions and, of course, overall weight (mass) may or may not be an important factor. Thus, any approach is sensitive to built-in assumptions and it is best to adopt one that has a degree of mathematical rigour and uses consistent, easily understood assumptions. This chapter, therefore, sets up comparisons based on the performance of structures with identical linear dimensions. Note that, with the use of lattice structures and the design freedoms offered by additive manufacturing, it is becoming increasingly possible to easily alter the shape and dimensions of a design, but this chapter will mainly assume conventional manufacturing is being utilised (see Syam et al. 2017 for an example of early work on the use of lattice structures in precision design).

By allowing the materials under study to be the only change allowed, modelling to extract useful property groups can be simplified by the use of the theory of dimensional similarity; be careful to differentiate between two different uses of 'dimension', which here relates to the underlying basic physical units of an expression, not to size (see Chapter 2). Design choices tend to focus on a small number of archetypical requirements such as the need to carry specified loads without failure or excessive elastic deflection. Consider, for example, setting up equations (mathematical models) that describe how a point on structures of various degrees of internal complexity will deflect under the action of an applied force at a specified location. Using the mass–length–time scheme for dimensional analysis, all these models result in a displacement (L) and have a force (MLT^{-2}) as the independent input. The displacement is a manifestation of strains and so depends on internal stresses and the elastic properties of the material, while the stresses depend on the geometry of the structure. All systems in which, say, axial or bending stresses dominate behaviour must have a net effect of turning the input force into a stress ($ML^{-1}T^{-2}$), and so must ultimately have the same dimensional influence $\times(L^{-2})$. Similarly, the materials factors have a net dimensional influence $\times(M^{-1}LT^2)$ to convert the stress into a strain and a final geometric factor $\times(L)$ to generate the displacement. The overall conclusion is that exploring only the very simplest structure that reflects the type of behaviour of interest will reveal how the material affects the function (with respect to that behaviour) in all systems of any complexity.

The simplest archetypal structure is an axially loaded rod, or strut. The axial stress σ relates simply to the strain ε by Hooke's law,

$$\sigma = E\varepsilon \qquad (12.1)$$

or, in terms of axial displacement δ and applied force F,

$$\delta = \frac{Fl}{EA}, \qquad (12.2)$$

where A is the cross-sectional area and l is the length. There is commonly a desire to minimise any change in size of (or strain in) a metrology loop and in many other practical situations. This leads to the, unsurprising, guideline that a high Young's modulus tends to be preferred. There are though a few important cases, for example in some flexure mechanisms, where a low stiffness material might be better. Nevertheless, Young's modulus (or elastic modulus E) is a valid property group on its own. In any comparison, a higher value than the reference (model material) is generally preferred, and this is indicated by an upward-pointing bar on the property profile, as feels natural. This style will be used as a general convention: for all property groups, the condition most commonly considered to be better will take a larger value. It is inevitable that many structural members will need to carry significant loads and the integrity of the loops (including to some extent in terms of creep and fatigue effects) requires that their stresses do not approach the elastic limit. So, considering Equations 12.1 and 12.2 and substituting σ by Y, the yield stress (or, if preferred, a safety-factor scaled version of it) shows that Y is also a useful property group on its own.

Consider, now, the bending behaviour of a uniform cantilever beam under lateral forces, using the linear, small deflection theory covered in all basic structural mechanics texts (see, for example, Beer and Johnston 2004). The end-deflection resulting from an end force F is

$$\delta = \frac{Fl^3}{3EI},$$ (12.3)

where l is its length and I the second moment of area of its cross-section. If this cantilever is placed horizontally with no load other than its own weight, the end-deflection will vary as

$$\delta \propto \frac{\rho A g l^4}{EI},$$ (12.4)

where ρ is the density, A the cross-sectional area (and so its volume is Al) and g is the acceleration due to gravity. The constant in the denominator of Equation 12.3 arises from the specific boundary conditions for the situation modelled, whereas Equation 12.4 is reduced to a proportionality by omitting the equivalent constant. This has been done to emphasise the underlying physical similarities between these two models. Each comprises the same geometrical term and a force term, which in Equation 12.4 itself happens to comprise a materials property (density) and geometrical parameters; factors such as physical constraints and point or distributed load functions are absorbed into a boundary-value constant. The proportional relationship, combined with the nature of this constant, further demonstrates the dimensional similarity of the models that allows the materials-specific behaviours to be extracted from the simplest of archetypes (often without need to fully solve even those forms).

Continuing in this vein, the maximum bending moment that the beam can carry without any yielding is

$$M_{\text{max}} = \frac{2YI}{d}$$ (12.5)

with d being the depth of the beam in the plane of bending and Y the allowable (yield) strength in tension, or compression if lower (which is fairly uncommon, but can occur with a few materials of potential interest for high-precision applications). The maximum bending moment will depend on the types and positions of the loads applied, but the message for materials choice is exactly the same as it was for axial loads: Y taken alone is a useful generic signifier of load-bearing capacity. Also, because it will often be important to control self-weight distortion of metrology loops, Equation 12.4 suggests that the group E/ρ should preferentially be large.

To further illustrate how potentially useful property groups can be extracted, the static behaviour of beams can be expressed in terms of an effective linear stiffness. For example, noting that for a rectangular cross-section of $b \times d$, $I = bd^3/12$, Equation 12.3 rearranges to give the stiffness of a transversely loaded beam as varying according to

$$\lambda \propto \frac{Ebd^3}{l^3}. \tag{12.6}$$

There might be situations where a beam must span a fixed distance and it is inconvenient to vary its breadth, but there is scope for adjusting its depth. The depth strongly influences the stiffness and linearly affects the overall weight. For some pre-specified stiffness, the required depth would be $d_\lambda \propto E^{-1/3}$. Then, the weight would be proportional to $\rho/E^{1/3}$, and selecting for a low value of this group would be optimal in this scenario. Following the convention that higher values should correspond to better performance in the common cases, the property group $E^{1/3}/\rho$ would be a good candidate for inclusion in a selection profile. However, while important in fields such as aerospace, it less frequently reflects constraints on precision design and so will not be included in the basic set used here. Similarly, there are cases, including certain conditions relating to the (Euler) buckling of columns, where selecting for high $E^{1/2}/\rho$ would be sensible; the line corresponding to constant values of this buckling criterion is plotted in Figure 12.3. Ashby (1989) offers an original discussion of many such examples, where it is perhaps more easily accessible than in that group's later extended publications.

The discussion so far has concerned static behaviour. This is because, despite some notable exceptions such as sub-systems of diamond turning machines and microcircuit steppers, many precision systems run under low-speed and low-acceleration conditions where inertial loads do not impose major constraints on design choices. Nevertheless, some non-static behaviour should be captured within a general property profile. An interesting intermediate case (perhaps classified as 'quasi-static') arises with flexure mechanisms. Flexures are often applied as a means of obtaining precisely repeatable movement, which implies that there must be no hint of plastic behaviour in their ligaments or hinges (see Chapter 7 for a fuller discussion on flexures). On the other hand, it is also helpful to get a good range of motion compared to their overall size, implying a large strain in their elastic elements. Inserting the yield stress into Equation 12.1 indicates that maximum allowable strain is governed by increasing the value of Y/E, making this a useful property group. As discussed in Chapter 7, resilience Y^2/E is also an interesting property for flexures, although it is not included in the base set of groups used here. Noting the general arguments that it is good to keep the reaction forces in loops associated with drives relatively low, flexure systems of modest stiffness might be preferred, providing they are adequately stable. The previous discussion around stiffness then implies a materials selection towards low elastic modulus E. This is a criterion opposite in sense to most cases where E is used alone, but it relates to just one scenario and has to be treated as one of a few unavoidable anomalies likely to occur in a general-purpose set of property groups; if flexural considerations dominated a particular design application, it would be reasonable to use $1/E$ in place of E in order to follow the convention of increasing values being preferred. In terms of controlling dynamic forces associated with accelerating a flexure mechanism, it seems generally desirable to keep its mass reasonably small for the size required, implying simply that lower density materials are attractive. Thus, higher $1/\rho$ could be used as a selector; it is, however, not that often of high relevance to precision engineering and so is not included in the base set used here.

Vibration is one type of dynamic behaviour of very definite relevance to precision engineering. In the absence of other design requirements, it is good practice to seek relatively high natural frequencies for the major structural and metrology loops of a system (see Chapter 13). Energy transfer from the environment tends to couple less well at higher frequencies, so reducing the chance of significant resonance. Again, the simplest archetype is the first natural frequency (fundamental) of a uniform cantilever, which is widely quoted as

$$\omega_n \simeq 3.52 \left[\frac{EI}{\rho A l^4} \right]^{1/2}. \tag{12.7}$$

Higher natural frequency maps against larger values of the group E/ρ (or its square root). An analysis of elastic wave equations would show that the less sensitive grouping $(E/\rho)^{1/2}$ is also the longitudinal wave velocity, or speed of sound, in a long, thin rod. Variations of this ratio occur in several basic scenarios, so E/ρ is a definite candidate for the general-purpose set of property groups. Some applications (for example, some types of vibration sensors) require a low natural frequency, in which case materials selection is for a low value of this group; this is considered here as another anomaly of the general-purpose set.

It seems right that there should be some representation of limiting inertial effects even in a general set of property groups focused on precision engineering. A uniform rod rotating about its end provides a characteristic example. Restricting, for simplicity, to constant speed conditions, each infinitesimal section of the rod will experience a centripetal acceleration depending on its radius from the pivot and a consequent centripetal force (stress). All these elementary forces have to be reacted at the pivot and so the stress in the rod integrates to a maximum there, given by

$$\sigma = \frac{\rho \omega^2 l^2}{2}, \tag{12.8}$$

where ω is the angular velocity and l the length of the rod. For maximal rotation rates (or other high accelerations), substituting a maximum stress Y for σ shows that a high value of the group Y/ρ would be preferable.

This section has demonstrated how simple archetypical models of mechanical systems can be used to extract materials property groups that will have functional significance for various applications and so can form the basis of a property profile to aid materials selection. It is important to emphasise that the discussion can be treated as an illustration of how to construct bespoke profile sequences tuned to defined ranges of application. The discussion has also generated a set of property groups that are likely to occur commonly in the mechanical behaviour of precision engineering systems. These will be included in the general-purpose profile set used in this chapter: E, Y, E/ρ, Y/E and Y/ρ. They will be presented consistently in this order when tabulated or plotted, on the grounds that it reflects a general progression from typically more 'static' to typically more 'dynamic' behaviour as the sequence progresses.

12.3.3 Thermal Properties and Behaviour

Thermal disturbances to mechanical loops are a particular concern in high-precision design; indeed, it is often observed that every precision instrument or machine is also a thermometer. So, some thermal properties will often be very relevant to materials selection and feature strongly in the general set of property groups. The approach in this section will follow closely that of Section 12.3.2. Generally, the final indication of potential thermal disturbance to the dimensional stability of a loop is likely to be in terms of expansion caused by temperature changes and temperature gradients. These changes can arise from many sources, commonly involving several different properties to relate an undesirable but unavoidable energy source to a temperature effect on the stability of loops within a device or machine. Thermal disturbances can be imposed by environmental changes, by localised internal sources such as motors and other actuators, or even by such things as heat from an operator's hand (typically having a surface temperature around 30 °C, whereas machines tend to operate nearer to 20 °C). Some effects might be virtually unchanging (consider a technical decision to run a high-quality calibration room at a non-standard but well-controlled temperature and to compensate for it post-measurement) or tending only to slow drift. Other sources might be one-off or repeating transients. So, a general-purpose set of thermal property groups should reflect the same notion of covering 'static', 'quasi-static' and 'dynamic' behaviours, as was applied in the previous section when considering stress-related properties.

Putting to one side some very demanding applications, such as those in space science, it is reasonable to assume that environmental changes do not occur too rapidly in situations where high-precision devices are being used, provided some care is taken to avoid radiative sources such as direct sunlight. If so, passive bodies of moderate size will remain in approximate thermal equilibrium with their environment. To reduce susceptibility to temperature changes, the essential requirement will then be to place a limit on the direct thermal expansion of the body. If the body is located kinematically (see Chapter 6), so that small changes in size can be accommodated without the generation of significant forces, or is essentially free (as with a gauge block, for example), this control is simply for a low coefficient of linear thermal expansion or expansivity (coefficient of thermal expansion [CTE] is also commonly used). To maintain the convention that larger values reflect more preferred performance, the appropriate property group is, therefore, $1/\alpha$, where α is the expansivity. Often, however, a body will be more tightly constrained, perhaps as a sub-structure of a nominally rigid reference loop. In the limit, no change in length would be permitted and so internal compressive stress (with matching end-reaction forces) must develop to exactly compensate elastically for any natural thermal expansion. The free expansion under a change of temperature can be expressed as a strain of $\alpha\Delta T$ and, as long as it remains in the elastic region, the stress needed to cancel this strain is

$$\sigma = -E\alpha\Delta T. \tag{12.9}$$

Since the general guideline is to reduce loop stresses where possible, materials selection should, therefore, be guided by higher values for $1/\alpha E$.

Another idealised model of quasi-static thermal behaviour considers the effect of a dissipative energy source, perhaps a motor, that runs (more or less) continually and at a steady power. Some of the heat energy will be conducted along structural members, implying a raised local temperature and expansion. The simplest archetype is one-dimensional steady heat flow q along a uniform rod of length L and cross-sectional area A. Fourier's law for conduction shows that this requires a temperature differential between the ends of the rod such that

$$q = \frac{kA\Delta T}{L}, \tag{12.10}$$

where k is the thermal conductivity. A uniform rod will have a constant temperature gradient, so, assuming that the end that is remote from the source is at ambient temperature, its average temperature rise will be $\Delta T/2$. The overall expansion of the rod, expressed as a strain, will be

$$\varepsilon = \frac{\Delta T}{2}\alpha = \frac{qL}{2kA}\alpha. \qquad (12.11)$$

Low strain is, therefore, associated with high values for the property group k/α when the constraints on the body concerned do not restrict expansion. Following the argument used in Equation 12.9, a high value of $k/\alpha E$ would be the preferred criterion in order to minimise internal stresses in bodies physically constrained to remain at constant length.

Should there be a non-steady, or transient, thermal disturbance, a rapid return to steady conditions is usually desirable. In other words, it is desirable that the heat energy diffuses rapidly. Again, taking the simplest case of the one-dimensional heat equation, a spatially distributed temperature profile $\theta(x)$ will, in the absence of additional energy sources (or sinks), vary in time and space as

$$\frac{\partial \theta}{\partial t} = \frac{k}{c\rho}\frac{\partial^2 \theta}{\partial x^2}, \qquad (12.12)$$

where t is time, x is position and $c\rho$ is the volumetric specific heat. The specific heat c is always associated with density in these relationships because the size of an object directly influences how much heat must be gained, lost or moved internally to cause a particular change in a temperature distribution. The group $k/c\rho$ is known as the thermal diffusivity. Equation 12.12 is first order in time and so clearly describes an exponential characteristic settling to an equilibrium state with a reciprocal time constant of $k/c\rho$. Generally, a high value of this group is preferred.

Diffusivity itself has no effect on the dimensional disturbance of the reference loops that might occur. It is, therefore, always associated with a temperature-based scenario to elicit a net thermal strain. One of the two archetypical situations is a temperature profile directly imposed, with the exchange of whatever heat energy is needed to achieve it, such as when a small object has been held between an operator's fingers for a while. The other is when a fixed amount of heat energy is transferred relatively quickly, perhaps from an actuator operating briefly, and a temperature profile builds up in response. In the fixed temperature profile case, the expansion will still depend on an average temperature above ambient, even though the temperature profile is likely to be non-linear. The selection criteria for free and fully constrained bodies are the same as those discussed earlier, with the simple added preference for high diffusivity so that the expansion effects die away rapidly. In the fixed heat transfer case, the temperature change will be inversely proportional to the volumetric specific heat and then the net dimensional effect will be the average of that change scaled by the expansivity. Hence, a high value of $c\rho/\alpha$ or $c\rho/\alpha E$ would be preferred for the free and fixed bodies, respectively. Note, in passing, that the interactions of heat capacity, net expansivity and diffusivity also lie at the heart of the good practice requirements that machine tools and measuring instruments should be powered up a considerable time before they are used and their workpieces should be allowed time to soak in the operational environment before being processed. Manufacturers of such equipment will generally offer their own guidelines on this matter.

Perhaps more so than with stress-related properties, the thermal behaviour of materials in a precision engineering context is governed by a small number of parameters that group in several different ways according to the situation; the most obvious illustration is that thermal expansion is regularly scaled by elastic modulus to account for stresses in heavily constrained parts of structural loops. It could become distracting to include all these variations in a general-purpose property profile, while it is easy to select the most relevant ones when constructing an application-specific profile. So, a fairly minimal set that captures instances of all the behaviour patterns will be used here, along with this reminder to consider which of them could be used in combination to account for other situations. The chosen thermal property groups are $k/c\rho$, $c\rho/\alpha$, k/α, $k/\alpha E$, $1/\alpha$ and $1/\alpha E$. This set will be presented consistently in this order, an arrangement chosen to mirror that selected in Section 12.3.2 for the stress-related property groups, by reflecting a progression in relevance from typically more 'dynamic' towards more 'static' situations.

12.4 Materials for Precision Engineering: Common and Interesting Choices

Section 12.3 discussed methods for comparing materials, and so guiding the materials selection process for a design. While requirements for a specific application or the existence of a previous design will often provide an obvious reference point for such comparisons and give a clear indication of which property groups should be given priority, it also argues for a general-purpose set of property groups focused on loop stability in precision mechanical devices. It also suggested a 'model' material in terms of a set of basic properties that appear quite attractive for common situations in precision engineering and so could be used as a reference when there is no obvious existing choice. This approach is now adopted to explore the merits of a range of materials, from the most commonly adopted ones to a few of specialised interest. It is important to note that all values given here are intended only for guidance to typical behaviour patterns. This is suitable for outline designing intended to identify preferred types of material, but it remains crucial that the selection is confirmed by detail design calculations based on suppliers' data. In order to emphasise this need, values quoted here are nearly always normalised to the model material, so preventing them being applied without some further thought. Numerical data in this form, covering both the base properties and property groups, is provided for a selection of materials in Table 12.A.1 through Table 12.A.4 in the Appendix. Comparisons are obviously much easier if the property groups are consistently tabulated and plotted in the same order, although the actual order is arbitrary. Practical experience suggests that ordering according to the scheme

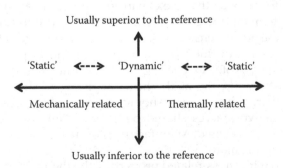

FIGURE 12.5
Schematic for the layout used in this chapter for the property group profiles relating to mechanical stress and thermal behaviours.

summarised in Figure 12.5 works conveniently for plotting overall property group profiles along a horizontal axis and it will be used throughout this chapter. Drawing justification from the discussions in Section 12.3, the thermally related set is placed to the right of the mechanical stress-related one, such that all the more dynamically related groups appear in the central region with those more related to static behaviours separated towards the left and right.

12.4.1 Metallic Materials

A survey of commercially produced precision mechanical systems will confirm that steel is a very widely used material; it is, for example, especially common within the main structures of machine tools and coordinate measuring systems, even though aluminium alloys and other materials have been used in some cases. This contrasts a little with the production of experimental systems or ones made in relatively small numbers for more specialised purposes, where steel is certainly used regularly but a wider range of other materials come into play. The reasons for this are pragmatic ones. Steel is quite strong and stiff, very readily available, and costs less than many alternatives. Steel machines well and almost every machine shop will have experience working with it. The many variations of steel alloys and heat treatments provide considerable control over strength, hardness, long-term stability, corrosion resistance and so on. These features create considerable incentives for using it in mainstream production environments. Compared to the model material for precision mechanical applications, conventional steels tend to be indifferent performers; although not at all bad in some mechanical aspects, since two of the model properties are based on mild steel and strength is typically higher for other steels. Figure 12.6 illustrates the property profile for a typical spring steel, which is similar to mild steel in all characteristics except yield strength; it should be no surprise that the property groups associated with flexure applications stand out. Thermal performance is quite poor in some areas.

Other widely used ferrous materials include cast iron, which, as indicated in Table 12.A.3, looks to be quite a poor performer but has the useful properties of being easily cast into large pieces, such as machine bases, and tending to have higher internal damping than most metals because graphite phases in its matrix help to dissipate vibrational energy. Stainless steels,

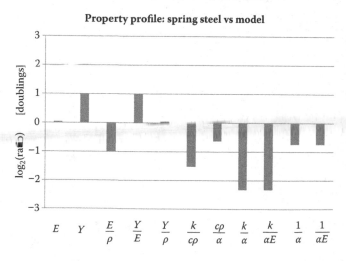

FIGURE 12.6
Property group profile for a typical spring steel, relative to the model material in Table 12.1. It uses data from Table 12.A.3.

which have large proportions of chromium (commonly around 20%) and nickel (usually around 10% or more), provide the obvious advantage of high corrosion resistance, but this may not be so critical in the controlled environments common to many precision applications. As Figure 12.7 indicates, the profile for stainless steel is inferior to other steels in most respects, particularly some thermal behaviours. It is also a lot more difficult to machine cleanly than steels with lower levels of alloying. However, its corrosion resistance is often a significant benefit and there are important applications where the use of stainless steel is almost essential, often related to the relative chemical neutrality of its surface oxide layer: biomedical, food-related and vacuum applications are obvious examples. Alloying commonly affects phase transition temperatures and many stainless steels contain enough nickel and chromium to make them austenitic under ambient conditions. Only the ferritic phase of iron is ferromagnetic and so an austenitic steel provides a 'non-magnetic' option that is not so different to mild steel except in a couple of the thermal property groups.

Aluminium and copper alloys are the other most commonly used metals, based on general arguments similar to those for steels. They mostly machine easily so overall manufacturing costs remain reasonable even if the raw material is rather more expensive, offer heat-treatable variations in strength and stability, and tend to have good recycling value. As a very broad comparison, on a per unit mass basis, stainless steel and aluminium alloy are likely to cost around two to three times as much as mild steel, with copper alloys a little more expensive still. Brasses (essentially copper with around 30% or more zinc) were the traditional base materials for instrument makers in previous generations, but the balance of economic and technical factors makes them now less attractive for many tasks. Bronzes are alloys of copper with mainly tin (typically of the order of 10%) and are also no longer so attractive in general. Both bronze and, less so, brass are used in sliding bearings, often in alloys containing other softer elements such as lead (see Chapter 7). Especially given easy availability as thin sheets, phosphor bronze (fairly low tin plus a few percent of phosphorus) and beryllium copper (around 2% beryllium) are good candidates for springs and ligaments in flexure devices. Figure 12.8 illustrates the property profile for a typical

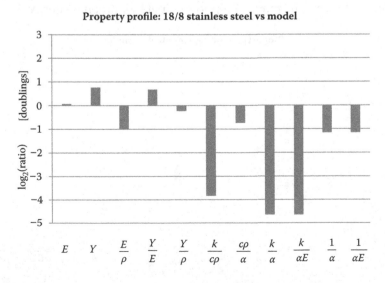

FIGURE 12.7
Property group profile for a typical stainless steel, relative to the model material in Table 12.1. It uses data from Table 12.A.3.

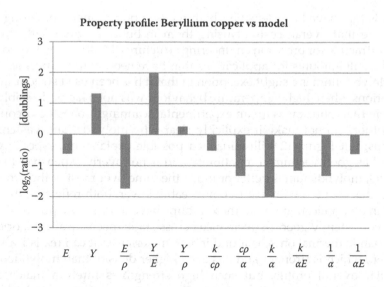

FIGURE 12.8
Property group profile for a typical beryllium copper, relative to the model material in Table 12.1. It uses data from Table 12.A.3.

beryllium copper, which, because of its high yield strain in its hardened state, might be compared to the spring steel shown in Figure 12.6; other brasses and bronzes have quite similar profiles, except for the influence of different yield strengths. High-purity copper, especially as 'oxygen free copper', has important applications in the diamond turning of moulds having very high surface quality, based on its high ductility, thermal response and chemistry. Pure aluminium is also attractive for diamond turning, for creating metallic (electrically conducting) layers in microsystems and for thin foils. Otherwise pure aluminium is usually too soft and alloys with typically a few percent of copper (and small amounts of other elements) are used; the everyday use of the word 'aluminium' often actually relates to a 'Duralumin' type of alloy with around 4% copper. Because several of the features of aluminium alloys were discussed in the context of setting up property groups in Section 12.3.1, a full plot is not provided here but all the trends can be seen clearly in Table 12.A.3. As would be expected, many other aluminium alloys, often including silicon or magnesium, are used for superior performance in terms of strength, fatigue, machinability and so on. Noting the potential implications of their quite low density and high conductivity and expansivity, aluminium alloys are reasonable general choices for monolithic flexures and for similar components in many instruments, but perhaps less so in, for example, machine tools where the structural loops will be more heavily loaded. It should be noticed that the thermal performance of all these alloys is (at least on the measures used here) poorer than those of pure aluminium and copper. Adding just a little copper to aluminium confers the mechanical advantages of alloying, but seriously reduces its thermal conductivity in particular because the larger copper atoms distort the lattice, so impeding the free movement of electrons as transporters of thermal (i.e. kinetic) energy and electrical energy. Pure copper, and to a lesser extent pure aluminium, provides a high thermal diffusivity which is attractive if mechanical loops need to return rapidly to equilibrium after unavoidable heat disturbances, but, again, alloying reduces this parameter. Copper, having quite high density and conductivity, is also a good choice for heat sinks.

Those metals that have high densities, high melting points and offer enough challenges for manufacture that overall costs of using them in bulk are quite high do not appear immediately attractive for precision engineering structures. Mostly, they would be expected to be used in small amounts for applications that have need of their specific highly useful properties. Molybdenum is a slight exception in that it has been used structurally in at least a few applications where high temperature behaviour was not a factor. Possibly the earliest recorded use in this context was for an experimental scanning probe microscope (Fein et al. 1987). The authors do not make it explicitly clear why molybdenum was chosen for the reference loops, but Figure 12.9 illuminates a possible explanation, especially when also noting that it has good vacuum compatibility and is not overly expensive. In comparison with the model, molybdenum is better or much the same over most of the property groups being used in this chapter, with just two noticeably lower, both reflecting the influence of high density in an application where these groups have a fairly low practical significance. This level of consistently good performance across static mechanical categories, and both static and dynamic thermal ones, is found in few materials that can realistically be used in bulk. Tungsten, which is stronger and has even higher density than molybdenum, offers a broadly similar overall profile, but such high strength is rarely a major criterion for instrument reference loops and the extra challenges of manufacture make it an unlikely practical alternative. It does, though, have some interesting special applications, for example, high stiffness combined with its thermal behaviour (and also its crystal lattice) makes tungsten useful as a thin-film buffer layer in microsystems, to avoid thermal stresses being induced in semiconductor layers by other metallic layers such as conductors. The 'precious metals', platinum, silver and gold can be conveniently added to this group, although they are rather easier to work. In precision engineering contexts, all appear almost always as thin layers or coatings to provide a particular surface layer functionality, be that electrical, optical, electro-chemical, bio-compatible or others. The stability and low shear strength of gold has even led to occasional use as a thin-film solid state lubricant in extremely demanding environments such as deep space missions (where lead alloys might be used in more normal circumstances).

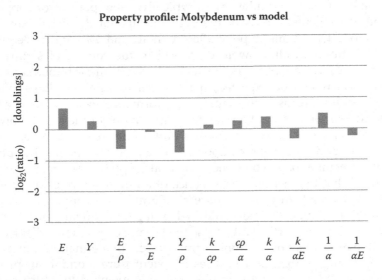

FIGURE 12.9
Property group profile typical of molybdenum, relative to the model material in Table 12.1. It uses data from Table 12.A.3.

A few low-density, fairly strong metals get mentioned often as higher technology alternatives to aluminium alloys across a range of engineering applications. Mostly, though, the reasoning focuses on specific properties not necessarily of high concern within precision engineering. Beryllium has a generally quite attractive property group profile, especially where stiffness and low density are needed, but has a major downside of oxidising very readily and that oxide being highly toxic, and a hazard even to touch. It should, therefore, only be considered when there is no feasible alternative to exploiting its specific properties. (Note, though, that beryllium copper, see earlier, where a relatively small amount of beryllium is tightly bound within the alloy, is not regarded as hazardous.) Titanium has high strength, which combined with low stiffness and low density, provides good values for the more dynamically oriented mechanical property groups, but its thermal characteristics are poor for a metal. It is also more challenging to machine than its obvious rival alloys, partly because of its thermal behaviour. It is, therefore, generally more expensive to use than an aluminium alloy and its modestly better performance in some mechanical contexts are likely to win out only when they must be pushed towards the limits (for example, in some aerospace applications). This is unlikely to be the case in precision engineering, where even the high value of Y/E is unlikely to be enough for selecting titanium for a conventional flexure mechanism, once the manufacturing challenges are taken into account; this decision might change if, say, minimising the overall size of a device is a dominant design driver. Magnesium alloys (usually with a few percent of aluminium and other elements in smaller amounts) are relatively easy to machine, and can be relatively strong but have (for a metal) very low Young's modulus and high expansivity. They offer overall property group profiles not so different to those of aluminium alloys, better in some groups (notably Y/E) but rather poorer in others. Their use in precision engineering is fairly uncommon.

As happens in many fields, some alloys have been developed specifically to focus on the needs for precision references. For example, one of the most important is Invar, an iron alloy with about 36% nickel and small amounts of other elements. Originally developed by Charles Edouard Guillaume for chronometers and length standards, it has most properties similar to mild or stainless steels but with much lower expansion. Many alloys of ferromagnetic metals such as iron, nickel or cobalt display magnetostrictive behaviour in which a small change of dimension (arising from internal strains) occurs when an imposed magnetic field changes strength. The thermal properties of Invar depend on a quite complex set of internal interactions of magnetic and elastic effects, broadly meaning that thermally induced strain (expansivity) causes changes to the internal magnetic states that tend to generate a counter-strain. With careful heat treatment, the expansivity can be reduced to 10^{-6} K^{-1}, or a little less, over a usefully wide range of temperatures around room temperature. Few other materials can better this and those that do tend to be brittle. With machining characteristics similar to a difficult stainless steel and the need for accurate heat treatment, It tends to be used only when low expansion and toughness are needed. Quite small changes to the composition of Invar (mainly increasing from 36% to around 42% nickel content) lead to another alloy, Elinvar, which has slightly larger expansivity, but the interesting property that its Young's modulus holds constant over a range of temperatures around ambient; this constancy is of particular benefit for reference springs, for example in chronometers. Generally, materials become less stiff as temperatures rise and thereby loops might progressively distort even if applied or internally generated loads are constant. It is worth a final comment that all materials properties are determined experimentally and tend to vary to some degree with temperature; strictly, then, uncertainty budgets for metrological loops should include not just the direct effect of a particular property or property group but also, ideally, the measurement and environmental uncertainty in the value used. In practice, this uncertainty in the 'constants' is normally neglected but, as with all second-order effects, it is wise to be aware that it exists and to treat the values with caution when considering the most demanding applications (see Chapter 9 for more detail on uncertainty budgets).

12.4.2 Ceramics and Glassy Materials

In summary, ceramics and glasses tend to be brittle, which can restrict their practical capability to carry tensile and bending loads. Many are inherently strong, but this strength is not generally realised within components made from these materials because small cracks and other surface defects act as stress raisers, and lead to the rapid propagation of failure surfaces. A catastrophic failure mode is commonly thought a disadvantage compared to the more gradual collapse likely with ductile materials (metals, in particular) but is sometimes useful in precision systems. If a reference loop is accidentally subject to abusive loading it might be better that it is physically broken (acting like the mechanical equivalent of an electrical fuse) rather than suffering unobserved plastic distortion that could introduce an error into subsequent operations of the device. In any case, modern manufacturing methods with final machining or polishing from near net-shape formation can provide very good surfaces and will ensure the continued growth in applications for these materials.

The ceramics of most interest for precision engineering tend to be oxides, nitrides or carbides of aluminium, silicon or a few other elements. Figure 12.10 shows a property group profile typical of an oxide ceramic, in this case alumina (oxide ceramics tend to be identified by an 'a' or 'ia' ending). It is quite similar to, on average a little better than, the model material across the mechanical groups and not bad thermally except in groups that involve its low thermal conductivity. Zirconia is another common engineering ceramic, offering higher strength but otherwise tending to be inferior to alumina across the property groups used here (see Figure 12.11), implying that it would only occasionally be preferred for precision applications. Ceramics are electrical insulators that are perhaps still seen as applied mostly to small, often intricate, components. However, they have been applied very

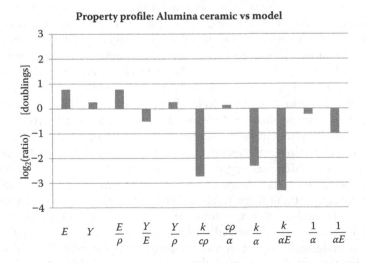

FIGURE 12.10
Property group profile for a typical alumina engineering ceramic, relative to the model material in Table 12.1. It uses data from Table 12.A.4.

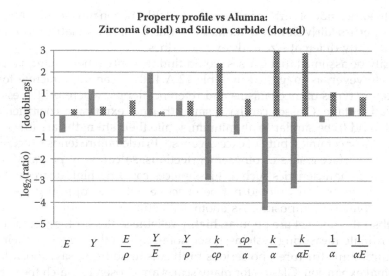

FIGURE 12.11

Property group profiles of typical zirconia and silicon carbide ceramics, relative to that for the typical alumina shown in Figure 12.10. It uses data from Table 12.A.4.

successfully to large loop structures, even as machine tool beds since at least the 1980s (Ueno 1989); the design rules are, of course, different to those for cast iron or steel.

Nitride and carbide ceramics offer a mix of properties of some specific interest for precision engineering. Silicon nitride is important in microsystems technology (especially socalled surface micro-machining, which is actually a lithographic process) where strong, insulating thin films are used both structurally and as part of lithographic masking processes. As a bulk material, silicon nitride is sometimes worth the extra expense to exploit a strong ceramic similar in stiffness to alumina, but otherwise out-performing it on every property group of the set used here; only those groups involving its low thermal conductivity are below the model values. As illustrated by Figure 12.11, bulk silicon carbide performs better overall than other common ceramics (alumina is used as the reference material in this plot) and compares quite favourably to the model material because its thermal conductivity is high for a ceramic. Its high stiffness might, though, be a slight disadvantage for some types of constrained loop elements. Care must be taken, though, because silicon carbide can form a large number of distinct crystal phases and so some aspects of its overall performance can depend on how it is processed. It is a tough, hard material, which adds to the challenges, including the costs, of manufacturing complex shapes, but it can still be machined (mainly using grinding processes) and polished. Tungsten carbide is well known as a material of exceptional strength and stiffness, characteristics that make it very attractive for applications such as cutting tool tips or gauge blocks. These properties and its very high melting point also make it difficult to work with and so it is otherwise quite rarely seen in precision applications. In practice, tungsten carbide tools or other parts are often sintered with significant amounts of a metal such as cobalt to act as a binder; many of its properties will be affected by such binders. The cubic crystal form of boron nitride is also much used for cutting and grinding tools, offering a performance often not much inferior to natural diamond tips and superior to it for cutting steels, for which diamond is chemically unsuited. This is not such a surprising use once the technology was developed to produce it reliably. Being composed of the elements immediately to either side of carbon in the periodic table, boron nitride might be expected to have some properties fairly similar to diamond, while the inherent lattice strain induced by the atoms of slightly different size adds extra hardness.

It might easily be assumed that glasses would find use only when high-quality optics are being made. However, as suggested in Table 12.A.4, they can be attractive for structural members in some types of mechanisms and reference loops, where brittleness is not necessarily a disadvantage. Glasses can have small thermal expansion coefficients and their elastic moduli tend to be similar to aluminium, while their strengths are low compared to most metal alloys or ceramics, but not excessively so. Brittle failure tends to occur in practice from the sites of surface cracks or other imperfections, so carefully prepared surfaces can lead to surprising characteristics such as glass pieces carrying high degrees of bending; if this were not so, glass fibres would not be effective either as optical guides or as reinforcement in polymeric composites. As another illustration, high-quality optical gratings have glass substrates and yet are popular, highly reliable scales for high-precision machine tools and coordinate measuring systems; given some care in the design of their mountings to steel or aluminium structures, no serious problems arise from, say, shock loads or differential thermal expansion. Glasses for many scales are chosen to match the expansivity of cast iron.

A fundamental characteristic of a glass is that it is amorphous, having no long-range crystal structures. A possible consequence could be that a glassy material might creep somewhat more readily than crystalline ones, which has led to questions of its long-term stability for use in reference loops. In practice, the glasses used in precision applications do not cause problems in this respect, but it would be wise to take extra care in unusual circumstance such as regular use at elevated temperatures. On the other hand, glasses can be moulded, fused, ground and polished; they are not that difficult to machine, even by conventional good-quality tools. Ordinary silicate glasses (for example, Crown glass) are used from time to time in precision engineering, especially for one-off experimental devices but even as monolithic flexure mechanisms in commercial instruments (Mollenhauer et al. 2006). However, glasses having low expansivity tend to be of greater interest for precision applications. The properties of borosilicate glass are similar to other silicate glasses except that borosilicate glass expands typically around three times less than does Crown glass and it is a basis for the formulation of the ultra-low expansion compositions known as ULE. Fused silica (silicon dioxide) and fused quartz are close to the notional ideal of a glass and offer excellent stability alongside an expansivity well under one-tenth that of Crown glass and comfortably below the part-per-million boundary where there are few rivals. Silicon dioxide is also commonly used in MEMS, both structurally and as a sacrificial layer during lithographic manufacture.

If low thermal expansion is the dominant requirement in an application, then options are fairly limited to materials such as Invar or fused silica, where other general performance and manufacturing factors will be sacrificed to the specific need. The ultimate example of this approach lies with ultra-low expansion ULE glass-ceramics. At a basic level, they comprise a quite conventional glass to which various additions have been made, notably titanium dioxide. This leads to a distribution of tiny crystallites (ceramic phase) in the glass matrix and, with careful heat treatment, it can be arranged that the transition between these phases, which have different densities, at temperatures near ambient almost exactly compensates for the normal thermal expansion. Local expansivity below 10^{-8} K^{-1} can be achieved, while the material can in other respects be treated much like an ordinary glass. A common composition known as 'Zerodur' (see, for example, Brehm et al. 1985) is optically transparent but has a characteristic golden-yellow tint caused by scattering from the crystallites. It is relatively expensive and very much a material that is likely to be used only when its special characteristics address a very specific need, but see also Section 12.4.5 for further discussion of the instrument shown in Figure 12.1. For example, while low expansivity is desirable to minimise environmental and handling uncertainties in a length standard, the relative softness of glass leaves it vulnerable to scratching and ULE is unlikely to be used for this task except in controlled facilities.

12.4.3 Polymeric and Composite Materials

It would be quite surprising to find thermoplastic or thermoset polymers used directly as major components of the mechanical loops in precision machines or instruments. Compared to the other materials classes already considered here, they are strongly insulating and have low density (usually well below 2000 kg·m^{-3}). They tend, however, to have low elastic modulus (typically a few gigapascal), at best modest strength (some tens of megapascal), high thermal expansivity and are sometimes susceptible to creep and to water absorption (including from humid atmospheres), all of which could reduce loop stability. Nevertheless, there is widespread use of many different polymers for functionally important roles within precision devices. Of course, polymeric materials also feature widely in such things as cosmetic, non-structural outer casings, shielding against airflow, safety screens and so forth, but such uses will not be further considered in this chapter. There is limited value in comparing polymer properties to those of the model material used in this chapter, because some values are grossly different, as are the typical applications. Nevertheless, for consistency, illustrative values for the normalised properties of a few are included in Tables 12.A.2 and 12.A.4. When normalised this way, common engineering polymers appear to have quite similar properties, but the actual differences between them become potentially important once a design decision in favour of polymers has been taken.

Two major groups of engineering polymeric resins can be broadly summarised as follows. Thermoplastic materials can, once made, be chopped up, heated and melted back into a single piece having much the same properties as the original. Hence, they are then suited to processes such as extrusion and injection moulding. A wide range of thermoplastics is used across general engineering applications, with those commonly found in precision applications including acrylics, polycarbonates and polyacetals. Thermoset materials are formed by the chemical reaction of two (or more) components which cure together (increasingly cross-linking the molecular structure) over time, often, but not always, at moderately elevated temperatures. The precursor resin is generally liquid and so these materials are well suited to casting processes. Typical examples are epoxies and polyesters. Since many thermosets and thermoplastics will flow readily under modest pressures and temperatures, they are excellent at reproducing fine detail from the surfaces of moulds while using relatively simple manufacturing equipment. Thermoplastics can also be pressed under gentle heat to form precise surface topographies. A commonplace example of this is the almost zero cost of producing each individual copy when manufacturing polycarbonate compact discs or DVDs. The micrometre-scale pits that encode the data can be reproduced with uncertainties as low as a few nanometres. The analogue track on a vinyl record is an exemplar from a previous generation of technology.

Thin layers of polymers are frequently used at interfaces between major subsections of mechanical loops. Their low shear-strength provides good low-friction layers for dry sliding bearings with, for example, PTFE, polyacetal and polyamides (nylon) being common choices. Polymers are also used very successfully as adhesives, providing that the loading does not lead to a high shear stress. The technical details of these applications are extensive and beyond the scope of this chapter. It can be noted, though, that concerns about low dimensional stability might not be so relevant when only films are used. Even a significant strain in such a small length of the entire loop could well be negligible overall. Care still needs to be taken that any such strains do not affect the loop shape, whereby their effect could be magnified by Abbe effects (see Chapter 10).

Polymer use for defining structural loops tends to be limited to very small devices where loads and consequent strains will be small. They can be attractive for applications in which large numbers of components with small, well-aligned features are needed, as are found in connectors for arrays of single optical fibres or very fine electrical contacts. The combination of low stiffness and low density might sometimes be exploitable for flexure devices and some types of diaphragms for pressure (acoustic) sensing. Occasionally, they might be exploited for proof-of-principal prototyping, because of the ease of working with them. Polymethyl methacrylate (PMMA), which is available in large sheets and blocks that are readily machined, and various epoxies are commonly used. PMMA and epoxy resins are easy to cast or mould from liquid precursors. Note that these two materials are also the basis of many of the photoresists used in lithography for microelectronics and MEMS. This photo technology can also be used to generate functional structures in the polymer layer itself, which can either remain anchored to a substrate or be 'floated' off by a wet chemical process. PMMA (in particular) can be used for LIGA-like processes in which a high-energy lithographic step produces quite deep structures upon which a galvanic process then makes replicas in a metal such as nickel. Polyethyl ethyl ketone (PEEK) is another relatively stiff and strong polymer that has (for a polymer) reasonably high temperature tolerance and low shrinkage after injection moulding, giving it increasing technological relevance in many areas. Typical uses are in lightweight gears for power transmission, which suggests also potential for application in some types of precision mechanisms.

Polymers that can be made from liquid precursors are very attractive as the matrix material for engineering composites that offer much improved stiffness and strength. Of most importance in the present context are glass-fibre and carbon-fibre reinforced polymers, which often provide properties directly comparable to the model material. Although care must be taken over thermal and some other properties, the ease with which complex shapes can be made can make them useful in precision structural applications. The underlying idea is that adding, for example, a stronger material that bonds well to the polymer provides extra tensile and bending strength while preserving many of the other useful properties of the matrix. The reinforcement can be in small pieces randomly distributed. Greater advantages can be obtained by using longer fibres laid down in a more organised way along one of more directions. This results in anisotropic properties, which is interesting because the actual loading in most loop structures is also anisotropic. Note that anisotropy applies across all properties, with some resultant effects of potential significance. For example, differential thermal expansion will lead to internal stresses, compressive in the matrix as temperature rises. This effect might sometimes lead to concerns about loop distortion or even de-bonding between the matrix and fibre, but it can have positive outcomes. The net linear expansivity of the composite can be much reduced along the fibre direction, with a consequently higher value laterally. Some types of carbon fibre can show negative expansion around room temperature (see, for example, Kulkarni and Ochoa, 2006) and so the expansivity along one axis of some epoxy composites can reduce to almost zero, making them ideal for such things as low-mass pushrods. For general purposes, the most common layouts tend to have layers of fibres in two directions at 90° in order to provide a good compromise between tensile and shear behaviour. Materials failure in fibre reinforced polymers could be exacerbated by internal voids arising from poor preparation, but is mostly through failures of bonding to the fibres under tensile loading, including delamination, where a failure plane can cause a peeling away from a fibre-rich layer. How the reinforcement affects compressive strength is even more difficult to specify and crushing failure of the matrix might occur under higher loads on compressive surfaces. The actual behaviour of a particular fibre composite is so specific to the quality of the bonding, to the type, amount and distribution of the fibres, and to the exact matrix that no further discussion is warranted in an overview such as this.

Many engineering polymers are actually particulate composites rather than fibre composites, where filler materials such as clays, chalk or silica powders are mixed into the matrix. The reasons for doing this can range from improving wear resistance to simply reducing the cost of bulk material. Metal matrix composites are also of occasional interest in the present context, especially cemented carbides (cermets), such as tungsten carbide in a cobalt matrix, considered earlier. Further exploration of all these issues can be found in standard texts, for example, Askeland (1996).

By no means are all technological composite materials fibre reinforced polymers. Many of their important applications are less relevant to precision engineering; for example, the uses of woods and special reinforced metals. Reinforced concrete is an obvious example, where the concrete (itself a composite of cement and stone) provides an easily processed (pourable) approximation to stone, with steel bars adding tensile strength as a second composite function. So-called 'polymer concrete' is a development of this idea aimed at applications such as machine bases. Comprising typically an epoxy resin heavily loaded with granite chips, it might be used to approximate a granite base in a castable material that polishes to a reasonably smooth and durable surface and can easily accommodate inset fittings for slots, tapping points and so on. Like cast iron, it offers good damping properties, but can be used in a wider range of situations. It can be poured into major structural loops fabricated from steel tubes to provide damping and some extra rigidity (giving, overall, an analogue to reinforced concrete).

Another family of composites is created by bonding together thin layers of different materials with the intention of generating asymmetrical behaviour. Simple temperature sensors and thermally driven actuators have long been made by effectively welding together thin sheets of two metals having different coefficients of thermal expansion: a bimetallic device. Set up as a beam, the upper surface expands relative to the lower one (or vice versa) as the temperature changes, which can be described in terms of tensile and compressive strains relative to the neutral axis and, in the absence of heavily constraining external forces, causes a change of curvature and consequent lateral deflection. Note, in passing, that bimorph actuators apply the same principle with the differential expansion driven by effects such as piezo-electricity. An axial variation of this concept can be used to hold a datum position against environmental thermal changes. Conceptually, a rod of length L_1 is attached to a base and then a shorter rod, length L_2, attached alongside via a bracket at its free end. The free end of the second rod then provides a reference point a distance $(L_1 - L_2)$ from the base. A temperature change of $\Delta\theta$ will cause a net change in this distance of $(\alpha_1 L_1 - \alpha_2 L_2)\Delta\theta$. So, by selecting the lengths in inverse ratio to the expansivities of the rods, the positional variation can be reduced in principle to zero. Formerly, this idea was used within the compound pendulums for high-precision clocks, but it would now be more usual to make a simpler design from, say, Invar. Manufacturing methods associated with microsystems technology allow the production of novel types of composite, but comments on them will be deferred to the following section.

12.4.4 Esoteric and Emerging Materials

If materials-related costs are often of somewhat less concern in precision applications than in others, it becomes possible to consider the use of materials that might at first sight seem highly unlikely candidates. Indeed, by using some of the methods of microsystems technology, it is possible to create materials that do not exist in bulk form. This section examines just a few of these esoteric options, intending thereby to encourage designers to look more widely for others.

Given the very sophisticated technology developed for microelectronics processing, silicon provides much of the underpinning for the current generations of MEMS devices. Single crystal and polycrystalline silicon, its oxide and nitride, processed by lithographic methods known as 'surface micromachining' and 'bulk micromachining' (Gardner et al. 2001; Chapters 5 and 6 provide an accessible overview of these processes), dominate the basic structures of micro-devices such as accelerometers (for example, for airbag sensors) and also millimetre-scale systems such as pressure sensors or inkjet nozzles. Note that the microsystems field commonly refers, potentially confusingly, to 'polysilicon' as a shortened version for 'polycrystalline'; there is no implication here of a polymerisation process. The properties of single crystal material will generally be anisotropic, different along different crystal axes or planes. For example, silicon has a diamond lattice structure (which can also be interpreted as an FCC structure) and its elastic modulus is maximum, around 190 GPa, along the most closely packed directions, but only around 130 GPa normal to them. Unsurprisingly, polysilicon generally shows intermediate values. There is, though, a larger challenge in obtaining good-quality data for properties at the micro-scale. One reason is that very small structures cannot contain sizable flaws and so may well be comprised of material notably closer to theoretical ideals than are found in bulk measurements. On the other hand, any remaining imperfections might have an unusually large influence. Also, it is extremely challenging to measure the properties at the truly relevant scale and much of the available data has been inferred from the observed behaviour of complete microsystems, for which there might, among other factors, be large uncertainties in some of the cross-sectional dimensions. Consequently, no values will be quoted here and it is advised that published information is treated with caution. For example, some very high strength values can be found in the literature, which might be valid in that specific application but not readily transfer to a different context.

The interesting mechanical potential of silicon was first expressed strongly by Peterson (1982), but even this classic paper did not appreciate the potential of single-crystal silicon for large mechanical loops. In the subsequent decades, the availability of very large silicon crystals, demanded for microelectronics, has opened up new possibilities. This material also acts as a warning example about relying too directly on general sources for property data. Tables 12.A.2 and 12.A.4 include a set of values for (111) silicon, showing it to be comparable to the model material mechanically and superior thermally, but it is possible to find other sources of data which show the same trend much more weakly and can be up to a factor of three different in some of the property groups. While data about common metals tends to be quite consistent, property values reported for these more esoteric materials can be very sensitive to how testing was undertaken and it is wise, if possible, to design on the basis of a testing approach that reflects the specific application. Single-crystal silicon is produced with almost no crystal defects and so it is almost ideally elastic-brittle and has extremely low damping. Its use for large-scale devices is mainly restricted to cases where specific other properties offer major advantages: for example, x-ray interferometer monoliths that exploit diffraction at crystal planes to execute picometre-precision displacement measurements can include quite complicated arrangements of flexure mechanisms and kinematic locators (see, for example, Chetwynd et al. 1990). Being very brittle, silicon can be notoriously fragile if surface defects become stressed, but it can nevertheless be worked without undue difficulty by a combination of various types of diamond grinding followed by etching to remove the surface damaged layer. This is facilitated by its low density which makes it relatively buoyant when immersed in typical liquid etchants so that it can be gently 'tumbled' for uniform etching without experiencing large forces. It is, therefore, capable of wider application, although it is likely to remain a clear choice only for specialised applications.

The property group values given for diamond in Table 12.A.4 reflect those of an outstanding choice of material for many precision applications. It is as good as the model across all the groups and sufficiently exceeds it in most to the extent that a profile plot is not greatly helpful. It is very hard and offers the interesting and unusual combination of excellent thermal conductivity with electrical insulation. It is, of course, inconceivable as a material for such things as macroscopic loop structures, for reasons of availability even before excessive costs are considered. It appears in bulk at the millimetre-scale in a few special cases such as natural diamond cutting tools and blades for microsurgery and microtomes. The situation is very different when thin films are considered. Various vapour deposition techniques can put down layers of almost pure diamond phase carbon across large areas of substrate. Such layers do not provide properties as good as those of pure single crystal material and so are almost always referred to as 'diamond-like coatings' (DLCs). These coatings nevertheless offer properties approaching those of diamond and they are increasingly exploited for producing smooth, hard, low-wear functional surfaces. As methods improve, it is plausible that high-quality diamond-phase carbon could be grown on MEMS as part of a surface micro-machining lithographic production with parts of the substrate then etched away; perhaps, then, one day will see some application for stand-alone diamond micro-devices.

Other allotropes of carbon are also of interest. The hexagonal crystal structure of graphite leads to shear properties that make it a very attractive solid lubricant. However, it can be made into very large pieces that are readily machined; indeed, extremely large pieces are commonly used in nuclear reactors. In a precision engineering context, its general properties balanced against cost and so forth are not so attractive for structural applications. However, as it can be machined and polished and is slightly porous to gases, it finds applications such as within the structure of special air bearings, not just as a powder lubricant. Graphene, which can be thought of as a monolayer of graphite containing the hexagonal lattice, is by its nature a nanotechnological material still under rapid development but likely to offer important applications across electronics, sensing and so on. It is, though, not a mechanical material useful for large-scale structures and so it will not be further discussed here. The same argument applies to other allotropes such as fullerenes (including the famous C_{60} buckminsterfullerene 'football like molecule') and the whole family of carbon nanotubes (CNTs). These materials are, though, being considered for reinforcements for composites, within metallic as well as polymer matrices. New high-performance materials of great interest to precision engineering might arise over the coming years.

Another field likely to grow is the use of processes such as vapour deposition methods (both physical, such as sputtering, and chemical) to produce materials at the micro-scale that cannot be made conventionally. Given good enough process control, it is now practical to place layers of different materials atop of each other to create a crystal-like structure that will not grow naturally and consequently has unusual properties. There are now very many such 'meta-materials' being researched and developed, especially, but not only, for their optical properties, but they are well beyond the scope of this chapter. Even without such levels of control, it is possible to produce layers of effectively novel materials. Sputtering simultaneously from two or more source materials can allow a more-or-less uniform coating of non-stoichiometric material. For example, chromium and nitrogen will form a hard, tribologically attractive layer (referred to as Cr-N, because it is not a compound in the normal sense) that is used in applications such as protective coatings for hard-drive discs. Simple control of sputtering parameters can produce a range of distinctly different surface morphologies with differing physical properties (Gerbig et al. 2007).

12.4.5 Final Comments: Properties and Practical Design Choices

The previous sections have discussed and compared the properties of a wide range of materials, using property group profiles, and sometimes graphical methods, to illustrate their more and less attractive features for application to precision mechanical structures. The appendix provides tables of properties and property groups normalised against an artificially conceived model material considered to represent typically good behaviour for precision mechanical applications. It is very important to stress that these tables are illustrative, drawing on rounded values typical of those reported in the wider literature, which can be considered adequate for initial planning and the screening of conceptual designs but lack the necessary authority to be used directly in detailed designing. Similarly, it should be reiterated that the approach to graphical comparisons is offered as a background technique to be adopted and adapted as individual users see fit; hence, further exploration of them is left as an exercise. A few final observations about practical realisations are offered here.

Consider again the examples of surface topography instruments illustrated by Figure 12.1. The majority of general-purpose micro-topography instruments continue to have major loops based on steel, just as they did fifty years previously. The addition of, for example, granite (or granite-like) bases or columns appears often to be as much for aesthetic and marketing reasons as of genuinely direct functional importance. Such instruments provide adequate performance for the majority of industrial needs and remain highly effective in terms of such things as robustness and cost, including the costs of change to legacy designs. By the 1960s, it was becoming clear that some applications needed higher precision than was available from these general designs, but only for a restricted range of workpiece shapes and sizes. The first commercial system to address this need was the Talystep 1, introduced by Rank Taylor Hobson Limited, primarily for measuring step heights over relatively short traverse lengths. It derives this short but extremely repeatable probe scan by means of a very deep flexure hinge. Thus, the overall loop dimensions remained quite large (or the order of 300 mm) despite it accommodating workpieces of no more than around 25 mm deep. A predominantly cast design led to a low number of interfaces in the main loops. It was anticipated that use would always be within fairly well-controlled metrology rooms or clean rooms, where overall temperature would not change by much, especially over the time of a typical measurement, but there might still be significant airflow and consequent thermal gradients. High thermal conductivity to better maintain a constant temperature within the instrument was considered more important than low expansivity because thermally induced change of the metrology loop shape tends to generate larger errors than does uniform expansion or contraction. The loop design was revolutionary and so could not draw on much legacy design. Taking these and other arguments together, the final decision was to make the majority of the instrument from aluminium alloy and it has proved very successful. Note, for interest, that the small loops within the probe head of the instrument included several quite esoteric materials intended to introduce thermal compensation in the one region where short-term expansion was a potentially serious problem.

By the 1990s there was a need to accommodate a somewhat wider range of ultra-precision workpieces and new materials were available. In particular, work at the UK National Physical Laboratory had by then demonstrated the first of a whole series of special calibration topography instruments based on the extensive use of optically polished ultra-low-expansion glass ceramic and special sliding bearings using thin films of polymers (typically PTFE). This was adopted in the commercial Nanostep 1 instrument, allowing a loop design rather more like that of a conventional surface profilometer that could readily incorporate longer scan ranges. The use of a material having such low expansivity ensured excellent loop stability and this sole factor could dominate the materials selection issue: it was the obvious choice provided only that no other property actually prevented proper operation under reasonable conditions. In practice, some smaller, critical components in the probe head were difficult to make using standard ULE ceramic stock and fused silica was substituted in these cases. These materials choices led to a relatively expensive system, though, and both designs continue to be used. Garratt and Bottomley (1990) give a good description of how the concept was developed into commercial form from the NPL system of Lindsey et al. (1988).

The preceding example helps to emphasise that even in precision engineering there are relatively few occasions on which choice of a less common material is a simple, direct matter of meeting a functional requirement for, say, structural loops. Commonly, design briefs will provide a reasonable level of flexibility over choices of dimensions, which, taken alongside less tangible factors associated with manufacture and costs, leads to the use of common engineering materials that are in a formal sense sub-optimal in their property group profiles. Mostly, this will be a perfectly rational design compromise. Inverting this observation, it appears that the rational approach is to start with one of a small number of commonly used material choices (steel much in evidence, if not dominant in this set) and considering a move to less common, less well understood, or more expensive materials only when particular challenges arise in satisfying a design brief or there are prospects for a step-change in overall system cost-effectiveness. Commercial flexure mechanisms offer another illustration. Steels and aluminium alloys have quite similar profiles across the property groups typically relevant to flexures, which are reasonable but by no means the best amongst the material discussed in this chapter. Steel is very common in applications using fabricated leaf-spring flexures, perhaps with copper alloy ligaments. One factor here is the ready availability of a large selection of sheet stocks from which the ligaments can be easily cut out or etched to produce almost arbitrarily complex shapes. One-off monolithic flexure designs often use aluminium because they are easy to machine from plate stock; also, recycling value might become a factor if large amounts of material are being removed. However, commercial high-precision stages tend to use steel (at least in standard products). High out of-plane stiffness, to reduce parasitic motion errors deriving from external forces, might benefit simply from high E and lead to this taking a higher weighting than other property groups. Steel might then become a really good option for a dimensionally compact design, which is likely to be a good marketing feature for a standard product. Actually, many modern monolithic flexures are made by combining a number of very narrow through-cuts created predominantly by wire erosion techniques that involve little material removal; again, this can and will influence the materials choice.

It is fitting to end this chapter with comments about manufacturing because it is usually a key factor in a successful design. More excitingly, it seems quite probable that many of the near-future breakthroughs in precision engineering will arise from novel combinations of materials and non-conventional manufacturing processes. Additive manufacture (AM) offers particularly good prospects. AM was originally conceived as a 'rapid prototyping' tool in which, for example, successive thin layers of a solid piece could be built up by successively spreading layers of a suitable powder and then selectively fusing them to the underlying structure that builds up. The particle bonding was rarely strong (perhaps comparable to a 'green state' ceramic) and the resulting components were too fragile to function but still useful for early checking for design errors such as geometric interferences. Alternative methods that are conceptually similar but involve selective curing of polymeric resins were similarly restricted. However, with improved technology and a much wider choice of better-prepared material feedstocks, AM is now a viable small-batch production technique. Larger-scale systems can, for example, produce complex 3D shapes in many metals and plastics. There is clear potential for medical applications such as custom designs for titanium prosthetics to replace damaged bone or for biodegradable polymers that provide 'scaffolds' to encourage tissue growth. The ability to make effectively monolithic 'open' metallic structures (visually similar to 3D trusses or foams, according to scale) offers new prospects for high strength-to-weight structures of obvious interest in aerospace and quite likely application to structural loops in precision machines (Syam et al. 2017). At smaller scales, micro-stereo-lithography offers the possibility of building objects by selectively curing liquid resins at a resolution governed ultimately only by the size of achievable optical resolution. There is no reason in principle why functionally active fine powders should not be incorporated into such resins to provide novel composites. Currently, AM methods perhaps do not deliver enough control to feature much in true precision engineering production, although their capabilities are growing steadily (Gardner et al. 2001, Gibson et al. 2015). What is certainly the case is that the AM machines themselves will make considerable and growing use of the principles covered in this book.

Appendix

TABLE 12.A.1

Some Indicative Mechanical and Thermal Properties for Selected Metals, Normalised to the Model Material Given in Table 12.1

Material	E	Y	ρ	c	k	α
Aluminium	0.36	0.40	0.68	1.22	1.58	3.43
Beryllium	1.59	1.15	0.46	2.43	1.34	1.71
Copper	0.65	0.77	2.24	0.51	2.57	2.37
Molybdenum	1.63	1.53	2.55	0.34	0.92	0.71
Titanium	0.60	2.33	1.13	0.70	0.14	1.27
Tungsten	2.06	4.50	4.83	0.18	1.11	0.64
Cast iron	0.75	0.70	1.83	0.69	0.33	1.57
Mild steel	1.05	1.00	1.97	0.56	0.37	1.57
Steel, spring	1.03	2.00	1.97	0.53	0.37	1.64

(Continued)

TABLE 12.A.1 (CONTINUED)

Some Indicative Mechanical and Thermal Properties for Selected Metals, Normalised to the Model Material Given in Table 12.1

Material	E	Y	ρ	c	k	α
Steel, hard	1.05	3.33	1.97	0.56	0.23	1.57
Invar	0.73	1.33	2.00	0.67	0.14	0.16
18/8 Stainless	1.03	1.67	1.98	0.68	0.10	2.29
Elinvar	0.85	1.23	2.00	0.61	0.07	0.57
Mg alloy	0.20	0.83	0.44	1.40	0.78	3.79
Brass 70/30	0.53	1.50	2.14	0.49	0.73	2.79
Bronze 90/10	0.65	2.00	2.23	0.48	0.33	2.43
Phosphor bronze	0.55	1.67	2.23	0.48	0.47	2.43
Beryllium copper	0.63	2.50	2.06	0.47	0.67	2.43
Duralumin	0.37	1.00	0.70	1.20	0.98	3.29

TABLE 12.A.2

Some Indicative Mechanical and Thermal Properties for Selected Non-Metallic Materials, Normalised to the Model Material Given in Table 12.1

Material	E	Y	ρ	c	k	α
Silicon (111)	0.95	0.63	0.58	0.94	1.05	0.33
Diamond	6.00	10.0	0.88	0.51	3.93	0.17
Silicon carbide	2.05	1.50	0.78	1.33	0.84	0.54
Silicon nitride	1.55	3.33	0.80	0.73	0.22	0.50
Alumina	1.66	1.15	0.95	1.40	0.23	1.19
Zirconia	1.03	2.73	1.44	0.62	0.021	1.46
Tungsten carbide	3.60	11.1	3.75	—	0.35	1.04
Fused silica	0.35	0.23	0.54	1.13	0.014	0.071
Fused quartz	0.35	0.23	0.55	1.12	0.010	0.071
Crown glass	0.35	0.23	0.63	0.93	0.007	1.14
Zerodur	0.46	0.32	0.63	1.10	0.011	0.007
Epoxy-granite	0.18	0.07	0.63	1.28	0.011	1.71
PTFE	0.002	0.08	0.55	1.40	0.002	11.4
PMMA	0.014	0.25	0.30	1.96	0.001	10.0
Polycarbonate	0.011	0.23	0.30	1.84	0.001	9.43
Polyester	0.012	0.18	0.33	3.07	0.001	14.3
PEEK	0.018	0.31	0.33	—	0.002	6.71
Epoxy	0.016	0.24	0.29	2.53	0.002	8.57

TABLE 12.A.3

Indicative Property Groups for a Selection of Metals, Normalised to the Model Material in Table 12.1

Material	E	Y	E/ρ	Y/E	Y/ρ	$k/c\rho$	$c\rho/a$	k/α	$k/\alpha E$	$1/\alpha$	$1/\alpha E$
Aluminium	0.36	0.40	0.52	1.13	0.59	1.92	0.24	0.46	1.30	0.29	0.82
Beryllium	1.59	1.15	3.44	0.72	2.49	1.19	0.66	0.78	0.49	0.58	0.37
Copper	0.65	0.77	0.29	1.18	0.34	2.25	0.48	1.09	1.67	0.42	0.65
Molybdenum	1.63	1.53	0.64	0.94	0.60	1.08	1.19	1.29	0.79	1.40	0.86
Titanium	0.60	2.33	0.53	3.89	2.07	0.18	0.62	0.11	0.19	0.79	1.31
Tungsten	2.06	4.50	0.43	2.19	0.93	1.29	1.33	1.72	0.84	1.56	0.76
Cast iron	0.75	0.70	0.41	0.93	0.38	0.26	0.81	0.21	0.28	0.64	0.85
Mild Steel	1.05	1.00	0.53	0.95	0.51	0.33	0.70	0.23	0.22	0.64	0.61
Steel, spring	1.03	2.00	0.52	1.95	1.02	0.35	0.64	0.22	0.22	0.61	0.59
Steel, hard	1.05	3.33	0.53	3.17	1.70	0.21	0.70	0.15	0.14	0.64	0.61
Invar	0.73	1.33	0.36	1.84	0.67	0.10	8.54	0.89	1.23	6.36	8.78
18/8 Stainless	1.03	1.67	0.52	1.63	0.84	0.07	0.59	0.04	0.04	0.44	0.43
Elinvar	0.85	1.23	0.43	1.45	0.62	0.05	2.15	0.12	0.14	1.75	2.06
Mg alloy	0.20	0.83	0.46	4.17	1.90	1.27	0.16	0.21	1.03	0.26	1.32
Brass 70/30	0.53	1.50	0.25	2.86	0.70	0.70	0.38	0.26	0.50	0.36	0.68
Bronze 90/10	0.65	2.00	0.29	3.08	0.90	0.31	0.44	0.14	0.21	0.41	0.63
Phosphor bronze	0.55	1.67	0.25	3.03	0.75	0.44	0.44	0.19	0.35	0.41	0.75
Beryllium copper	0.63	2.50	0.31	3.97	1.21	0.69	0.40	0.27	0.44	0.41	0.65
Duralumin	0.37	1.00	0.52	2.74	1.43	1.17	0.26	0.30	0.82	0.30	0.83

Note: Metals tend to be reasonably stiff and strong, dense, ductile (to varying degrees) and tough. Production with metals is usually fairly straightforward, using many combinations of casting, forging, machining, polishing, ECM and powder metallurgy.

TABLE 12.A.4

Indicative Property Groups for a Selection of Non-Metallic Materials, Normalised to the Model Material in Table 12.1

Material	E	Y	E/ρ	Y/E	Y/ρ	$k/c\rho$	$c\rho/a$	k/α	$k/\alpha E$	$1/\alpha$	$1/\alpha E$
Silicon (111)	0.95	0.63	1.65	0.67	1.10	1.93	1.63	3.14	3.31	3.00	3.16
Diamond	6.00	10.0	6.86	1.67	11.4	8.80	2.61	22.9	3.82	5.83	0.97
Silicon carbide	2.05	1.50	2.65	0.73	1.94	0.81	1.90	1.55	0.75	1.84	0.90
Silicon nitride	1.55	3.33	1.94	2.15	4.17	0.38	1.17	0.44	0.28	2.00	1.29
Alumina	1.66	1.15	1.74	0.69	1.21	0.17	1.12	0.19	0.12	0.84	0.51
Zirconia	1.03	2.73	0.71	2.67	1.90	0.02	0.62	0.01	0.01	0.69	0.67
Tungsten carbide	3.60	11.1	0.96	3.09	2.97	–	–	0.33	0.09	0.96	0.27
Fused silica	0.35	0.23	0.65	0.67	0.43	0.02	8.53	0.20	0.56	14.0	40.0
Fused quartz	0.35	0.23	0.64	0.67	0.42	0.02	8.62	0.14	0.40	14.0	40.0
Crown glass	0.35	0.23	0.56	0.67	0.37	0.01	0.51	0.01	0.02	0.88	2.50
Zerodur	0.46	0.32	0.72	0.70	0.50	0.02	96.9	1.49	3.28	140.0	307.0
Epoxy-granite	0.18	0.07	0.28	0.42	0.12	0.01	0.47	0.01	0.04	0.58	3.33
PTFE	<0.01	0.08	<0.01	41.7	0.15	<0.01	0.07	<0.01	0.07	0.09	43.8

(Continued)

TABLE 12.A.4 (CONTINUED)

Indicative Property Groups for a Selection of Non-Metallic Materials, Normalised to the Model Material in Table 12.1

Material	E	Y	E/ρ	Y/E	Y/ρ	k/cp	cp/a	k/α	k/αE	1/α	1/αE
PMMA	0.01	0.25	0.05	17.9	0.84	<0.01	0.06	<0.01	0.01	0.10	7.14
Polycarbonate	0.01	0.23	0.04	21.2	0.78	<0.01	0.06	<0.01	0.01	0.11	9.64
Polyester	0.01	0.18	0.04	15.3	0.56	<0.01	0.07	<0.01	0.01	0.07	5.83
PEEK	0.02	0.31	0.06	16.8	0.93	–	–	<0.01	0.01	0.15	8.16
Epoxy	0.02	0.24	0.05	15.5	0.83	<0.01	0.08	<0.01	0.02	0.12	7.53

Note: Ceramics tend to be stiff, strong, fairly low density and brittle. They are commonly made by powder processes (pressing and sintering) or forms of casting. Glasses are less stiff and strong, and also brittle, although many machine quite well. Production tends to involve casting followed by machining and/or polishing. Polymers are less stiff and strong than the other materials considered here, have higher thermal expansions and lower general long-term stability. Production of polymer components often involves forms of injection moulding and extrusion, but some machine quite well.

Exercises

1. Look up the linear thermal expansion coefficients and elastic moduli for a few materials from classes, such as metals, ceramics and engineering polymers, and sketch a log-plot of them, in the broad style of an Ashby chart. What general trend appears? Suggest a physical explanation for this trend (a top-level description, not a mathematical one).

2. By considering the Euler buckling formulae for axially loaded struts or columns, propose a property group that could be optimised if a major design criterion were to carry a specified compressive load using a strut of minimum weight. Hence comment on which types of materials might be good choices for the application, by referring to Figure 12.3 or other data.

3. Research briefly the 'machinability' of each of the following general types of materials: brass, mild steel, stainless steel, Invar, titanium and silica glass. Rank them in terms of likely ease of production of good surface quality on a workpiece being made in a small, non-specialist machine shop. Do any property groups correlate closely with this ranking, and, if so, why?

4. The resilience ($Y^2/2E$) of a material is, in effect, a volumetric measure of its ability to store strain energy, which is desirable in some types of elastic (flexure) mechanisms. Demonstrate that this property group is dimensionally equivalent to strain energy. Then, determine its value, normalised to the model material used in this chapter, for a selection of materials that might be considered as candidates for a flexure mechanism. What observations would you make, taking account, also, of other property groups often significant for flexure designs?

5. Look up a range of the properties relevant to precision engineering applications for Guillame metal, better known as Invar (which was originally a commercial name). What is the usual composition of this alloy? Compare these properties and composition to those of the alloys usually known as Superinvar and Elinvar. Hence, comment on the

types of applications for which they would be good choices of material, noting also any disadvantages likely to restrict their wider use.

6. Generate a property group profile using Duralumin (aluminium alloy) as the 'reference material' for each of a magnesium alloy, titanium (alloy), silicon carbide, fused quartz and tungsten carbide. Hence, comment on which of their properties might be of interest for different types of application in precision engineering. Justify your observations by citing application areas for those you would recommend.

7. A laser interferometer is being used for length measurement in a room where the temperature cycles slowly over a ±2 K band. Its reference arm has both the beam-splitter and a retro-reflector firmly attached, 400 mm apart, to a steel bar of 50 mm square cross-section. There is, therefore, a thermally derived variability of the reference length. Assuming that this is not simply a mistake, why might this reference arm design have been chosen?

8. A production process requires measurement of a particular length, in the range 24.7 mm to 25.3 mm on steel components. Measurements must occur reasonably close to the manufacturing machine and are to be completed rapidly, almost immediately after removing a workpiece from the machine. A measurement standard deviation of ±1.0 μm is desired. To reduce the size of the effective metrology loop, it is proposed to take a differential measurement by contacting both the workpiece and a standard reference sphere or gauge block (nominally 25 mm) with a conventional, commercial displacement gauge of range ±0.5 mm. The reference gauge blocks or spheres will almost certainly be standard, commercially available ones. They tend to be made from materials such as tungsten carbide, steel, alumina and Zerodur. Why is this range available? Rank these materials, with brief explanations, in an order of preference for this application.

9. A monolithic leaf-spring flexure mechanism used in an aerospace instrument needs to be compact compared to its operational range of motion and to demand only modest energy to operate. Hence, a strong aluminium alloy (having $Y \sim 1.5$ relative to the model material used here) has been used. There is a proposal to use a titanium alloy instead. Investigate the advantages and disadvantages of this change and so recommend whether to adopt it.

10. While not so directly applicable to the design of precision structures, it is still useful to be aware of the implications of typical atomic-level behaviours in different classes of material. So, as an example, look up and comment upon the Wiedemann-Franz law.

References

Ashby M. F. (1989). On the engineering properties of materials, *Acta Metall.*, **37**:1273–93.

Ashby M. F. (1991). On material and shape, *Acta Metall.*, **39**:1025–39.

Askeland D. R. (1996). *The Science and Engineering of Materials*, 3rd S.I. edition (adapted by Haddleton F., Green P. and Robertson H), London: Chapman & Hall.

Beer F. P. and Johnston E. R. (2004). *Mechanics of Materials*, 3rd edition (SI units), Singapore: McGraw-Hill.

Brehm R., Driessen J. C., van Grootel P. and Gijsbers T. G. (1985). Low thermal expansion materials for high precision measurement equipment, *Precision Engineering*, **7**:157–60.

Cebon D. and Ashby M. F. (1994). Materials selection for precision instruments, *Meas. Sci Technol.*, 5:296–306.

Chetwynd D. G. (1987). Selection of structural materials for precision devices, *Precision Engineering*, 9:3–6.

Chetwynd D. G. (1989). Materials selection for fine mechanics, *Precision Engineering*, **11**:203–9.

Chetwynd D. G., Schwarzenberger D. R. and Bowen D. K. (1990). Two dimensional x-ray interferometry, *Nanotechnology* 1:19–26.

Fein A. P., Kirtley J. R. and Feenstra R. M. (1987). Scanning tunneling microscope for low temperature, high magnetic field and spatially resolved spectroscopy, *Rev. Sci. Instrum.* **58**(10):1806–10.

Gardner J. W., Varadan V. K. and Awedelkarim O. O. (2001). *Microsensors, MEMS and Smart Devices.* Chichester, U.K.: John Wiley & Sons.

Garratt J. D. and Bottomley S. C. (1990). Technology transfer in the development of a nanotopographic instrument, *Nanotechnology*, **1**:38–43.

Gerbig Y. B., Spassov V., Savan A. and Chetwynd D. G. (2007). Topographical evolution of sputtered chromium nitride thin films, *Thin Solid Films*, **515**:2903–20.

Gibson I., Rosen D. W. and Stucker B. (2105). *Additive Manufacturing Technologies: 3D Printing, Rapid Prototyping, and Direct Digital Manufacturing.* New York: Springer.

Griffiths A. A. (1920). The phenomena of rupture and flow in solids, *Proc. Roy. Soc. Lond.*, **221**:582–93.

Kulkarni R. and Ochoa O. (2006). Transverse and longitudinal CTE measurements of carbon fibers and their impact on interfacial residual stresses in composites, *J. Compos. Mater.*, **40**:733–54.

Lindsey K., Smith S. T. and Robbie C. J. (1988). Sub-nanometre surface texture and profile measurement with NANOSURF 2, *Ann. CIRP*, **37**:519–22.

Mollenhauer O., Ahmed S. I.-U., Spiller F. and Haefke H. (2006). High-precision positioning and measurement systems for microtribotesting, *Tribotest*, **12**:189–99.

Peterson K. E. (1982). Silicon as a mechanical material, *Proc. IEEE*, **70**:629–36.

Smith S. T. and Chetwynd D. G. (1994). *Foundations of Ultraprecision Mechanism Design*, Chapter 8, New York: Taylor Francis.

Syam W. P., Jianwei W., Zhao B., Maskery I. and Leach R. K. (2017). A methodology to design mechanically-optimised lattice structures for vibration isolation, *Precision Engineering*, in press.

Ueno S. (1989). Development of an ultra-precision machine tool using a ceramic bed, 5th Int. Precision Engineering Seminar (IPES-5), Monterey, California.

Waterman N. A. and Ashby M. F. (1991). *Elsevier materials selector*, London: Elsevier.

13

Environmental Isolation

Waiel Elmadih, Marwène Nefzi and Eric S. Buice

CONTENTS

ABSTRACT   The performance of a precision machine depends on the attenuation of the environmental disturbances acting upon it, and considerable time and expense is invariably expended to isolate external and internal noise sources. The magnitude of these disturbances also influences the design of process controls (see Chapter 14). This chapter discusses the shock, vibration, thermal and acoustic isolation needed to protect precision machines. Vibration arises from dynamic machine motion, can be generated within the machine itself, transmitted through the floor, coupled in with direct linkages, or might be a result of vibrations and impacts in one machine generating acoustical noise that is absorbed by another. The attenuation of the impact of these perturbations on the machine can be achieved either by isolation or energy dissipation. Isolation refers to the process where the excitation magnitude is prevented from transmitting to the structure of the precision machine, while energy dissipation attenuates the excitation magnitude by dissipating the energy through heat or friction. In this chapter, an understanding of external excitations is first developed by reviewing and classifying different excitation sources. The impact of these excitations on the dynamic behaviour of a precision machine is investigated. Following the dynamic section, the focus is shifted to thermal isolation, where passive and active thermal isolation methods are introduced. In closing, the chapter outlines approaches to acoustic isolation. Different techniques and methods are discussed throughout the chapter to design appropriate isolation systems that ensure the minimisation of external impact on the application of precision machines.

13.1 Introduction

Environmental isolation primarily aims at protecting and improving the performance of precision machines by mitigating the impact of environmental disturbances. These disturbances include a wide spectrum of excitations that range from mechanical, thermal and acoustic sources. This chapter primarily focuses on mechanical vibration, as many performance characteristics in precision engineering may be affected by these excitations. For example, the mechanical vibration of a lens could lead to undesired optical aberrations. Precision engineers have often faced the challenge of designing isolation systems to filter out vibration and reduce its impact on the whole system.

The main goal of this chapter is to provide designers with adequate approaches to

1. Model environmental excitations
2. Design isolation systems

The chapter will, therefore, cover the following:

- Determination of possible mechanical excitations
- Modelling of the mechanical excitation signals
- Determination of the excitation impact on precision machines
- The design of passive isolation systems for excitation of moderate magnitudes and frequencies
- The design of active isolation systems for excitations of high magnitudes and low frequencies

This list is rearranged into a flow chart (Figure 13.1) for a better understanding of the different steps that should be addressed when designing environmental isolation systems. It should be pointed out that these steps often involve iterations and redesigns.

The approaches discussed here are valid for a wide spectrum of machines. Figure 13.2 shows a photolithographic machine that is needed for the fabrication of integrated circuits in modern electronic devices. The manufacturing technology involves different processes that aim at building billions of transistors on a semiconductor substrate (for example, a silicon wafer).

The core process in transistor building is photolithography, which enables transistors and features of sizes down to 10 nm to be packed close to one another in a small area by printing the pattern of the semiconductors onto the surface of the wafer. As shown in Figure 13.3, the core process includes the projection of light onto a wafer through high-performance optics. At some point before projection onto the wafer, the light has to go through a mask (also called a reticle) that is effectively a plate patterned to shape the light to create the desired features when projected onto a thin photo-sensitive resist coating on the wafer surface. After projection of the first set of integrated circuits, the wafer is stepped to a new position by a certain index amount to allow for the projection of a new set of patterns (see Figure 13.3). These patterns are important for the creation of transistors and other features on the wafer which are installed in the following steps in the building process. Errors in the indexing of the wafer, the positioning of the optics mount, or any of the parts associated with the building of transistors are often traced back to the external excitations. The fact that the aforementioned processes are dynamic and highly sensitive implies the need for freedom of movement in different directions, which increases the vulnerability of the projector and lens mounts to vibration and thermal expansion in different degrees of freedom.

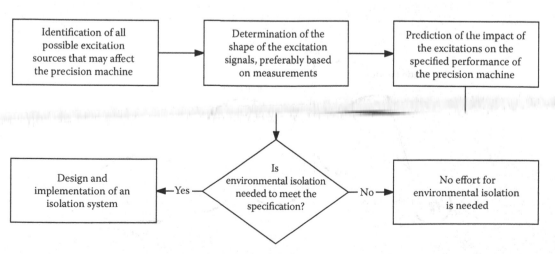

FIGURE 13.1
Simplified flow chart of showing the sub-systems needed for isolation design.

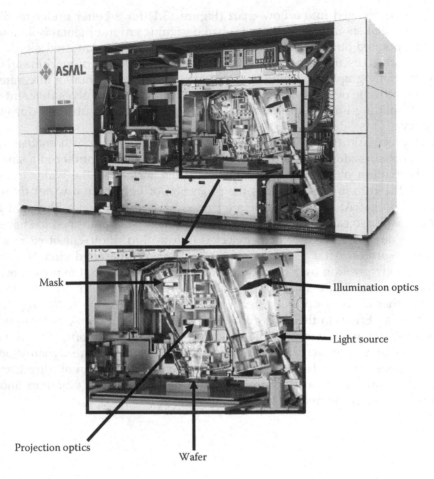

FIGURE 13.2
Photolithographic machine for producing semiconductor devices.

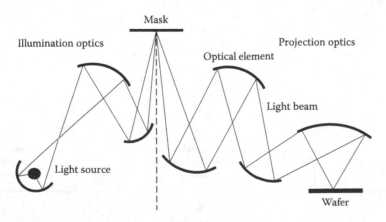

FIGURE 13.3
Schematic representation of a lithographic machine.

13.2 Sources of Shock and Vibration

In precision engineering, mechanical vibration originates from different sources, for example, ground-borne (or seismic) motion, fluid flow, imbalances of moving parts within the machine itself, and imbalances of moving parts of nearby machines. It is possible to classify these sources in terms of the excitation signal, for example, step function, harmonic signal and sweep (or chirp) signal. But first, it is necessary to make a distinction between two possible types of vibration that affect precision machines: free vibration and forced vibration. For each vibration type, numerous sources are discussed in the following sections. A typical signal type is assigned to each vibration source to cover a wide range of the environmental conditions to which precision machines are exposed.

13.2.1 Free Vibrations (The Transient Response)

Free vibrations result from an initial disturbance of a mechanical system and mainly arise when the system is put into operation, and allowed to freely vibrate. This is often the case when some movable components of the precision machine are set into motion, especially during position referencing, or due to an applied impulse. The system then vibrates at its natural frequency. Usually, the amplitudes of these vibrations decay quickly due to damping. Further methods for isolating a source of free vibrations may not be required if the resulting decay time is deemed to be adequate. For instance, the initialisation of the positions of the lenses of an optical system may induce optical aberrations, but such aberrations may or may not be significant.

13.2.2 Forced Vibrations

Forced vibrations arise when one component of a mechanism produces forces that induce components elsewhere in the mechanism to vibrate. This assumes that the two objects are connected in a way that energy can be transmitted between them. In this context, base and force excitations are discussed. Base excitation usually refers to the excitation of the foundations or the frame on which the precision machine is mounted. Forced excitation results from additional external forces that act on the machine. In the following sections, both excitation sources are addressed.

13.2.2.1 Base Excitation

When a precision machine is mounted on a frame or directly on the ground (in the following, the term *base frame* will be used), it is susceptible to base vibrations. Figure 13.4 depicts a typical example that is often encountered in precision engineering systems, which is the isolation of optical elements from external vibrations. The mounting of the optical element is modelled by a spring with stiffness k and a damper with a viscous damping coefficient c. In general, this simple model reveals the main behaviour of many mechanisms, including the mounting of optical elements considered here.

In optical systems, external excitations may displace the mounting of the optical elements, which leads to undesired optical aberrations. Fortunately, it is possible to improve the mounting by adjusting its mechanical characteristics, for example, the stiffness and viscous

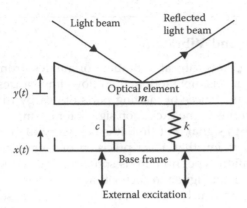

FIGURE 13.4
Base-excited optical element.

damping in order to obtain a response that does not affect the optical performance. The following question arises: What are the sources of this sudden motion?

Without being exhaustive, the following sources can be outlined:

- Neighbouring machines that are mounted on the same ground or frame to the precision machine
- Acoustic noise that makes the base vibrate
- Natural phenomena, such as slumping, wind, and so forth, that affect the whole building and thereby the base of the precision machine
- Vibrations from a truck, airplane, and so on during transport
- Motion of the frame during handling

Whereas the first three sources may affect the precision machine during operation, the last two sources mainly apply when it is out of operation. Even if the performance of the precision machine is not relevant when it is not in operation, the expected excitations should be assessed to evaluate possible damage to expensive parts. Furthermore, the induced vibrations may be amplified by the internal dynamics of the building, ground, frames and so on. In Exercise 1, only one dominant frequency has been considered. In reality, the base frame and the precision machine have other natural frequencies and corresponding modal behaviours.

13.2.2.2 Force Excitation

Figure 13.5 is an example where the position of an optical element in the vertical direction is sensed and corrected by a controlled actuator. The actuator exerts a force on the mounted optical element whose response depends on the stiffness and damping of the connection to the actuation frame.

Different sources could lead to a forced excitation of different parts of the precision machine. Some typical sources are listed here:

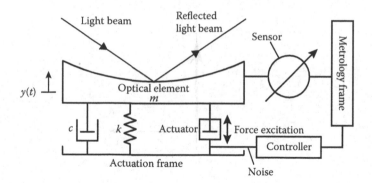

FIGURE 13.5
Force-excited optical element.

- The flow of cooling mediums needed to transfer heat in order to avoid high temperatures in sensitive parts
- Imbalances, for example, due to an inhomogeneous mass distribution
- Actuation forces
- Process forces that act directly on the precision machine, such as handling robots which are mounted on the machine
- Pumps that have to be mounted on the precision machine

13.3 Environmental Excitations

Once a vibration source and type is known, it is necessary to determine the shape of the excitation signal acting on the precision machine. This task is necessary for attenuation of the excitation and has to be addressed during the design process to be able to predict the system response, and to adequately design the connections and the mounting of all parts of the machine.

In this section, some generic mathematical functions are discussed which are often used in the analysis of environmental excitations. These are the unit impulse, the unit-step function, and harmonic and periodic functions. Finally, random excitations are briefly introduced.

13.3.1 The Impulse Function

The unit impulse $\delta(t)$ is defined as a function of time t that is zero everywhere, except at the origin where it takes an infinite value:

$$\delta(t)\begin{cases} = 0, \ t \neq 0 \\ \to \infty, \ t = 0. \end{cases} \tag{13.1}$$

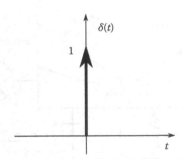

FIGURE 13.6
Unit-impulse function.

However, the time integral of the unit impulse satisfies the unity condition, given by

$$\int_{-\infty}^{+\infty} \delta(t)dt = 1. \tag{13.2}$$

The unit impulse is also called the delta function, or the Dirac function, and is represented by an arrow of height equal to 1 at the origin, as shown in Figure 13.6. The impulse function corresponds to the time integral on the whole time axis. The magnitude of unity is, in fact, the area under the impulse and not its height. If the unit impulse function is scaled by a constant A, the time integral from $-\infty$ to $+\infty$ becomes A.

Not many physical situations can be practically modelled as an impulse function. Generally, abrupt changes in physical quantities, such as velocity, acceleration and force, are bounded (not infinitely large as is the case for the impulse function) and they last for a given time duration (not for an infinitesimally small time as is the case for the impulse). Why is it then important to understand the impulse function when talking about environmental isolation? There are three main reasons:

1. The impulse function can be considered to model environmental excitations that have a very short duration and a very large amplitude as in the case of a collision; the impulse function quite effectively models the instantaneous change of parts colliding with each other, which is often the case when precision machines are handled or transported. In such cases, the determination of the amplitude and time duration of collisions is usually challenging, so the impulse function can be used as an approximation.

2. Investigating the impulse response of a precision machine may also help to capture its main dynamical behaviour, especially when different natural frequencies of the machine are expected. Indeed, the spectral analysis of an impulse function, that is, a horizontal line over the whole frequency axis, makes clear that this function equally excites all frequencies of a system. Hence, the impulse response of a system allows for a fast investigation of its dynamical behaviour.

3. The third reason for the importance of the impulse function is related to the study of linear time-invariant (LTI) systems. Each arbitrary excitation input can be decomposed to a continuous train of amplitude-modulated impulses. Hence, the response of LTI systems to any input can be derived from the impulse response. Time invariance indicates that the output of a system will not change, as long as the input is applied on the system now or τ seconds from now, except when the input

time is delayed by an amount τ. Precisely, the response of an LTI system $y(t)$ to any input $f(t)$ can be obtained by the integral in Equation 13.3 (called a convolution integral), if the impulse response $h(t)$ is known:

$$y(t) = \int_0^t f(\tau)h(t - \tau)d\tau$$

$$= \frac{1}{m\omega_d} \int_0^t f(\tau)e^{-\zeta\omega_n(t-\tau)} \sin(\omega_d(t - \tau))d\tau. \tag{13.3}$$

The second integral in Equation 13.3 contains the impulse response for a single degree of freedom spring mass damper system which is discussed in details in Section 13.4.1.

13.3.2 The Step Function

The unit step is defined as

$$u(t) = \begin{cases} 0, & t < 0 \\ 1, & t \geq 0. \end{cases} \tag{13.4}$$

The unit-step function can be used to represent sudden changes of variables, especially when the change persists for a specific amount of time. The step function can be used to model well-known physical situations as in the case of a constant force acting suddenly on a mass or the case of a body exposed to constant temperature.

The step function helps in obtaining information about the dynamic stability of the function, that is, determining whether the system is able to reach a stationary state when starting from another state, which is especially important for active isolation systems that have to be stable in different states. Moreover, the step response provides an insight into the dynamical behaviour of complex systems by means of characteristics that can directly be determined from the excitation plot, for example, settling time, overshoot or rise time.

13.3.3 Further Functions for Modelling Sudden Changes

In practice, there are further functions that are often used to model sudden changes. The following are examples of these functions.

13.3.3.1 Rectangular Function

In contrast to the impulse function, the time duration of a rectangular pulse is not infinitesimally small. Its amplitude is also bounded. The mathematical expression is given by

$$f(t) = \begin{cases} 0, & t < 0 \\ 1, & 0 \leq t \leq \Delta T, \\ 0, & t > \Delta T \end{cases} \tag{13.5}$$

where ΔT is the maximum time boundary at which the rectangular function has a magnitude of unity.

13.3.3.2 Half-Sine Function

While a rectangular function is useful in filtering signals and separating the used signals from unused signals, a half-sine function is more appropriate for computational purposes. This is because a half-sine function is smoother than a rectangular function. The mathematical expression is given by

$$f(t) = \begin{cases} 0, & t < 0 \\ \sin\left(\frac{\pi}{\Delta T}t\right), & 0 \le t \le \Delta T. \\ 0, & t > \Delta T \end{cases} \tag{13.6}$$

It should be noted that half-sine functions are often used to determine response spectra of dynamical systems.

13.3.3.3 Sinc Function

The sinc function is noteworthy, not only because it may be used to model external excitations, but because it corresponds to the continuous inverse Fourier transform of a rectangular spectrum of width π that, in turn, can be used with suitable coordinate transformations to model ideal low pass filters.

The mathematical expression is given by

$$f(t) = \begin{cases} \dfrac{\sin(\pi t)}{\pi t}, & t \ne 0 \\ 1, & t = 0. \end{cases} \tag{13.7}$$

13.3.3.4 Decaying Sine Function

The mathematical expression of a decaying sine function is given by

$$f(t) = A \cdot e^{-\lambda t} \cdot \sin(\omega t + \Phi), \tag{13.8}$$

where A is the amplitude, ω is the frequency of the signal, λ is a parameter that describes the damping in the isolation and is a measure of the decay rate of the oscillations, and Φ is the phase of the sine function.

13.3.4 Harmonic Excitation

The mathematical expression is given by

$$f(t) = A\sin(\omega t + \phi) = \text{Im}\left\{Ae^{j(\omega t + \phi)} = Ae^{j\omega t}e^{j\phi}\right\}, \tag{13.9}$$

where A is the amplitude, ω is the angular frequency and Φ is the phase of the signal. The steady-state harmonic response of a system is useful for modelling vibration isolation

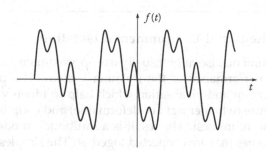

FIGURE 13.7
Periodic excitation.

and can readily be obtained using the complex exponential representation of a harmonic signal.

13.3.5 Periodic Excitation

The French mathematician Joseph Fourier found that any periodic signal can be decomposed into an infinite sum of sines and cosines with each term in the sum being exact integer multiples of the fundamental frequency. This sum is called the Fourier series and is discussed in more detail in Chapter 3. Figure 13.7 shows an example of a periodic function.

By extension, the same decomposition can be applied to arbitrary functions using a continuous distribution of frequencies. This is known as Fourier transform (again, see Chapter 3). With spectral analysis, it is possible to determine the frequency content of any input signal, including those that are not periodic (see Addison 2017).

13.3.6 Random Stationary Vibrations

Random vibration is a non-deterministic dynamic motion, in which different frequencies, that do not comprise rational integer multiples of frequency, exist concurrently. The amplitudes and phases of the frequency components composing the signal are random. A graph of a generic, relatively broad bandwidth, random vibration is illustrated in Figure 13.8.

For such a signal that extends to infinity, the Fourier transform is not defined. Power spectral density (PSD) is used instead to characterise these excitations and describes the energy distribution spanning a band of frequencies. For a white noise signal, the PSD is a continuous horizontal line over all frequencies.

FIGURE 13.8
Random excitation.

13.4 Specifying the Required Environmental Isolation

The designers of environmental isolation usually use approximate physical and mechanical models to specify the performance of the isolation system. The modelling of complex systems involves abstraction and idealisation which help to identify the relevant features, such as the number, nature (whether rigid or deformable) and coupling between the bodies that build up the system of interest. The result is a multibody model, that is, a system of rigid and deformable bodies that are connected together. The simplest model of a precision machine and its isolators is a rigid mass m mounted on springs of stiffness k and dampers of capacity c (Figure 13.9) that is effectively identical to the lens mechanism model of Figure 13.4. Clearly, this idealisation is an oversimplification but is adequate for illustrating the principles of shock and vibration isolation. Designers should keep in mind that assumptions are made to reduce the complexity of a process to a simple model, for example, the internal dynamics of the system components are neglected.

The required performance of isolation can be described in terms of the characteristics of the machine response. In this context, two types of responses are distinguished: the steady-state and the transient response of the machine. The steady-state response is relevant when the external excitations are of long duration, as is the case of periodic and random vibrations that occur during the operation of the machine (see Sections 13.3.5 and 13.3.6). The transient response is relevant when the machine is subject to short-duration shocks that can occur (sometimes when the machine is not in operation during handling, transport or when being installed). Steady-state responses can be characterised by the transmissibility, which is defined as the ratio of the response amplitude of a system in the steady state to the excitation amplitude. Transmissibility is, therefore, the complement of isolation; the lower the transmissibility, the higher the isolation will be. For transient responses, shock response spectra (SRS) are used to evaluate the performance of the isolation, which is a plot of the maximum response to an applied shock. Transmissibility and shock responses are discussed in Section 13.4.1.

13.4.1 Vibration Transmissibility

The transmissibility of a system can be derived from its transfer function. For the one degree of freedom (DOF) system shown in Figure 13.9, the transfer function can be obtained by

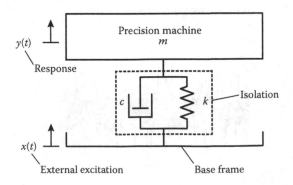

FIGURE 13.9
Simple model of the precision machine and its isolator.

taking the Laplace transform of the equation of motion of the precision machine, that is, the equation that governs its motion (see Chapter 4). Newton's second law of motion leads to the following ordinary differential equation

$$m\ddot{y} + c\dot{y} + ky = c\dot{x} + kx,$$ (13.10)

where m denotes the mass of the precision machine, c is the viscous damping of the connection and k is its stiffness. The Laplace transform of Equation 13.10 is

$$s^2 mY(s) + scY(s) + kY(s) = scX(s) + kX(s),$$ (13.11)

where s is the Laplace variable. The transfer function, that is, the ratio between the output $Y(s)$ and the input $X(s)$, is given by

$$H(s) = \frac{Y(s)}{X(s)} = \frac{cs + k}{ms^2 + cs + k}.$$ (13.12)

The Laplace variable s is often replaced by $s = \pm j\omega$ to obtain the steady state frequency response function (FRF), where ω can be interpreted as the excitation frequency. The FRF of the Laplace function in Equation 13.12 is given by

$$H(j\omega) = \frac{cj\omega + k}{-m\omega^2 + cj\omega + k},$$ (13.13)

where $H(j\omega)$ is a complex function. The graphical representation of $H(j\omega)$ is given in a Bode plot (Figure 13.10). As an example, take a vibration system of mass $m = 100$ kg, stiffness $k = 2 \times 10^6$ N·m^{-1}, and $c = 1 \times 10^3$ N·s·m^{-1}. The resulting natural frequency of the system can be calculated as $f_0 = \frac{1}{2\pi}\sqrt{\frac{k}{m}} = 22.5$ Hz, and the resulting damping ratio is calculated as

$$\zeta = \frac{c}{2\sqrt{km}} = 0.035 = 3.5\ \%.$$

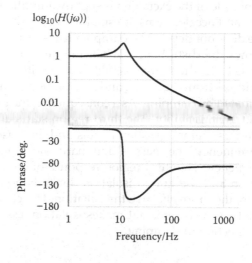

FIGURE 13.10
Frequency response function of a 1-DOF system.

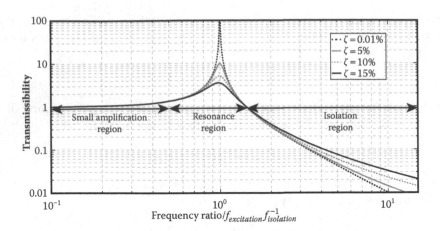

FIGURE 13.11
Transmissibility of 1-DOF system.

The abscissa of the Bode plot of Figure 13.10 is the excitation frequency, while the ordinate corresponds to the magnitude $|H(j\omega)| = \left|\dfrac{Y(j\omega)}{X(j\omega)}\right|$ and phase of the output with respect to the input signal. By means of a Bode plot, it is possible to calculate the response of the precision machine to each harmonic excitation, whose amplitude and frequency are known. It should be pointed out that for LTI systems, the output and input signals always possess the same frequency. Nevertheless, the amplitude and phase of the output differ from that of the input signal, depending on the excitation frequency. For each frequency, the ratio between the output and input signals and the phase of the output signal with regard to the input signal can be determined.

The transmissibility of a single DOF system corresponds to the magnitude plot $|H(j\omega)|$ of Figure 13.10. In Figure 13.11, and in contrast to the Bode plot, the abscissa of the transmissibility is the ratio between the excitation and the undamped natural frequency of the isolation. Three regions may be identified in this plot. In the first region, the frequency ratio is less than 0.5, and the amplitude of the excitation is slightly amplified by the isolation. In this region, because velocities and accelerations are small it is only the effect of the spring that dominates. The response does not depend on damping (by a factor of 1 to 1.5). The isolated machine essentially follows the disturbance. When the ratio between the excitation and isolation frequency is close to 1,[*] the transmissibility reaches its maximum. This is the resonance region where the forces from the spring and acceleration of the mass tend to cancel so that the response is dominated by damping. As a consequence, the larger the damping ratio, the lower the maximum transmissibility. The third region starts after the crossover at the frequency ratio of $\sqrt{2}$. In this region, isolation begins and, because acceleration increases with the square of the frequency, the mass is dominant and the attenuation varies with the inverse square of frequency. The third region response of the machine is less than the amplitude of the excitation and the rate of isolation depends on the damping ratio. The smaller the damping ratio, the more effective the isolation. Hence, increasing the damping reduces the transmissibility at resonance but makes isolation less effective. A trade-off is, therefore, needed in the selection of damping.

---

[*] For undamped single DOF systems, it is exactly 1. For damped single DOF systems, it is close to 1.

The magnitude transmissibility of a single DOF system can also be calculated using the following:

$$T_d = \left|\frac{F_T}{F_A}\right| = \left|\frac{Y}{X}\right| = \sqrt{\frac{1 + (2\zeta r)^2}{(1 - r^2) + (2\zeta r)^2}}, \tag{13.14}$$

where T_d is the transmissibility, ζ is the damping ratio, r is the ratio of the input frequency to the undamped natural frequency of the spring and mass, F_A is the amplitude of the force applied to the mass, F_T is the magnitude of the force transmitted to the base or ground of the spring mass damper, Y is the motion of the ground and X is the amplitude of the response at the mass. Apparent from Equation 13.14, transmissibility can be used to calculate the forces transmitted to the ground by a system that is generating a harmonically varying force or, in the opposite direction, the displacement of the system due to motion of the ground.

Most isolation systems are designed based on these factors. Given the excitation frequency and the desired maximum allowed transmissibility, it is possible to specify the isolation frequency.

Most precision machines are placed on vibration isolation tables to protect them from external excitations. Figure 13.12 shows a model of a precision machine mounted on a vibration isolation table. The table is isolated from the ground by a spring of stiffness k_1 and a damper of capacity b_1.

In addition to ground vibration, the moving parts/unbalanced parts of the precision machine, for example the tools in a cutting machine, will produce forces in the form of additional vibrations that disturb the machine (not considered in this analysis). The susceptibility to disturbances is modelled with a spring of stiffness k_2 and a damper of capacity b_2. In this model $M_1 = M_T + M_b$ is the combined mass of the vibration isolation table and machine base, x_1 is the table displacement due to external vibration, M is the effective mass of the supporting structure of the precision machine, and ξ_2 is the relative displacement between the components of interest in the precision machine. The equations governing motion of this system are

$$M_2\ddot{x}_1 + M_2\ddot{\xi}_2 + \frac{1}{2}b_2\dot{\xi}_2 + k_2\xi_2 = 0, \tag{13.15}$$

$$(M_1 + M_2)\ddot{x}_1 + M_2\ddot{\xi}_2 + b_1\dot{x}_1 + k_1x_1 = b_1\dot{y} + k_1y. \tag{13.16}$$

The displacement of the vibration isolation table x_1 and the machine disturbance ξ_2 are expressed in terms of the frequency response functions H_{1Y} and H_{2Y}, respectively

$$x_1 = Ye^{j\omega t}\, H_{1Y}(j\omega), \tag{13.17}$$

$$\xi_2 = Ye^{j\omega t}\, H_{2Y}(j\omega), \tag{13.18}$$

where H_{1Y} is the motion of the isolation table relative to the ground movement, and H_{2Y} is the machine disturbance relative to the ground movement. Substituting Equations 13.17 and 13.18 into Equations 13.15 and 13.16 results in the linear matrix equation

$$\begin{bmatrix} (-j\omega^2(M_1 + M_2) + j\omega b_1 + k_1) & -\omega^2 M_2 \\ -\omega^2 M_2 & -\omega^2 M_2 + j\omega b_2 + k_2 \end{bmatrix} \begin{Bmatrix} H_{1Y} \\ H_{2Y} \end{Bmatrix} = \begin{Bmatrix} 0 \\ j\omega b_1 + k_1 \end{Bmatrix},$$

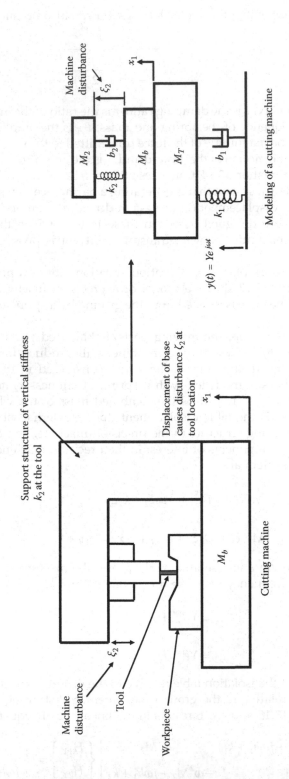

FIGURE 13.12
Precision machine mounted on an isolation table.

or

$$\begin{bmatrix} e_{11} & e_{12} \\ e_{21} & e_{22} \end{bmatrix} \begin{Bmatrix} H_{1Y} \\ H_{2Y} \end{Bmatrix} = \begin{Bmatrix} b \\ 0 \end{Bmatrix}. \tag{13.19}$$

The frequency responses in Equation 13.19 can be solved using Cramer's rule (see Chapter 4):

$$H_{1Y}(j\omega) = \frac{\begin{vmatrix} b & e_{12} \\ 0 & e_{22} \end{vmatrix}}{\begin{vmatrix} e_{11} & e_{12} \\ e_{21} & e_{22} \end{vmatrix}}, \tag{13.20}$$

$$H_{2Y}(j\omega) = \frac{\begin{vmatrix} e_{11} & b \\ e_{21} & 0 \end{vmatrix}}{\begin{vmatrix} e_{11} & e_{12} \\ e_{21} & e_{22} \end{vmatrix}}. \tag{13.21}$$

At low-frequency ground vibrations, the table motion is similar to the ground's motion, which is in turn transferred to the base of the precision machine mounted on the isolation table. Although the table does not isolate these low frequencies, the high stiffness and relatively low mass of the precision machine does. However, significant issues occur when the frequencies of the isolator become higher and cause the table to vibrate at frequencies close to the natural frequency of the precision machine. Optimally, the configuration should have low damping in the table to maximise attenuation at higher frequencies and high damping in the precision machine to attenuate resonant peaks. High damping also turns out to be desirable for optimal dynamics in high-speed machining.

As an example, consider the plot of the two responses given by Equations 13.20 and 13.21, shown in Figure 13.13 based on the model of Figure 13.12 for which the axes are plotted on a base 10 logarithmic scale. For both the isolator and machine, the natural frequencies and

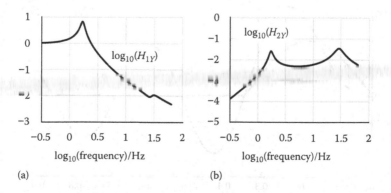

(a) (b)

FIGURE 13.13
Steady-state frequency responses for the model of Figure 13.12: (a) response of isolation table to ground vibration, and (b) machine disturbance due to ground vibration. The parameter values are $M_T = 400$ kg, $M_2 = 50$ kg, $M_b = 35$ kg, $M_1 = M_T + M_b$, $b_1 = 700$ N·s·m^{-1}, $b_2 = 2500$ N·s·m^{-1}, $k_1 = 50$ kN·m^{-1}, $k_2 = 1.5$ MN·m^{-1}.

damping ratios are 0.078 and 1.78 Hz, and 0.144 and 27.57 Hz respectively. When the machine is mounted onto the isolation table, the system will then have two DOF with corresponding eigenvalues (these are the roots of the system and can be thought of as the characteristic values about each resonance of the form $-\zeta\omega_n \pm j\omega_d$) of $-0.804 \pm j\, 10.689$ and $-28.33 \pm j\, 182.0$ respectively. In terms of damping ratio and undamped natural frequency, these eigenvalues correspond to 0.070 and 1.82 Hz, and 0.137 and 32.93 Hz. Comparing the combined system to the isolator and machine on their own, it is apparent that the dynamics of each is not significantly changed. Figure 13.13a shows the response of the isolation table to ground motion. It is apparent that the table moves with the ground at low frequency and, for frequencies above resonance, rapidly attenuates the ground motion. Figure 13.13b represents the ratio of vibration disturbance between the tool and workpiece relative to ground motion and presumably would correlate to the surface texture machined into the part. The maximum disturbance occurs at the two resonance peaks with values of -1.57 and -1.46. Because of the log 10 scale, this corresponds to around 3.5% of the ground amplitude being transmitted into the workpiece at the resonant frequency of the machine and 2.5% at that of the isolator. Outside of these two peaks, there is substantial attenuation at low and high frequencies and typically less than 1% between them.

13.4.2 Shock Transmissibility

As discussed in Section 13.4.1, the task of vibration isolation is to design isolators with natural frequencies that are well below the disturbing frequency. This leads to 'soft' isolators that are vulnerable to short-duration shocks that may occur, not only during start-up and shutdown of the machine but also during handling and transport.

Shock response spectra (SRS) are used to characterise shock transmissibility. For a better understanding of SRS, the transient response of a precision machine mounted on the isolators found in the previous section is plotted in Figure 13.14. The excitation signal is a half sine extracted from a sine wave of frequency 1.47 Hz corresponding to a pulse width of 0.34 s. As can be seen in Figure 13.14, the amplitude of the input signal is amplified by a factor of 1.6. It is now interesting to know how the transient response of the machine varies

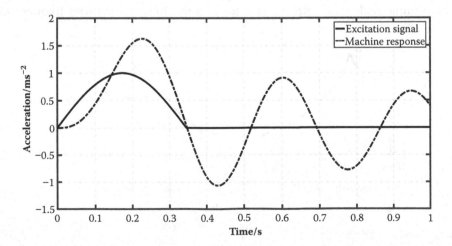

FIGURE 13.14
Machine response to a half-sine with a pulse width of 0.34 s.

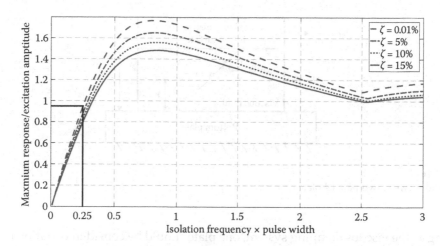

FIGURE 13.15
Shock response spectrum.

depending on the pulse width of the half sine. In other words, which shock signals can be attenuated?

To address this question, an SRS is given in Figure 13.15. The abscissa of this graph is the product of the frequency isolation with the pulse width of the half sine. The ordinate is the ratio between the maximum response and the excitation amplitude. Clearly, isolation is only possible when the product of the frequency isolation with the pulse width is smaller than 0.25.

13.4.3 Viscous and Non-Viscous Damping

Damping represents the mechanism by which vibration energy is dissipated. The damping force is described by the relative velocity between two parts of a viscous damping system, as discussed in Section 13.4.3.1. It is assumed for modelling purposes that the damping element is massless and does not contribute stiffness to the system, and is represented as one of the models described in the following two sections.

13.4.3.1 Viscous Damping

Viscous damping is the most commonly applied damping mechanism for reducing the amount of vibration energy induced in the system. In viscous damping, the damping force is considered to be proportional to the relative velocity between two bodies. Typically this is achieved using eddy currents or by using motion of the body to force fluid to flow (typical fluids being water, oil or air). Examples include fluid flow between a piston and a cylinder wall; or fluid present between two sliding surfaces, or fluid being squeezed out of two parallel surfaces (called squeeze film damping); and fluid present between a bearing and a journal.

Viscous dampers can be constructed in different ways. For instance, viscous damping can be obtained by the presence of a viscous fluid between two parallel plates, as shown in Figure 13.16. In Figure 13.16, the two parallel plates have a distance h and a fluid of viscosity μ between them. The two plates can move parallel to each other in the same direction with different velocities v, or in opposite directions with equal or unequal velocities. For simple

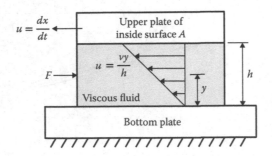

FIGURE 13.16
Viscous fluid between two parallel plates.

modelling of the viscous damping system, one plate should be considered stationary, while the other plate is moved with a relative velocity v to the stationary plate.

The fluid particles in contact with the stationary plate do not have any velocity ($v = 0$), while those in contact with the upper moving plate (as shown in the example in Figure 13.16) travel with the same speed v as the plate. It is acceptable to linearly model the velocity of the particles forming the intermediate layers of the fluid, although other approaches can be used (Rao 2016). Newton's second law for viscous flow, indicates that the shear stress τ of a fluid particle in a layer of distance y from the lower fixed plate can be expressed as

$$\tau = \mu \frac{du}{dy}. \tag{13.22}$$

The differential term is the velocity gradient. The resisting force developed on the inside surface of the upper plate is the product of shear stress and area

$$F = \tau A = \mu \frac{Av}{h}, \tag{13.23}$$

where A is the surface area of the moving plate. Since μ, A and h are all constants, Equation 13.23 can be written as

$$F = cv, \tag{13.24}$$

where $c = \mu A/h$ is the damping constant. More identification models of viscous damping can be found in Adhikari (2014).

13.4.3.2 Non-Viscous Damping (Coulomb and Structural, or Material, Damping)

In practice, a damping system includes both viscous and non-viscous properties. A damping system that does not depend on the relative velocity between the moving components in the damping system is called a non-viscous damping system. Two common non-viscous damping models are discussed: Coulomb damping and hysteretic damping.

Coulomb damping occurs when a mass slides on a dry surface and is produced by dissipating energy through friction. According to Coulomb's law of dry friction, the damping force is proportional to the normal force which acts on the contact plane, giving a sliding force F of

$$F = \mu N = \mu W = \mu mg, \tag{13.25}$$

where N is the normal force acting on the contact plane, which is sometimes equal to the weight of the sliding object; and μ is the coefficient of friction, which depends on the surface condition and the material in contact but is otherwise considered to be independent of load, speed, surface texture and apparent contact area (known as the modified Amonton-Coulomb laws). The damping force direction is opposite to the displacement direction and does not depend on the displacement magnitude nor the velocity; it depends on the force acting normally between the sliding surfaces.

Consider the single DOF non-viscous damping system shown in Figure 13.17. The mass m slides on the surface, and the spring causes the displacement to have two values of $+x$ and $-x$ with regard to the initial position. This movement is influenced by the stiffness k of the spring.

To analyse the example in Figure 13.17, two cases should be considered. The first case, as shown in Figure 13.18, is when the displacement x is either positive or negative, but the first derivative $\dfrac{dx}{dt}$ is only positive. This means that the mass in the example in Figure 13.17 accelerates to the right from the very left end in the first half of the system oscillation. At the initial point, x is negative and decreases in an absolute value until it reaches the equilibrium point where $x = 0$. The mass then keeps accelerating to the right after the equilibrium point, where x starts to pick up a positive value. Newton's second law gives the following equation of motion:

$$m\ddot{x} = -kx - \mu N. \tag{13.26}$$

Solving the non-homogeneous second-order differential equation as it appears in Equation 13.26 gives

$$x(t) = A_1 \cos \omega_n t + A_2 \sin \omega_n t - \frac{\mu N}{k}, \tag{13.27}$$

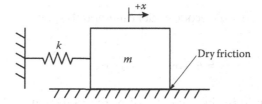

FIGURE 13.17
Single degree of freedom Coulomb damping system.

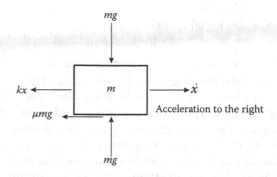

FIGURE 13.18
Analysis of half cycle with movement to the right.

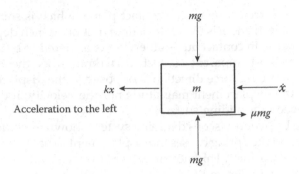

FIGURE 13.19
Analysis of half cycle with movement to the left.

where $x(t)$ is the amplitude in the first half of the vibration cycle, ω_n is the vibration frequency and A_1 and A_2 are constants that depend on the initial non-viscous damping system conditions.

The second case, as shown in Figure 13.19, is when the displacement x is either positive or negative, but the first derivative $\dot{x}$ is only negative. This means that the mass in the example in Figure 13.17 accelerates to the left from the very right end in the second half of the system oscillation. At the initial point, x is positive and it decreases in absolute value until it reaches the equilibrium point. The mass then keeps accelerating to the left after the equilibrium point where x starts to pick up a negative value. Newton's second law gives the following equation of motion:

$$- kx + m\ddot{x} = -\mu N. \tag{13.28}$$

Solving this non-homogeneous second-order differential equation as it appears in Equation 13.28 gives

$$x(t) = A_3 \cos \omega_n t + A_4 \sin \omega_n t + \frac{\mu N}{k}, \tag{13.29}$$

where $x(t)$ is the amplitude in the second half of the vibration cycle, ω_n is the vibration frequency and, A_3 and A_4 are constants that depend on the initial non-viscous damping system conditions. The constant $\frac{\mu N}{k}$ describes the spring displacement under the presence of the damping force μN. Table 13.1 shows the equations for calculating the constants A_1, A_2, A_3 and A_4, where x_0 is the initial motion amplitude.

When the two cases are combined to give the total amplitude of the system in one cycle, the amplitude is found to be decreasing by an amount $\frac{4\mu N}{k}$ in each cycle. This can be proven by summing the amplitudes of the first and second cases to give the full cycle amplitude.

The second type of non-viscous damping to be discussed here is hysteretic damping or material damping, and for the purpose of computing steady-state frequency responses is typically modelled using a complex elastic modulus or shear modulus $E' + jE''$ and $G' + jG''$ respectively. When a body deforms, the deformation energy is dissipated or absorbed by the body through a number of internal loss mechanisms within the material as a consequence of the deformation. The stress–strain curve of a vibrating body with material damping produces a hysteresis loop. The area of this hysteresis loop represents the energy

TABLE 13.1

Constants of Coulomb Friction

Constant	Equation
A_1	$x_0 - \dfrac{3\mu N}{k}$
A_2	0
A_3	$x_0 - \dfrac{\mu N}{k}$
A_4	0

loss per cycle per unit volume of the body as a result of material damping. Experimentally, the energy loss in each vibration cycle with this damping is reasonably modelled to be proportional to the square of the amplitude (Rao 2016).

Consider the representation shown in Figure 13.20, where the stiffness property of a body is represented by a spring of stiffness k, and the damping is assumed to be hysteretic with a hysteretic damping constant h. Often isolators are produced by suspending or supporting an isolation table with a viscoelastic material having relatively high losses such as the typically silicone-based viscoelastic damping polymers. Energy dissipation for these damping polymers is a function of frequency and, more significantly, temperature, with the complex moduli being provided by the manufacturers of this material. For many isolation systems these are used at constant temperature and at over limited frequency bandwith, usually a few to tens of hertz. At such conditions the properties of the damping polymers can be assumed constant. A support made from this material comprises a cylinder of length L and cross-sectional area A that is subject to longitudinal deformation in the x direction. The stiffness k of this support is therefore

$$k_h = \frac{E' + jE''}{L} A = \frac{E'(1 + j\beta)}{L} A = k + jh. \tag{13.30}$$

The dimensionless coefficient β is called the loss factor; the reason for this will be apparent shortly. Using Equation 13.30, the equation governing motion for the model of Figure 13.20 is

$$m\ddot{x} + (k + jh)x = m\ddot{x} + k(1 + j\beta)x = F, \tag{13.31}$$

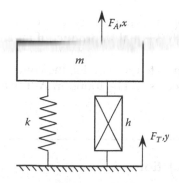

FIGURE 13.20
Hysteretic damping.

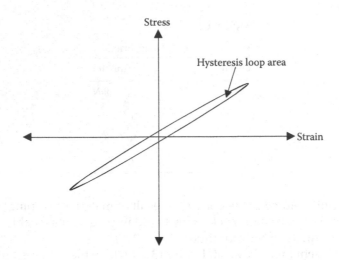

FIGURE 13.21
Hysteresis loop.

from which the frequency response is

$$H(j\omega) = \frac{1/k}{1 - \dfrac{\omega^2}{\omega_n^2} + j\beta},$$

$$|H| = \left|\frac{x}{F}\right| = \frac{1/k}{\left(\left(1 - \dfrac{\omega^2}{\omega_n^2}\right) + \beta^2\right)^{1/2}}, arg(H) = -\tan^{-1}\left(\frac{\beta}{1 - \dfrac{\omega^2}{\omega_n^2}}\right). \tag{13.32}$$

From the last of Equation 13.32 it is apparent that the phase shift contains only the frequency independent factor β that determines the lag between the displacement and force that in turn correspond to stress and strain in the material of the spring. Without hysteretic damping, $\beta = 0$ and the system becomes ideally elastic without losses, hence β represents a loss factor. Figure 13.21 shows the stress (force) and strain (deflection) during the loading and unloading of the hysterically damped body, when external vibrations deform the body.

The area of the hysteresis loop gives the energy loss for each loading–unloading cycle per unit volume of the body, which is equal to product of πh and the square of the vibration amplitude.

For a spring and damper connected in parallel, the complex stiffness is given by $\beta = h/k$. More non-viscous damping models can be found in Adhikari (2014).

13.5 Passive and Active Isolation

Specifications for isolation systems are determined using the models discussed in Section 13.4. The main difference between passive and active isolation systems is their working limits. Conventional passive systems provide isolation from high frequencies down to approximately 1 Hz. Advances in sensor technology enable active systems to isolate frequencies as low as 0.1 Hz (Ryaboy 2005).

13.5.1 Passive Isolation

Passive isolation provides vibration isolation when it is placed between the source of vibration and the sensitive precision machine. Passive isolation takes many forms depending on the isolation mechanism and has a wide range of performance characteristics. Possibly the simplest form of a passive isolator is an elastomer isolator, which is a soft material, such as rubber, placed between the source of vibration and the system to be isolated. The elastomer isolation combines damping and isolation properties. Another basic passive isolation mechanism is the use of metal springs, which provide an isolation mechanism but lack damping. Metal springs may be effective in avoiding the resonance region of a machine. They can also be used for transmitting frequencies that do not lie within the machine's natural frequency region.

13.5.1.1 Pneumatic Isolator

An important type of passive isolation system for use in precision applications is the pneumatic isolator. These provide low natural frequency isolation (down to 0.7 Hz) in comparison to metal springs and rubber isolators (Rivin 2003). Figure 13.22 shows a typical example of a pneumatic element used for low-frequency isolation. In Figure 13.22, the pneumatic element comprises a piston of cross-sectional area A moving inside a cylinder. The cross-sectional area can take any shape, but it is usually square or round in cross section. Consider v_i to be the instantaneous initial volume of the cavity between the piston and the cylinder which a fluid occupies.

With the settings in Figure 13.22, the initial absolute pressure in the cavity P_i is given by

$$P_i = P_a + \frac{mg}{A},\qquad(13.33)$$

where P_a is the atmospheric pressure at the level where the pneumatic mechanism is installed and m is the mass of the precision machine to be isolated. Assuming the

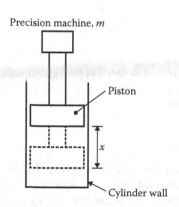

FIGURE 13.22
Example of a pneumatic element.

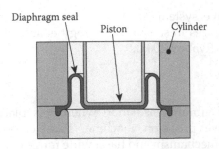

FIGURE 13.23
Representation of a diaphragm seal in a pneumatic isolator.

compression process of the fluid in the cavity to be adiabatic (no energy is transferred between the internal cavity and the surrounding cavity)

$$P_i V_i^n = P_x V_x^n,\tag{13.34}$$

where P_x and V_x are the absolute pressure and the cavity volume respectively, as a result of a linear movement of the piston by a displacement x; and n is the ratio of specific heats, which depends on the fluid occupying the cavity. n is 1.4 when air is used in adiabatic conditions.

Pneumatic isolators have smaller sizes than other passive isolation types used to isolate the same loads and frequencies. The gas volume provides low stiffness and supports the load with pressure. However, pneumatic isolators have relatively high vibration transmissibility for high frequencies, which is a shortfall that can be overcome by isolating the piston from the hard cylinder by a rolling diaphragm seal, as shown in Figure 13.23. A rolling diaphragm seal reduces the high-frequency transmissibility of the pneumatic isolation mechanism. But the rolling diaphragm also brings its own shortfall, as it has a relatively high stiffness and very low damping properties in the horizontal direction. A method for identifying the design parameters of these diaphragms can be found in Chen and Shih (2007).

The stiffness of a pneumatic isolator k_0 can be calculated as the first derivative of the pressure $\frac{dP}{dx}$ and is given by

$$k_0 = \frac{dP}{dx} = \frac{n P_i A^2}{V_i} \left[\frac{1}{1 - \left(\frac{A}{V_i}\right)x} \right]^{n+1}.\tag{13.35}$$

The stiffness value of a pneumatic isolator is proportional to the initial pressure inside the cavity. The initial pressure itself is influenced by the mass of the precision machine to be isolated, thus $k_0 \propto m$ (Rivin 2003).

13.5.2 Active Isolation Systems

Active isolation systems operate by means of external force actuators designed to oppose the disturbing force and hold the isolated mass motionless or within stated limits. A typical active isolation system with balanced mass, for example, the one DOF system shown in Figure 13.24, consists of a sensor to measure the deflection of the precision machine relative to an initial reference point, a signal processor to amplify and analyse the signal, and

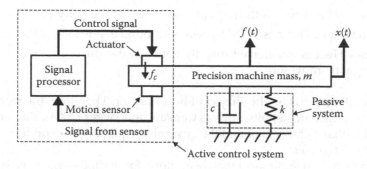

FIGURE 13.24
A typical active vibration isolation system.

actuators to apply the compensation in the form of force or motion. The compensation force or motion would then compensate the vibration. An example of an active vibration isolation system with a rotating unbalanced mass can be found in Rao (2016).

The force $f(t)$ shown in Figure 13.24 varies in magnitude and causes the precision machine of mass m, on which the force is applied, to vary in both the magnitude $x(t)$ and the direction of the deflection. The deflection is measured by the sensor, which transfers the signal to a signal processor (usually a computer). The signal processor produces a signal to the actuator, commanding it to develop a proportional force or motion. The type of sensors depends on the form of the feedback signal to be produced. Sometimes, the signal processor is linked to a passive system, such as a mechanical linkage or, an active system such as an electromagnetic, a hydraulic or a pneumatic network, so that the processor can perform functions such as integration, attenuation or amplification (see Chapter 14 for more in-depth information about control). The actuator which receives the signal can be a mechanical system, for example, fluidic, rack and pinion; a piezoelectric system; or an electromagnetic system.

Optimisation of the isolation is then dependent upon implementation of a particular control strategy with those of particular relevance to precision applications discussed in Chapter 14.

13.6 Thermal Isolation

In addition to the vibration and shock isolation discussed in Sections 13.4 and 13.5, it is also desirable to isolate any thermal loads that may be present in or around a precision machine. Actuators, friction due to sliding components, light, temperature of the environment and simply the presence of operators are all examples of heat loads that have an impact on the performance of a precision machine (see Chapter 11). Identifying the sources of a heat load and the cause of the error can be difficult, as illustrated in the wafer stepper example discussed in Chapter 10 that suffered from thermal drift caused by a focusing actuator (Stone 1989).

Before discussing thermal isolation, it is worthwhile to briefly introduce the basics of heat transfer. Heat transfer occurs from three distinct modes as follows (definitions from Carslaw and Jaeger 2004):

- Conduction—'Heat passes through the substance of the body itself'.
- Convection—'Heat is transferred by relative motion of portions of the heated body'.
- Radiation—'Heat is transferred directly between distant portions of the body by electromagnetic radiation'.

Extensive literature exists on the subject of heat transfer. Therefore, the reader is referred to common heat transfer textbooks, such as Carslaw and Jaeger (2004), Kaviany (2011), and Lienhard and Lienhard (2017) to obtain the governing equations for conductive, convective and radiative heat transfer.

However, the basic one-dimensional heat flow for a flat surface is still useful for approximate evaluations pertaining to thermal isolation. The total heat flux is given by

$$q_{Tot} = q_{cond} + q_{conv} + q_r, \tag{13.36}$$

where q_{cond}, q_{conv} and q_r are the heat flows based on conduction, convection and radiation modes of heat transfer respectively. The conduction, convection and radiation heat fluxes are given respectively by

$$q_{cond} = -kA\frac{\Delta T}{d}, \tag{13.37}$$

$$q_{conv} = hA(T_s - T_0), \tag{13.38}$$

$$q_r = \varepsilon\sigma A(T_c^4 - T_0^4), \tag{13.39}$$

where k is the thermal conductivity, A is the surface area, ΔT is the temperature difference, d is the cross-sectional depth, h is the convective heat transfer coefficient, ε is the emissivity, σ is the Stefan-Boltzmann constant, T_c is the absolute temperature of the component emitting the radiative heat and T_0 is the absolute temperature of the surrounding environment (note that these are absolute temperatures so the unit must be kelvin). It should be noted that Equation 13.38 is for a forced convection, which is a common condition experienced in a clean room (or temperature-controlled) environment where precision systems are utilised. For multi-dimensional conductive, radiative and convective equations (with additional boundary conditions), the reader is referred to common heat transfer textbooks, such as Carslaw and Jaeger (2004), Kaviany (2011) and Lienhard and Lienhard (2017).

A common method, used to perform a fast analysis of the heat flow, is the use of a thermal resistance model. The thermal resistance model is the analogy of an electrical resistance model. The thermal resistance terms for conduction, convection and radiation are given respectively by

$$R_{cond} = \frac{d}{kA} \tag{13.40}$$

$$R_{conv} = \frac{1}{hA} \tag{13.41}$$

and

$$R_r = \frac{1}{h_r A} = \frac{1}{\varepsilon \sigma (T_c + T_o)(T_c^2 + T_o^2)},$$ (13.42)

where R_{cond}, R_{conv} and R_r are the thermal resistance based on conduction, convection and radiation modes of heat transfer, respectively, and h_r is the radiation heat transfer coefficient. If the thermal resistances are in series, as shown in Figure 13.25a, then the total resistance R_{tot} of n numbers of resistances in series is given by

$$R_{tot} = R_1 + R_2 + \dots + R_n.$$ (13.43)

If the thermal resistances are in parallel, as shown in Figure 13.25b, then the total resistance R_{tot} of n number of resistances in parallel is given by

$$\frac{1}{R_{tot}} = \frac{1}{R_1} + \frac{1}{R_2} + \dots + \frac{1}{R_n}.$$ (13.44)

Knowing the total resistance, the heat flow can be calculated by

$$q = \frac{T_f - T_i}{R_{tot}},$$ (13.45)

where T_i and T_f are the temperatures at the initial and final points respectively. Referring to Figure 13.25, the heat flow with the resistance paths in series is

$$q = \frac{T_f - T_i}{R_1 + R_2},$$ (13.46)

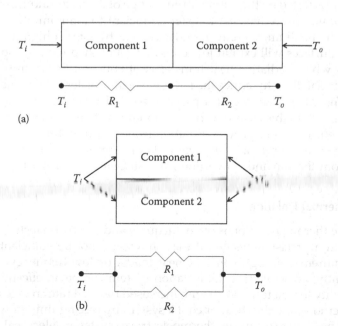

(a)

(b)

FIGURE 13.25
A simple thermal resistance model: (a) system in series and (b) system in parallel.

and for the resistance paths in parallel

$$q = \frac{(T_f - T_i)(R_1 + R_2)}{R_1 R_2}.$$ (13.47)

Note that the preceding equations are not time dependent, but similar equations can be formulated to include the time dependence term. Having these basic principles, an approximate calculation can be made to determine adequate thermal isolation. For multi-dimensional thermal analysis, it is advisable to use finite element analysis to ease the effort in determining the thermal effects on a precision system.

13.6.1 Passive Thermal Isolation

Passive thermal isolation is a method that is equivalent to the use of thermal insulation in a house. The goal of the thermal insulation inside a house is to provide a thermal resistive path (reduce heat flow) such that the internal temperature is isolated from the external temperature. Of course, if the exterior temperature is increased and kept constant for an infinite length of time, the interior temperature will become equal to the exterior temperature. With this in mind, passive thermal isolation is most effective when temperature changes are cyclic and at high frequencies.

Utilising a passive thermal isolation system in a design includes incorporation of materials with low thermal conductivities. Ceramics, glasses and plastics are good choices of materials with low thermal conductivity and are commonly used in precision systems. Air, if stationary, can also provide good insulation and can either be utilised as an intermediate layer between solid panels or by using porous structures, such as ceramic foams. A vacuum environment also exhibits very low thermal conductivity; thermal conductivity can be assumed to be zero for ultra-high vacuum environments. In precision systems that are predominantly affected by the temperature change of the surrounding environment, it is desirable to keep the convective heat transfer coefficient to a minimum by minimising the flow velocity of the medium. To reduce radiative heat transfer, a high reflective surface is desired, as these surfaces will exhibit low emissivity. Polished metals, especially gold, are excellent choices where radiative heat transfer is of concern. For example, gold foils are commonly used in satellites to reduce the radiative heat transfer (Maini and Agrawal 2006).

Figure 13.26 shows an example of a passive isolated positioning stage operating in a vacuum environment. In this example, the positioning system is housed inside two passive thermal-isolated enclosures to provide isolation from, not only the heat sources (electrical rack), but also the room serves as a thermal isolator from the building exterior conditions (e.g. radiation from the sun and convection from air being blown past the building).

13.6.2 Active Thermal Isolation

Although passive thermal isolation is often adequate and is a cost-effective way to limit the thermal effects on a precision system, it may, however, not be sufficient with increased performance requirements, high constant heat loads, or low-frequency thermal fluctuations. Consequently, an active thermal isolation system can be an effective alternative for providing the ability to isolate a system with increased thermal rate changes. Active thermal isolation uses methods to either heat or cool a system by forcing fluid flow (water cooling, ventilation fan, etc.) and/or using thermoelectrical systems (electrical heaters, Peltier coolers, etc.). The use of an air conditioner in a house is an example of an active thermal

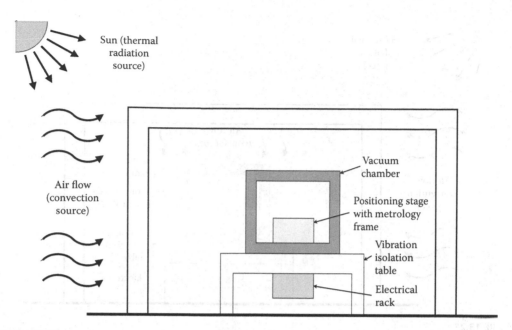

FIGURE 13.26

An example of a passive thermal isolation method applied to a positioning stage. The positioning stage is housed inside a chamber to isolate the heat flux emitted by the electrical rack, which is mounted to the underside of the vibration isolation table. Additionally, the building also acts as a passive thermal isolation chamber, isolating the radiation of the sun and convection generated by the wind flowing past the building.

isolation system used to maintain a cooler interior temperature than is present outside the house, or to compensate for larger thermal sources that are present inside the house, such as an oven. An example of an active thermal isolation is shown in Figure 13.27, which builds on Figure 13.26 by adding a cooling/heating system that can maintain a constant temperature regardless of the changes that may occur on the outside of the building.

Some companies can provide temperature control within 0.1 °C for metrology facilities at considerable cost. Such facilities are often maintained at positive pressures with close attention to the flow into and out of the room. Typically, the air is fed from the ceiling and exits at the floor, with the air returning back up through a cavity in the wall.

When implementing an active thermal isolation system, care is required to not offset the positive attributes gained from the thermal isolation at one point to only provide a negative impact at another point. For instance, the optical element shown in Figure 13.4 absorbs a portion of the light beam causing the optical element temperature to rise. To minimise the temperature change of the optical element and keep the base frame at a constant temperature, a water-cooled system is used to remove the heat caused by the absorption of the light. The return line of the water-cooled system will be at an elevated temperature, since the returning water has removed heat from the optical element. As seen in Chapter 11, it would be a poor design choice to have the return line in near proximity to the metrology frame because this would cause a direct measurement error, due to thermal expansion of the frame.

Further error sources can be in the form of flow-induced vibrations and acoustics. To minimise the flow-induced vibrations, the fluid flow should be kept to a minimum. However, reducing the fluid flow can limit the effectiveness of the thermal isolation, such that an appropriate compromise has to be found. Additionally, the frequency of the water pump

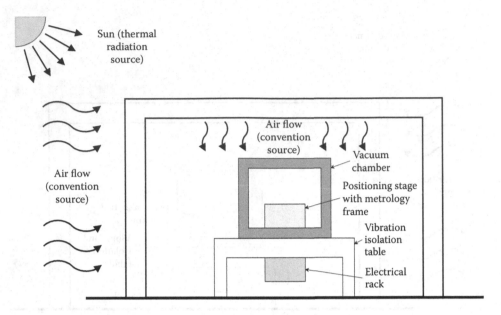

FIGURE 13.27
An example of an active thermal isolation method applied to the system shown in Figure 13.26. A cooling/heating system is added to the building to keep the internal environment of the building at a constant temperature. Note that a passive isolation is still present between the positioning stage and electrical rack.

can also cause a vibrational disturbance. The following are examples of precision machines using active thermal isolation to minimise machining errors due to thermal changes.

Lawrence Livermore National Laboratory (LLNL) has developed several diamond turning machines (DTMs) using active thermal isolation. One of the prominent examples of LLNL precision machine designs is the large optics diamond turning machine (LODTM). LODTM utilises active controlled airflow and water-cooling. The airflow is used to control the overall environment to within 0.005 °C peak-to-valley (PV), while critical machine elements are gravity-fed* water cooled to within 0.0005 °C PV. By introducing the active thermal isolations, LODTM was capable of machining parts with a surface accuracy of better than 25 nm root mean square, while turning optical components with diameters of up to 1.2 m (Saito et al. 1993).

The DTM #3 was another diamond turning machine developed at LNLL, which utilised a low-mass oil shower with a flow rate of approximately 1500 per minute to, not only maintain critical machine elements at a constant temperature of ±0.0025 °C, but also the part being machined at a constant temperature. This provided the DTM #3 with the capability of machining parts up to a diameter of 2.3 m with an accuracy of better than 1 μm (Klingmann and Sommargren 1999, Kobayashi 1984).

---

* Gravity-fed water cooling was chosen to avoid the pressure fluctuations that can be commonly observed by water pumps, which is an additional error source that would have been an additional error term in the LODTM error budget.

13.7 Acoustic Isolation

During the development of precision systems, acoustic isolation is often overlooked as a potential error source. The sound pressure levels generated by fans, actuators, pumps and even technicians talking can cause errors in highly sensitive precision systems. The sound pressure level is expressed in decibels (dB); a logarithmic unit used to express the ratio between two quantities of which one is a reference quantity. The sound pressure level L_p is given by

$$L_p = 10 \log \left(\frac{P(t)}{P_e} \right)^2, \tag{13.48}$$

where P_e is the reference sound pressure of the environment (for airborne sound $P_e = 20\ \mu\mathrm{Pa}$) and $P(t)$ is the instantaneous sound pressure given by

$$P(t) = P_o \sin(2\pi f)t, \tag{13.49}$$

where P_o is the instantaneous sound pressure above or below atmospheric pressure, f is the frequency and t is time (Vér and Beranek 2006). Specific equations (estimates) related to common components, such as pumps and fans, that are implemented into precision systems can be found in Barron (2003).

Since sound waves can be reflected by any hard surface, modelling the disturbance caused by a noise source can be complex. If the noise levels need to be lowered by approximately 10 dB, then it is common practice to implement an acoustic isolation chamber either directly to a noise source (such as a pump) or to the entire machine. Figure 13.28 shows a cross-section of an acoustic isolation chamber wall having a rigid exterior wall that reflects the acoustic sound wave ($P_3 = P_4$ assuming stiffness $k = \infty$) and an interior wall lined with an acoustic absorbing material (conical foams) to reduce the amplitude of the reflected acoustic sound wave ($P_1 > P_2$, P_2 would be zero if the acoustic absorbing material was perfect in absorbing the acoustic sound wave). Acoustic isolation chambers can achieve a reduction in noise up to

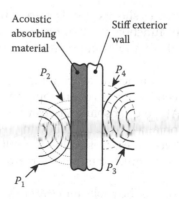

FIGURE 13.28
Cross section of an acoustic isolation chamber wall having an interior wall lined with an acoustic absorbing material that absorbs the initial acoustic sound wave P_1, and since the acoustic absorbing material is not perfect, a reflected acoustic sound wave P_2 with a lower amplitude is reflected. A rigid exterior wall reflects the acoustic sound wave P_4 of the initial acoustic sound wave P_3 ($P_3 = P_4$).

30 dB with appropriate design and use of appropriate material. For high-precision applications, even up to 50 dB is achievable (Barron 2003). It should also be noted that when implementing an acoustic enclosure, is important to ensure that heat sources are adequately cooled or ventilated, as heat can be trapped inside the enclosure (i.e. the enclosure acts as thermal insulation). For further detailed information, the following references may be of interest: Crocker (2007), Beranek (1993) and Vér and Beranek (2006).

Exercises

1. Consider a precision machine used to measure the surface texture of car engines on an engine building line. Due to transferred vibrations from the building environment, the base frame on which the precision machine is mounted displaces. An isolator is introduced with the aim of limiting the displacement transmissibility to $T_d = 3$. Determine the required damping ratio of the isolator assuming the system is of 1 DOF. Comment on the calculated value.

2. A precision machine of a 65 kg mass is used for manufacturing electronic circuits of laptop memory chips. The machine is mounted on an isosymmetric pentagon table which receives transmitted vibrations from the external environment causing all five sides of the table to vibrate at 1200 rpm. If five springs, each of $\xi = 0.02$, are to be introduced to link every table corner to the precision machine, determine the deflection on each spring if not less than 92% of the table vibration is to be isolated.

3. An optical instrument is to be isolated from its vibrating frame. The mass of the machine is 35 kg and it has a natural angular frequency of 26 rad·s^{-1}. The vibration needs to be controlled with a damping ratio of $\xi = 0.8$ for the optical instrument to take acceptable measurements. Considering an active isolation to be more expensive than a passive one, and that the passive system only uses a dashpot in the range of $0 \leq c \leq 450$ N·s·m^{-1}, design the most economical and effective isolation system. Justify your choice.

4. A precision machine experiences shocks of 9 Hz during transportation. Considering the excitation to form a half-sine function and that the machine can only tolerate excitations of 5.4 Hz, what would be the damping ratio and the isolation frequency of an appropriate shock isolator?

5. A damping mechanism for a surface measuring instrument can be approximated as a thin film of lubricant between two plates of an upper moving plate with surface area of 0.1 m^2. The lubricant is SAE 30 of absolute viscosity 0.3445 Pa·s and could provide a damping force F, which is obtained from the following relation: $F = 0.2v^2 - v$, where v is the velocity between the two plates. Determine the minimum height of the lubricant film that can provide the maximum damping force achievable with these damping settings.

6. The structural frame, supporting the probe and the entire measurement mechanism in a precision machine, is assumed to have a damping property that can be represented as Coulomb damping. External excitation causes the frame to make five vibration cycles in 0.5 seconds. Assume the initial position of the frame to be 10 μm away from the frame's equilibrium position and that the final position after five oscillations was measured to be 1 μm from the equilibrium position. Determine the Coulomb damping coefficient of friction for these settings.

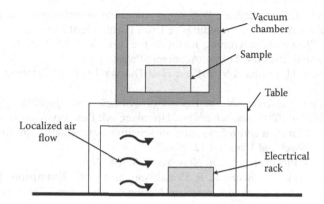

FIGURE 13.29
A sample is mounted inside a vacuum chamber, which is mounted to a vibration isolation table. Under the table is an electrical rack, which acts as a heat source to the table. To ensure that the electrical rack does not overheat, a fan is added to provide convection.

7. A precision machine of mass 10 kg is to be isolated pneumatically using a piston–cylinder mechanism. The piston has a diameter of 40 cm and travels a maximum displacement of 90 cm. Assuming the cylinder to have an internal length of 150 cm and that the mechanism uses air as a working fluid, what would be the best value of stiffness achievable with these settings?

8. If the isolator in Exercise 7 is to provide an isolation angular frequency of 62.4 rad·s^{-1}, why would it fail to provide such an isolation? If d is the diameter of the piston and P_i is the initial pressure inside the cylinder, determine the appropriate value $P_i d^2$ of the piston–cylinder mechanism that will allow this isolation frequency. Assume $\left(\frac{A}{V_i}\right)x = 0$.

9. Determine the thermal resistance equation (R_{tot} from the electrical rack to the sample of a passive thermal isolation system (Figure 13.29). A fan is used to provide additional cooling for the electrical rack.

10. List a minimum of two thermal isolation improvements that can be made to the system shown in Figure 13.29.

References

Adhikari, S. 2014. *Structural Dynamic Analysis with Generalized Damping Models: Identification*. Wiley.

Addison, P. S. 2017. *The Illustrated Wavelet Transform Handbook: Introductory Theory and Applications in Science, Engineering, Medicine and Finance*, 2nd edition. CRC Press.

Barron, R. F. 2003. *Industrial Noise Control and Acoustics*. New York: Marcel Dekker.

Beranek, L. L. 1993. *Acoustics*. Acoustical Society of America, 1844207.

Carslaw H. S., and J. C. Jaeger. 2004. *Conduction of heat in solids*. Oxford: Oxford University Press.

Chen, P.-C., and M.-C. Shih. 2007. "Modeling and robust active control of a pneumatic vibration isolator." *Journal of Vibration and Control* **13**(11): 1553–71.

Crocker, M. J. 2007. *Handbook of Noise and Vibrational Control*. Hoboken, NJ: John Wiley & Sons.

Kaviany, M. 2011. *Essentials of Heat Transfer: Principles, Materials, and Applications*. Cambridge: Cambridge University Press.

Klingmann, J. L., and G. E. Sommargren. 1999. "Sub-nanometer interferometry and precision turning for large optical fabrication." Proceedings of Ultra Lightweight Space Optics, Napa, California.

Kobayashi, A. 1984. "Precision machining methods for ceramics." In *Advanced Technical Ceramics*, edited by S. Somiya, 261–314. Tokyo: Academic Press.

Lienhard IV, J. H. and J. H. Lienhard V. 2017. *A Heat Transfer Textbook*. Cambridge, MA: Phlogiston Press.

Maini, A. K., and V. Agrawal. 2006. *Satellite Technology: Principles and Applications*. John Wiley & Sons.

Rao, S. S. 2016. *Mechanical Vibrations, 6th edition*. Hoboken, NJ: Pearson Education.

Ryaboy, V. M. 2005. "Vibration control systems for sensitive equipment: Limiting performance and optimal design." *Shock and Vibration* 12: 37–47.

Rivin, E. I. 2003. *Passive Vibration Isolation*. New York: ASME Press.

Saito, T. T., R. J. Wasley, I. F. Stowers, R. R. Donaldson, and D. C. Thompson. 1993. Precision and Manufacturing at the Lawrence Livermore National Laboratory. The Fourth National Technology Transfer Conference and Exposition. Anaheim, CA: NASA.

Stone, S. W. 1989. "Instrument design case study flexure thermal sensitivity and wafer stepper baseline drift." Edited by T. C. Bristow and A. E. Hatheway, *Proceedings of SPIE: Precision Instrument Design*, vol. 1036.

Vér, I. L., and L. L. Beranek. 2006. *Noise and Vibration Control Engineering: Principles and Applications*. Hoboken, NJ: John Wiley & Sons.

14

Control Systems for Precision Motion

Stephen Ludwick

CONTENTS

ABSTRACT Control algorithms are a key component of precision machine design. Strategies that employ both feedforward and feedback elements address the dual requirements of trajectory following with minimal vibration and robust disturbance rejection. Classical, frequency-domain based design techniques apply to both linearised lumped parameter models of a motion system as well as measured frequency responses that capture behaviour not included in the model. Idealised linear lumped parameter models that replace motors with unlimited applied forces and slideways with frictionless inertial masses are widely used in developing the initial control strategy, while direct measurements are used for model refinement and final tuning. Precision motion control problems commonly include structural flexibilities, sensitivities to actuator and sensor location, multiple sensors and actuators, axis coupling and multi-loop designs. This chapter provides a starting point for students who have taken a traditional introductory controls course to identify issues of particular interest for precision controller design. References giving detailed developments of specific controller strategies are included for the student desiring to address particular precision motion control applications.

14.1 Introduction

Servo control is another critical element in the overall design of precision systems. It enables rapid point-to-point positioning and the ability to track programmed contours, reduces the sensitivity to component variations, and greatly improves the rejection of external disturbances by effectively increasing the stiffness of a mechanism within particular frequency ranges. These improvements come at the cost of increased system complexity and the potential for instability with a poorly chosen servo control algorithm, but the trends towards lower-cost and higher-performing sensors and microprocessors suggest that servo control will become an increasingly critical part of precision mechatronic system designs.

Differences between the design of controllers for precision systems and conventional ones mirror those for mechanical design problems. The fundamental design principles remain the same for both, but designing for precision systems requires that a philosophy of determinism is embraced in rigorously investigating, identifying and appropriately compensating for apparent randomness in the performance metrics until the uncertainty in the results is below the requirements.

Several characteristics distinguish a high-precision motion control problem from a more conventional one. The absolute magnitude of the displacement (or any controlled variable) is not necessarily a defining factor. Control systems for precision motion must typically address several of the following problems:

- Compliance in the mechanical components that leads to performance-limiting vibrational modes
- Low-frequency vibrations resulting from motion of a base, vibration isolation system or machine frame
- A difference between measurements available for use in a control algorithm and the actual performance variables
- Nonlinear responses over some portion of the operation
- Operating at the limits of models that can be economically developed
- Cross-axis dynamic coupling in multi-axis systems

Additional constraints are the same as for conventional control design problems. The control algorithm must be reliable and robustly stable over the entire operating life of the system and must not depend on elaborate, manual, so-called expert adjustments. The highest precision systems are often used in advanced manufacturing operations where downtime becomes expensive and intolerable. This leads to a preference for a rigorous application of well-established techniques.

14.1.1 Two Degree of Freedom Control System Design

Precision systems are often controlled through a so-called two degree of freedom methodology that divides the algorithms into feedforward and feedback components, each with different objectives. Figure 14.1 shows the location of feedforward and feedback blocks in a typical control loop. The feedback algorithm, which operates on the tracking error e, is

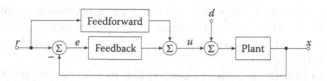

FIGURE 14.1
A two degree of freedom control system design includes a feedback controller that acts on an error signal and a feedforward controller that acts only on the reference command.

designed primarily for rejecting the external disturbance d for robustness to plant variations and for attenuation of unmodelled dynamics. The feedforward algorithm is used to track the reference command r with an acceptable level of following error. The plant is the physical system whose output variable is controlled by adjusting one or more inputs.

Feedforward control is a critical part of the design of control systems for precision systems in practice, but one that is often given minimal attention in most introductory textbooks on servomechanisms. Precise modelling of the plant is critical to high-performance feedforward controller designs (Iwasaki et al. 2012). Modelling allows more of the compensation to be generated in an open-loop feedforward sense and less via feedback, where it can be corrupted by sensor noise and faces the algebraic limitations imposed by the Bode sensitivity integral (also known as the 'waterbed effect') (Stein 2003). Sensitivity plots are described in more detail in Section 14.2.2.

Feedback control algorithms for precision systems use measurements of the present state of the system to determine the commands to use to adjust the plant output. These techniques respond to disturbances and correct for residual errors that result from inaccuracies in the feedforward command. The basic techniques of feedback control are generally introduced in an undergraduate-level control systems text (for example, Franklin et al. 2015) and are almost unchanged from their origins at Bell Labs in the 1930s and 1940s (Graham 1946). These techniques are collectively referred to as classical control techniques and conceptually amount to adding frequency-dependent stiffness to the system dynamics. The common and effective proportional–integral–derivative (PID) controller is a well-known example, and remains the starting point for many (and perhaps most) control algorithm designs.

Modern control design techniques, also known as state-space techniques, use time-domain differential equations that describe the dynamic response of the system. These techniques become more useful as the model order and number of inputs and outputs increase. However, far from being divergent techniques, classical and modern controls are better seen as different paths to the same (or similar) algorithms, and the choice of which to use depends on the particulars of the problem at hand.

14.1.2 Data-Based versus Model-Based Control

Two complementary approaches to control system design are model-based control and data-based control. The first step in a model-based control design is to generate a dynamic model of the system, often in the form of differential equations, or the Laplace transforms of them if in continuous time, or difference equations and Z-transforms of them, if operating in discrete time. The control system is then designed to meet the performance objectives around the model of the systems, with an assumption that the plant model represents the true plant to within a suitable level of uncertainty (Hou and Wang 2013). The root locus design technique in classical controls and pole-placement techniques in modern controls are examples of model-based design techniques. There are inevitably unmodelled dynamics and parametric uncertainty in the model, which the field of robust control attempts to manage. However, the lack of adequate, practical uncertainty descriptions limit their utility in many cases (Gevers 2002). Precision systems often operate at tolerances which are at the limits of readily designed mathematical models, which can ultimately limit the performance of strictly model-based designs.

Data-based control design techniques use explicit measurements of the plant response, often in the form of a frequency response. The measurement itself becomes the model on which the controller is designed. Data-based techniques are appealing for the control of precision systems because the designer is attempting to achieve performance at the limits of what a mathematical model can readily predict. The two techniques clearly reinforce each other, as increasingly sophisticated models can be created to explain the measured data, and thus ultimately influence the electro-mechanical design of the plant. Control techniques, namely classical frequency-domain techniques, are preferred for the control of precision systems as they operate equally well with both model-based and data-based design problems.

Data-based design techniques rely heavily on measured frequency response data. The core idea for linear systems is deceptively simple. The frequency response quantifies the changes in magnitude and phase that occur when a sinusoidal input signal is acted upon by a dynamic system. If the system is mathematically linear, there will be no change in frequency at the output (see Chapter 13). Implementation details become extremely important, however, and the reader is referred to a comprehensive review of frequency-domain based system identification techniques given in Pintelon and Schoukens (2012). Recent advances in the frequency-domain technique improve the performance when measuring systems contain an integrator (Widanage et al. 2015), when operating in closed-loop (Pintelon and Schoukens 2013), when the systems contain multiple inputs and outputs (Dobrowiecki et al. 2006) and perhaps most important in the presence of nonlinearities (Schoukens et al. 2016, Rijlaarsdam et al. 2017). Also important are techniques for extracting transfer function data directly from measured frequency response plots. That information can be used in root-locus plots that encourage a blending of data-based and model-based design techniques (Hoogendijk et al. 2015).

High-precision motion systems often contain several distinct axes of motion and multiple inputs and outputs. They also typically contain lightly damped resonances, requiring a fine frequency spacing to discriminate the actual resonant frequency. This creates the potential for an untenably large set of high-density frequency responses, from each input, to each output, with each axis, potentially over a grid of locations. Thus, there is the need to develop a strategy for planning the measurement process using a priori system knowledge (van der Maas 2016), as well as for efficiently processing the measured responses (Bruijnen and van der Meulen 2016).

The assumption of linearity in system response is useful, but usually only true to a first approximation. With well-designed electrical and mechanical systems, the influence of nonlinear dynamics might be small, but they are likely to be observable in a precision motion control design problem. The most readily apparent consequences of these nonlinearities are amplitude-dependent responses and the generation of harmonics of the input signal. It is important that nonlinearities be quantified for the designer to make an informed decision about whether a linear model will be adequate for the application or whether a more involved nonlinear modelling effort is required (Schoukens et al. 2014). Efficient and accurate identification of nonlinear behaviour usually begins with a careful selection of the excitation frequencies and subsequent analysis of frequency content in the response that was not present in the excitation signal (Vanhoenacker et al. 2001, Rijlaarsdam et al. 2010).

This measure is related to the coherence of the response, which is often low at low frequencies and near resonances where damping forces, that are often nonlinear, dominate. Typical sources of nonlinearities in motion stages include bearing friction, seal and cable hysteresis, actuator hysteresis and force (or torque) ripple, shifting modal masses over travel, and amplifier distortion.

14.2 Feedback Control Techniques

This section presents commonly used frequency-domain techniques for applying feedback control algorithms to precision motion systems. These techniques apply to both model-based, and data-based designs. A dynamic model consisting of two masses connected by a damped spring provides insight into the performance limitations imposed by the mechanical design, guidance on the effect of possible design changes and is the basis for a series of case studies. Rotary systems can equivalently be modelled by a series of rotational inertias connected by torsional springs and acted upon by applied torques. Systems that include both linear and rotational motion can be readily modelled using the techniques outlined in Chapter 4.

This section also reviews the use of frequency response measurements for characterising the response of a dynamic system. These responses may be used directly as part of a data-driven control design methodology, or used to derive and validate a parametric model for use in a model-based design methodology. A loop transmission or open-loop frequency response is the single most useful measurement available for characterising the dynamic performance of a servomechanism.

14.2.1 Loop Shaping and PID Control

Classical, frequency-domain design techniques are useful in designing control systems for precision systems and are more than capable of enabling high-speed, high-accuracy motion in demanding precision manufacturing applications (Butler 2011). The most common technique is the PID controller. Often presented in textbooks in a parallel form

$$C(s) = K_P + K_I \frac{1}{s} + K_D s, \tag{14.1}$$

with K_P as the proportional gain, K_I as the integral gain and K_D as the derivative gain. A more convenient and intuitive representation is the lead lag form

$$C(s) = K \left(\frac{s + 2\pi f_I}{s} \right) \left(\frac{s + 2\pi f_D}{2\pi f_D} \right) \left(\frac{(2\pi f_{lp})^2}{s^2 + 2\zeta_{lp}(2\pi f_{lp})s + (2\pi f_{lp})^2} \right) \tag{14.2}$$

that allows the designer to independently set the gain and controller zero locations. Figure 14.2 shows a block-diagram representation of the lead-lag controller. The lag portion is defined by the integrator and zero at f_I, and the lead portion is defined by the zero at f_D, along with a second-order low-pass filter at f_{lp}. This second-order low-pass filter attenuates the high-frequency response of the lead component, and should generally be included in any PID or lead-lag controller implementation.

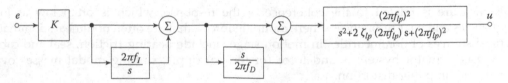

FIGURE 14.2
The lead-lag form of a PID controller allows for more intuitive tuning through independent and direct setting of the overall controller gain and derivative and integrator zero locations. The lead (or derivative) component is usually attenuated with a low-pass filter.

The desired bandwidth, or speed of response, largely determines the placement of the controller zeros in the transfer function. This process for selecting the parameters can be automated through numerical optimisation techniques (Yaniv and Nagurka 2004, van Solingen et al. 2016, Bruijnen et al. 2006), but some intuition can be developed by following the techniques presented in Butler (2011).

The plant usually has a predominantly low-pass characteristic, with a larger magnitude response at low frequencies. A mass driven by a force (or torque) has a transfer function from force to displacement that approximates a double-integrator at low frequencies. A flexure-based stage may be better modelled as a spring–mass–damper system, with a predominantly spring-like behaviour at low frequencies. The controller for these types of systems is often augmented with one or more additional integrators so that it has a large magnitude response at low frequencies.

Selecting initial values for the controller parameters is based largely on the low-amplitude response. By selecting a target bandwidth substantially below the first resonant frequency of the plant, a frequency ratio α (nominally around two to five) can be defined for the location and spacing of the controller zeros. As shown in Figure 14.3, the derivative zero is located at f_D, a factor of α below the bandwidth. The integrator zero is located at f_I, an additional factor of α below the derivative zero. The low-pass filter frequency f_{lp} is a factor

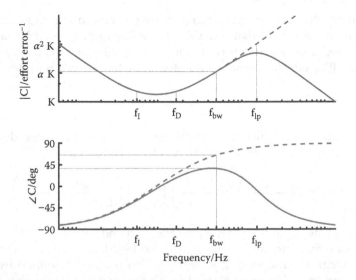

FIGURE 14.3
The high-frequency response of a PID controller (shown in dashed lines) is typically attenuated by a low-pass filter placed after the intended crossover frequency. The separation factor α is the spacing between the indicated frequencies.

of α (or more) greater than the bandwidth. The overall gain K is set so that the magnitude of the loop gain equals unity at the desired crossover frequency.

14.2.2 Characterising Servo Performance

Servo performance is characterised through the use of frequency response functions, but the particulars of which frequency response functions are most relevant depends on the application. A simplified representation of a typical motion control system, shown in Figure 14.4, identifies the multiple inputs, outputs and the overall system of a servomechanism. The main systems are the plant P, the feedback controller C, the feedforward controller C_{ff}, and a transformation from measured variables to performance variables P_{zx}. The design, modelling and analysis of the plant occupy a substantial part of this text, and included in the plant are the sensors for measuring variables of interest. There are four inputs to this system: a reference command r (presumably known a priori), a process disturbance d_1, an output disturbance d_2 and sensor noise n. Examples of process disturbances include any forces acting directly on the system, such as bearing friction, cable drag, machining or cutting forces, and acoustic disturbances. Ground floor vibration is an example of an output disturbance. The plant state x is corrupted by sensor noise n before becoming available to the controller as measurement y.

The actual point of interest for a precision motion application likely differs from the point where a measurement is available. It may be impractical or impossible to place a sensor exactly at the location of interest, and compliance of the internal machine structure cannot be neglected. There needs to be a distinction between the performance variables, indicated by the variable z, and the measured variables, indicated by y. A suitably accurate model can be used to infer the performance variables from the measured ones, and the interested reader is referred to Oomen et al. (2015) for details of the implementation.

Sensitivity functions characterise the ability of feedback control systems to reject disturbances. Assuming single-input, single-output systems, the output responds to the inputs according to

$$x = \frac{PC_{ff} + PC}{1 + PC}r + \frac{P}{1 + PC}d_1 + \frac{1}{1 + PC}d_2 + \frac{-PC}{1 + PC}n \qquad (14.3)$$

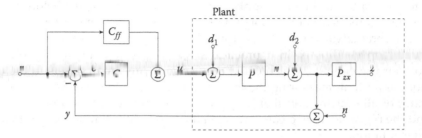

FIGURE 14.4
A servomechanism typically has multiple inputs and interconnected systems. The outputs of the feedback block C and feedforward block C_{ff} combine with disturbance d_1 to form the input to the plant P, which is often proportional to a force. The plant output is corrupted by disturbance d_2 and measurement noise n before being available to the controller as a measurement y. Notice that the controlled variable x may differ from the intended performance variable z if there is compliance between the measurement and the point of interest for the application represented by the system P_{zx}.

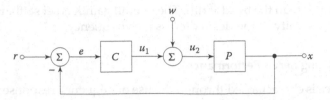

FIGURE 14.5
A typical single-axis loop gain measurement is performed by injecting a known disturbance signal between the output of the controller and the input to the plant. This often takes the form of an additive current command (proportional to force or torque) in the case of an electromagnetic actuator.

from which the sensitivity function S, the complementary sensitivity function T and the process sensitivity function S_p are defined. These are given as

$$S = \frac{1}{1 + PC}, \quad T = \frac{PC}{1 + PC} \quad \text{and} \quad S_p = \frac{P}{1 + PC} \tag{14.4}$$

in their single-input, single-output forms. These functions share the loop gain $L = PC$, and the frequency domain design method is based on setting closed-loop performance characteristics for each sensitivity function by adjusting the loop gain characteristics (Franklin et al. 2015).

Bode plots and Nyquist diagrams of the loop gain together determine the system crossover frequency (loosely referred to as bandwidth), phase margin and gain margin. A more comprehensive measure of stability robustness is the modulus margin. This value is inverse of the peak magnitude (or infinity norm) of the sensitivity function, and is the minimum distance between the loop gain and the critical −1 point on a Nyquist plot (Garcia et al. 2004). Whereas the more familiar gain margin and phase margin quantify the nearness to instability if either the loop gain or loop phase change independently, the modulus margin quantifies the effect of a simultaneous change—a worst-case change that would lead to instability.

A frequency response measurement showing the loop transmission, sometimes called an open-loop frequency response, is the single most valuable measurement that a control systems designer can take. With deference to the care that must be taken in the choice of excitation signals for specific cases, in general, a functional measurement can be taken by injecting a disturbance signal immediately at the output of the control algorithm and before it becomes an input to a plant, as shown in Figure 14.5. This control effort is largely proportional to force or torque in many applications. The ratio of the signal before and after the injection point is a measure of the loop gain L, which is a function of frequency.

There are several advantages to measuring the loop gain, or so-called 'open-loop' gain under closed-loop conditions. In a practical sense, this eliminates the problem of axes drifting to their travel limits, which could occur due to any small offsets to the excitation signal. More important, measuring the loop gain in a closed-loop configuration quantifies the influence of all systems and their computational time delays around the loop. Those delays, and the phase lag they create, are often the bandwidth-limiting factor in a discrete-time implementation.

14.2.3 Data-Driven Loop Shaping

This section demonstrates data-driven tuning techniques based on measured frequency response data. The first step usually involves closing the servo loop with a stabilising, but low-performing, set of gain parameters through autotuning or even empirical tuning techniques. With the system stabilised, the loop transmission is measured as described in the discussion in Section 14.2.2, and by removing the known controller response, arrives with a measured plant frequency response, as shown in Figure 14.6. Notice that a properly presented plant response contains the engineering units of the system; in this case, units of compliance.

The next step is to design a controller in the frequency domain. This is often performed directly on the system using vendor-supplied tuning tools, but can also be done offline using the measured frequency response. In the case shown in Figure 14.6, the plant response showed a resonant peak at 75 Hz, leading to the choice of a desired crossover frequency (or bandwidth) of 35 Hz. Using the lead-lag form of a PID controller as given in Equation 14.2, and adjusting the gain for unity magnitude at 35 Hz, results in a feedback controller with the transfer function

$$C(s) = 0.375 \frac{(s + 2\pi 17.5)}{2\pi 17.5} \frac{(s + 2\pi 8.75)}{s} \frac{(2\pi 140)^2}{s^2 + 2(0.707)(2\pi 140)s + (2\pi 140)^2} \quad [\text{N} \cdot \mu\text{m}^{-1}]. \quad (14.5)$$

Notice that the controller also has units, and one of the more tedious, but necessary, steps in a controller design is to make the unit conversions required of the specific implementation. This may include scale factors resulting from analogue-to-digital conversion, encoder quantisation, motor force constants, amplifier scale factors and discrete-time equivalences. It is good practice to include the units of all signals on block diagrams of controller implementations in order to limit possible confusion.

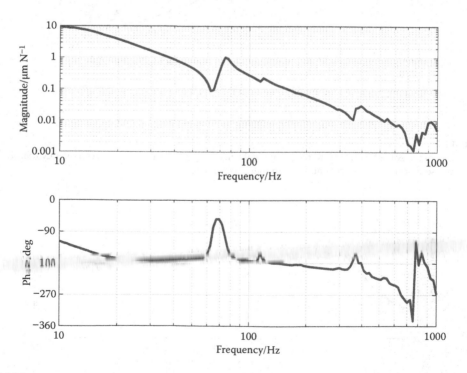

FIGURE 14.6
This measured frequency response of a generic linear positioning system with a payload shows both the general trend of a two-decades-per-decade slope expected of an inertial mass as well as the indication of several resonant peaks.

The resulting loop transmission has unity gain at the crossover frequency, and generally acceptable levels of gain and phase margin. Figure 14.7 shows the loop gain $L = PC$ with a crossover frequency of 35 Hz, a phase margin of 38° and a gain margin of 8.5 dB. Further application-specific tuning may be necessary from here, but the system is now under robustly stable closed-loop control.

The sensitivity plots should also always be reviewed. The peak of the sensitivity magnitude is 6 dB, as shown in Figure 14.8. This gives a modulus margin of 0.5 and is again a

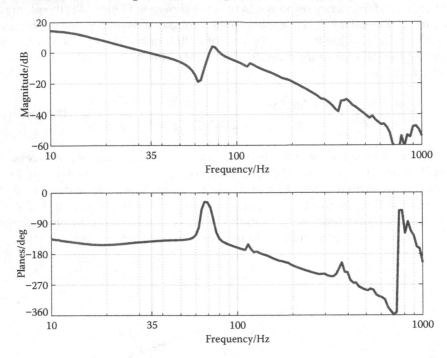

FIGURE 14.7
The addition of a feedback controller to the plant results in a loop gain with a crossover frequency (bandwidth) of 35 Hz, a phase margin of 38 degrees and a gain margin of 8.5 dB.

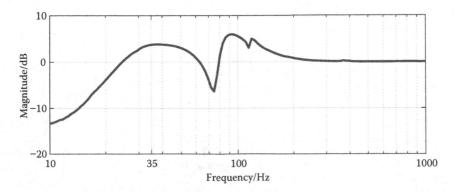

FIGURE 14.8
The sensitivity plot of the closed-loop system shows a peak value of 6 dB, which quantifies the nearness to instability under any plant or controller changes, as well as showing the amplification of sensor noise.

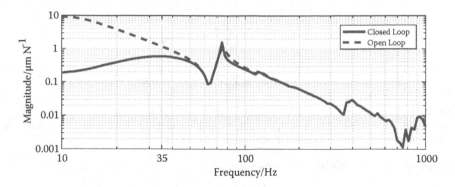

FIGURE 14.9
The process sensitivity illustrates how the control algorithm has effectively added stiffness (or reduced compliance), thus reducing the displacement resulting from a disturbance force at the plant input compared to the open-loop system. This additional stiffness at low frequencies came at the cost of higher compliance in the area of the resonant frequency.

generally acceptable value. Attempts at higher bandwidth will reduce the modulus margin, indicating a system that is closer to instability, and has worse in-position performance through the amplification of sensor noise at any frequency range where the sensitivity is greater than unity (0 dB). Also interesting is the plot of the process sensitivity, shown in Figure 14.9. This plot in particular emphasises the role of feedback control in adding frequency-dependent stiffness. Notice that the low-frequency compliance has decreased compared to the open-loop system. This indicates reduced displacement in response to a disturbance force applied at the plant input. The cost is additional compliance in the range of frequencies close to the resonant peak. At high frequencies, where the control is largely inactive, the open-loop and closed-loop responses are essentially identical.

14.2.4 Systems with Compliant Payloads

The control systems for high-precision mechatronic systems need to compensate for mechanical resonances in the structure. The general technique of modal decomposition (Munnig Schmidt et al. 2014) characterises linear system dynamics as a summation of an arbitrary number of second-order systems. The participation factor of each mode depends on where system inputs are applied and where measurements are taken. The following examples derive control strategies for a typical system that acts as a rigid-body mass at low frequencies, but has some internal flexibility that results in a vibration mode at higher frequencies.

The linear dynamics of precision motion systems can be modelled as a series of mass–spring–damper systems, with their behaviour predicted by second-order linear differential equations. A two-body system, as shown in Figure 14.10, is a useful example. The Laplace transform of the equations of motion take the form

$$
\begin{bmatrix} X_1 \\ X_2 \end{bmatrix} = \frac{1}{s^2 \left(s^2 + \dfrac{b}{m_c} s + \dfrac{k}{m_c} \right)} \begin{bmatrix} \dfrac{1}{m_1} \left(s^2 + \dfrac{b}{m_2} s + \dfrac{k}{m_2} \right) & \dfrac{1}{m_2} \left(\dfrac{b}{m_1} s + \dfrac{k}{m_1} \right) \\ \dfrac{1}{m_1} \left(\dfrac{b}{m_2} s + \dfrac{k}{m_2} \right) & \dfrac{1}{m_2} \left(s^2 + \dfrac{b}{m_1} s + \dfrac{k}{m_1} \right) \end{bmatrix} \begin{bmatrix} F_1 \\ F_2 \end{bmatrix} \tag{14.6}
$$

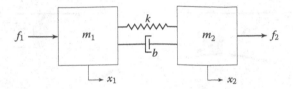

FIGURE 14.10
A model of two masses connected by a damped compliance demonstrates common dynamic responses of precision motion systems and forms the basis for more complex systems.

when presented in a multiplicative sense, and

$$\frac{X_1}{F_1} = \frac{1}{(m_1 + m_2)s^2} + \frac{\dfrac{m_2^2}{(m_1 + m_2)^2}}{m_c s^2 + bs + k} = \frac{1}{m_1 + m_2}\left(\frac{1}{s^2} + \frac{\dfrac{m_2}{m_1}}{s^2 + \dfrac{b}{m_c}s + \dfrac{k}{m_c}}\right), \quad (14.7)$$

$$\frac{X_2}{F_1} = \frac{1}{(m_1 + m_2)s^2} + \frac{\dfrac{-m_1 m_2}{(m_1 + m_2)^2}}{m_c s^2 + bs + k} = \frac{1}{m_1 + m_2}\left(\frac{1}{s^2} + \frac{-1}{s^2 + \dfrac{b}{m_c}s + \dfrac{k}{m_c}}\right) \quad (14.8)$$

and

$$\frac{X_1}{F_2} = \frac{1}{(m_1 + m_2)s^2} + \frac{\dfrac{-m_1 m_2}{(m_1 + m_2)^2}}{m_c s^2 + bs + k} = \frac{1}{m_1 + m_2}\left(\frac{1}{s^2} + \frac{-1}{s^2 + \dfrac{b}{m_c}s + \dfrac{k}{m_c}}\right), \quad (14.9)$$

$$\frac{X_2}{F_2} = \frac{1}{(m_1 + m_2)s^2} + \frac{\dfrac{m_1^2}{(m_1 + m_2)^2}}{m_c s^2 + bs + k} = \frac{1}{m_1 + m_2}\left(\frac{1}{s^2} + \frac{\dfrac{m_1}{m_2}}{s^2 + \dfrac{b}{m_c}s + \dfrac{k}{m_c}}\right) \quad (14.10)$$

in an additive form. The variable $m_c = m_1 m_2 / (m_1 + m_2)$ is called the combined mass term. Notice that the poles of the transfer function (roots of the denominator) are identical regardless of where a measurement is taken. Their locations are fixed and do not change. Zeros (roots of the numerator) on the other hand depend on the particular inputs and outputs of a system. Their location is often neglected in specifications that call only for plant dynamics with a first natural frequency above a certain value.

A review of the limiting cases gives insight into the dynamic response. If the second mass is vanishingly small ($m_2 \ll m_1$), the behaviour reverts to a double integrator with a mass approximately equal to m_1. This is true wherever the measurement is made. When the masses have roughly comparable magnitudes, then the dynamic responses will depend on where the measurement is made. Figure 14.11 shows the frequency response plots of the two-mass system, driven at mass 1, and measured at both mass 1 and mass 2. The response measured at the same point that the force is applied is commonly called the collocated response, as the drive and measurement are located coincidental with each other. The response from a drive on mass 1 to a measurement on mass 2 is referred to as non-collocated.

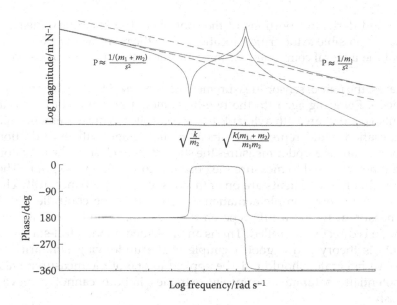

FIGURE 14.11

Frequency responses of collocated and non-collocated systems show identical pole locations, but the appearance of a complex conjugate set of zeros form a notch in the collocated case. This is typically the more common response for well-designed mechanical systems with a relatively tight structural loop between the actuator and feedback sensor.

14.2.5 Dual-Loop Systems

Measurements taken at sensor locations near the actuator are often not adequate to fully evaluate dynamic performance at the point of interest. Precision systems often include feedback from multiple sensors. In a dual-loop system, one feedback sensor may be located close to the actuator, with a second located as close as possible to the point of interest. The control system designer can now choose between using the second sensor primarily for low-frequency accuracy improvement or potentially to directly control a flexible mode of the structure.

The most common use for an additional feedback device is for accuracy improvement by taking a measurement as close as possible to the point of interest (see Chapter 10). As modelled in Figure 14.12, the primary feedback measurement y_1 is usually taken as close as possible to the actuator, thus creating a collocated dynamic system, and is used for the

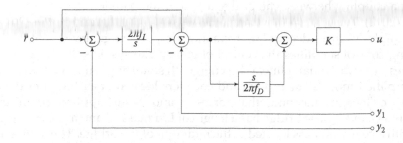

FIGURE 14.12

This dual-loop implementation of a lead-lag type PID controller uses a measurement near the point of interest for the integral action, and measurement made close to the actuator for the proportional and derivative components.

proportional and derivative portions of the controller. The secondary measurement y_2 is made as close as possible to the critical location (often the point where a tool and workpiece interact), and the integral component of the controller acts on the error between it and the reference r.

In some cases, the pair of sensor measurements can be used to directly attenuate the main vibration mode. Referring again to the two-mass model in Figure 14.10, assume that the displacements x_1 and x_2 are both available to the controller as measurements y_1 and y_2. This type of idealisation could represent a linear motion stage with an additional angular pitching mode. A linear encoder measures the stage motion close to the actuator, whereas a laser interferometer is used to measure the position close to the workpoint. The important observation is that measurements are on both sides of the compliance, fully characterising the resonance. A preferred implementation is now a PD-type controller on the primary collocated measurement and a PID-type on the secondary workpoint measurement. The vibration mode is directly controlled. This is an implementation of full-state feedback from modern controls theory and a good example of the underlying similarity between the approaches. An integrator should only be applied to one of the measurements, otherwise each can potentially saturate at opposite extremes if both cannot achieve zero error simultaneously.

14.2.6 Decoupling Control–Gantry Systems

Precision motion control systems are often multi-axis, may not have collocated sensors and actuators, or may have redundant sensors and actuators. This creates a multivariable control problem for which classical control techniques, designed around single-input, single-output models, are difficult, or at least non-intuitive, to apply. Details of full multi-variable control system design practices are provided in Skogestad and Postlethwaite (1996). However, classical techniques can be applied to a class of multiple-input, multiple-output control problems through the application of axis decoupling techniques. These decoupling techniques are coordinate transformations—applied to actuator inputs and sensor measurements—that substantially reduce a coupled plant model and control problem into a diagonalised series of single-input, single-output models. The formalised technique is very similar to the modal coordinate transformation frequently used in vibration analysis (see Chapter 13).

Decoupling control techniques are often applied to gantry control problems. This style of mechanism, also called a parallel, or H-bridge axis, uses two actuators and two sensors acting substantially at the ends of a single bridge axis. The bridge axis itself also usually contains a motor used to drive a payload offset from the centre of stiffness of the bridge axis, resulting in a coupling between the bridge axis and the parallel gantry axes. Gordon and Erkorkmaz (2012), Teo et al. (2007), García-Herreros et al. (2013) and Butler (2011) all show detailed examples of these decoupling techniques applied to gantry systems.

Decoupling control simplifies the control of gantry-like systems. In this case, the gantry bridge moves in a substantially linear direction with small angular deviations. Figure 14.13 shows a simplified model. The motors and encoders located at each end of the bridge are substantially collocated, meaning that forces f_1 and f_2 are applied collinear with the displacements x_1 and x_2. The bridge has a length of L, a mass m, and a polar second moment of mass J with a centre of mass located at the midpoint of the bridge. The stiffness k_θ (N · m · rad^{-1}) and damping b_θ (N · m · s · rad^{-1}) terms are the equivalent rotational values of the compliance in the guiding bearings when the bridge rotates through and angle θ. Given two

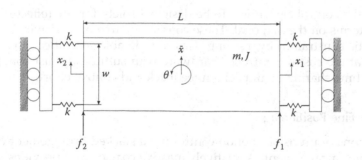

FIGURE 14.13
A gantry bridge has two dominant degrees of freedom; one linear and one rotational. Forces f_1 and f_2 are applied at both ends of the bridge, and are largely coincident with the sensors reading displacements x_1 and x_2. The bridge has a mass of m and a rotational inertia about the centre of mass of J.

bearings on each side, spaced by a distance w, they are related through $k_\theta = w^2 k$. This leads to a transfer function matrix from motor forces to measured positions according to

$$
\begin{bmatrix} X_1 \\ X_2 \end{bmatrix} =
\begin{bmatrix} \dfrac{1}{ms^2} + \dfrac{L^2/4}{Js^2 + b_\theta s + k_\theta} & \dfrac{1}{ms^2} - \dfrac{L^2/4}{Js^2 + b_\theta s + k_\theta} \\ \dfrac{1}{ms^2} - \dfrac{L^2/4}{Js^2 + b_\theta s + k_\theta} & \dfrac{1}{ms^2} + \dfrac{L^2/4}{Js^2 + b_\theta s + k_\theta} \end{bmatrix}
\begin{bmatrix} F_1 \\ F_2 \end{bmatrix}.
\tag{14.11}
$$

Notice the main challenge in designing the controllers: any single applied force leads to a displacement at two different locations. This is a particular case of the two-mass problem described earlier, except that instead of two linear degrees of freedom, there is a linear and a rotary. This example ignores the change in rotational inertia that occurs with motion of the bridge axis.

A coordinate change decouples the control problem. Notice that the state of the axis can be described equally well by a displacement (at the centre of the gantry) and an angle, and the control effort can be described equally well as an average force and a torque. Making the substitutions

$$
\begin{bmatrix} X_1 \\ X_2 \end{bmatrix} =
\begin{bmatrix} 1 & L/2 \\ 1 & -L/2 \end{bmatrix}
\begin{bmatrix} \overline{X} \\ \theta \end{bmatrix},
\tag{14.12}
$$

$$
\begin{bmatrix} F \\ T \end{bmatrix} =
\begin{bmatrix} 1 & 1 \\ L/2 & -L/2 \end{bmatrix}
\begin{bmatrix} F_1 \\ F_2 \end{bmatrix}
\tag{14.13}
$$

and simplifying yields the decoupled form of the transfer function matrix

$$
\begin{bmatrix} \overline{X} \\ \theta \end{bmatrix} =
\begin{bmatrix} \dfrac{1}{ms^2} & 0 \\ 0 & \dfrac{1}{Js^2 + bs + k} \end{bmatrix}
\begin{bmatrix} F \\ T \end{bmatrix}.
\tag{14.14}
$$

This allows the control algorithms to be designed solely for the (effective) single-input, single-output terms on the diagonal. These consist of a free mass driven by a force, and a rotational inertia constrained by a spring. This considerable simplification in the algorithm design comes at the cost of controller hardware with sufficiently fast communications to allow the real-time sharing of all position feedback and control efforts.

14.2.7 Coarse–Fine Positioning

Some applications of precision motion control use a stacked arrangement of axes in a so-called coarse–fine arrangement. A relatively massive coarse stage provides the long travel, while a smaller fine stage provides high-frequency corrections. Some examples include fast tool servos for turning non-rotationally symmetric optics or a fast focusing system for optics. Figure 14.14 shows a lumped parameter model of one such arrangement that stacks a fine flexure-based stage on a coarse long-travel stage. This creates a two-input, two-output transfer function matrix for the plant model

$$
\begin{bmatrix} X_1 \\ X_2 \end{bmatrix} =
\begin{bmatrix}
\dfrac{\dfrac{1}{m_1}\cdot\left(s^2+\dfrac{b}{m_2}s+\dfrac{k}{m_2}\right)}{s^2\left(s^2+\dfrac{b}{m_c}s+\dfrac{k}{m_c}\right)} & \dfrac{-\dfrac{1}{m_1}}{\left(s^2+\dfrac{b}{m_c}s+\dfrac{k}{m_c}\right)} \\[4ex]
\dfrac{\dfrac{1}{m_1}\cdot\left(\dfrac{b}{m_2}s+\dfrac{k}{m_2}\right)}{s^2\left(s^2+\dfrac{b}{m_c}s+\dfrac{k}{m_c}\right)} & \dfrac{\dfrac{1}{m_2}}{\left(s^2+\dfrac{b}{m_c}s+\dfrac{k}{m_c}\right)}
\end{bmatrix}
\begin{bmatrix} F_1 \\ F_{12} \end{bmatrix},
\tag{14.15}
$$

where m_1 represents the mass of the coarse stage and m_2 represents the mass of the fine stage. The variable $m_c = m_1 m_2/(m_1 + m_2)$ is again the combined mass term, and k and b are the stiffness and damping respectively. The positions of the coarse and fine stage are x_1 and x_2, the force applied by the coarse-stage motor is f_1, and the force applied by the fine-stage motor is f_{12}. Notice that the force from the fine stage applies equally to masses m_1 and m_2, but in opposite directions. Equation 14.15 generalises to

$$
\begin{bmatrix} X_1 \\ X_2 \end{bmatrix} =
\begin{bmatrix} P_{cc} & P_{cf} \\ P_{fc} & P_{ff} \end{bmatrix}
\begin{bmatrix} F_1 \\ F_{12} \end{bmatrix}.
\tag{14.16}
$$

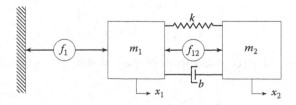

FIGURE 14.14
This model of a precision motion system shows a flexure-based fine stage mounted to a coarse stage. Control efforts are applied between the coarse stage and ground, and between the fine stage and coarse stage. The only measured position available to the controller is between the fine stage and ground.

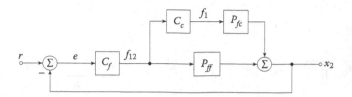

FIGURE 14.15
The block diagram of one possible coarse–fine control implementation uses measurements of the fine-stage position to generate control efforts for both the coarse-stage and fine-stage actuators.

The objective is to design a control strategy that positions the fine stage to follow a reference command by using commands to both the coarse- and fine-stage actuators. In addition, when settled, the position of the fine stage should be at its centre of travel relative to the coarse stage. Assume that only a single measurement is available: one that measures the position of the fine stage relative to the fixed base. Figure 14.15 shows a block diagram of the servo control problem. The fine-stage controller C_f generates a control effort in response to the measured position of the fine stage relative to a fixed reference, and the coarse-stage controller C_c generates a control effort in response to the fine-stage effort.

The coarse-stage controller is designed to simplify the effective plant to be controlled by the fine controller. The open-loop transfer function of the system is given by

$$L = (P_{fc}C_c + P_{ff})C_f \tag{14.17}$$

and the term in parentheses simplifies with a selection of C_c as a scalar multiplier of $(m_1 + m_2)/m_2$. The new effective plant becomes

$$P_f = P_{fc}C_c + P_{ff}$$

$$= \frac{\dfrac{1}{m_1 m_2}(bs + k)}{s^2\left(s^2 + \dfrac{b}{m_c}s + \dfrac{k}{m_c}\right)}\frac{(m_1 + m_2)}{m_2} + \frac{\dfrac{1}{m_2}}{s^2 + \dfrac{b}{m_c}s + \dfrac{k}{m_c}}$$

$$= \frac{\dfrac{1}{m_2}\left(\dfrac{b}{m_c}s + \dfrac{k}{m_c}\right)}{s^2\left(s^2 + \dfrac{b}{m_c}s + \dfrac{k}{m_c}\right)} + \frac{\dfrac{1}{m_2}}{s^2 + \dfrac{b}{m_c}s + \dfrac{k}{m_c}} \tag{14.18}$$

$$= \frac{1}{m_2 s^2}$$

With this simplification, the fine stage controller C_f is designed as a conventional PID or lead-lag compensator.

14.2.8 Systems with Frame Motion

The machine bases and frames used to support precision motion systems are not infinitely rigid, and the dynamics of their motion must be included in the system analysis. The machine base may take the form of a substantial mass (for example, a granite slab) mounted to a well-defined vibration isolation system or may be a structural frame, in which case the

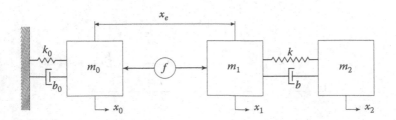

FIGURE 14.16
A precision motion system often includes a machine base or frame. The displacement measured by the feedback sensor is usually relative to this base.

local structure determines an effective modal mass and stiffness (see Chapters 11 and 13). Any forces applied to accelerate the payload of a motion system also accelerate the machine base or frame in the opposite direction. Figure 14.16 shows a lumped parameter model of this configuration. The mass of the base m_0 is typically much higher than that of the stage and moving elements, and so its acceleration is proportionally lower for the same applied force. It is usually not negligible, in the context of precision systems, however. It is also important to note that the available displacement measurement is usually between a point on the machine frame and the moving carriage, and so the motion of the machine base manifests itself as a residual vibration in the measurement at the end of the move.

The simplified model for the motion of a carriage on a machine base adds an additional mass m_0 to the flexible-payload model developed in Equation 14.5. Denoting x_1 as the absolute displacement of the carriage, x_0 as the absolute displacement of the base, and x_e as their difference, the set of transfer functions

$$\frac{X_0}{F} = \frac{-1}{m_0 s^2 + b_0 s + k_0} \tag{14.19}$$

$$\frac{X_1}{F} = \frac{1}{(m_1 + m_2)s^2} + \frac{\dfrac{m_2^2}{(m_1 + m_2)^2}}{m_c s^2 + bs + k} \tag{14.20}$$

$$\frac{X_e}{F} = \frac{1}{(m_1 + m_2)s^2} + \frac{\dfrac{m_2^2}{(m_1 + m_2)^2}}{m_c s^2 + bs + k} + \frac{1}{m_0 s^2 + b_0 s + k_0} \tag{14.21}$$

describe the dynamic behaviour.

The frequency response between the applied force and the position at the sensor again contains a complex-conjugate set of poles and zeros, but in this case, they typically fall below the expected crossover frequency. The result can be a low-amplitude but long duration residual vibration during settling.

14.2.9 Systems with Rolling Friction

The dynamic response of systems with rolling-element bearings changes dramatically at small displacements (see Chapter 7). Friction in this so-called pre-sliding or pre-rolling regime is not simply a function of velocity, but instead depends on the history (direction and distance) of past moves, becoming in effect a hysteretic spring. At small displacements (generally a micrometre and below), the bearings act like springs before eventually breaking free and rolling over longer distances (Futami et al. 1990). There are a number of descriptive models for this behaviour (Armstrong-Hélouvry et al. 1994, Al-Bender et al. 2005, Feldman et al. 2016), but most remain empirical and less directly applicable in the mechanical design phase of a project. The development of predictive models remains an active research topic (Al-Bender and Swevers 2008, De Moerlooze et al. 2011).

There are several adverse effects to this spring-like frictional behaviour. The amplitude of the low-frequency loop gain is reduced when compared to the same response in the large-travel regime (Otsuka and Masuda 1998). A lower-magnitude loop gain increases the magnitude of the sensitivity response, indicating a reduced ability to reject disturbances and track reference commands. This is most noticeable as extended settling times compared with what would be expected based on the large-amplitude linear model. In addition, the combination of a spring-like behaviour of the bearings carrying a mass creates a resonant peak in the frequency response (Yoon and Trumper 2014). This peak, often called a Dahl resonance after the inventor of one of the first descriptive models (Dahl 1968), can lead to a low modulus margin (i.e. a high sensitivity peak) and sustained oscillations while the axis is attempting to hold position.

Researchers have proposed numerous control techniques for addressing this non-linearity, and the breadth of approaches suggests that a universal solution remains elusive. The design problem is complicated by the sensitivity of this behaviour to small changes in lubrication, bearing preload (including thermally induced changes) and recent operating history. Figure 14.17 shows a measured frequency response from motor force to measured position of a commercially available linear motor stage. Note the wide variation in effective spring constant over even a small number of samples. If possible to determine early in the

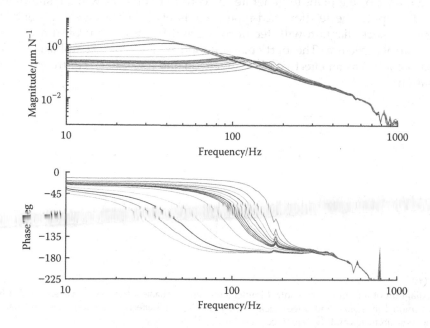

FIGURE 14.17
Measured frequency response plots of a linear motion stage that uses rolling elements bearings taken at different levels of input amplitude show the variable spring-like behaviour of rolling-element bearings in their pre-rolling regime.

mechanical design phase, the best strategy may be to select bearing preload and lubrication such that the maximum Dahl resonance is below the intended crossover frequency (Butler 2011). This ensures that overall loop stability can be maintained at all operating conditions.

14.3 Feedforward Control Techniques

Feedforward algorithms form the second part of a two degree of freedom controller design for precision systems. A true feedforward controller does not use any real-time measurements, but instead operates solely open-loop, using a priori information about the expected dynamic response of a system to generate a control effort that will have the desired response. Araki and Taguchi (2003) present a tutorial-style article showing the functional equivalence between several different feedforward control structures. Two that are of interest here are the feedforward type, which is designed to be an approximate plant inversion, and the set-point filter type, which typically alters the frequency content of the reference command for vibration suppression. Lee et al. (2011) present a design methodology for two degree of freedom control principles applied to the nanopositioning systems used in scanning probe microscopy (see Chapter 5).

14.3.1 Model-Inverting Feedforward

The feedforward-type filter is used to approximately invert the expected response of the plant. That is, given a model of a plant and a desired output, the feedforward filter uses an inverse model of the plant to generate the control effort that will substantially lead to the desired output. The feedforward-type filter is effectively open-loop control. Figure 14.18 shows a block diagram with feedforward and feedback control in a two degree of freedom control structure. The particular location of the blocks emphasises the role of feedforward control as an effective plant inversion and a feedback control for disturbance compensation.

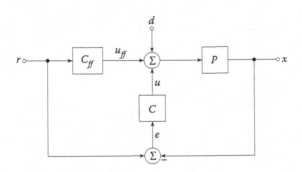

FIGURE 14.18
This block diagram of a feedforward control implementation emphasises the role of a feedforward block as an approximate model inversion, and a feedback block to manage the effects of disturbances and residual errors between the measurements and the reference.

Setting the feedforward filter to an approximate plant inversion reduces the servo errors in response to a change in reference. Following the terminology of Figure 14.18 results in the transfer function

$$\frac{e}{r} = \frac{1 - PC_{ff}}{1 + PC} \tag{14.22}$$

and the closer that the term PC_{ff} is to unity at each frequency, the lower the error in response to a reference command. There are of course limits to the effectiveness of model-inverting feedforward. Plant uncertainty varies with frequency and defines the regions over which feedforward control can be used to improve the response to reference commands (Devasia 2002). In addition, non-minimum phase zeros in the plant response, which are relatively common in discrete-time representations of plant models or when mode shapes include both linear and angular components, become unstable poles in a plant inversion. This is mathematically feasible as an end-to-end pole-zero cancellation, but unusable in any practical system due to the presence of unstable intermediate signals.

Many motion systems exhibit predominantly rigid-body behaviour. That is, the low-frequency response can be effectively modelled as inertial mass, with higher-frequency dynamics described by a summation of mode shapes

$$P(s) = \frac{1}{ms^2} + \sum_{i=1}^{N} \frac{k_i}{m\left(s^2 + 2\zeta_i \omega_i s + \omega_i^2\right)}. \tag{14.23}$$

The numerators k_i are the modal influence coefficients, and depend on where in the system the forces are applied and where in the system the displacements are measured. Thus, a model inversion becomes a double-differentiator multiplied by the approximate mass, $\hat{m}$, of the system. A feedforward controller of

$$C_{ff}(s) = \hat{m}s^2 \tag{14.24}$$

produces a control effort proportional to acceleration if the reference command is given in units of displacement. This is the acceleration feedforward term common and relatively effective in most industrial motion controllers. Likewise, for systems with pronounced viscous friction, a velocity feedforward term adds a control effort proportional to the velocity to compensate for particular frictional effects.

Continuing the model-inversion process at higher frequencies becomes more difficult in practice. High-frequency dynamics often have a higher amount of uncertainty and typically consist of a series of complex-conjugate pole and zero pairs. These have the characteristics of rather rapid changes in phase, and attempts to invert the response can lead to large errors if the model matching is not close enough. In addition, the higher-order dynamics are also more likely to change over time and to vary with the system mass and particular location of all of the axes in a system. These variations can of course be measured and added to an increasingly complex plant and feedforward model, but other techniques exist.

The frequency content in a reference command is usually low relative to the structural dynamics in a well-designed precision motion system. Step changes are popular in analysis, but are almost never applied as a reference in any practical motion system (van Dijk and

Aarts 2012). This means that the feedforward controller needs only to invert the plant over the range of frequencies that are contained in the reference command. Jerk derivative control compensates for the low-frequency content of higher-frequency resonant modes providing the benefit of better tracking of the frequencies typically contained in a reference command, but reducing the sensitivity to modelling errors or plant changes. Following the derivations in Boerlage (2006) and Boerlage et al. (2004), for the plant model given by Equation 14.23 (and with negligible damping, typical for a mechanical system), the appropriate value for jerk-derivative gain is

$$
\delta = \frac{-m \sum_{i=1}^{N} k_i \prod_{j \in \{1,..,N|j \neq i\}} \omega_j^2}{\prod_{i=1}^{N} \omega_i^2}
\tag{14.25}
$$

and the overall feedforward controller, including both acceleration and jerk-derivative terms, becomes

$$
C_{ff}(s) = \hat{m}s^2 + \delta s^4.
\tag{14.26}
$$

It is important to recognise that most feedforward control calculations take place in discrete time and typically operate on a sequence of position commands. Thus, the underlying trajectory generator must produce commands that are sufficiently differentiable to enable higher-order feedforward control designs (Chang and Hori 2006). A fourth-order feedforward compensator requires at least a fourth-order polynomial reference trajectory (Boerlage et al. 2003). In addition, there is usually some delay between a feedforward command and the dynamic response caused by analogue-to-digital conversion times and often multiple different discrete-time calculations. This time delay can cause large errors at higher frequencies and generally should also be compensated (Butler 2012). A generalised implementation of model-inverting feedforward, implemented in discrete time, can take the form of a finite impulse response (FIR) filter. The values of these filter coefficients can then be modified via an optimisation technique that uses iterative trials to converge on a feedforward filter that best minimises tracking error by compensating for the differences between the model and the actual system (van der Meulen et al. 2008).

14.3.2 Setpoint Filtering

Setpoint filtering feedforward strategies strategically alter the frequency content in a reference command. A step response contains high-amplitude content across all frequencies, and it has already been stated that step commands are rarely used in practice (due to both the high-frequency content and to the likelihood of amplifier saturation). Instead, the command is smoothed to limit the peak velocity, acceleration or jerk. This smoothing, done as part of the trajectory generation, creates a profile with a given frequency content. The frequency response of the servo system determines how well this profile is followed.

The shape of the reference command determines its frequency content, and thus, the degree to which structural resonances on the mechanics are excited by the motion commands. Short-duration move commands contain higher frequencies, leading to the counterintuitive result that the fastest move-and-settle times are often achieved by using a slower move command that does not excite structural resonances. Low-pass filters or notch filters are sometimes added to modify the frequency content of the command, or the command can be defined as a high-order polynomial with a carefully shaped frequency spectrum (Sencer and Tajima 2017). So-called tuning problems are often actually problems with the shape of the commanded trajectory.

14.3.3 Command Shaping

Command shaping is a type of feedforward control strategy that improves settling time in a system by modifying the reference command and creating opportunistic self-cancelling vibrations. It is important to note that these techniques are only effective in reducing the level of vibrations caused by motion commands. Vibrations excited by other inputs (environmental disturbances, for example) are not attenuated in any way. One such command shaping technique is a sparse FIR filter with the times and magnitudes of the filter values designed such that transient vibrations created by the first part of a move command are cancelled by subsequent scaled and time-shifted copies. The easiest to understand conceptually is the Posicast filter (Tallman and Smith 1958) applied to a lightly damped second-order system. The impulse response of this system is an exponentially decaying sinusoid. A second impulse, timed to begin a half period after the first, and slightly reduced in magnitude, will create a scaled and time-shifted copy of the initial impulse response. The superposition of their responses cancels all but the first half-period of the oscillation. Figure 14.19 shows the simulated result of applying a two-tap command shaping filter to a step response of a lightly damped second-order system. Notice that the oscillations created by the delayed step cancel those created by the initial step, and their superposition is vibration-free.

Extended command shaping techniques generally vary the magnitude and timing of the impulses to minimise the amount of residual vibration present given uncertainty in the natural frequency. The longer the length of the command shaping filter, the more the move command time is extended and the greater the insensitivity to changes in natural frequency. The technique presented here follows the derivations in Singer and Seering (1990) and Singhose and Seering (2011), and can be effective at reducing residual vibrations in motion systems with lightly damped mechanical resonances.

The design process for these command shaping filters begins with measurements of the natural frequency and damping ratio of the problematic mechanical resonance. Estimates of

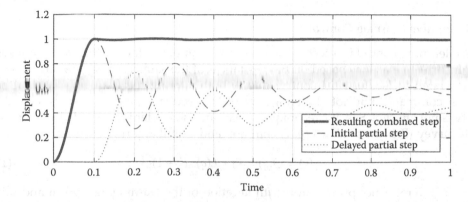

FIGURE 14.19
Example of a command shaping filter implementation showing that delaying and correctly scaling a portion of the command can create self-cancelling out-of-phase resonances.

TABLE 14.1

Impulse Times and Magnitudes for a Vibration-Reducing Command Shaping Filter

Time	2-Tap Filter Coefficient	3-Tap Filter Coefficient	4-Tap Filter Coefficient
0	$\dfrac{1}{1+K}$	$\dfrac{1}{1+2K+K^2}$	$\dfrac{1}{D}$
ΔT	$\dfrac{K}{1+K}$	$\dfrac{2K}{1+2K+K^2}$	$\dfrac{3K}{D}$
$2\Delta T$		$\dfrac{K^2}{1+2K+K^2}$	$\dfrac{3K^2}{D}$
$3\Delta T$			$\dfrac{K^3}{D}$

these values taken from time trace data are often sufficient for an initial trial. In the following relationships, the frequency ω_0 is given in units of radians per second and the damping ratio ζ is unitless. Defining as intermediate variables

$$K = e^{-\dfrac{\zeta\pi}{\sqrt{1-\zeta^2}}}, \tag{14.27}$$

$$\Delta T = \dfrac{\pi}{\omega_0\sqrt{1-\zeta^2}}, \tag{14.28}$$

$$D = 1 + 3K + 3K^2 + K^3. \tag{14.29}$$

and the coefficients of the preshaping filter values for two-, three- and four-tap filters are shown in Table 14.1. Note here some practical limitations imposed by a discrete-time implementation. The filter delay ΔT must be approximated by an integer multiple of the servo sample period. This means that the filter is likely to be perfectly tuned for a frequency near, but not exactly at, the resonant frequency. Also, the filter coefficients should be rounded carefully so that they sum to one, otherwise the final position setpoint will not accurately equal the intended final value. This is only one example of a command shaping filter, and the interested reader is encouraged to visit the references for a fuller survey of the field.

14.3.4 Iterative Learning Control

Iterative learning control is a technique designed to improve the performance of a system through corrective actions based on past instances of the same (or very similar) tasks. It is loosely related to feedforward control due to both altering the frequency content of the reference command, but not changing any of the feedback gains. As such, closed-loop stability is not affected. Specific techniques vary, and the reader is referred elsewhere for an overall survey (Bristow et al. 2006). A common and useful algorithm is

$$r_{j+1}(k) = Q(q)[r_j(k) + L(q)e_j(k+1)], \tag{14.30}$$

where r_j is the reference profile on the jth iteration of the learning algorithm and e_j is the error between the measured and desired outputs. $Q(q)$ and $L(q)$ are the Q-filter (typically with low-pass characteristics) and the learning function (with proportional and rate-dependent characteristics) respectively, with q denoting the forward shift operator.

14.4 Summary

The design of controllers is not fundamentally different for a precision system versus a conventional system any more than are mechanical design techniques. The distinction is again in the additional number of error-creating influences that the designer of a precision system needs to address. An appropriate strategy is to use a two degree of freedom control structure with feedforward control to provide the bulk of the control effort in response to setpoint changes, and feedback control to respond to disturbances and eliminate residual errors. The designer can deterministically achieve the closed-loop performance requirements by using frequency-domain techniques applied to a model of the system, followed by fine-tuning based on direct measurements.

Exercises

1. Converting the lead-lag form of a controller to a parallel PID form: Controllers developed in the lead-lag form often need to be implemented in a parallel PID form to fit the existing parameterisation of a commercially available PID implementation. Given a feedback controller with the transfer function

    $$C(s) = 10 \cdot \frac{(s + 50)(s + 150)}{s}, \tag{14.31}$$

    determine the appropriate values of K_P, K_I and K_D for an equivalent controller expressed as

    $$C_{PID}(s) = K_P + K_I \frac{1}{s} + K_D s. \tag{14.32}$$

2. Designing a controller for an air bearing stage: Design a lead-lag controller for a linear-motor-driven air-bearing stage system. The moving mass m is 10 kg, the force constant of the motor K_m is 25 N/A and the encoder used for measuring displacement has a resolution of 10 nm per count. The units of the controller should be amperes per count, with a design that achieves a 50 Hz crossover frequency with a phase margin between 50° and 60°.

3. Model parameter estimation: The frequency response shown in Figure 14.6 is quite typical for that of a precision motion system. There will always be some mechanical compliance, as indicated by a zero-pole doublet, but the underlying mechanical design has placed the sensor and actuator close together so that the resulting resonance is of the collocated type. Fit the parameters of a two-body model given by

    $$P(s) = \frac{m_2 s^2 + bs + k}{s^2(m_1 m_2 s^2 + (m_1 + m_2)bs + (m_1 + m_2)k)} \tag{14.33}$$

    to the measured frequency response shown in Figure 14.6.

4. Converting a plant model to a summation of modes: An algorithmic model-fitting routine has returned the following polynomial expression of a plant transfer function:

$$P(s) = \frac{5s^2 + 502.654824574s + 6316546.816697190}{100s^4 + 12566.3706144s^3 + 157913670.4174297s^2} \quad \text{mN}^{-1}. \quad (14.34)$$

Using a partial fraction expansion, convert this multiplicative expression into a summation of individual vibration modes. Report the total moving mass, the mass decoupling ratio and the first natural frequency. What would be a reasonable target for a closed-loop bandwidth?

5. Command shaping filter design: Design a three-tap command shaping filter to attenuate a 150 Hz vibration observed in the closed-loop response of a positioning system. Time-domain plots indicate a damping ratio of 0.02. Express the answer in a form that can be implemented on a discrete-time system with a 5000 Hz update rate.

6. Frequency responses of two degree of freedom systems: Given the transfer function model of a compliant mass driven by a force,

$$P(s) = \frac{1}{m}\left(\frac{1}{s^2} + \frac{\alpha}{s^2 + \omega_n^2}\right), \quad (14.35)$$

plot the frequency response curves for four different values of the modal participation factor α:

Case 1: $\alpha > 0$

Case 2: $-1 < \alpha < 0$

Case 3: $\alpha = -1$

Case 4: $\alpha < -1$

For graphing purposes, assume a mass of $m = 1$ kg, negligible damping and a natural frequency of $\omega_n = 1000$ rad s^{-1}. Describe the physical significance of each case.

7. Systems with dynamic coupling: Multi-axes mechanical systems may sometimes exhibit unwanted dynamic coupling between axes. That is, a control effort commanded to a primary axis results in motion in multiple axes. This could be due to mechanical misalignments or cable management systems or could be inherent in the mechanical design (for example, stages for generating vertical motion often use a wedge-style design and may mounted on an additional horizontal axis; acceleration of the horizontal axes couples into the vertical direction). In this exercise, begin with a set of plant models that have unwanted dynamic coupling between them,

$$X_1 = P_1(U_1 + k_{12}U_2)$$
$$X_2 = P_2(U_2 + k_{21}U_1). \quad (14.36)$$

The variables x_1 and x_2 represent the displacement of the axes, u_1 and u_2 are the control efforts, P_1 and P_2 are the plant models that convert effort to displacement, and k_{12} and k_{21} represent the axis coupling. Determine an appropriate decoupling strategy that allows the two controllers to be designed independently of each other.

8. Modelling a gantry axis with an off-centre load: The model developed for the gantry system shown in Figure 14.13 assumes that the centre of mass of the gantry is coincident with the geometric centre of the gantry bridge. The more general case is that the centre of mass is offset from the geometric centre and may in fact change quite significantly as the other axes move through their travel. Derive the equations of motion of the gantry

bridge for the case where the centre of mass is located a distance of a_1 from the point where force f_1 is applied, and a distance a_2 from the point where force f_2 is applied (noting that $a_1 + a_2 = L$).

9. Additive and multiplicative forms of plant models: Derive the additive forms of the two-body system dynamic equations (as given in Equation 14.6) starting with the multiplicative forms (as given in Equation 14.5).

10. Model-inverting feedforward filter design: Design a model-inverting feedforward controller in the form

$$C_{ff}(s) = A_{ff}s^2 + J_{ff}s^4 \qquad (14.37)$$

for a plant that is approximated by the transfer function

$$P(s) = \frac{1}{20}\left(\frac{1}{s^2} + \frac{0.4}{s^2 + 20s + 1000000} + \frac{0.2}{s^2 + 40s + 4000000}\right) \ [\text{mN}^{-1}]. \qquad (14.38)$$

References

Abir J, Longo S, Morantz P, Shore P 2017 Virtual metrology frame technique for improving dynamic performance of a small size machine tool *Precision Engineering* **48** 24–31

Al-Bender F, Lamport V, Swevers J 2005 The generalized Maxwell-slip model: A novel model for friction simulation and compensation *IEEE Transactions on Automatic Control* **50** 1883–7

Al-Bender F, Swevers J 2008 Characterization of friction force dynamics *IEEE Control Systems Magazine* **28** 64–81

Araki M, Taguchi H 2003 Two degree-of-freedom PID controllers *International Journal of Control, Automation, and Systems* **1** 401–11

Armstrong-Hélouvry B, Dupont P, de Wit CC 1994 A survey of models, analysis tools and compensation methods for the control of machines with friction *Automatica* **30** 1083–138

Boerlage M 2006 MIMO jerk derivative feedforward for motion systems Proceedings of the 2006 American Control Conference 3892–7

Boerlage M, Steinbuch M, Lambrechts P, van de Wal M 2003 Model-based feedforward for motion systems Proceedings of the 2003 IEEE Conference on Control Applications 1158–63

Boerlage M, Tousain R, Steinbuch M 2004 Jerk derivative feedforward control for motion systems Proceedings of the 2004 American Control Conference 5 4843–8

Bristow DA, Tharayil M, Alleyne AG 2006 A survey of iterative learning control *IEEE Control Systems Magazine* **26** 96–114

Bruijnen D, van de Molengraft R, Steinbuch M 2006 Optimization aided loop shaping for motion systems Proceedings of the 2006 IEEE International Conference on Control Applications 255–60

Bruijnen D, van der Meulen S 2016 Faster computation of closed loop transfers with frequency response data for multivariable loopshaping *IFAC-PapersOnLine* **49**(13) 87–92

Butler H 2011 Position control in lithographic equipment [Applications of control] *IEEE Control Systems Magazine* **31** 28–47

Butler H 2012 Feedforward signal prediction for accurate motion systems using digital filters *Mechatronics* **22** 827–35

Chang B-H, Hori Y 2006 Trajectory design considering derivative of jerk for head-positioning of disk drive system with mechanical vibration *IEEE/ASME Transactions on Mechatronics* **11** 273–9

Dahl P 1968 A solid friction model *DTIC Document*

De Moerlooze K, Al-Bender F, Van Brussel H 2011 Modeling the dynamic behavior of systems with rolling elements *International Journal of Non-Linear Mechanics* **46** 222–33

Devasia S 2002 Should model-based inverse inputs be used as feedforward under plant uncertainty? *IEEE Transactions on Automatic Control* **47** 1865–71

Dobrowiecki T, Schoukens J, Guillaume P 2006 Optimized excitation signals for MIMO frequency response measurements *IEEE Transactions on Instrumentation and Measurement* **55** 2072–9

Feldman M, Zimmerman Y, Gissin M, Bucher I 2016 Identification and modeling of contact dynamics of precise direct drive stages *Journal of Dynamic Systems, Measurement, and Control* **138** 071001

Franklin GF, Powell JD, Emami-Naeini A 2015 *Feedback control of dynamic systems*, 7th ed. Pearson

Futami S, Furutani A, Yoshida S 1990 Nanometer positioning and its microdynamics *Nanotechnology* **1** 31–7

Garcia D, Karimi A, Longchamp R 2004 Robust PID controller tuning with specification on modulus margin *Proceedings of the American Control Conference* **4** 3297–302

García-Herreros I, Kestelyn X, Gomand J, Coleman C, Barre P-J 2013 Model-based decoupling control method for dual-drive gantry stages *Control Engineering Practice* **21** 298–307

Gevers M 2002 Modelling, identification and control, in *Iterative Identification and Control* Springer 3–16

Gordon DJ, Erkorkmaz K 2012 Precision control of a T-type gantry using sensor/actuator averaging and active vibration damping *Precision Engineering* **36** 299–314

Graham RE 1946 Linear Servo Theory *Bell Labs Technical Journal* **25** 616–51

Hoogendijk, R, van de Molengraft MJG, den Hamer AJ, Angelis GZ, Steinbuch M 2015 Computation of transfer function data from frequency response data with application to data-based root-locus *Control Engineering Practice* **37** 20–31

Hou Z-S, Wang Z 2013 From model-based control to data-driven control: Survey, classification, and perspective *Information Sciences* **235** 3–35

Iwasaki M, Seki K, Maeda Y 2012 High precision motion control techniques: A promising approach to improving motion performance *IEEE Industrial Electronics* **6** 32–40

Lee C, Mohan G, Salapaka S 2011 *2DOF Control design* in *Control technologies for emerging micro and nanoscale systems* Springer

Munnig Schmidt R, Schitter G, Rankers A, van Eijk J 2014 *The design of high performance mechatronics* 2nd ed Delft University Press

Oomen T, Grassens E, Hendricks F 2015 Inferential motion control: Identification and robust control framework for positioning an unmeasurable point of interest *IEEE Transactions on Control Systems Technology* **23** 1601–10

Otsuka J, Masuda T 1998 The influence of nonlinear spring behaviour of rolling elements on ultra-precision positioning control systems *Nanotechnology* **9** 85–92

Pintelon R, Schoukens J 2012 *System identification: A frequency domain approach* 2nd ed John Wiley & Sons

Pintelon R, Schoukens J 2013 FRF measurement of nonlinear systems operating in closed loop *IEEE Transactions on Instrumentation and Measurement* **62** 1334–45

Rijlaarsdam D, Nuij P, Schoukens J, Steinbuch M 2017 A comparative review of frequency domain methods for nonlinear systems *Mechatronics* **42** 11–24

Rijlaarsdam D, van Loon B, Nuij P, Steinbuch M 2010 Nonlinearities in industrial motion stages— Detection and classification *Proceedings of the American Control Conference* 6644–9

Schoukens J, Marconato A, Pintelon R, Rolain Y, Schoukens M, Tiels K, Vanbeylen L, Vandersteen G, Van Mulders A 2014 System identification in a real world *IEEE 13th International Conference on Advanced Motion Control (AMC)* 1–9

Schoukens J, Vaes M, Pintelon R 2016 Linear system identification in a nonlinear setting *IEEE Control Systems Magazine* **36** 38–69

Sencer B, Tajima S 2017 Frequency optimal feed motion planning in computer numerical controlled machine tools for vibration avoidance *Journal of Manufacturing Science and Engineering* **139** 011006-1-13

Singer NC, Seering WP 1990 Preshaping command inputs to reduce system vibration *Journal of Dynamic Systems, Measurement, and Control* **112** 76–82

Singhose W, Seering W 2011 *Command generation for dynamic systems* Lulu

Skogestad S, Postlethwaite I 1996 *Multivariable feedback control: Analysis and design* John Wiley & Sons

Stein G 2003 Respect the unstable *IEEE Control Systems* 23 12–25

Tallman GH, Smith OJM 1958 Analog study of dead-beat Posicast control *IRE Transactions on Automatic Control* 4 14–21

Teo C, Tan K, Lim S, Huang S, Tay E 2007 Dynamic modeling and adaptive control of an H-type gantry stage *Mechatronics* 17 361–7

van der Maas R, van der Maas A, Dries J, de Jager B 2016 Efficient nonparametric identification for high-precision motion systems: A practical comparison based on a medical X-ray system *Control Engineering Practice* 56 75–85

van der Meulen SH, Tousain RL, Bosgra, OH 2008 Fixed structure feedforward controller design exploiting iterative trials: Application to a wafer stage and a desktop printer *Journal of Dynamic Systems, Measurement, and Control* 130 051006

van Dijk J, Aarts R 2012 Analytical one parameter method for PID motion controller settings IFAC Conference on Advances in PID Control WeC2.4

van Solingen E, van Wingerden JW, Oomen T 2016 Frequency-domain optimization of fixed-structure controllers *International Journal of Robust and Nonlinear Control*

Vanhoenacker K, Dobrowiecki T, Schoukens J 2001 Design of multisine excitations to characterize the nonlinear distortions during FRF-measurements *IEEE Transactions on Instrumentation and Measurement* 50 1097–102

Widanage WD, Omar N, Schoukens J, Van Mierlo J 2015 Estimating the frequency response of a system in the presence of an integrator *Control Engineering Practice* 32 1–11

Yaniv O, Nagurka M 2004 Design of PID controller satisfying gain margin and sensitivity constraints on a set of plants *Automatica* 40 111–6

Yoon JY, Trumper DL 2014 Friction modeling, identification, and compensation based on friction hysteresis and Dahl resonance *Mechatronics* 24 734–41

Index

Page numbers followed by f and t indicate figures and tables, respectively.

S

Saddle effect, 284
Sample mean, defined, 69
Samples, defined, 65
Sampling distributions, 69
Scalar quantities, defined, 57
Scalars, defined, 58
Scanners, wafer, 110
Scanning probe microscopes (SPMs), 190
Scraping
 of cast iron ways, 396
 slideway, dry bearings, 283, 283f
Screw displacement, 234, 235f
Screw drives, 350–352, 351f
Screw motion, 208, 210, 235, 236f
Screw parameters, defined, 235
Screw theory, 234
Screw thread, 220
 measurement, 440–441, 441f, 441t
Seals, 288
Self aligning bearing, 239
Self-correction/calibration, 9
Semiconductors, 530–531
Sensors
 contact displacement, 160–162
 indicators, 160, 161f
 LVDTs, 160, 162
 non-contact displacement, 162–164
 capacitive sensors, 163
 inductive (eddy-current) sensors, 164
 overview, 162
Serial kinematic structure, 251–254
 benefits and limitations, 252, 254
 characteristic features, 251, 252, 253f
Servo control, 602
 characterising performance, 607–608,
 607f–608f
Setpoint filtering feedforward strategies, 622–623
Shear plate, 480–481, 481f
Shear strains, defined, 76
Shear stress, 78, 86
Shock; *see also* Vibrations
 sources of, 569–571
 transmissibility, environmental isolation
 and, 582–583, 582f–583f
Shock response spectra (SRS), 576,
 582–583, 583f
Shot noise, defined, 19
Signal conditioning component, of circuit, 45
Silicon carbide, 245
Silicon nitride, 125, 245, 549

Simple model involving linear and angular
 motion, 116–122
Simple pendulum, 121–122
Simulations
 required number of, Monte Carlo
 method, 426
Sinc function, 574
Sine error(s), 456–457, 456f
 combination with Abbe and cosine errors,
 457, 458f
 one-dimensional depiction of, 456–457, 456f
Single-value uncertainty evaluation
 mean, standard deviation and standard
 deviation of mean, 415–418, 416t,
 417f, 417t
 standard uncertainty concept, 422–423
 uncertainty distributions and confidence
 intervals, 418–422, 419f–420f, 421t, 422f
Singularities, geometric, 224–226
SIOS Nanomeasuring Machine (NMM), 182
Skewness, characterisics, 67
Slideway, dry bearings
 design considerations, 284–287, 284f–286f
 double V, 285, 285f
 flat, 284–287, 284f–285f
 manufacturing, 281–284, 283f
 scraping, 283, 283f
 symmetric spindle design, 286–287, 286f
 V-groove, 284–287, 284f–285f
Sliding bearings, *see* Dry (sliding) bearings
Sliding pair, 277
 wear mechanism maps, 280, 280f
Slip and stick regions, in contact area,
 87–88
Slocum, 247
SmartScope CNC 500 vision system, 178
Snell's law, 91, 92f
Software, CMS, 175–176
Software compensation method, 393–395, 453
 algorithms, 395
 calibration routines, 393–394
 computer control, 394–395, 394f
Solenoid actuators, 325–330, 326f
Solid mechanics, concepts, 75–90
 beam theory, 79–81
 Hertz contact theory, 82–87
 contact between two parallel cylinders,
 86, 87
 contact between two spheres, 83–86
 overview, 82–83
 Hooke's law for linear elastic, isotropic
 materials, 78–79

Printed in the United States
by Baker & Taylor Publisher Services